Ecology and Evolution of Flowers

Ecology and Evolution of Flowers

Edited by

LAWRENCE D. HARDER
Department of Biological Sciences, University of Calgary, Calgary, Alberta, Canada T2N 1N4

SPENCER C. H. BARRETT
Department of Botany, University of Toronto Toronto, Ontario, Canada M5S 3B2

OXFORD
UNIVERSITY PRESS

OXFORD
UNIVERSITY PRESS

Great Clarendon Street, Oxford OX2 6DP

Oxford University Press is a department of the University of Oxford.
It furthers the University's objective of excellence in research, scholarship,
and education by publishing worldwide in

Oxford New York

Auckland Cape Town Dar es Salaam Hong Kong Karachi
Kuala Lumpur Madrid Melbourne Mexico City Nairobi
New Delhi Shanghai Taipei Toronto

With offices in

Argentina Austria Brazil Chile Czech Republic France Greece
Guatemala Hungary Italy Japan Poland Portugal Singapore
South Korea Switzerland Thailand Turkey Ukraine Vietnam

Oxford is a registered trade mark of Oxford University Press
in the UK and in certain other countries

Published in the United States
by Oxford University Press Inc., New York

© Oxford University Press 2006

The moral rights of the authors have been asserted
Database right Oxford University Press (maker)

First published 2006

All rights reserved. No part of this publication may be reproduced,
stored in a retrieval system, or transmitted, in any form or by any means,
without the prior permission in writing of Oxford University Press,
or as expressly permitted by law, or under terms agreed with the appropriate
reprographics rights organization. Enquiries concerning reproduction
outside the scope of the above should be sent to the Rights Department,
Oxford University Press, at the address above

You must not circulate this book in any other binding or cover
and you must impose the same condition on any acquirer

British Library Cataloguing in Publication Data

Data available

Library of Congress Cataloging in Publication Data

Ecology and evolution of flowers / edited by Lawrence D. Harder, Spencer C. H. Barrett.
    p. cm.
  ISBN-13: 978–0–19–857086–8 (pbk. : alk. paper)
  ISBN-10: 0–19–857086–4 (pbk. : alk. paper)
  ISBN-13: 978–0–19–857085–1 (alk. paper)
  ISBN-10: 0–19–857085–6 (alk. paper)
 1. Plants, Flowering of. 2. Plants ecology. 3. Plants—Evolution. I. Harder, Lawrence D. II. Barrett, Spencer C. H.
    QK830.E26 2006
    575.6—dc22   2006022878

Typeset by Newgen Imaging Systems (P) Ltd., Chennai, India
Printed in Great Britain
on acid-free paper by
Antony Rowe Ltd.,
Chippenham, Wiltshire

ISBN 0–19–857085–6   978–0–19–857085–1
ISBN 0–19–857086–4 (Pbk.)   978–0–19–857086–8 (Pbk.)

10 9 8 7 6 5 4 3 2 1

Dedicated to the memory of David G. Lloyd (1937–2006), natural historian and founder of the modern theory of plant reproduction

Preface

The reproductive organs and mating biology of angiosperms exhibit greater variety than those of any other group of organisms. Flowers and inflorescences are also the most diverse structures produced by angiosperms. Why should organs that serve but one main purpose, reproduction, evolve such matchless variety? The answer seems to lie in the interactions of plants with their pollen vectors, which are necessary to overcome the fundamental mating problem confronted by plants, namely their immobility. The significance of these interactions is apparent in the observation that most of the reproductive diversity of flowering plants involves features that function during pollination and mating. Thus, the search for explanations of this diversity should logically focus on mating as a process and an outcome, its ecological and genetic context, its consequences for variation in maternal and paternal success within and between populations, and its evolution within species and lineages.

Analysis of the function and evolution of flowers has changed considerably during recent decades. The traditional descriptive, natural-history approach, which focused almost exclusively on flowers themselves, is increasingly accompanied by conceptually motivated experiments in both the laboratory and field, formal mathematical theory, genetic analysis, and studies of pollinator behaviour. In addition, the scope of topics considered has broadened significantly as the long-standing, artificial division between ecological studies of pollination and genetic studies of mating has eroded, so that the reproductive biology of plants is fully integrated into evolutionary ecology. This expansion of approaches and perspectives continues to expose new questions and stimulate reanalysis of enduring questions. For example, during the decade since the publication of the previous general research volume on floral biology[1], topics that have attracted increased interest include: pollen limitation of seed production; the nature and strength of selection on floral traits; inflorescence function; the relative incidence and consequences of specialization and generalization in plant-pollinator interactions; the occurrence, causes and implications of hybridization; the community context of plant reproduction, including the effects of herbivory on floral evolution; and the phylogenetics and comparative biology of floral adaptation. Moreover, the scale at which some of these topics are now being investigated has expanded beyond the local population to encompass a broader geographical context, including metapopulations and regional assemblages. Given the dynamic nature of this discipline, a review of recent developments seemed timely, which prompted us to organize this volume.

Ecology and Evolution of Flowers includes 18 chapters written by internationally recognized authorities, which both review recent advances in floral biology and present new results. The chapters are organized into an introductory chapter and four major sections that consider different aspects of floral ecology and evolution. The first chapter reviews the seminal contributions of David G. Lloyd to the strategic analysis of plant reproduction. We have chosen to highlight Lloyd's contributions, because his penetrating functional analysis, until illness ended his research career in 1993, identified the current research agenda for major areas in plant evolutionary biology. The remaining chapters build on Lloyd's legacy and illustrate the ecological function of flowers, the

---

[1] Lloyd DG, and Barrett SCH (1996). *Floral biology: studies on floral evolution in animal-pollinated plants*. Chapman and Hall, New York.

alternate evolutionary solutions that enable plants to reproduce successfully in varied environments, and the resulting process of adaptive diversification. In all cases, we encouraged authors both to provide a synopsis of current knowledge of their chosen topic, and to use their own empirical or theoretical studies to illustrate current issues and prospects for future analysis in more detail.

The major sections of *Ecology and Evolution of Flowers* in turn examine functional aspects of floral traits, the ecological influences on reproductive adaptation, the evolutionary ecology of mating and sexual systems, and the role of floral biology in angiosperm diversification. The first section, "Strategic perspectives on floral biology", presents current theoretical perspectives on the relations of fitness to reproductive allocation and floral traits and their implications for floral adaptation and mating-system evolution. The second section, "Ecological context of floral function and its evolution", illustrates that the evolution of reproductive strategies does not occur in isolation, but instead depends on population characteristics and the influences of other species in the community that act as mutualists, competitors or predators. This section also considers the consequences of anthropogenic disturbance for floral function. The third section, "Mating strategies and sexual systems" considers the functional consequences of the diverse deployment of female and male organs within flowers and within and among plants. This variety in organ deployment contributes to the diverse patterns of mating within plant populations and characterizes the remarkable variety of sexual systems in angiosperms. The final section, "Floral diversification", examines the genetic opportunities and constraints associated with floral adaptation and considers the role of this floral evolution in speciation, species integrity and lineage diversification. All sections of the volume are supported by a glossary to assist readers in understanding relevant terminology and concepts of plant reproduction, ecology and evolution. Overall, this integrated treatment illustrates current understanding of the role of floral function and evolution in the generation of angiosperm biodiversity and offers many suggestions for the next phase of research on this intriguing subject.

Compilation of such a volume requires the involvement and cooperation of many participants. We are grateful to numerous individuals who contributed to the content and production of this book. First and foremost, we acknowledge the essential contributions of the authors, who provided the necessary content, worked to enhance synergy among chapters, and tolerated the idiosyncrasies of the editors. Second, we greatly appreciate the continued enthusiasm for this project of Ian Sherman (Commissioning Editor, Biology, Oxford University Press) and his decisiveness in moving the volume along. We also thank the many individuals who reviewed individual chapters, thereby enhancing the volume as a whole, including: Lynn Adler, Paul Aigner, Scott Armbruster, Camille Barr, Christine Caruso, Andrea Case, Pierre-Olivier Cheptou, Mitch Cruzan, Marcel Dorken, Tim Holtsford, Linley Jesson, Pedro Jordano, Peter Klinkhamer, Josh Kohn, Claire Kremen, Elizabeth Lacey, David McCauley, David Moeller, Bill Morris, Sally Otto, Emmanuelle Porcher, Shane Richards, Loren Rieseberg, Ophélie Ronce, Satoki Sakai, Doug Schemske, Dan Schoen, Stacey Smith, John Stinchcombe, James Thomson, Jana Vamosi, Diego Vázquez, Arthur Weis, Colin Webb, Steven Weller, and Paul Wilson. The assistance of Bill Cole, Chris Eckert, and Matt Routley was invaluable in facilitating reviews and interaction among authors. In addition, Stefanie Gehrig (Assistant Commissioning Editor, Biology, Oxford University Press) aided considerably during the final production of this book. Finally, we thank Dale Hensley and Suzanne Barrett for tolerating our compulsive fascination with flowers and for their forbearance while we were immersed in the writing and editing of this volume.

Lawrence D. Harder
Calgary
Spencer C.H. Barrett
Toronto
January 2006

Contents

List of contributors	xvii
1 David G. Lloyd and the evolution of floral biology: from natural history to strategic analysis	**1**
Spencer C. H. Barrett and Lawrence D. Harder	
1.1 Introduction	1
1.2 Biographical sketch	2
1.3 Self- and cross-fertilization in plants	5
1.3.1 Early investigations on mating systems	5
1.3.2 Integration of pollination and mating	6
1.4 Gender strategies	8
1.4.1 Early investigations of plant sexual diversity	8
1.4.2 Theories on the evolution of sexual systems	10
1.4.3 Gender concepts and theory	10
1.5 Allocation strategies	11
1.5.1 Allocation to competing functions	12
1.5.2 Size-number compromises	13
1.5.3 Application to specific problems	14
1.6 Floral mechanisms	14
1.6.1 Sexual interference	15
1.6.2 The evolution of heterostyly	15
1.7 Lloyd's evolution	17
References	18
Part 1 Strategic perspectives on floral biology	**23**
2 Selection on reproductive characters: conceptual foundations and their extension to pollinator interactions	
Martin T. Morgan	25
2.1 Introduction	25
2.2 Phenotypic selection on reproductive strategy	26
2.2.1 Phenotypic selection and evolutionary stable strategies	26
2.2.2 Phenotypic selection and inheritance of quantitative characters	27
2.2.3 Measuring phenotypic selection	28
2.3 Plant and pollinator interaction	29
2.3.1 Specialist or generalist?	30
2.3.2 Pollination as trade	32
2.3.3 Ecological dynamics	32
2.4 Self-fertilization, phenotypic selection, and reproductive assurance	34

	2.4.1 Ecological consequences of selfing mode	35
	2.4.2 Evolutionary consequences of selfing mode	35
	2.4.3 Variable pollination environments	36
2.5	Discussion	37
	References	38

3 Evolutionarily stable reproductive investment and sex allocation in plants
Da-Yong Zhang — 41

3.1	Introduction	41
3.2	The classic model of sex allocation for outcrossing species	42
	3.2.1 Gain curves	43
	3.2.2 ESS sex allocation	44
	3.2.3 Evolution of sexual systems	46
3.3	Plant size and sex allocation	47
	3.3.1 Budget effects of plant size on fitness gains	48
	3.3.2 Direct effects of plant size on fitness gains	50
	3.3.3 Why are male and female function often displaced temporally in cosexual plants?	51
3.4	Joint evolution of reproductive effort and sex allocation in perennial plants	53
	3.4.1 Size-dependent reproductive effort	55
3.5	Discussion and conclusion	56
	References	57

4 Pollen and ovule fates and reproductive performance by flowering plants
Lawrence D. Harder and Matthew B. Routley — 61

4.1	Introduction	61
4.2	Pollen fates and ovule fates	62
	4.2.1 Pollen dispersal	62
	4.2.2 Pollen-tube growth	66
	4.2.3 Ovule fertilization	66
	4.2.4 Seed development	66
	4.2.5 Seedling establishment and parental fitness	68
4.3	Limits on seed production	68
	4.3.1 Pollen limitation	68
	4.3.2 Ovule limitation	69
	4.3.3 Resource limitation	70
4.4	Examples of the roles of pollen and ovule fates in floral and mating-system evolution	71
	4.4.1 Improvements in pollen export	71
	4.4.2 Simultaneous, autonomous self-pollination without pollen discounting	72
	4.4.3 Facilitated self-pollination	74
	4.4.4 Delayed self-pollination	76
4.5	Concluding discussion	76
	4.5.1 The asymmetry of pollen and ovule fates	76
	4.5.2 Limits on seed production and reproductive evolution	77
	References	78

Part 2 Ecological context of floral function and its evolution — 81

5 Models of pollinator-mediated gene dispersal in plants
James E. Cresswell — 83

5.1	Introduction	83

5.2 Three iconic patterns of pollinator-mediated gene dispersal	84
5.3 A historical perspective on the theory of pollinator-mediated gene dispersal	85
5.4 Qualitative generalizations from the portion-dilution model	90
5.5 In pursuit of quantitative predictions from the portion-dilution model	93
5.6 Evolutionary biology of the paternity shadow	95
5.7 Prospects for the theory of pollinator-mediated gene dispersal	97
References	98

6 Pollinator responses to plant communities and implications for reproductive character evolution — 102
Monica A. Geber and David A. Moeller

6.1 Introduction	102
6.2 Properties of plants and communities	104
6.3 Plant-community effects on pollinator responses	105
6.3.1 Pollinator behavioural responses	105
6.3.2 Pollinator demographic responses	105
6.3.3 Pollinator-community structure	107
6.4 Consequences of pollinator responses for selection on plant reproductive traits	107
6.4.1 Heterospecific pollen transfer	108
6.4.2 Pollinator visitation rate	108
6.4.3 Pollinator-community composition	111
6.5 Community context and mating-system evolution in *Clarkia*	111
6.5.1 Study system	112
6.5.2 Pollinator responses to *Clarkia* communities	112
6.5.3 Consequences of community context for selection on the mating system	113
6.5.4 Biogeographic patterns of mating-system variation and plant-community diversity	113
6.6 Conclusions and future directions	115
References	116

7 Non-pollinator agents of selection on floral traits — 120
Sharon Y. Strauss and Justen B. Whittall

7.1 Introduction	120
7.2 Selection on reproductive traits by non-pollinator agents	123
7.2.1 Petal colour	125
7.2.2 Flower shape	126
7.2.3 Flower size and display size	127
7.2.4 Nectar	128
7.2.5 Sexual systems	129
7.2.6 Flowering phenology	130
7.3 Relative strengths of pollinator and non-pollinator agents of selection	131
7.3.1 A case study of *Raphanus sativus*	132
7.4 Synthesis of ecological and genetic observations	134
7.5 Community context of trait evolution	135
References	135

8 Flowering phenologies of animal-pollinated plants: reproductive strategies and agents of selection — 139
Gaku Kudo

8.1 Introduction	139

	8.2 Components of flowering phenologies	**140**
	8.2.1 Individual phenology	**140**
	8.2.2 Population phenology	**141**
	8.2.3 Community phenology	**142**
	8.3 Flowering phenologies as reproductive strategies	**143**
	8.3.1 Abiotic influences on phenology	**143**
	8.3.2 Pollination	**144**
	8.3.3 Herbivory	**146**
	8.3.4 Seed dispersal and germination	**146**
	8.3.5 Sexual system	**147**
	8.3.6 Limits on phenological evolution	**148**
	8.4 Flowering phenology and selection on plant reproduction: case studies of alpine plants	**149**
	8.4.1 Flowering-time effects on pollination success and seed quality	**149**
	8.4.2 Interspecific competition for pollination along snowmelt gradients	**149**
	8.4.3 Metapopulation structure and phenological separation	**152**
	8.5 Concluding remarks for future research	**153**
	References	**156**
9	**Flower performance in human-altered habitats**	**159**
	Marcelo A. Aizen and Diego P. Vázquez	
	9.1 Introduction	**159**
	9.2 General effects of disturbance and their reproductive consequences	**161**
	9.3 Effects of human-caused perturbations on plant attributes	**162**
	9.4 Effects of human-caused perturbations on pollinator attributes	**164**
	9.5 Relation of pollination to modified plant attributes	**165**
	9.6 Relation of pollination to modified pollinator attributes	**167**
	9.7 Relation of plant reproduction to modified pollination	**168**
	9.8 Translation into a path-analysis framework: an example	**169**
	9.9 Modulators of plant reproductive response	**171**
	9.10 Anthropogenic disturbance and the structure of pollination interaction networks	**172**
	9.11 Prospects	**175**
	References	**176**
Part 3	**Mating strategies and sexual systems**	**181**
10	**Reproductive assurance and the evolution of uniparental reproduction in flowering plants**	**183**
	Christopher G. Eckert, Karen E. Samis, and Sara Dart	
	10.1 Introduction	**183**
	10.2 Reproductive assurance and self-fertilization: theoretical context	**184**
	10.3 Reproductive assurance and self-fertilization: empirical approaches	**186**
	10.3.1 Reproductive assurance and the evolution of mixed-mating systems	**187**
	10.3.2 Exploiting intraspecific variation to test the reproductive assurance hypothesis	**190**
	10.4 Asexual reproduction: a neglected mechanism of reproductive assurance	**196**
	10.4.1 Challenges and opportunities for investigating the evolution of asexuality	**198**
	10.5 More experiments needed	**199**
	References	**200**

11 The evolution of separate sexes: a focus on the ecological context — 204
Tia-Lynn Ashman

11.1 Introduction — 204
11.2 The gynodioecy pathway to dioecy — 205
    11.2.1 Hermaphroditism to gynodioecy — 205
    11.2.2 Gynodioecy to dioecy — 206
    11.2.3 Predicted relations of female frequency and key model parameters — 206
    11.2.4 Comparisons of observed and predicted female frequencies — 208
11.3 Importance of ecological context — 209
    11.3.1 Harsh abiotic environment — 209
    11.3.2 Harsh environments alter pollination, pollen movement, and mating system — 214
    11.3.3 Enemies—analogous to harsh environments? — 215
    11.3.4 Multi-species interactions and little-studied aspects of ecological context — 217
11.4 Spotlight on unresolved issues — 217
    11.4.1 Subdioecy: reflective of constraints or adaptation? — 218
    11.4.2 Pollinators: facilitators or inhibitors of sexual-system evolution? — 219
References — 219

12 Effects of colonization and metapopulation dynamics on the evolution of plant sexual systems — 223
John R. Pannell

12.1 Introduction — 223
12.2 Single-event versus recurrent colonization — 224
12.3 Effects of single-event colonization on the sexual system — 224
    12.3.1 Long-distance dispersal to oceanic islands — 224
    12.3.2 Long-distance dispersal and the evolution of dioecy in *Cotula* — 225
12.4 Evolution of sexual systems in a metapopulation — 226
12.5 Evolution of dominant versus recessive traits in a metapopulation — 229
12.6 Modes of selfing and the evolution of geitonogamy — 231
    12.6.1 Model of geitonogamy in a metapopulation — 232
    12.6.2 Model results and discussion — 232
12.7 Conclusions — 236
References — 236

13 Floral design and the evolution of asymmetrical mating systems — 239
Spencer C. H. Barrett and Kathryn A. Hodgins

13.1 Introduction — 239
13.2 The evolution and functional basis of floral and sexual-system diversity — 242
13.3 Mating in monomorphic populations — 243
13.4 Mating in dimorphic populations — 244
    13.4.1 Evolution and maintenance of stigma-height dimorphism — 245
    13.4.2 Asymmetrical mating and biased morph ratios — 246
13.5 Mating in trimorphic populations — 246
    13.5.1 Evolution of morph ratios — 247
    13.5.2 Variation and evolution of sexual organs — 251
13.6 Discussion — 252
References — 254

Part 4 Floral diversification — 257

14 Ecological genetics of floral evolution — 260
Jeffrey K. Conner

14.1 Introduction — 260
14.2 Simple polymorphisms: floral colour — 261
14.3 Ecological genetics of quantitative traits — 263
14.4 Natural selection on floral traits — 266
14.5 Independent evolution of correlated traits in radish — 268
  14.5.1 Microevolution — 268
  14.5.2 Macroevolution — 272
14.6 Future directions — 273
References — 274

15 Geographical context of floral evolution: towards an improved research programme in floral diversification — 278
Carlos M. Herrera, María Clara Castellanos, and Mónica Medrano

15.1 Introduction — 278
15.2 Representation of geographical variation in pollination studies — 279
  15.2.1 Patterns: how much attention has geographical variation in plant traits and pollinators received? — 280
  15.2.2 Processes: how much do we know about geographical variation in selection on pollination-related traits? — 281
15.3 Outcomes and limitations of geographically informed studies — 282
  15.3.1 Outcomes — 282
  15.3.2 Limitations and a proposal — 282
15.4 A case study: clinal variation of *Lavandula latifolia* flowers and pollinators — 284
  15.4.1 Methods — 284
  15.4.2 Step 1: Characterize geographical variation in pollinators — 285
  15.4.3 Step 2: Demonstrate pollinator-mediated selection on floral traits — 286
  15.4.4 Step 3: Assess geographical divergence in selection — 287
  15.4.5 Step 4: Evaluate the match between divergent selection and phenotypic divergence — 288
  15.4.6 Step 5: Genetic basis of population differences in floral traits — 289
  15.4.7 Interpretation and caveats — 289
15.5 Concluding remarks: towards an improved research programme in floral diversification — 290
References — 291

16 Pollinator-driven speciation in plants — 295
Steven D. Johnson

16.1 Introduction — 295
16.2 Why flowers evolve — 297
16.3 The geographic pollinator mosaic — 298
16.4 Pollination ecotypes — 300
16.5 The scale of gene flow in plants — 301
16.6 Geographic modes of pollinator-driven speciation — 302

16.7 Identifying pollinator-driven speciation	303
16.8 Pollinators and reproductive isolation	304
16.9 Reinforcement of isolating barriers	305
16.10 Adaptive radiation	305
16.11 Conclusions	306
References	306

17 Floral characters and species diversification — 311
Kathleen M. Kay, Claudia Voelckel, Ji Y. Yang, Kristina M. Hufford, Debora D. Kaska, and Scott A. Hodges

17.1 Introduction	311
17.2 How might floral traits affect diversification?	312
17.3 Common tests for key innovations	314
17.4 Methods	316
17.4.1 Trait datasets	316
17.4.2 Analyses	317
17.5 Results	317
17.5.1 Pollination mode	317
17.5.2 Floral symmetry	317
17.5.3 Floral nectar spurs	318
17.5.4 Dioecy	319
17.6 Discussion	319
References	322

18 Floral biology of hybrid zones — 326
Diane R. Campbell and George Aldridge

18.1 Introduction	326
18.2 Genetic architecture of species differences; what do hybrid flowers look like?	328
18.3 Floral traits and the frequency of mating between species and hybrids	329
18.3.1 Ethological isolation	329
18.3.2 Mechanical isolation	331
18.4 The relative importance of ethological and mechanical isolation	332
18.4.1 Floral traits in *Ipomopsis aggregata* and *I. tenuituba* and pollinator behaviour	332
18.4.2 Simulation model of ethological and mechanical isolation	333
18.4.3 Geographical variation in pre-mating isolation	335
18.5 Post-mating isolation	337
18.6 Floral traits and the fitness of hybrids	340
18.7 Conclusions and future directions	341
References	342

Glossary	**346**
Index	**353**

List of contributors

Marcelo A. Aizen
Laboratorio Ecotono, CRUB
Universidad Nacional del Comahue
Quintral 1250
8400 Bariloche,
Río Negro
Argentina (marcito@crub.uncoma.edu.ar)

George Aldridge
Rocky Mountain Biological Laboratory
PO Box 519
Crested Butte, CO 81224
USA (contracow@yahoo.com)

Tia-Lynn Ashman
Department of Biological Sciences
University of Pittsburgh
Pittsburgh, PA 15260
USA (tia1@pitt.edu)

Spencer C. H. Barrett
Department of Botany
University of Toronto
Toronto, Ontario M5S 3B2
Canada (barrett@botany.utoronto.ca)

Diane R. Campbell
Department of Ecology and Evolutionary Biology
University of California
Irvine, CA 92697
USA (drcampbe@uci.edu)

María Clara Castellanos
Estación Biológica de Doñana, CSIC
Avenida de María Luisa s/n
E-41013 Sevilla
Spain (mcastel@ebd.csic.es)

Jeffrey K. Conner
Kellogg Biological Station and Department
of Plant Biology
Michigan State University
3700 East Gull Lake Drive
Hickory Corners, MI 49060
USA (connerj@msu.edu)

James E. Cresswell
School of Biosciences
University of Exeter
Exeter
UK (j.e.cresswell@ex.ac.uk)

Sara Dart
Department of Biology
Queen's University
Kingston, Ontario K7L 3N6
Canada (darts@biology.queensu.ca)

Christopher G. Eckert
Department of Biology
Queen's University
Kingston, Ontario K7L 3N6
Canada (eckertc@biology.queensu.ca)

Monica A. Geber
Department of Ecology and Evolutionary Biology
Cornell University
Ithaca, NY 14850
USA (mag9@cornell.edu)

Lawrence D. Harder
Department of Biological Sciences
University of Calgary
Calgary, Alberta T2N 1N4
Canada (harder@ucalgary.ca)

Carlos M. Herrera
Estación Biológica de Doñana, CSIC
Avenida de María Luisa s/n
E-41013 Sevilla
Spain (herrera@cica.es)

Scott A. Hodges
Department of Ecology, Evolution and Marine Biology
University of California
Santa Barbara, CA 93106
USA (hodges@lifesci.ucsb.edu)

Kathryn A. Hodgins
Department of Botany
University of Toronto
Toronto, Ontario M5S 3B2
Canada (hodgins@botany.utoronto.ca)

Kristina M. Hufford
Ecology and Evolution
321 Steinhaus Hall
University of California
Irvine, CA 92697
USA (huffordk@uci.edu)

Steven D. Johnson
School of Biological and Conservation Sciences
University of KwaZulu-Natal
P.Bag X01, Scottsville
Pietermaritzburg 3209
South Africa (johnsonsd@ukzn.ac.za)

Debora D. Kaska
Department of Ecology, Evolution and Marine Biology
University of California
Santa Barbara, CA 93106
USA (kaska@lifesci.ucsb.edu)

Kathleen M. Kay
Department of Ecology, Evolution and Marine Biology
University of California
Santa Barbara, CA 93106
USA (kkay@lifesci.ucsb.edu)

Gaku Kudo
Graduate School of Environmental Earth Science
Hokkaido University
Sapporo 060–0810
Japan (gaku@ees.hokudai.ac.jp)

Mónica Medrano
Estación Biológica de Doñana, CSIC
Avenida de María Luisa s/n
E-41013 Sevilla
Spain (monica@ebd.csic.es)

David A. Moeller
Department of Plant Biology
University of Minnesota
St. Paul, MN 55108
USA (moell021@tc.umn.edu)

Martin T. Morgan
Washington State University
Pullman
WA 99164–4236
USA (mtmorgan@comcast.net)

John R. Pannell
Department of Plant Sciences
University of Oxford
South Parks Road
Oxford OX1 3RB
UK (john.pannell@plants.ox.ac.uk)

Matthew B. Routley
Department of Biological Sciences
University of Calgary
Calgary, Alberta T2N 1N4
Canada (routley@ucalgary.ca)

Karen E. Samis
Department of Biology
Queen's University
Kingston, Ontario K7L 3N6
Canada (samisk@biology.queensu.ca)

Sharon Y. Strauss
Section of Evolution and Ecology
University of California
One Shields Ave, 2320 Storer Hall
Davis, CA 95616
USA (systrauss@ucdavis.edu)

Diego P. Vázquez
Instituto Argentino de Investigaciones de las Zonas Áridas
Centro Regional de Investigaciones Científicas y Tecnológicas
5500 Mendoza
Argentina (dvazquez@lab.cricyt.edu.ar)

Claudia Voelckel
Department of Ecology, Evolution and Marine Biology
University of California
Santa Barbara, CA 93106
USA (voelckel@lifesci.ucsb.edu)

Justen B. Whittall
Section of Evolution and Ecology
University of California
One Shields Ave, 2320 Storer Hall
Davis, CA 95616
USA (jbwhittall@ucdavis.edu)

Ji Y. Yang
Department of Ecology, Evolution and Marine Biology
University of California
Santa Barbara, CA 93106
USA (albertoyang@yahoo.com)

Da-Yong Zhang
MOE Key Laboratory for Biodiversity Science and Ecological Engineering
College of Life Sciences
Beijing Normal University
Beijing 100875
P. R. China (zhangdy@bnu.edu.cn)

CHAPTER 1

David G. Lloyd and the evolution of floral biology: from natural history to strategic analysis

Spencer C. H. Barrett[1] and Lawrence D. Harder[2]

[1] Department of Botany, University of Toronto, Ontario, Canada
[2] Department of Biological Sciences, University of Calgary, Alberta, Canada

Outline

David G. Lloyd's scholarly contributions provide the conceptual foundation for many aspects of plant reproductive biology. Here we provide a biographical sketch of Lloyd's life, trace his intellectual development, and highlight the main research problems that he tackled during his 30-year career. Our review reveals how Lloyd started as a botanist making natural history observations in the classic Darwinian tradition. As his career progressed, Lloyd embraced the optimality approach of evolutionary ecology, enabling him to provide some of the first strategic analyses of plant reproductive adaptations. Many of Lloyd's ideas have influenced contemporary research and we identify four major areas in which he made seminal contributions: self- and cross-fertilization, gender strategies, allocation strategies and floral mechanisms. Lloyd's work on plant mating introduced the concept of "modes of self-pollination" and demonstrated how they influence whether selfing can evolve. Lloyd pioneered the concept of plant gender and was the foremost authority on the evolution of plant sexual systems. Lloyd analyzed the diversity of evolutionary stable strategies in plants and recognized that they all involve either allocation among competing functions or size-number compromises. His work in floral biology emphasized the significance of male reproductive success and he championed the idea that intra-sexual selection to increase the proficiency of pollen dispersal primarily guides floral evolution. Lloyd's strategic perspective often allowed him to consider topics beyond the plant kingdom and when illness ended his career he was working on a book on the evolution of social behaviour. The extensive body of concepts that Lloyd developed through keen observation, incisive intellect, and realistic theory established him as the founder of the theory of plant reproduction and comprise his enduring legacy.

1.1 Introduction

After lying largely dormant since the end of the nineteenth century, studies of the ecology and evolution of plant reproduction revived during the 1960s (Baker 1979). Subsequent developments expanded on the rich perspectives contributed during the 19th century stimulated by Darwin's insightful explorations of plant reproduction (1862, 1876, 1877) and identified new perspectives that remain the focus of research today. Among the individuals who contributed to advances during the modern age of plant reproductive biology, David G. Lloyd stands out for his conceptual synthesis. Lloyd's scholarly work laid the foundation for much of today's research on the ecology and evolution of flowers, as well as for several other fields of evolutionary biology. Indeed, editors of influential books in plant reproductive biology (Lovett Doust and Lovett Doust 1988;

Geber *et al.* 1999) have dedicated their volumes to Lloyd because of his seminal work in this field. However, neither the significance of Lloyd's contributions, nor their impact has been reviewed previously. Because of our strong conviction that Lloyd is one of the pre-eminent plant evolutionary biologists of the modern era, ranking alongside the venerable G. Ledyard Stebbins (Crawford and Smocovitis 2004), Verne Grant (Rieseberg and Wendel 2004) and Herbert Baker (Barrett 2001), we decided to highlight Lloyd's contributions as a tribute to his legacy and a fitting beginning to this volume.

The goal of our review is to illustrate how Lloyd's research contributions provide the concepts and tools for solving many outstanding questions in floral ecology and evolution. We begin with a short biography and then review the main problems on which Lloyd worked during his career, in roughly chronological order. This review reveals how Lloyd's ideas evolved, masterfully linking natural history and the strategic analysis of reproductive adaptations. Lloyd's ability to combine observations of the natural world in the classic Darwinian tradition with penetrating theoretical insights distinguishes him from his peers and marks him as a leading evolutionary ecologist of our time.

1.2 Biographical sketch

David Graham Lloyd was born on June 20, 1937 at Manaia, Taranaki on the North Island of New Zealand, with his identical twin brother Peter Lloyd (Fig. 1.1a). Apparently, the twins were so similar that only their mother could tell them apart reliably. Unfortunately, she died of cancer when they were eight leaving their father, a dairy farmer at Taranaki, to raise the twins, their brother Trevor and sister Judith. We are particularly indebted to Peter and Trevor Lloyd for providing many of the details for this biographical sketch of David Lloyd.

David had a fairly happy childhood at Manaia with the farm routine of twice daily milking of cows and visits to his grandfather's farm next door to see the poultry and bees. David later worked during holidays in the local cheese factory, which his father supplied with milk, and in the "gut room" of the slaughter house stripping by hand the contents of animal intestines before they were processed into sausage casings. According to Trevor, this was one of the worst jobs locally "and regarded as the bottom of the heap socially and work-wise." However, swimming in local rivers, the town swimming pool and a nearby beach were great childhood pleasures, although their mother's death cast a large shadow over the children's early lives.

By all accounts, David excelled at school, attending Manaia primary school where he and Peter skipped a grade because they were brighter than most students in their cohort. From 1950 to 1954, David attended New Plymouth Boys High School along with brother Peter. New Plymouth was a boarding school known mostly for rugby, rather than academic excellence: a "boarding school for country kids, no more." However, David was taught by several exceptional teachers, fostering an early interest in the sciences. While at New Plymouth, David was also head prefect of Pridham House and, despite being small in stature, was an outstanding athlete (sprinting and long jump) and Rugby football player. After completing high school, he and Peter both obtained Taranaki scholarships to assist in their subsequent university education.

David's family upbringing and school experiences provide no obvious indication that he was destined to become one of New Zealand's most influential biologists. David was the only member of his family to choose biology as a profession and, although at school he was strongly attracted to the sciences, especially physics, he was not an especially outdoors type, rarely went camping, and had no particular affinity for plants and animals. According to Peter, "both brothers were attracted by theory and abstract reasoning, rather than a childhood fascination with frogs." From Peter's perspective it is "uncertain why David chose botany at university," although according to the molecular evolutionist David Penny, who also attended New Plymouth and went on to Canterbury, the enlightened analytical botanical teaching of W. R. Philipson at Canterbury, rather than the traditional descriptive approach, was probably formative. The interests of the Lloyd twins

Figure 1.1 Images of David G. Lloyd from throughout his life. (a) The twins Peter (P) and David (D) at an early age. (b) As a young man prior to departure to the United States to start doctoral studies at Harvard. (c) At the start of his academic career in the glasshouse at the University of Canterbury with his *Cotula* (*Leptinella*) collection. (d) Conducting field studies of *Narcissus* in Andalucía, Spain, during March 1990. (e) In his office at Canterbury at the time of his election in 1992 to the fellowship of the Royal Society of London.

diverged during high school, as Peter was attracted to the humanities and David to the sciences.

With scholarships in hand, the Lloyd twins went off to university in 1955, with David entering Canterbury College (now University of Canterbury) at Christchurch and Peter going to Victoria University College, Wellington (now Victoria University). According to Peter, despite this divergence the two "remained very similar in intellect" over the years. At Canterbury, David was the first person to complete the new 4-year B.Sc. honours degree, finishing with first-class honours (1959). Peter was also awarded first-class honours in a Master of Arts degree, and the brothers then both applied for the prestigious Frank Knox Memorial Scholarship and were ranked first (David) and second (Peter), apparently causing some confusion in the application process. As for many young and ambitious students, it was time to leave the academic isolation of New Zealand and obtain "OE" (overseas experience). Both brothers chose to conduct their doctoral studies in the United States, with David going to Harvard University and Peter to Duke University. Today, Peter Lloyd is Emeritus Professor of Economics at the University of Melbourne, Australia, having specialized in international trade. He has some regrets that "I did not collaborate as much as I would have liked with David during our careers, despite the similarities between economic and evolutionary theories." Although they did not publish together, it is probably no coincidence that some of David Lloyd's most interesting theoretical contributions were based on ideas from economic theory, for which he has acknowledged Peter (see Lloyd and Venable 1992).

In 1959 David left New Zealand for the first time (Fig. 1.1b), travelling by boat through the Panama Canal to New York and then on to Boston. At Harvard he initially intended to work on maize genetics with Paul Mangelsdorf, but soon became more attracted to studies of variation and evolution in wild plant populations. For his doctoral dissertation David chose to work with the systematist Reed Rollins on *Leavenworthia*, a genus of annual crucifers rich in floral and mating system diversity. His Ph.D. thesis is a classic study of the causes and consequences of the evolution of selfing from outcrossing and was published in 1965 as a massive, 131-page article in the *Contributions from the Gray Herbarium of Harvard University*. This article is widely cited and contains many of Lloyd's early ideas on selective mechanisms. After completing his thesis in 1964, David returned to New Zealand, taking up a research fellowship at the University of Canterbury. In 1967, David became employed as a lecturer in botany at the University of Canterbury, where he remained for the rest of his career, eventually becoming Professor of Plant Science in 1986. During David's time at Canterbury a generation of researchers in plant ecology and evolution made the long trek to New Zealand to work with him as graduate students, post-doctoral fellows or visiting professors. International visitors enabled Lloyd to keep abreast of the latest developments in the burgeoning field of plant reproductive biology, particularly during the 1970s and 1980s.

David Lloyd's publications during his tenure at the University of Canterbury reveal several features of his character and intellectual development. Lloyd's career was based on ideas, and he believed strongly that the originators of ideas should be acknowledged. Consequently, Lloyd's papers are characterized by reviews of relevant intellectual precedent and he viewed papers by others who ignored this precedent to be disrespectful. Lloyd's publications during the first decade of his career describe his functional interpretations of plant reproduction based on direct observation. These descriptions are largely unsupported by statistical analysis and indeed Lloyd used such approaches sparingly throughout his career. However, Lloyd's conceptual approach to biology eventually led him to formulate his ideas mathematically, allowing him to manipulate concepts in a formal manner that often revealed unexpected conclusions. Although theoretical approaches had dominated population genetics since its inception during the 1930s and were being applied increasingly by animal ecologists led by R. H. MacArthur, ecological and evolutionary botany had remained largely immune to mathematical analysis (but see Lewis 1941; Crosby 1949). Consequently, Lloyd's initial theoretical publications in 1974 (Lloyd 1974a, b, c) were among the first to apply the power of mathematics to the conceptual analysis of plant reproduction. Lloyd was particularly attracted to the optimality approaches being applied in the developing field of evolutionary ecology and used the analogy between constrained optimization and natural selection to great effect in his analysis of reproductive strategies. After convincing himself that simpler phenotypic models often identified the same optima as more complex genetic models (Lloyd 1977), Lloyd focused on phenotypic traits in his mathematical analysis of the evolution of plant reproduction. Interestingly, although Lloyd continued publishing descriptive papers on plant adaptations, he tended to present such observations separately from his theoretical papers. Nevertheless, Lloyd's theory was always motivated biologically and incorporated his intimate knowledge of reproductive mechanisms. The extensive body of concepts that Lloyd developed through keen observation, incisive intellect and realistic theory established him as the founder of the theory of plant reproduction and comprise his enduring legacy.

David Lloyd received many awards and distinctions during his career. He was elected to the Royal Society of New Zealand in 1984 and in 1993 he was made a foreign honorary member of the American Academy of Arts and Sciences. In 1992 he became only the seventh scientist resident in New Zealand to be elected to the Royal Society of London (Fig 1.1e). Signing the charter book in London and seeing Charles Darwin's signature also in the book was of special significance for Lloyd. Darwin had been a major influence on his work, particularly Darwin's pioneering work on plant sexual systems (Darwin 1877). Lloyd's citation from his election certificate to the Royal Society succinctly summarizes his main contributions:

Distinguished for his elegant experimental and theoretical studies of sexuality in flowering plants and of its costs and benefits. He has analyzed the ways in which natural selection may influence the allocation of resources between the sexes and the conflict of interest between maternal investment in the numbers and size of progeny. He has also made major contributions to understanding the special features of the flora of New Zealand and outlying islands.

Tragically, soon after this highpoint of his scientific career, Lloyd's life changed forever. On December 17, 1992 he was admitted to hospital with a mysterious ailment and soon lost his vision and went into a coma. Although David revived from the coma, paralysis effectively ended his career. When illness struck, the hand-written manuscript of Lloyd's *magnum opus*, a volume on evolutionary strategies, was only partially complete and the scientific community was deprived of a major synthesis of his theories on evolution and selection. As discussed below, several papers from this planned book (Lloyd 2000a, b, c) have been published subsequently owing to the efforts of several of Lloyd's closest colleagues, Lynda Delph, Curtis Lively, and Colin Webb. After a long and heroic struggle, David Lloyd died peacefully on May 30, 2006 at his home in Christchurch, with his wife Linda Newstrom-Lloyd, also a reproductive biologist, and several family members by his side.

1.3 Self- and cross-fertilization in plants

We will not understand the evolution of self-fertilization properly until we know more about its functional dimensions as well as the genetic aspects. (Lloyd and Schoen 1992, p. 367)

1.3.1 Early investigations on mating systems

The relative advantages of selfing and outcrossing and their consequences for mating-system evolution intrigued Lloyd throughout his academic career. Lloyd's interest in this topic began with his Ph.D. project on the evolution of self-compatibility in two *Leavenworthia* (Brassicaceae) species, both of which include self-incompatible and self-compatible geographic races (Lloyd 1965). As a result of three field seasons and glasshouse trials, Lloyd recognized 15 races of *Leavenworthia crassa* and four of *Leavenworthia alabamica*, with 13 of the 19 races restricted to a 6.4×7.2 km^2 area. Lloyd concluded that self-compatibility evolved several times within both species and that "there is only one adaptive peak in *Leavenworthia* for those species and races which rely on cross-pollination, but numerous adaptive combinations of the same characters have been adopted by those races which are frequently or predominantly self-pollinated" (p. 82). He also identified 12 trends in floral and inflorescence characteristics associated with increased auto-fertility among races, including a change from outward- to inward-facing anthers and decreases in flower number, petal length, pistil length, pollen:ovule ratio, and mass of individual seeds. Except for flower number, the reduction in floral traits in self-compatible populations is not associated with inbreeding depression (Busch 2005), so that these trends seem to reflect adaptations to selfing. Some of these trends had been observed previously (notably by Darwin 1876), but the observation that pollen:ovule ratio declined with increased auto-fertility (Fig. 1.2) was novel and significant, given Lloyd's conclusion that "...the evolution of self-compatibility and the reduction in the pollen:ovule index have largely preceded other changes" (p. 78). Following an idea from Darwin (1876), Lloyd proposed that "...trends towards decreases in the anther lengths and pollen:ovule indices perhaps reflect increased efficiency in (self-) pollination in these races..." (p. 68). Cruden (1977) later elaborated this idea with extensive comparisons of pollen:ovule ratios in selfing and outcrossing species.

Lloyd used his observations to discriminate between an ecological and a genetical hypothesis for the evolution of self-compatibility in *Leavenworthia*. On the one hand, "earlier workers, including Darwin and Herman Müller, generally attributed the evolution of self-compatibility to the need for an adequate seed set under conditions where cross-pollination was insufficient for this purpose" (Lloyd 1965, p. 128). In contrast, emphasizing selection on the regulation of recombination, Darlington, Mather, and Stebbins had proposed that because of higher homozygosity "a self-fertilizing plant can become more 'closely adapted' to its immediate environment," even though "...self-fertilizing plants achieve an increase in immediate fitness at the expense of a decreased flexibility" (p. 129). Lloyd found support for the reproductive-assurance hypothesis in two results: a qualitative difference in pollinator abundance among large and small populations, and fertilization failure of up to 16% of flowers in populations. In contrast, he argued that the evolution of self-compatibility could not provide an immediate genetic advantage, because his comparison of germination by self- and outcrossed seeds from a largely self-incompatible population indicated strong inbreeding depression. Throughout his career Lloyd was sceptical of population genetic arguments for the evolution of selfing, favouring instead the "retrieval of the cost of meiosis, more assured fertilisation, and easier colonisation" (Lloyd 1979a, p. 604) to account for this frequent transition.

1.3.2 Integration of pollination and mating

Lloyd's emphasis on the ecological context of mating remained a prevalent theme in his research on mating-system evolution. In particular, Lloyd was struck by the paradox that "many topics of floral ecology have been rejuvenated by innovative studies of reproductive strategies for deploying adaptive mechanisms, but the new paradigm has had hardly any impact on the traditional topic of cross- versus self-fertilization" (Lloyd and Schoen 1992, p. 358). In response, he developed an insightful perspective on mating systems, which recognizes that plants can self-pollinate in several ways. Initially, Lloyd (1975a, 1979b) recognized three "modes" of self-pollination for plants with chasmogamous (open) flowers, based on when self-pollination occurs relative to cross-pollination: prior, competing, and delayed. Later,

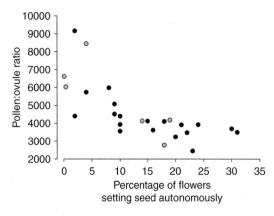

Figure 1.2 Relation of pollen:ovule ratio to the proportion of flowers setting seed autonomously in a glasshouse for populations of *Leavenworthia alabamica* (grey symbols) and *L. crassa* (black symbols). Based on data from Lloyd (1965).

he sub-divided self-pollination that occurs simultaneously with cross-pollination to include: (1) self-pollination within a flower that occurs without the action of a pollen vector (competing self-pollination), or (2) as a result of vector activity (facilitated self-pollination), and (3) self-pollination between flowers on the same or different ramets (geitonogamy) (Lloyd and Schoen 1992, Lloyd 1992a). As Lloyd pointed out, these modes of self-pollination differ in their selective advantages because of differences in their dependence on pollinators, the extent to which they reduce pollen export (pollen discounting) and production of outcrossed seeds (seed discounting), and their ability to provide reproductive assurance (Table 1.1).

Lloyd used this framework to demonstrate mathematically that the contrasting features of the different modes of self-pollination influence the selection of self-fertilization in self-compatible species (Lloyd 1979b, 1992a), leading him to three main conclusions. First, he proposed that modes of self-pollination that result from interaction with pollen vectors "are unavoidable by-products of adaptations for cross-pollination in the sense that they cannot be reduced below certain levels without sacrificing outcrossing" (p. 377). Second, he concluded that the primary advantage of autonomous self-pollination, especially delayed self-pollination, results from reproductive assurance when cross-pollination is inadequate. Lloyd's (1992a) discussion of the relative consequences of limited cross-pollination for the benefits and costs of different selfing modes is a particularly valuable demonstration of the dependence of plant mating on the prevailing pollination environment, which has been demonstrated empirically only recently (Elle and Carney 2003; Kalisz et al. 2004; Moeller and Geber 2005). This recognition of the context-dependent nature of mating led Lloyd to his third conclusion, namely, that superiority of outcross pollen relative to self-pollen in competition for ovule fertilization ("prepotency," Darwin 1876) may commonly allow prior and simultaneous self-pollination to provide reproductive assurance with little penalty. This suggestion remains largely untested.

Lloyd's emphasis on the mating costs of self-pollination significantly influenced subsequent studies of floral biology. Although pollen discounting had been recognized previously as a potentially important consequence of self-pollination (Nagylaki 1976; Holsinger et al. 1984), Lloyd was the first to consider its relation to floral mechanisms (Lloyd 1992a; Lloyd and Schoen 1992). This clarification accelerated a broadening of pollination studies, which had historically focused on the function of individual flowers, to incorporate the function of entire floral displays (Harder

Table 1.1 Principal features of the modes of self-pollination for flowers, after Lloyd (1975a, 1979b, 1992a)

Mode of self-pollination	Timing relative to cross-pollination	Requires action of a pollen vector?	Pollen discounting	Seed discounting during fertilization (zygote discounting)	Provides reproductive assurance?
Prior	Before	No	Limited	Limited to considerable, depending on pollen limitation	Yes
Autonomous, simultaneous[a]	Simultaneous	No	None to considerable	Limited to considerable, depending on pollen limitation	Yes
Facilitated intrafloral[b]	Simultaneous	Yes	Considerable	Limited to considerable, depending on pollen limitation	Yes
Geitonogamy	Simultaneous	Yes	Complete, if transfer within and among plants are identical	Limited to considerable, depending on pollen limitation	No, unless associated with more pollinator visits
Delayed	After	No	None	None to limited	Yes

[a]Lloyd's "competing" self-pollination.
[b]Lloyd's "facilitated" self-pollination.

et al. 2004). In addition, Lloyd was the first to refer specifically to the displacement of outcrossed seeds by self-fertilization as seed discounting. Although this negative relation is implicit in most genetic models of mating-system evolution (e.g., Nagylaki 1976), which typically assume complete fertilization, the consequences of the weakening of this tradeoff when cross-pollination is insufficient for complete fertilization had not been explored. Finally, the alternate modes of selfing that Lloyd recognized have been measured for several species (e.g., Eckert 2000; Johnson et al. 2005), their implications for the evolution of plant mating have received additional theoretical analysis (e.g., Schoen et al. 1996; Morgan and Wilson 2005; Chapter 4) and the differential consequences for maternal fitness have since been demonstrated by Herlihy and Eckert (2002).

Lloyd's mating-system models consistently predicted either exclusive outcrossing or complete selfing, which contrasted with a growing body of evidence that many plants produce mixtures of selfed and outcrossed seeds (Barrett and Eckert 1990; Goodwillie et al. 2005). Although Lloyd (1992a) was aware of several genetic models that predicted mixed mating, he doubted whether the proposed mechanisms could explain its prevalence, given the restricted conditions that they required. In the absence of a general functional explanation, Lloyd concluded that mixed mating generally resulted from combinations of the inevitability of geitonogamy for plants that display multiple flowers simultaneously, and the advantages of reproductive assurance in the face of insufficient pollen dispersal. In contrast, more recent theoretical analysis have identified factors that may select for mixed mating even when seed production is not pollen limited (Goodwillie et al. 2005; Chapter 4).

The role of pollination in governing mating patterns should be self-evident to even the most casual observer of flowers. Nevertheless, during most of the twentieth century research conducted on these two fundamental aspects of plant reproduction took separate courses, with remarkably little cross-fertilization. Lloyd's most important contribution to studies of mating-system evolution was to introduce a functional dimension to the topic by forcing researchers to consider how and why self-pollination occurs, and the demographic and environmental context in which mating takes place (Lloyd 1979b, 1980a, 1992a). This ecological perspective has balanced population genetic approaches, which have traditionally dominated research in this area (e.g., Nagylaki 1976; Charlesworth 1980; Lande and Schemske 1985). The integration of ecological and genetic aspects stimulated by Lloyd has been incorporated increasingly in studies of plant mating (Holsinger 1996; Barrett and Pannell 1999). The functional linkage between pollination and mating provided the theme for one of Lloyd's last projects: an edited volume on floral biology (Lloyd and Barrett 1996) to commemorate the bicentenary of the publication of C. K. Sprengel's (1793) pioneering book on floral adaptations promoting cross-pollination.

1.4 Gender strategies

Morphological descriptions of sex tend to rely on appearance not function. Moreover, they ignore the fact that the sexual performance of a flower or plant depends not only on its own nature, but also on the gametes produced by other flowers and plants in the same population. (Lloyd 1980b, p. 104)

1.4.1 Early investigations of plant sexual diversity

On returning to New Zealand from Harvard, Lloyd finished writing several additional papers on *Leavenworthia* from his doctoral research (Lloyd 1967, 1968a, b, 1969) and then turned his attention to local research problems. Because Lloyd's *Leavenworthia* studies had established his interest in reproductive biology, particularly the evolution of mating systems, it is not surprising that he now became interested in another major aspect of this topic: the evolution of gender strategies. His colleague Eric Godley had helped to establish that the New Zealand flora was particularly rich in species with unisexual flowers (dicliny), including many gynodioecious and dioecious taxa (reviewed in Godley 1979; Webb et al. 1999). Lloyd therefore chose *Cotula* (now *Leptinella*), which is particularly rich in sexual diversity, to investigate the functional basis of this variation (Fig. 1.1c). For the next 15 years, in addition to taxonomically revising *Cotula* section Leptinella (Lloyd 1972a), Lloyd focused much of his energies on three main topics: the evolution and maintenance of sexual systems, sex ratios in gynodioecious and dioecious populations, and the concept of gender and its measurement in populations. This work was published in both empirical and theoretical papers from 1972 to 1984 and established Lloyd as the foremost authority on the ecology and evolution of plant sexual systems. Later in his career he also collaborated with his Ph.D. student Lynda Delph on her studies of gender dimorphism in New Zealand *Hebe* (Delph and Lloyd 1991, 1996) and with Mark Schlessman and colleagues on the evolution of sexual systems in New Caledonian Araliaceae (Schlessman *et al.* 1990a, b, 2001).

The four papers on *Cotula* (Lloyd 1972b, c, 1975b, c) published by Lloyd at the start of his professional career are of historical interest, because they contain several themes, which he later developed more fully. For example, he recognized in *Cotula* that the traditional morphological categories used to classify sexual systems were insufficient for evolutionary studies because they ignored function. Moreover, the typology involved failed to recognize that sex expression often varies quantitatively in plants (e.g., Fig. 1.3) and that sex inconstancy is common, especially in diclinous populations. Rank-frequency curves of the sex expression of individuals are featured in the *Cotula* papers and presage Lloyd's later development of quantitative measurements of functional gender (e.g., Lloyd 1980b). Also, his discussion of energy expenditure on floral adaptations that promote outcrossing represent the beginning of his interest in allocation strategies in outcrossing and selfing plants (e.g., Lloyd 1984, 1987a). Lloyd's comparative and functional analyses of sex expression in *Cotula* species indicated that there was considerable evolutionary lability of sexual systems in the genus. He proposed transitions from gynomonoecy to monoecy and then to dioecy, but with reversions to monoecy: inferences that can now be tested using phylogenetic methods. His finding of an association between polyploidy and dioecy in *Cotula* (Lloyd 1975b) anticipated current work on this topic (Miller and Venable 2000) and his studies of sex ratios paved the way for his later theoretical treatments of this topic.

Sex ratios in dioecious populations should generally be close to unity after the period of parental investment, as a result of negative frequency-dependent selection (Fisher 1930). Lloyd was interested in examining this proposition and determining the mechanisms that could account for biased sex ratios. With Ph.D. student Colin Webb he undertook surveys of sex ratios in sexually dimorphic genera of Umbelliferae in New Zealand and found that male-biased sex ratios occurred most commonly (Lloyd 1973, Webb and Lloyd 1980). They proposed that differences in the costs of reproduction between females and males could explain male-biased sex ratios, particularly

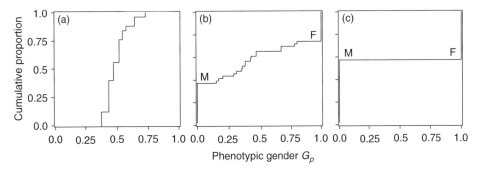

Figure 1.3 The use of Lloyd's method for illustrating gender variation within and among populations. Variation in phenotypic gender for three populations of *Sagittaria latifolia* (Alismataceae): (a) a completely hermaphroditic population, (b) a population containing females, hermaphrodites, and males, and (c) a dioecious population. Based on data from Sarkissian *et al.* 2001.

in long-lived species for which repeated episodes of reproduction magnify differential costs. Their major review of this topic (Lloyd and Webb 1977), which today is Lloyd's most cited publication, documented differences between the sexes and highlighted how their distinct roles in sexual reproduction influence the evolution of sex ratios. Lloyd also became interested in the mechanisms responsible for female-biased sex ratios, which are reported less often (Lloyd 1974b). In this paper he effectively disposed of previous group-selection arguments based on maximizing seed production at the population level and argued that the common occurrence of female bias in species with sex chromosomes could result from differential fertilization of ovules by female- versus male-determining gametophytes as a consequence of the genetic differentiation of sex chromosomes. This interesting idea has some support, although the mechanism(s) responsible are still unclear (Conn and Blum 1981; Stehlik and Barrett 2004).

1.4.2 Theories on the evolution of sexual systems

Lloyd made important theoretical contributions to our understanding of the evolution and maintenance of dimorphic sexual systems. He developed analytical models to explain the maintenance of gynodioecy and androdioecy based on survival, seed fertility, modes of selfing, and inbreeding depression with both nuclear and cytoplasmic modes of inheritance (Lloyd 1975a). The models provided insights into the strikingly different frequencies of these two sexual systems among angiosperm species and the common occurrence of gynodioecy on islands, such as Hawaii and New Zealand. His model predicting sex ratios in gynodioecious species based on the transmission of genes via pollen and ovules (Lloyd 1976) redefined our concept of this dimorphic condition. Lloyd demonstrated that although gynodioecy was generally viewed as a type of sexual system, its boundaries are in fact not distinct, with populations merging into hermaphroditism and dioecy at opposite extremes.

One of Lloyd's most controversial decisions arising from this work was to refer to hermaphroditic plants in gynodioecious and sub-dioecious populations as "males" (inconstant or fruiting males), even though they produce ovuliferous flowers. This decision recognized that the fitness of fruiting males in such populations results largely from contributing pollen to female plants. However, because the notion that males can produce seed is counterintuitive, Lloyd's terminology was not adopted in a recent volume on gender and sexual dimorphism in plants (Geber *et al.* 1999; Sakai and Weller 1999, pp. 7–8), despite many loyal adherents.

Finally, Lloyd (1982) introduced the concepts of pollen and seed shadows for understanding the selection of dioecy from cosexuality. This idea has been extended recently to demonstrate theoretically that the separation of sexes increases the variance in successful pollen and seed dispersal, which reduces mean offspring recruitment compared with hermaphrodites (Heilbuth *et al.* 2001; Wilson and Harder 2003). As a consequence, dioecious species are less competitive than hermaphroditic species unless they live at high density, or have specific mechanisms that expand pollen and/or seed shadows.

1.4.3 Gender concepts and theory

The hallmark of Lloyd's work on plant sexual systems was his elaboration of the concept of gender. In contrast to the term "sex," which reflects phenotype, "gender" describes the relative genetic contribution of individuals to the next generation as female and male parents, or their functional "femaleness" or "maleness" (Lloyd 1979b). Lloyd was the first to recognize that despite the complexity of plant sexual diversity, virtually all species can be classified into two distinct gender strategies, depending on whether populations are monomorphic or dimorphic for gender (Lloyd 1980b, c). Populations with "gender monomorphism" show quantitative (unimodal) variation in gender (Fig. 1.3a) and usually comprise individuals that produce offspring through both ovules and pollen (cosexes). Alternatively, populations with "gender dimorphism" show strong bimodality in gender, with two distinct sexual morphs that function primarily, but often not exclusively, as female or male parents (Fig. 1.3c).

Lloyd developed quantitative methods for describing plant gender (Lloyd 1979b, 1980b, c; Lloyd and Bawa 1984). Ideally the functional gender of individual i would be measured by the proportion of all genes that it transmits as a female (f_i) and male parent (m_i),

$$G_f = \frac{f_i}{f_i + m_i}. \tag{1}$$

However, paternal genes are difficult to track in populations, so that Lloyd proposed a phenotypic measure,

$$G_p = \frac{o_i}{o_i + p_i E}, \tag{2}$$

in which o_i and p_i are the numbers of ovules and pollen grains produced by individual i, respectively, and E is the ratio of ovules to pollen grains in the population as a whole. Therefore, phenotypic gender considers the production of pollen and ovules (or seeds) of individual plants relative to the average ratio of expenditure in the population. Note that Lloyd (1979b, 1980b, c) originally referred to G_p as functional gender, but changed terminology in his 1984 review with K. S. Bawa to that given in eq. 1 and 2. Values of G_p can range from 0 to 1, denoting in the extremes strictly male and female plants, respectively (Fig. 1.3c). G_p has now been measured to describe gender strategies in a wide range of flowering plants (e.g., Fig. 1.3: also see Thomson and Barrett 1981; Lloyd and Bawa 1984; Wolfe and Shmida 1997; Vaughton and Ramsey 2002). However, truly functional measurement of gender based on mating success using genetic markers is still in its infancy (Elle and Meagher 2000; Morgan and Conner 2001). Moreover, the problem with relative measurements of performance, as used in Lloyd's gender formulation, rather than absolute measures of performance remains a thorny issue for determining fitness in plant populations in which gender varies with plant size (Sarkissian et al. 2000; Chapter 3).

In addition to clarifying gender, Lloyd (1984) considered theoretical explanations for the strongly biased allocation of reproductive resources to female function in animal-pollinated cosexual plants. Previous analyses of this problem had identified a variety of mechanisms that could cause such unequal allocation (e.g., Charlesworth and Charlesworth 1981; Charnov 1982); however they had not assessed the relative importance of these mechanisms. Based on his own analysis, Lloyd concluded that "(a)n upper limit on paternal fitness offers the most promise of explaining the observed deviations emphasizing female expenditure" (p. 298). With characteristic perceptiveness, Lloyd then recognized that such limits would invoke "intra-sexual selection to increase the proficiency of pollen donation," resulting in "increasing precision of pollination ... and morphological trends towards reduction and fusion of floral parts and zygomorphy" (p. 300). Thus Lloyd viewed gender allocation and the evolution of floral mechanisms as integrated components of the reproductive strategies of plants.

1.5 Allocation strategies

The central consideration of adaptive strategies is how individual organisms are selected to deploy their limited resources among various structures and behaviors.
(Lloyd 1989, p. 185)

The allocation of time, energy and nutrients governs many aspects of organisms' lives, including phenology, gender and sex ratio, fecundity, growth, defence, and intrinsic longevity. Indeed the life history represents a series of allocations (e.g., age at first reproduction, schedule of reproductive effort) that fundamentally determine fitness. The life history mediates the effects of all physiological, morphological, and behavioural traits on fitness (Roff 1992), so that selection on most traits depends on their influences on allocation patterns.

Given the biological importance of allocation, Lloyd's active analysis of optimal allocation patterns, particularly those affecting reproduction, is not surprising. His interest in allocation seems to have developed from his early studies of self- versus cross-fertilization, gender, and sex ratio. In February 1979, Lloyd presented the ground plan for his subsequent studies of allocation at an international symposium on "Reproduction in Flowering Plants" in Christchurch, New Zealand, identifying six "parental strategies" which commanded his attention for the remainder of his academic career: gender strategies; relative maternal and paternal expenditures; size-number compromises; the temporal control of maternal investment; and breeding patterns, including the relative incidence of sexual versus asexual reproduction and of self- and cross-fertilization (Lloyd 1979a). During the next decade he developed the powerful insight that, instead of six strategies, "(t)he diversity of evolutionarily stable strategies (ESSs) can be classified into two major superfamilies, allocation strategies and size-number strategies, which involve additive and multiplicative expenses respectively" (Lloyd 1989, p. 185). Lloyd's analysis of these strategic superfamilies broadened his scientific contributions from botanical subjects to the entire scope of evolutionary ecology.

Before reviewing the two classes of allocation strategies, we point out that Lloyd's basic conception of allocation problems was not unique (e.g., see Smith and Fretwell 1974; Charnov 1982). However, his approach differs in two ways from previous analyses of allocation. First, in Lloyd's opinion mathematical "procedures used (previously) are abstruse to all but the most numerate biologists" (1984, p. 281), prompting him to adopt a more intuitive approach, which we describe below. Second, previous analyses of allocation problems had largely considered specific situations (e.g., sex allocation, Charnov 1982), whereas Lloyd developed a general framework that emphasized the conceptual unity of allocation problems, rather than their functional diversity.

1.5.1 Allocation to competing functions

Organisms must commonly divide a limited resource among two or more functions, each of which affect fitness. As examples, Lloyd (1985) mentioned the following: the production of the different kinds of structures that perform the same function in different circumstances (e.g., shade versus sun leaves), diet composition and habitat use of foraging animals, division of altruistic acts among relatives, and production of different sterile castes by social insects. To the extent that investment in these functions is drawn from a common resource pool, increased investment in one function must decrease investment in others. In this situation, natural selection favours the allocation pattern that maximizes overall fitness, rather than performance of any individual function. As Lloyd demonstrated in a series of papers, this optimal allocation pattern satisfies a specific set of characteristics.

As a simple example, consider two functions, A and B (such as the production of ovules versus pollen), which are financed by the same resource pool. Suppose that proportion a of these resources is allocated to A and the remainder, $b = 1 - a$, is allocated to B. These functions contribute to fitness (w) according to $w \propto f(a)$ (Fig. 1.4a, dashed line)

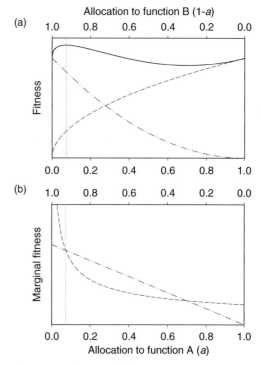

Figure 1.4 An example of aspects of allocation to competing functions. Panel (a) illustrates the relation of fitness contributions by two functions, A (dashed line) and B (dashed–dotted line), and total fitness (solid line) to allocation in the competing functions. Panel (b) depicts the first derivatives of the fitness contributions (marginal fitnesses) by functions A (dashed line) and B (dashed-dotted line). The vertical dotted line indicates the allocation that maximizes total fitness, at which the marginal fitnesses are equal (and $\partial^2 w/\partial a^2 < 0$).

and $w \propto g(1-a)$ (Fig. 1.4a, dashed dotted line), with total fitness equal to the sum of the fitness contributions,

$$w = f(a) + g(1-a) \quad (3)$$

(Fig. 1.4a, solid line). Lloyd (1988) showed that at the optimal allocation ($\hat{a}$) the fitness gain through one function associated with a tiny change in allocation exactly equals the fitness loss through the other function, so that

$$\frac{\partial w}{\partial \hat{a}} = -\frac{\partial w}{\partial [1-\hat{a}]} = \frac{\partial w}{\partial \hat{b}} \quad (4)$$

(Fig. 1.4b), as long as fitness is maximized at some intermediate allocation ($0 < \hat{a} < 1$: note that $\hat{a}$ is an optimum only if $\partial^2 w/\partial \hat{a}^2 < 0$). This equality of marginal fitnesses at the optimal allocation occurs even though the absolute fitness contributions by each function may differ considerably (Fig. 1.4). Furthermore, equality of marginal fitnesses occurs for any number of functions that compete for the same resource (see Lloyd and Venable 1992; Venable and Lloyd 2004).

1.5.2 Size–number compromises

Organisms often engage in processes that involve reiteration, either simultaneously (e.g., production of many pollen grains or ovules in individual flowers), or sequentially (e.g., catching prey in a series of patches during a single foraging bout). In such cases, a single pool of R resources is divided into n units of relatively equal size, s. In general, an individual's fitness is a positive function of both the number of units produced [$n = f_n(s)$] and their size [$f_s(s)$], so that

$$w = f_n(s) f_s(s) \quad (5)$$

(Lloyd 1987b). Notice that, in contrast to the allocation of resources among competing functions (eq. 3), fitness now depends on a product, rather than on a sum. Resource limitation creates an inverse relation between the number and size of units that can be produced, $n = R/s$ (Fig. 1.5a), so selection cannot maximize unit size and number simultaneously. Instead, selection favours the combination of unit size and number that

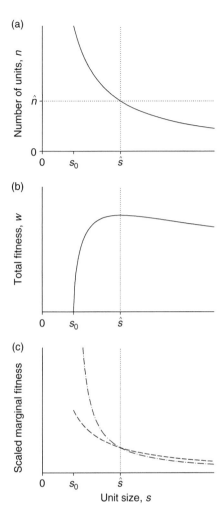

Figure 1.5 An example of aspects of size–number compromises. Panel (a) depicts the underlying tradeoff between unit size (s) and number (n) caused by resource limitation. Panel (b) illustrates the total fitness that an individual derives from all units produced given that fitness increases linearly with unit number and increases in a decelerating manner with unit size above a minimum viable unit size (s_0). Panel (c) illustrates the influences of unit size on the first derivatives of the fitness contributions (marginal fitnesses) by unit size (dashed line) and number (dashed–dotted line) divided by the absolute fitness contribution (the scaled marginal fitness for unit number has been multiplied by -1). The horizontal and vertical dotted lines indicate the unit number and size, respectively, that maximizes total fitness.

maximizes the total fitness resulting from all units, given the size–number tradeoff.

Lloyd's formal analysis of size–number compromises began in 1987, when he published a mathematical model that generalized Smith and Fretwell's (1974) graphical model of offspring size (Lloyd 1987b). According to this model, when individuals produce units of the optimal size ($\hat{s}$)

$$\frac{\partial f_s(\hat{s})}{\partial \hat{s}} \cdot \frac{1}{f_s(\hat{s})} = -\frac{\partial f_n(\hat{s})}{\partial \hat{s}} \cdot \frac{1}{f_n(\hat{s})}, \quad (6)$$

where $\hat{n} = R/\hat{s}$ (Fig. 1.5c). This result resembles eq. 4 in that at the optimal size–number compromise the fitness gain resulting from a tiny increase in one fitness component (e.g., unit size) is counteracted by the fitness loss resulting from the reduction in the other component (e.g., unit number). However, in this case the marginal fitnesses for unit size and number at the optimum are divided by their corresponding absolute fitness contributions, rendering the scaled marginal fitnesses dimensionless, so that they can be compared directly. As Lloyd noted, when fitness varies linearly with unit number eq. 6 simplifies to

$$\hat{s} = \frac{f_s(\hat{s})}{\partial f_s(\hat{s})/\partial \hat{s}}, \quad (7)$$

which is equivalent to the well-known marginal-value solution that Charnov (1976) derived for the optimal patch-residency time for a foraging animal. As eq. 7 demonstrates, in this special case the optimal unit size does not depend on unit number, so that analysis can focus on production of a single unit (or visit to a single patch).

1.5.3 Application to specific problems

Equations 4 and 6 are powerful for two reasons. First, by recognizing that allocation to competing functions and size–number compromises are recurring and pervasive themes in the lives of organisms, Lloyd provided two general solutions that illustrate the unity of biological processes. For example, this insight allowed Lloyd (1989) to analyze the seemingly unrelated topics of self-incompatibility and sterile castes in eusocial insects as facets of the same general problem. Second, eq. 4 and 6 illustrate that the optimal solutions to all allocation problems depend primarily on how fitness changes as allocation is modified (marginal fitness), rather than on the absolute fitness realized from any specific allocation pattern. Therefore, analysis of any allocation problem should focus on identifying the allocation pattern for which a small change in allocation is accompanied by exactly counterbalancing changes in marginal fitness.

Given the practical difficulties of measuring fitness contributions and of modifying allocation patterns to characterize marginal fitness, Lloyd's general predictions (eq. 4 and 6) are difficult to test directly. However, as Lloyd illustrated with many theoretical examples (e.g., Lloyd 1983, 1984, 1985, 1987a, b, 1988, 1989, 1992b), inclusion of details concerning the influences on fitness leads to specific predictions about allocation patterns that allow indirect tests of the equality of marginal fitnesses.

1.6 Floral mechanisms

(I)ntra-sexual selection to increase the proficiency of pollen donation, particularly the number of visitors that can remove pollen from a flower, is the major selective force guiding floral evolution. (Lloyd 1984, p. 300)

Understanding the function of flowers was a major focus for much of Lloyd's research career. He recognized that the two most significant events in the history of floral biology were the realization that many features of flowers facilitate cross-fertilization (Sprengel 1793), and that the progeny from cross-fertilization generally perform better than those from self-fertilization (Knight 1799). As Darwin had shown from extensive studies of floral mechanisms that promote outcrossing (Darwin 1862, 1877) and experiments on inbreeding depression (Darwin 1876), these two concepts could explain the function of flowers. As indicated by the quotation preceding this paragraph, Lloyd embraced the developing perspective (e.g., Willson 1979, 1994) that traits controlling mating in angiosperms are subject to sexual selection, even though most species are hermaphroditic. At the time, reproductive botany was being transformed by the recognition that plants may also be subject to Bateman's (1948) Principle, namely that male success is typically limited by mating opportunities, whereas female success is often limited by resource availability (although see Burd 1994; Wilson *et al.* 1994). Although not the initiator of sexual selection theory for plants (a source of some consternation), Lloyd helped develop this perspective through his consideration of how floral traits influence pollen export (Lloyd and Yates 1982; Lloyd 1984; and see Section 1.4.3), including the implications of different modes of self-pollination for pollen discounting (see Section 1.3.2). We now review the two aspects of floral mechanisms that Lloyd considered most fully.

1.6.1 Sexual interference

Following Darwin, most workers during the twentieth century interpreted floral traits, such as self-incompatibility, heterostyly, dicliny, dichogamy, and herkogamy, as mechanisms that function solely to promote outcrossing (Richards 1997). However, Lloyd recognized difficulties with this as a universal explanation, because many species possess several of these mechanisms and this redundancy seemed unnecessary. To resolve this paradox, Lloyd developed an alternative hypothesis that recognized the seemingly obvious fact that most flowering plants are hermaphroditic and thus acquire fitness as both maternal and paternal parents. While not minimizing the significance of anti-selfing mechanisms, he proposed that some floral traits reduce conflict between female and male function, which he called pollen–stigma interference (Lloyd and Yates 1982, p. 904). Using protandrous *Wahlenbergia albomarginata* as an example, Lloyd and Yates postulated that secondary pollen presentation in this species (and other members of the Campanulaceae) was a mechanism for segregating pollen and stigma function, thus avoiding the sexual interference that is an inevitable consequence of hermaphroditism.

In two classic publications with Webb in the *New Zealand Journal of Botany* on the functions of dichogamy (Lloyd and Webb 1986) and herkogamy (Webb and Lloyd 1986) the concept of sexual interference was developed more fully and its significance for floral evolution outlined. "We postulate that selection to avoid pollen-stigma interference is virtually universal in outcrossing flowering plants, and that such selection is responsible in whole or in part for diverse floral features" (Lloyd and Webb 1986, p. 138). In these papers, Lloyd and Webb provided the first functional classification of the diverse forms of dichogamy and herkogamy, using as examples many species from the New Zealand flora. They argued that as a consequence of Bateman's (1948) Principle the avoidance of self-interference usually increases the proficiency of pollen dispersal, thereby benefiting paternal fitness more than maternal fitness. Recent experimental studies support this hypothesis (Fetscher 2001; Routley and Husband 2003). Intriguingly, Lloyd and Webb (1986) considered the consequences of dichogamy only for within-flower interference, even though Darwin (1877) had clarified its role in inflorescence function and Lloyd (1979a) had previously considered some consequences of geitonogamy, which can be viewed as female interference with pollen export. The role of dichogamy in reducing geitonogamy and the associated pollen discounting have since been demonstrated empirically (Harder *et al.* 2000; Routley and Husband 2003) and its influence on inflorescence architecture has been considered theoretically (Jordan and Harder 2006).

Lloyd and Webb (1986) were careful to distinguish sexual interference in stigma and anther function from sexual conflict in organisms with separate sexes, a topic of considerable current interest (Arnqvist and Rowe 2005). During the interaction between flowers and pollinators, the same position for stigmas and anthers may typically maximize pollen import and export, respectively, in the absence of interference. However, because of the interference that would result if these organs occupied the same position, selection favours temporal or spatial separation of female and male organs, resulting in dichogamy or herkogamy, respectively. In contrast, with sexual conflict females and males have different optima, creating discord because the sexes rely on each other.

1.6.2 The evolution of heterostyly

Lloyd next turned his attention to the long-standing puzzle of the evolution and adaptive significance of heterostyly. Darwin's (1877) volume on plant sexual systems was devoted largely to heterostyly and several prominent evolutionary biologists from the U.K. (e.g., R. A. Fisher, J. B. S. Haldane, K. Mather, D. Lewis, B. Charlesworth, and D. Charlesworth) had worked on these floral polymorphisms. However, questions remained concerning the evolution of heterostyly. It did not escape Lloyd's attention that unlike his research on gender strategies, which are exceptionally well represented in the New Zealand flora, he was now studying a problem for which there were no representative species native to New Zealand (Godley 1979). Perhaps this was an advantage, as it enabled him to approach heterostyly in a fresh way, unencumbered by the details of a particular group, such as *Primula* on which much previous literature was concentrated (reviewed in Richards 1997). However, Lloyd did study heterostyly outside of New Zealand and his interest in *Narcissus* took him to the Iberian Peninusla in 1990 (Fig. 1.1d) in an effort to settle a long-standing controversy concerning the nature of sexual polymorphisms in this genus (Lloyd et al. 1990; Barrett et al. 1997). This work is discussed further in Chapter 13.

Lloyd's contributions on heterostyly extended earlier research with Jocelyn Yates (Lloyd and Yates 1982) and Webb (Webb and Lloyd 1986) on sexual interference. These ideas were developed more fully in two book chapters with Webb on the evolution and selection of distyly (Lloyd and Webb 1992a, b). These chapters made several novel contributions that continue to influence research on this topic. Significantly, Lloyd and Webb argued that the prevailing hypothesis for the order of establishment of morphological and physiological traits in the heterostylous syndrome (Yeo 1975; Ganders 1979) was probably incorrect. Earlier theoretical work by Charlesworth and Charlesworth (1979) proposed that the diallelic incompatibility system that occurs in most distylous species evolved first, setting up conditions that favoured selection for the reciprocal arrangement of stigma and anther heights (reciprocal herkogamy) that characterizes heterostylous species. In the Charlesworths' model, inbreeding avoidance is the primary selective mechanism resulting in the establishment of incompatibility in populations. Selection for efficient cross-pollination then leads to the establishment of reciprocal herkogamy.

Comparative data and a phenotypic selection model based on pollen transfer led Lloyd and Webb to propose the opposite sequence from that depicted in the Charlesworths' model. They argued that reciprocal herkogamy was likely to evolve first through selection for more proficient cross-pollination and that incompatibility evolves subsequently due to a combination of specialization for legitimate pollination and active selection to restrict self-fertilization (Fig. 1.6). This sequence revived Darwin's (1877) original proposal that the reciprocal herkogamy of heterostylous taxa preceded the evolution of diallelic incompatibility. It also supports Darwin's interpretation that heterostyly evolves principally to promote legitimate cross-pollination, rather than to avoid inbreeding.

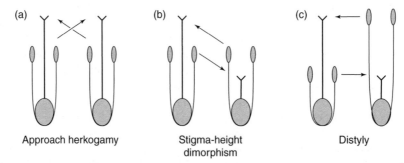

Figure 1.6 The Lloyd and Webb (1992a, b) model for the evolution of distyly. (a) The ancestral condition involves a monomorphic population of animal-pollinated plants with approach herkogamy. (b) A short-styled variant invades, creating a polymorphic population containing two morphs that differ in style length, but not anther height (stigma-height dimorphism). (c) Selection to improve the proficiency of cross-pollination results in the establishment of an anther-height polymorphism and hence the evolution of distyly (reciprocal herkogamy). Arrows indicate the principal types of pollen transfer postulated for each condition.

Recent comparative and experimental work has supported Lloyd and Webb's models (Kohn and Barrett 1992; Stone and Thomson 1994; Graham and Barrett 2004); however, it is too soon to exclude the possibility that some heterostylous groups followed the opposite path.

To test whether heterostyly promotes legitimate pollination Lloyd and Webb (1992b) introduced a novel means of analyzing empirical data on the composition of pollen loads on stigmas of the floral morphs. During the preceding 20 years, many workers had sought direct evidence for the Darwinian hypothesis based on pollen-load data, with limited success (reviewed in Ganders 1979). Lloyd and Webb (1992b) reanalyzed published data for distylous *Jepsonia heteranda* (Ganders 1974) and tristylous *Pontederia cordata* (Barrett and Glover 1985) using a different method of calculation that allows pollen loads to be examined in terms of male function, as well as the more conventional female function. They used this approach because their pollen-transfer models indicated stronger selection for more proficient cross-pollination on male than on female fitness. Their empirical analysis revealed that the average proficiencies of legitimate donation and receipt were approximately twice those of the corresponding illegitimate combinations, a surprising finding given the mixed results of previous studies. Therefore, instead of there being little support for Darwin's hypothesis, as most authors had concluded, Lloyd and Webb's reanalysis provided convincing evidence for the function of heterostyly. Indeed, the two-fold advantage they detected represents strong selection, even for polymorphic traits. As they concluded, "few selection hypotheses can claim this degree of support" (Lloyd and Webb 1992b, p. 202).

1.7 Lloyd's evolution

David Lloyd's work spanned 30 years and has significantly shaped understanding of the ecology and evolution of flowers and several other areas of plant reproductive biology. His papers have been cited over 5000 times and stimulate much current theoretical and empirical work in this field. Of the 89 journal articles and book chapters that Lloyd wrote during his career, 48 were solo-authored, indicating that he often worked independently and that many of the insights presented in Lloyd's papers are uniquely his own. Lloyd collaborated increasingly as his career developed, although this interaction primarily involved empirical, rather than theoretical projects. While a faculty member at Canterbury, Lloyd trained few post-graduate students (3 M.Sc. and 5 Ph.D. students), at least by North American standards, although several (Lynda Delph, Philip Garnock-Jones, Alastair Robertson, Colin Webb) have gone on to establish their own reputations in evolutionary biology and systematics. The burgeoning growth of reproductive biology during the late 1970s resulted in Lloyd being invited frequently to North America and Europe as a visiting scholar and his lectures and seminar courses strongly influenced a generation of students. Although Lloyd was not naturally gregarious and at times could be quite reserved, he greatly enjoyed chatting with eager graduate students about their work and providing constructive criticism if that was appropriate. Visits abroad were critical for keeping Lloyd abreast of the latest developments and as his reputation grew he took full advantage of travel away from the relative isolation of New Zealand. Nevertheless, that isolation may have distanced Lloyd from academic fads, facilitating his independent development of novel approaches and perspectives.

It is clear from Lloyd's academic career that he began as a well-trained botanist with keen observational skills and a background in systematics, in the tradition of the time. As his career progressed he became increasingly drawn to the works of W. D. Hamilton and G. C. Williams, rather than botanical icons, such as Stebbins, Grant, and Baker, who had influenced his early work. After Darwin, Hamilton was Lloyd's evolutionary hero and his visit to Oxford to meet Hamilton was especially significant to him. Indeed, the last lecture that Lloyd attended, blind, paralysed and in a wheelchair, was to hear Hamilton when he visited Christchurch in 1995 and he was thrilled when Hamilton visited him at Burwood Hospital later for a chat. During the 1980s, Lloyd increasingly embraced the optimality approach of evolutionary ecology, which zoologists had borrowed from economics, and applied this general framework to examine the adaptive nature of diverse plant characteristics, ranging from the function of specific floral traits, such as the retractile hairs on *Wahlenbergia* styles, to the strategic deployment of resources within plants that govern their life history.

Towards the end of his career, Lloyd's strategic perspective increasingly allowed him to consider topics beyond the plant kingdom. Indeed, when tragic illness ended his career, Lloyd was working on a book on the theory of natural selection in which he was to recast arguments about phenotypic models of selection and replace the role of kin selection with family selection in many circumstances. Although this unfinished work concerned the evolution of social behaviour, Lloyd, because of his strong love of natural history, found remarkable adaptive parallels for traits and behaviours between animal and plant kingdoms. David Lloyd not only provided the conceptual foundation for floral biology, but he also contributed fundamentally to the theoretical underpinnings of evolutionary ecology.

Acknowledgements

We thank the many colleagues who have shared their experiences and knowledge of David Lloyd with us. In particular, we thank Linda Newstrom-Lloyd, Peter Lloyd, Trevor Lloyd, Colin Webb, Linley Jesson, and Alastair Robertson for comments on this chapter.

References

Arnqvist G and Rowe L (2005). *Sexual conflict*. Princeton University Press, Princeton, NJ.
Baker HG (1979). Anthecology: Old Testament, New Testament and Apocrypha. *New Zealand Journal of Botany*, **17**, 431–40.
Barrett SCH (2001). The Baker and Stebbins era comes to a close. *Evolution*, **55**, 2371–4.
Barrett SCH and Eckert CG (1990). Variation and evolution of mating systems in seed plants. In S Kawano, ed. *Biological approaches and evolutionary trends in plants*, pp. 2229–54. Academic Press, London.
Barrett SCH and Glover DE (1985). On the Darwinian hypothesis of the adaptive significance of tristyly. *Evolution*, **39**, 766–74.
Barrett SCH and Pannell JR (1999). Metapopulation dynamics and mating-system evolution in plants. In PM Hollingsworth, RM Bateman, and RJ Gornall, eds. *Molecular systematics and plant evolution*, pp. 74–100. Taylor and Francis, London.
Barrett SCH, Cole WW, Arroyo J, Cruzan MB, and Lloyd DG (1997). Sexual polymorphisms in *Narcissus triandrus* (Amaryllidaceae): is this species tristylous? *Heredity*, **78**, 135–45.
Bateman AJ (1948). Intra-sexual selection in *Drosophila*. *Heredity*, **23**, 349–68.
Burd M (1994). Bateman's principle and plant reproduction: the role of pollen limitation in fruit and seed set. *Botanical Review*, **60**, 83–139.
Busch JW (2005). Inbreeding depression in self-incompatible and self-compatible populations of *Leavenworthia alabamica*. *Heredity*, **94**, 159–65.
Charlesworth B (1980). The cost of sex in relation to mating system. *Journal of Theoretical Biology*, **84**, 655–71.
Charlesworth D and Charlesworth B (1979). A model for the evolution of distyly. *American Naturalist*, **114**, 467–57.
Charlesworth D and Charlesworth B (1981). Allocation of resources to male and female functions in hermaphrodites. *Biological Journal of the Linnean Society*, **15**, 57–74.
Charnov EL (1976). Optimal foraging, the marginal value theorem. *Theoretical Population Biology*, **9**, 129–36.
Charnov EL (1982). *The theory of sex allocation*. Princeton University Press.
Conn JS and Blum U (1981). Sex ratio of *Rumex hastatulus*: the effect of environmental factors and certation. *Evolution*, **35**, 1108–16.
Crawford DJ and Smocovitis VB, eds. (2004). *The scientific papers of G. Ledyard Stebbins (1929–2000)*. ARG Gantner Verlag, Ruggell, Liechtenstein.
Crosby JL (1949). Selection of an unfavorable gene-complex. *Evolution*, **3**, 212–30.
Cruden RW (1977). Pollen-ovule ratios: a conservative indicator of breeding systems in flowering plants. *Evolution*, **31**, 32–46.
Darwin CR (1862). *The various contrivances by which orchids are fertilised by insects*. Murray, London.
Darwin CR (1876). *The effects of cross and self fertilisation in the vegetable kingdom*. Murray, London.
Darwin CR (1877). *The different forms of flowers on plants of the same species*. Murray, London.
Delph LF and Lloyd DG (1991). Environmental and genetic control of gender in the dimorphic shrub *Hebe subalpina*. *Evolution*, **45**, 1957–64.
Delph LF and Lloyd DG (1996). Inbreeding depression in the gynodioecious shrub *Hebe subalpina* (Scrophulariaceae). *New Zealand Journal of Botany*, **34**, 241–7.

Eckert CG (2000). Contributions of autogamy and geitonogamy to self-fertilization in a mass-flowering, clonal plant. *Ecology*, **81**, 532–42.

Elle E and Carney R (2003). Reproductive assurance varies with flower size in *Collinsia parviflora* (Scrophulariaceae). *American Journal of Botany*, **90**, 888–96.

Elle E and Meagher TR (2000). Sex allocation and reproductive success in the andromonoecious perennial *Solanum carolinense* (Solanaceae). 2. Paternity and functional gender. *American Naturalist*, **156**, 622–36.

Fetscher AE (1999). Stigma behavior in *Mimulus aurantiacus* (Scrophulariaceae). *American Journal of Botany*, **86**, 1130–5.

Fisher RA (1930). *The genetical theory of natural selection*. Oxford University Press, London.

Ganders FR (1979). The biology of heterostyly. *New Zealand Journal of Botany*, **17**, 607–35.

Geber MA, Dawson TE, and Delph LF, eds. (1999). *Gender and sexual dimorphism in flowering plants*. Springer-Verlag, Berlin, Germany.

Godley EJ (1979). Flower biology in New Zealand. *New Zealand Journal of Botany*, **17**, 441–66.

Goodwillie C, Kalisz S, and Eckert CG (2005). The evolutionary enigma of mixed mating systems in plants: occurrence, theoretical explanations and empirical evidence. *Annual Review of Ecology, Evolution and Systematics*, **36**, 47–79.

Graham SW and Barrett SCH (2004). Phylogenetic reconstruction of the evolution of stylar polymorphisms in *Narcissus* (Amaryllidaceae). *American Journal of Botany*, **91**, 1007–21.

Harder LD, Barrett SCH, and Cole WW (2000). The mating consequences of sexual segregation within inflorescences of flowering plants. *Proceedings of the Royal Society of London, Series B*, **267**, 315–20.

Harder LD, Jordan CY, Gross WE, and Routley MB (2004). Beyond floricentrism: the pollination function of inflorescences. *Plant Species Biology*, **19**, 137–48.

Heilbuth J, Ilves KL, and Otto SP (2001). The consequences of dioecy for seed dispersal: modeling the seed-shadow handicap. *Evolution*, **55**, 880–8.

Herlihy CR and Eckert CG (2002). Genetic cost of reproductive assurance in a self-fertilizing plant. *Nature*, **416**, 320–2.

Holsinger KE (1996). Pollination biology and the evolution of mating systems in flowering plants. *Evolutionary Biology*, **29**, 107–49.

Holsinger KE, Feldman MW, and Christiansen FB (1984). The evolution of self-fertilization in plants: a population genetic model. *American Naturalist*, **124**, 446–53.

Johnson SD, Neal PR, and Harder LD (2005). Pollen fates and the limits on male reproductive success in an orchid population. *Biological Journal of the Linnean Society*, **86**, 175–90.

Jordan CY and Harder LD (2006). Manipulation of bee behavior by inflorescence architecture and its consequences for plant mating. *American Naturalist* **167**:496–509.

Kalisz S, Vogler DW, and Hanley KM (2004). Context-dependent autonomous self-fertilization yields reproductive assurance in mixed mating. *Nature*, **430**, 334–7.

Knight T (1799). Experiments on the fecundation of vegetables. *Philosophical Transactions Royal Society of London*, **89**, 195–204.

Kohn JR and Barrett SCH (1992). Experimental studies on the functional significance of heterostyly. *Evolution*, **46**, 43–55.

Lande R and Schemske DW (1985). The evolution of self-fertilization and inbreeding depression in plants. I. Genetic models. *Evolution*, **39**, 24–40.

Lewis D (1941). Male sterility in natural populations of hermaphroditic plants. *New Phytologist*, **40**, 56–63.

Lloyd DG (1965). Evolution of self-compatibility and racial differentiation in *Leavenworthia* (Cruciferae). *Contributions from the Gray Herbarium of Harvard University*, **195**, 3–134.

Lloyd DG (1967). The genetics of self-incompatibility in *Leavenworthia crassa* Rollins (Cruciferae). *Genetica*, **38**, 227–42.

Lloyd DG (1968a). Pollen tube growth and seed set in self-incompatible and self-compatible *Leavenworthia* (Cruciferae) populations. *New Phytologist*, **67**, 179–95.

Lloyd DG (1968b). Partial unilateral incompatibility in *Leavenworthia* (Cruciferae). *Evolution*, **22**, 382–93.

Lloyd DG (1969). Petal color polymorphisms in *Leavenworthia* (Cruciferae). *Contributions from the Gray Herbarium of Harvard University*, **198**, 1–40.

Lloyd DG (1972a). A revision of the New Zealand, Subantarctic and South American species of *Cotula* section Leptinella. *New Zealand Journal of Botany*, **10**, 277–372.

Lloyd DG (1972b). Breeding systems in *Cotula* I. The array of monoclinous and diclinous systems. *New Phytologist*, **71**, 1181–94.

Lloyd DG (1972c). Breeding systems in *Cotula* II. Monoecious populations. *New Phytologist*, **71**, 1195–202.

Lloyd DG (1973). Sex ratios in sexually dimorphic Umbelliferae. *Heredity*, **31**, 239–49.

Lloyd DG (1974a). Theoretical sex ratios of dioecious and gynodioecious angiosperms. *Heredity*, **32**, 11–34.

Lloyd DG (1974b). Female-predominant sex ratios in angiosperms. *Heredity*, **32**, 35–44.

Lloyd DG (1974c). The genetic contributions of individual males and females in dioecious and gynodioecious angiosperms. *Heredity*, **32**, 45–51.

Lloyd DG (1975a). The maintenance of gynodioecy and androdioecy in angiosperms. *Genetica*, **45**, 325–30.

Lloyd DG (1975b). Breeding systems in *Cotula* III. Dioecious populations. *New Phytologist*, **74**, 109–23.

Lloyd DG (1975c). Breeding systems in *Cotula* IV. Reversion from dioecy to monoecy. *New Phytologist*, **74**, 125–45.

Lloyd DG (1976). The transmission of genes via pollen and ovules in gynodioecious angiosperms. *Theoretical Population Biology*, **9**, 299–316.

Lloyd DG (1977). Genetic and phenotypic models of natural selection. *Journal of Theoretical Biology*, **69**, 543–60.

Lloyd DG (1979a). Parental strategies in angiosperms. *New Zealand Journal of Botany*, **17**, 595–606.

Lloyd DG (1979b). Some reproductive factors affecting self-fertilization in angiosperms. *American Naturalist*, **113**, 67–79.

Lloyd DG (1980a). Demographic factors and mating patterns in angiosperms. In OT Solbrig, ed. *Demography and evolution in plant populations*, pp. 67–88. Blackwell Scientific, Oxford, UK.

Lloyd DG (1980b). Sexual strategies in plants III. A quantitative method for describing the gender of plants. *New Zealand Journal of Botany*, **18**, 103–8.

Lloyd DG (1980c). The distribution of gender in four angiosperms illustrating two evolutionary pathways to dioecy. *Evolution*, **33**, 673–85.

Lloyd DG (1982). The selection of combined versus separate sexes in seed plants. *American Naturalist*, **120**, 571–85.

Lloyd DG (1983). Evolutionarily stable sex ratios and sex allocations. *Journal of Theoretical Biology*, **105**, 525–39.

Lloyd DG (1984). Gender allocations in outcrossing cosexual plants. In R Dirzo and J Sarukhán, eds. *Perspectives on plant population ecology*, pp. 277–300. Sinauer Associates, Sunderland, MA, USA.

Lloyd DG (1985). Parallels between sexual strategies and other allocation strategies. *Experientia*, **41**, 1277–85.

Lloyd DG (1987a). Allocations to pollen, seeds and pollination mechanisms in self-fertilizing plants. *Functional Ecology*, **1**, 83–9.

Lloyd DG (1987b). Selection of offspring size at independence and other size versus number strategies. *American Naturalist*, **129**, 800–17.

Lloyd DG (1988). A general principle for the allocation of limited resources. *Evolutionary Ecology*, **2**, 175–87.

Lloyd DG (1989). The reproductive ecology of plants and eusocial animals. In PJ Grubb and JB Whittaker, eds. *Towards a more exact ecology*, pp. 185–208. Blackwell, Oxford, UK.

Lloyd DG (1992a). Self- and cross-fertilization in plants. II. The selection of self-fertilization. *International Journal of Plant Sciences*, **153**, 370–80.

Lloyd DG (1992b). Evolutionarily stable strategies of reproduction in plants: who benefits and how? In R Wyatt, ed. *Ecology and evolution of plant reproduction*, pp 137–68. Chapman and Hall, New York.

Lloyd DG (2000a). The selection of social actions in families: I. A collective fitness approach. *Evolutionary Ecology Research*, **2**, 3–14.

Lloyd DG (2000b). The selection of social actions in families: II. Parental investment. *Evolutionary Ecology Research*, **2**, 15–28.

Lloyd DG (2000c). The selection of social actions in families: III. Reproductively disabled individuals and organs. *Evolutionary Ecology Research*, **2**, 29–40.

Lloyd DG and Barrett SCH (1996). *Floral biology: studies on floral evolution in animal-pollinated plants*. Chapman and Hall, New York.

Lloyd DG and Bawa KS (1984). Modification of the gender of seed plants in varying conditions. *Evolutionary Biology*, **17**, 255–338.

Lloyd DG and Schoen DJ (1992). Self- and cross-fertilization in plants. I. The modes of self-pollination. *International Journal of Plant Sciences*, **153**, 358–69.

Lloyd DG and Venable DL (1992). Some properties of natural selection with single and multiple constraints. *Theoretical Population Biology*, **41**, 90–110.

Lloyd DG and Webb CJ (1977). Secondary sex characters in seed plants. *Botanical Review*, **43**, 177–216.

Lloyd DG and Webb CJ (1986). The avoidance of interference between the presentation of pollen and stigmas in angiosperms. I. Dichogamy. *New Zealand Journal of Botany*, **24**, 135–62.

Lloyd DG and Webb CJ (1992a). The evolution of heterostyly. In SCH Barrett, ed. *Evolution and function of heterostyly*, pp. 151–78. Springer-Verlag, Berlin.

Lloyd DG and Webb CJ (1992b). The selection of heterostyly. In SCH Barrett, ed. *Evolution and function of heterostyly*, pp 179–207. Springer-Verlag, Berlin.

Lloyd DG and Yates JMA (1982). Intrasexual selection and the segregation of pollen and stigmas in hermaphrodite plants, exemplified by *Wahlenbergia albomarginata* (Campanulaceae). *Evolution*, **36**, 903–13.

Lloyd DG, Webb CJ, and Dulberger R (1990). Heterostyly in species of *Narcissus* (Amaryllidaceae) and *Hugonia* (Linaceae) and other disputed cases. *Plant Systematics and Evolution*, **172**, 215–27.

Lovett Doust J and Lovett Doust L, eds. (1988). *Plant reproductive ecology: patterns and strategies*. Oxford University Press, Oxford.

Miller JS and Venable DL (2000). Polyploidy and the evolution of gender dimorphism. *Science*, **289**, 2335–8.

Moeller DA and Geber MA (2005). Ecological context of the evolution of self-pollination in *Clarkia xantiana*: population size, plant communities, and reproductive assurance. *Evolution*, **59**, 786–99.

Morgan MT and Conner JK (2001). Using genetic markers to directly estimate male selection gradients. *Evolution* **55**, 272–81.

Morgan MT and Wilson WG (2005). Self-fertilization and the escape from pollen limitation in variable pollination environments. *Evolution*, **59**, 1143–8.

Nagylaki T (1976). A model for the evolution of self-fertilization and vegetative reproduction. *Journal of Theoretical Biology*, **58**, 55–8.

Richards AJ (1997). *Plant breeding systems*, 2nd edition. Chapman and Hall, New York.

Rieseberg LH and Wendel J (2004). Plant speciation—rise of the poor cousins. *New Phytologist*, **161**, 1–21.

Roff DA (1992). *The evolution of life histories: theory and analysis*. Chapman and Hall, New York.

Routley MB and Husband BC (2003). The effect of protandry on siring success in *Chamerion angustifolium* (Onagraceae) with different inflorescence sizes. *Evolution*, **57**, 240–8.

Sakai AK and Weller SG (1999). Gender and sexual dimorphism in flowering plants: a review of terminology, biogeographic patterns, ecological correlates, and phylogenetic approaches. In MA Geber, TE Dawson, and LF Delph, eds. *Gender and sexual dimorphism in flowering plants*, pp. 1–31. Springer-Verlag, Berlin, Germany.

Sarkissian TS, Barrett SCH and Harder LD (2001). Gender variation in *Sagittaria latifolia* (Alismataceae): is size all that matters? *Ecology*, **82**, 360–73.

Schlessman MA, Lloyd DG and Lowry PP (1990a). Evolution of sexual systems in New Caledonian Araliaceae. In G Gottsberger and GT Prance, eds. *Reproductive biology and evolution of tropical woody angiosperms. Memoirs of the New York Botanic Gardens*, **55**, 105–17.

Schlessman MA, Lowry PP, and Lloyd DG (1990b). Functional dioecism in the New Caledonian endemic *Polysias pancheri* (Araliaceae). *Biotropica*, **22**, 133–9.

Schlessman MA, Plunkett GM, Lowry PP, and Lloyd DG (2001). Sexual systems of New Caledonian Araliaceae: a preliminary phylogenetic reappraisal. *Edinburgh Journal of Botany*, **58**, 221–8.

Schoen DJ, Morgan MT, and Bataillon T (1996). How does self-pollination evolve? Inferences from floral ecology and molecular genetic variation. *Philosophical Transactions of the Royal Society of London, Series B*, **351**, 1281–90.

Smith CC and Fretwell SD (1974). The optimal balance between the size and number of offspring. *American Naturalist*, **108**, 499–506.

Sprengel CK (1793). *Das entdeckte Geheimniss der Natur im Bau und in der Befruchtung der Blumen* I. Vieweg sen., Berlin Reprint by J Cramer and HK Swann, Lehre, and by Weldon and Wesley, Codicote, New York, 1972.

Stehlik I and Barrett SCH (2005). Mechanisms governing sex-ratio variation in dioecious *Rumex nivalis*. *Evolution*, **59**, 814–25.

Stone JL and Thomson JD (1994). The evolution of distyly: pollen transfer in artificial flowers. *Evolution*, **48**, 1595–606.

Thomson JD and Barrett SCH (1981). Temporal variation of gender in *Aralia hispida* Vent. (Araliaceae). *Evolution*, **35**, 1094–107.

Vaughton G and Ramsey M (2002). Evidence of gynodioecy and sex ratio variation in *Wurmbea biglandulosa* (Colchicaceae). *Plant Systematics and Evolution*, **232**, 167–79.

Venable DL and Lloyd DG (2004). Allocation under multiple resource constraints. *Evolutionary Ecology Research*, **6**, 1109–21.

Webb CJ and Lloyd DG (1980). Sex ratios in New Zealand apioid Umbelliferae. *New Zealand Journal of Botany*, **18**, 121–6.

Webb CJ and Lloyd DG (1986). The avoidance of interference between the presentation of pollen and stigmas in angiosperms. II. Herkogamy. *New Zealand Journal of Botany*, **24**, 163–78.

Webb CJ, Lloyd DG, and Delph LF (1999). Gender dimorphism in indigenous New Zealand seed plants. *New Zealand Journal of Botany*, **37**, 119–30.

Willson MF (1979). Sexual selection in plants. *American Naturalist*, **113**, 777–790.

Willson MF (1994). Sexual selection in plants: perspective and overview. *American Naturalist*, **144** (Supplement), S13–39.

Wilson P, Thomson JD, Stanton ML, and Rigney LP (1994). Beyond floral Batemania: gender biases in selection for pollination success. *American Naturalist*, **143**, 283–96.

Wilson WG and Harder LD (2003). Reproductive uncertainty and the relative competitiveness of simultaneous hermaphroditism versus dioecy. *American Naturalist*, **162**, 220–41.

Wolfe LM and Shmida A (1997). The ecology of sex expression in a gynodioecious Israeli desert shrub (*Ochradenus baccatus*). *Ecology*, **78**, 101–10.

Yeo PF (1975). Some aspects of heterostyly. *New Phytologist*, **75**, 147–53.

PART 1

Strategic perspectives on floral biology

Natural selection of traits that affect direct or indirect interaction among individuals commonly depends on the composition of the population. In particular, the fitness consequences of traits involved in outcrossing and/or competition depend on the characteristics of potential mates and competitors. Because of this frequency dependence, the selection of traits that influence mating and/or competition results in the rise to prominence of a specific "strategy" for interaction that promotes fitness in the face of any other strategy (the evolutionary stable strategy, or ESS).

Reproductive performance of plants is commonly frequency dependent, because outcrossing success depends on the mating opportunities present in the population and several reproductive stages involve competition, including pollen grains competing to fertilize ovules, developing seeds competing for maternal resources, and seedlings competing for establishment sites. Consequently, a strategic perspective on plant reproduction commonly provides powerful insights into the evolution of reproductive traits, including allocation of resources to reproduction, floral and inflorescence characteristics, and the nature of mating and sexual systems. This perspective is incorporated formally in the mathematical search for evolutionary stable strategies, which Maynard Smith introduced and Charnov (1982) and Lloyd (Chapter 1) first applied to the analysis of plant reproduction, and this theory motivates many empirical studies (e.g., de Jong and Klinkhamer 2005). The three chapters in Part 1 introduce the strategic perspective on plant reproduction and its application to the analysis of diverse traits.

In Chapter 2, Martin Morgan introduces ESS analysis of plant reproduction and compares it with quantitative-genetic approaches. The ESS approach focuses on fitness differences among alternative phenotypes, which are assumed to be genetically determined, whereas quantitative-genetic analysis of selection emphasizes the genetic variation and covariation that govern phenotypic variation and determine the opportunities for genetic responses to phenotypic selection. These approaches offer complementary perspectives on floral evolution: the ESS approach identifies optimized phenotypes that promote reproductive function, whereas the quantitative genetic approach characterizes whether and how rapidly such an optimum might be approached. Through a series of examples concerning plant–pollinator interaction and the incidence of self- versus cross-fertilization, Morgan illustrates the development of ESS theory from general principles. In addition to identifying key features of the selection of reproductive traits, Morgan's analysis provides novel insights into the ecological dynamics of populations that accompany reproductive adaptation, including the risk of extinction. Based on his understanding of the theory of phenotypic and genetic evolution, Morgan also provides guidance for empirical studies of selection on reproductive characteristics.

In Chapter 3, Da-Yong Zhang addresses two aspects of plant reproduction that have received the most theoretical analysis from the strategic perspective, the allocation of resources to reproduction versus other functions (reproductive investment), and the allocation of resources to female versus male function (sex allocation). In contrast to most analyses of these allocation problems, which consider the proportions of resources invested in competing functions, Zhang focuses on absolute allocations. This approach allows Zhang to consider the consequences of size variation, either during individual growth or among individuals, for reproductive investment and sex allocation. In addition, Zhang introduces a graphical, rather than mathematical, approach for ESS analysis of allocation to competing functions, which provides straightforward solutions to problems that would otherwise be challenging, if not intractable. This combination of new approaches allows Zhang to address diverse topics, including the general life-history problem of allocation to survival versus reproduction, the evolution of combined versus separate sexes, and extended allocation to female function after male function.

Chapter 4, by Lawrence Harder and Matthew Routley, provides a detailed examination of the fates of pollen and ovules as they affect plant reproductive strategies, especially the evolution of selfing and outcrossing. Delineation of these fates provides a framework for understanding all aspects of the ecology and evolution of plant reproduction. Comparison of the relative magnitudes of these fates illustrates both the essential asymmetry of reproductive performance through female and male function, which underlies sexual selection, and the main functional constraints on reproductive success. Indeed, as Harder and Routley illustrate, explicit consideration of these fates exposes previously unrecognized aspects of reproduction, including ovule limitation of seed production, the dependence of resource limitation on the production of "excess" ovules, and the specific importance of post-dispersal inbreeding depression for mating-system evolution. Similar to the preceding chapters, Chapter 4 demonstrates the value of explicit strategic analysis in explaining many aspects of plant reproduction that elude intuition.

Selected key references

Charnov EL (1982). *The theory of sex allocation*. Princeton University Press, Princeton, NJ.

de Jong TJ and Klinkhamer PGL (2005). *Evolutionary ecology of plant reproductive strategies*. Cambridge University Press, Cambridge.

Lloyd DG (1984). Gender allocation in outcrossing cosexual plants. In R Dirzo and J Sarukhán, eds. *Perspectives on plant population ecology*, pp. 277–300. Sinauer Associates, Sunderland, MA.

Lloyd DG (1988). A general principle for the allocation of limited resources. *Evolutionary Ecology*, **2**, 175–87.

Lloyd DG (1992). Self- and cross-fertilization in plants. II. The selection of self-fertilization. *International Journal of Plant Sciences*, **153**, 370–80.

Maynard Smith J (1982). *Evolution and the theory of games*. Cambridge University Press, Cambridge.

CHAPTER 2

Selection on reproductive characters: conceptual foundations and their extension to pollinator interactions

Martin T. Morgan

Washington State University, Pullman, WA, USA

Outline

This chapter reviews conceptual insights and directions for understanding natural selection on reproductive characters. Evolutionary stable strategy (ESS) approaches are ideal for formulating concepts and thinking strategically about the phenotypic selection of reproductive characters. Quantitative genetic models (in the sense popularized by Lande) provide tools for characterizing phenotypic selection and inheritance on a microevolutionary scale, while relating closely to the conceptual insights of ESS analysis. Pollinators and the ecological context of pollination can be central to selection on reproductive characters, but often receive only implicit treatment in phenotypic selection models. Recent approaches begin to address this shortcoming, as illustrated by attempts to understand when generalist versus specialist pollination evolves. Equally important can be selection for reproductive assurance (selfing), especially under variable pollinator service. Such selfing influences the structure of genetic variation and evolutionary response to phenotypic selection.

2.1 Introduction

Angiosperm flowers awe the casual observer and naturalist alike with their subtle or, sometimes, elaborate morphology. Natural selection acting on small variations within populations is undoubtedly the creative process generating most diversity of form. And yet, angiosperm flowers pose unique challenges to understanding the action of natural selection. Unlike other life-history characters, an individual's success at outcrossing depends on reproductive strategies of others in the population and for this reason is inherently frequency dependent. Most plants are hermaphroditic, so floral adaptation involves a compromise between the conflicting requirements of female and male functions. Outcrossing plants rely on pollinators and produce seeds that contribute directly to population growth. Studies of natural selection on reproductive characters must therefore acknowledge that plant-pollinator interactions shape selection and that the action of selection affects population dynamics. Hermaphroditism allows self-fertilization, which also has consequences for selection of reproductive characters (e.g., diminishing the importance of pollinator attraction), population growth rates, and the structuring of heritable genetic variation. All of these evolutionary factors contribute to our understanding and appreciation of floral diversity.

This chapter broadly examines the operation of selection on reproductive characters, primarily emphasizing conceptual and theoretical understanding. The chapter considers three main topics. First, I review evolutionary stable strategy (ESS) and quantitative genetic approaches to understanding selection. Quantitative genetic approaches are particularly appropriate for documenting selection in natural populations, but require adequate measures of female and especially male fertility. Second, I explore consequences of plant and pollinator interactions and the population dynamic consequences of reproduction for understanding phenotypic selection on reproductive characters. Plant-pollinator interaction and population dynamics may be particularly important for phenotypic selection of generalist versus specialist modes of pollination, although investigation of this topic is incomplete. Finally, I describe insights that arise from consideration of population dynamics for the phenotypic selection of self-fertilization. A general theme is that ecological factors contribute as significantly as the genetic transmission advantage and inbreeding depression to the evolution of selfing rates.

2.2 Phenotypic selection on reproductive strategy

Two features of reproduction by hermaphroditic plants are particularly important for phenotypic selection: floral traits influence fertility (i.e., opportunities for genetic transmission) through both female and male functions, and reproductive success through either gender generally depends on the reproductive strategies of other individuals in the population. These features are conveniently encapsulated in the following characterization of the fitness of an individual with trait value z,

$$W(z) = \frac{1}{2}\left[\frac{W_f(z)}{\overline{W}_f} + \frac{W_m(z)}{\overline{W}_m}\right] \qquad (1)$$

(see Charnov et al. 1976; Lloyd 1977; Charlesworth and Charlesworth 1979). In eq. 1, $W_f(z)$, $W_m(z)$ are the female and male fertilities associated with the trait value z, and $\overline{W}_f$ and $\overline{W}_m$ are the corresponding population average fertilities. Terms such as $W_f(z)/\overline{W}_f$ are *relative fertilities* and, because average fertilities depend on trait values in the population, reflect the frequency dependence of sexual reproduction. The averaging of relative fertilities in eq. 1 is appropriate for autosomal genes of diploid organisms, ensuring that "everyone has exactly one father and one mother" (Charnov 1982, p. 8).

2.2.1 Phenotypic selection and evolutionary stable strategies

The consequences of eq. 1 can be examined by two approaches: *ESS* and *quantitative genetics*. An ESS identifies an individual strategy, z^*, that has higher fitness than any strategy z similar to z^*, when z^* is common. To identify the ESS, one seeks a maximum of relative fitness (eq. 1) when $z \approx z^*$, so that $\overline{W}_f \approx W_f(z^*)$ and $\overline{W}_m \approx W_m(z^*)$. Note again that the fitness of an individual adopting strategy z is frequency dependent, because $W(z) = (1/2)[W_f(z)/\overline{W}_f + W_m(z)/\overline{W}_m]$ depends on the strategy of other individuals in the population through the average fertilities. Fitness is maximized when $dW(z)/dz|_{z\approx z^*} = 0$, or

$$\frac{1}{2}\left[\frac{1}{\overline{W}_f}\frac{dW_f(z)}{dz}\bigg|_{z\approx z^*} + \frac{1}{\overline{W}_m}\frac{dW_m(z)}{dz}\bigg|_{z\approx z^*}\right] = 0, \qquad (2)$$

as long as the second derivative of the relative fitness function is negative. Terms of the form $(1/\overline{W}_f)dW_f(z)/dz|_{z\approx z^*}$ describe *marginal fitness returns* (Lloyd 1985). At the ESS, marginal returns through female function are equal in magnitude, but opposite in sign, to marginal returns through male function. The ESS identifies strategy z^* as a *local* maximum, but more deviant strategies (e.g., due to major-gene mutations, or sterility of one gender or the other) may increase in frequency when introduced into the ESS population. Although the ESS describes local stability, populations that are not at the ESS may *not* evolve towards the ESS under some circumstances. *Convergence stable strategies* (CSS) characterize situations in which populations actually evolve to the ESS. Novel evolutionary dynamics are possible when an ESS is not a CSS. The mathematical approach of *adaptive dynamics* (e.g., Geritz et al. 1998; see the extensive commentary in Waxman and Gavrilets 2005 *et seq.*) popularizes the convergence stability criterion.

A slightly more general perspective on phenotypic selection incorporates the *benefits* and *costs* of a reproductive strategy. Benefits are already incorporated in the form of relative fitness gains through male and female functions, e.g., $W_b(z) = (1/2)\bigl(W_f(z)/\overline{W}_f + W_m(z)/\overline{W}_m\bigr)$. Costs, $W_c(z)$, accrue from life-history trade-offs during other, typically earlier, phases of the life cycle. Fitness is the sum of costs and benefits, and marginal benefits and costs balance at an equilibrium, $(1/\overline{W}_b)dW_b(z)/d\,z|_{z\approx z^*} = -(1/\overline{W}_c)dW_c(z)/d\,z|_{z\approx z^*}$. Haig and Westoby (1988) used such a cost-benefit analysis to argue that optimal allocations to pollinator attraction and seed provisioning limit reproductive success simultaneously. Note that marginal gains contributing to the net benefit (including marginal gains through male and female functions) are no longer constrained to balance at equilibrium, for example, when $W_b(z) = (1/2)\bigl(W_f(z)/\overline{W}_f + W_m(z)/\overline{W}_m\bigr)$ there can be equilibria when eq. 2 is not satisfied. In this sense, including other life-history components breaks the trade-off between male and female reproductive strategies. Houle (1991) and others noted an important corollary of this, that overall constraints neither require nor imply genetic correlations between pairs of traits.

ESS analysis offers significant insight, for several reasons. The basic mathematics (eqs 1 and 2) involves elementary calculus, so analytic results are accessible to many botanists. Tractable solutions to the ESS often require simple expressions for fitness, nudging the practitioner to simplify complex biological situations to the features essential for understanding phenotypic selection. Astute assessment of the merits of ESS models relies on concordance between underlying assumptions and biological reality, rather than on mathematical prowess. For these reasons, ESS models are excellent tools to organize and reason strategically about complicated biological scenarios. The central role of early ESS analyses of plant reproductive characters in motivating contemporary theoretical and empirical research illustrates the utility of this approach. Examples include the phenotypic selection of self-fertilization (Lloyd 1979), separate sexes (e.g., Lloyd 1982), and reproductive allocation in cosexual plants (Lloyd 1984).

Despite facilitating strategic analysis of evolution under specific model assumptions, the ESS approach is often difficult to test with empirical observations of fitness and phenotypic selection. The main reasons for this shortcoming are the continuum of trait values characterizing populations, the action of phenotypic selection on several traits simultaneously, and the short-term and non-equilibrium time span available for observing the action of phenotypic selection.

2.2.2 Phenotypic selection and inheritance of quantitative characters

Quantitative genetic techniques introduced to evolutionary biology by Lande (1976) are well-suited to empirical assessment of fitness and short-term (microevolutionary) change in trait value due to phenotypic selection and inheritance (quantitative genetic models also offer significant insight into macroevolutionary change). Central to Lande's approach is the *phenotypic selection gradient*, β, defined in a manner very similar to marginal gains. The selection gradient depicts the change in average trait value caused by fitness differences among phenotypes, measured in units of phenotypic standard deviations. If phenotypic selection is not too strong (see, e.g., Abrams *et al.* 1993 for a more complete delimitation), the selection gradient describes the slope of the relation of relative fertility to the average trait value,

$$\beta_f \approx \frac{1}{\overline{W}_f}\frac{dW_f(z)}{dz}\bigg|_{z\approx z^*}, \quad \beta_m \approx \frac{1}{\overline{W}_m}\frac{dW_m(z)}{dz}\bigg|_{z\approx z^*}. \quad (3)$$

Comparison of eqs 1 and 3 reveals that net phenotypic selection on character z is the average of the phenotypic selection gradients through male and female function, $\beta = (1/2)(\beta_f + \beta_m)$. As in the ESS analysis, an equilibrium occurs when the female phenotypic selection gradient is equal in sign, but opposite in magnitude, to the male phenotypic selection gradient, $\beta_f = -\beta_m$. The quantitative genetic perspective lends itself to statistical analysis of phenotypic selection, because a phenotypic selection gradient equals the slope of the regression of relative fertility on trait value.

The quantitative genetic approach generalizes readily to multiple traits (Lande 1979; Lande and Arnold 1983), with a vector of phenotypic selection gradients, **β**, representing changes in multiple characters wrought by differences in relative fitness. These multivariate phenotypic selection gradients are estimated by multivariate regression of relative fitness on trait values. Note, however, that change in the average value of one character may occur solely because of phenotypic selection on a second, phenotypically correlated, character (see Chapter 14).

Equation 1, marginal gains, and phenotypic selection gradients describe phenotypic selection, but not the inheritance of reproductive traits. The ESS approach does not consider either change in allele frequency after introduction or complicated forms of inheritance. In contrast, Lande's quantitative genetic approach incorporates both genetic and environmental sources of phenotypic variation. In particular, the between-generation response resulting from phenotypic selection and inheritance of multiple traits is described by

$$\Delta \bar{z} = \mathbf{G}\boldsymbol{\beta} = \frac{1}{2}\left[\beta_f + \beta_m\right], \quad (4)$$

where bold face indicates vectors of trait values, $\mathbf{z}$, and the phenotypic selection gradients, $\boldsymbol{\beta}$, acting on traits z_i, $i = 1, 2, \ldots$. The symmetric matrix $\mathbf{G}$ describes additive genetic variances and covariances between trait (see Chapter 14: the details of genetic variation are interesting and controversial, and the interested reader is referred to, e.g., Turelli and Barton 1994; Zhang and Hill 2005). For a single trait, strong phenotypic selection (large absolute β) may nonetheless result in limited between-generation change, $\Delta \bar{z}$, if there is little additive genetic variation ($\mathbf{G} \approx 0$). An absence of genetic variation thus results in equilibrium, in the sense that the trait does not evolve, even though the population is not at a fitness maximum and within-generation change in trait mean caused by phenotypic selection remains large. For multiple traits, specific patterns of genetic covariation (specifically, when the determinant of $\mathbf{G} = 0$) cause a population to evolve so that between-generation change in average trait value eventually halts, despite strong phenotypic selection. More generally (i.e., when the determinant of $\mathbf{G} \neq 0$), genetic correlations deflect character evolution away from the direction of steepest ascent of the fitness surface, but do so without stopping character evolution (see Chapter 14).

Phenotypic selection gradient analysis has been particularly insightful when applied to situations that might involve contrasting phenotypic selection. For instance, Ashman (2003, 2005) and Delph (e.g., Delph et al. 2004) contrasted the roles of genetic constraint and phenotypic selection in shaping reproductive character change in genders of gynodioecious species. As another example, Nuismer and Cunningham (2005) documented patterns of phenotypic selection that suggest that reproductive characters in diploid and autotetraploid populations of *Heuchera grossularifolia* diverge in response to phenotypic selection for reduced intercytotype mating, rather than as an epigenetic consequence of increased ploidy.

2.2.3 Measuring phenotypic selection

Assessment of phenotypic selection gradients requires a reasonable measure of "fitness". Evolutionarily relevant fitness measures account for the change in gene frequencies from a particular stage of one generation to the comparable stage of the next generation (Charlesworth 1980). Such information is seldom available, and investigators typically measure *fitness components* during a portion of the lifespan. Fitness components reflect "true" fitness accurately if they are statistically independent of phenotypic selection acting during other periods. This independence might seem reasonable for floral traits (e.g., how can phenotypic selection on floral characters occur other than at flowering?), but such reasoning neglects the correlation between reproductive and other life-history characters emphasized in eq. 4 (also see Chapters 7 and 8). Such associations may be responsible for the implausibly strong phenotypic selection reported by many studies, including some that considered plant reproductive characters (Hereford et al. 2004). Nevertheless, necessity often requires assessment of fitness components to measure phenotypic selection on reproductive characters.

Female fertility—Studies of phenotypic selection through female function typically estimate phenotypic selection gradients by regressing relative seed (or fruit) production on floral reproductive characters (e.g., Chapter 15). Several factors may make such estimates unsatisfactory. Seed production may not have proportional effects on establishment, germination, and survival to reproduction in the next generation. Maternal effects may obscure the additive genetic contributions of trait value to fitness. Trade-offs with other life-history components, especially size-number trade-offs occurring *before* trait measurement, can negate or reverse the sign of the true phenotypic selection gradient (e.g., Houle 1991). These difficulties are not unique to measures of female fertility, and in that limited sense characterizing phenotypic selection through female function presents no unique technical or conceptual obstacles.

Male fertility—Estimation of male phenotypic selection gradients poses additional challenges, primarily because of the difficulty associated with measuring male reproductive success (Snow and Lewis 1993). A first approach is to use an easily measured proxy, such as pollen production or removal. This is particularly attractive in species such as *Asclepias* and orchids, for which pollinia are readily counted (e.g., Broyles and Wyatt 1990; Morgan and Schoen 1997). There are few guarantees, though, that pollen production or removal correlates with actual male fertility (e.g., Johnson *et al.* 2005). Alternatively, genetic markers can be used to infer male reproductive success based on assays of all, or most, potential males and females in the "parental" generation, and a sample of seeds in the "offspring" generation. Historically this approach considered allozymes, but it is now feasible to use single-nucleotide polymorphisms. Usually, candidate sets of an offspring and its putative parents (a triplet) are assessed to identify those with genotypes consistent with known offspring and parental genotypes. Various methods are then used to estimate phenotypic selection from the set of genetically possible male parents (Garant and Kruuk 2005). Exclusion and related methods use genetic data to eliminate all but one triplet. These methods are easiest to understand, but surprisingly prone to subtle biases. For example, parental genotypes differ in the probability of exclusion, and hence in fraction of offspring actually assigned (Brown 1990). In addition, these methods suffer from reduced statistical power when some offspring cannot be assigned parents uniquely and they estimate fitness parameters (i.e., individual fertilities), rather than the statistics of direct interest, namely the phenotypic selection gradients. For these reasons, methods that rely on specific models of genetic inheritance and phenotypic selection (e.g., Adams *et al.* 1992; Smouse and Meagher 1994; Morgan and Conner 2001; Burczyk *et al.* 2002) are preferred, even though they are conceptually and computationally more challenging.

2.3 Plant and pollinator interaction

How do pollinators fit into the phenotypic selection equations outlined above? Pollinators shape the gain curves that impose phenotypic selection on plant traits. For example, Lloyd and Webb considered how pollinator foraging and interference between male and female functions (Lloyd and Webb 1986; Webb and Lloyd 1986) drive the evolution of breeding systems such as heterostyly (Lloyd and Webb 1992). More concretely, Harder and Thomson (1989) studied transport of *Erythronium grandiflorum* pollen by nectar-collecting bumble bees. Longer floral visits and greater pollen removal increased pollen export, but at a diminishing rate. Floral traits influence how much pollen is removed, so that the relation of pollen deposition to these traits describes a gain curve that could be used in eq. 1. Additional floral and pollinator traits undoubtedly shape the relation of pollen export to removal. The multivariate formulation of eq. 4 can describe how the overall floral phenotype changes in response to phenotypic selection imposed by pollinators. In this case, the details of pollination biology provide essential information for formulating phenotypic selection (e.g., what specific mathematical expression best describes the relation of fertility to a trait), but does not fundamentally change how phenotypic selection contributes to floral character evolution (e.g., by enforcing the population average equality of male and female fertility contributions).

Implicit treatment of pollinators provides profitable insight into phenotypic selection on plant reproductive characters. However, in reality pollinators exhibit extensive diversity and evolutionary flexibility in their foraging strategies (e.g., Goulson 2003). Do some situations require a more complete treatment of coupled plant and pollinator evolutionary change?

2.3.1 Specialist or generalist?

Waser et al. (1996) stimulated considerable debate in the pollination literature by emphasizing that plant-pollinator interactions are typically generalized, involving many pollinator species per plant species and vice versa, rather than specialized, reciprocal mutualisms. Waser et al. supported this proposal with a review of empirical evidence and a model that considered the relation of plant fitness to visits from up to two pollinator species. Type i pollinators have abundance N_i, visitation rate per pollinator V_i, and "effectiveness" g_i, so that the fitness of a generalist plant exploiting both pollinators is $W_G = \sum_{i=1}^{2} N_i V_i g_i$. The authors then asked when the fitness of a "specialist" plant adopting a slightly different strategy that alters, for example, pollinator effectiveness $g_{S,1} = g_1 + \delta$, $g_{S,2} = g_2 - \delta$ exceeds that of the generalist. The fitness of the specialist plant is $W_S = \sum_{i=1}^{2} N_i V_i g_{S,i}$ and specialization is favoured (i.e. $W_S > W_G$) when $N_1 V_1 > N_2 V_2$. This condition leaves no scope for generalist pollination, except when $N_1 V_1 = N_2 V_2$ exactly. To resolve the discrepancy between theoretical prediction and empirical data, Waser et al. proposed that temporal variation in N_i, V_i, g_i and $g_{S,i}$ can, under some conditions, favour generalist pollinators. Note that temporal variation causes relations of fitness to trait value that are linear and positive on an arithmetic scale to become "decelerating" on the logarithmic scale. This transformation is important, because evolution in temporally variable environments occurs in response to differences in geometric mean fitness, and a logarithmic scale portrays geometric mean fitnesses accurately. The model of Waser et al. omitted many relevant features of plant reproduction, including gender, frequency dependence and non-linear relations between traits and fertility.

Aigner (2001) provided a useful extension to Waser et al.'s model by allowing for general functional relations between plant trait value and fitness through different pollinators, $W_i(z)$. With two pollinators, Aigner found that the optimal trait value z^* balances marginal gains $W_1'(z^*) = -W_2'(z^*)$ through each pollinator. In a graphical analysis, Aigner argued that predominance of specialization or generalization depends on the amount of curvature (roughly the strength of stabilizing phenotypic selection) and elevation (absolute contribution of each pollinator, as determined by pollinator abundance or visitation rate) of the fitness functions $W_i(z)$. The combination of these factors has unexpected consequences. For instance, as Fig. 2.1 illustrates, plants served by two equally "effective" pollinators (making the same absolute contribution to fitness) will nonetheless specialize on the pollinator exerting stronger stabilizing phenotypic selection on the plant reproductive character. This result contrasts with naive expectations of Stebbins' (1970) "most effective pollinator principle". Unfortunately, like Waser et al., Aigner's fitness function does not include male and female fitness components explicitly and the (graphical and mathematical) analysis does not include the gender-specific frequency dependence inherent in phenotypic selection of reproductive characters. To accept Aigner's insights into how phenotypic selection might shape generalist versus specialist

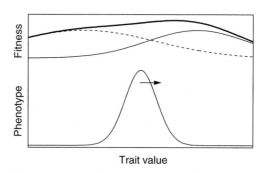

Figure 2.1 Aigner's characterization of selection for a generalized pollination system for a plant served by two pollinator species that impose contrasting selection on a floral trait. The upper curves relate the fitness contribution by each pollinator species (dashed and thin solid lines) and total fitness (thick line) to trait value. The two pollinators make comparable absolute fitness contributions (i.e., comparable fitness function elevations), but differ in strength of stabilizing selection (the selection "width", defined in Lande (1976), is twice as large for the dashed compared to solid line). Selection (arrow) drives the lower phenotypic distribution in the direction that maximizes total fitness (thick line), so that trait values match the pollinator that exerts stronger stabilizing selection more closely.

strategies, the fitness curves must be interpreted as implicitly summarizing male and female fitness components and frequency dependence.

Sargent and Otto (2006) presented perhaps the most satisfying model of trait specialization, as their model accounted for both pollen export and import explicitly. Rare plant species specialize on effective and common pollinators, whereas common species evolve generalization. Sargent and Otto also allowed for fitness trade-offs between pollinator attraction allocations (Fig. 2.2a). Specific forms of the trade-off promote generalization, much as specific trade-offs (between male and female functions, rather than pollinators) favour the evolution of hermaphroditism (Fig. 2.2b). Other forms of trade-off promote evolution towards specialization. A particularly intriguing scenario, illustrated in Fig. 2.2c, occurs when trade-offs initially drive the population towards a generalist strategy, but as the population approaches the generalist optimum the structure of the trade-offs promotes specialization and, potentially, speciation. Sargent and Otto thus formulated and analysed models of phenotypic selection appropriately, showing how observable empirical facts (the trade-offs of pollinator attraction) form assumptions that anticipate or predict evolutionary outcomes.

The progression from Waser et al.'s original model to Sargent and Otto's formulation clarifies the role of plant and pollinator abundance, pollen transfer, and gain curves in the phenotypic selection of generalist and specialist pollination; however, significant opportunities for conceptual and theoretical development remain. One important direction involves a more explicit coupling of plant and pollinator evolution, as illustrated in the next section. A second important direction involves coupling population and evolutionary dynamics. In Sargent and Otto's model, the numerical abundances of plants and pollinators are fixed parameters, whereas floral adaptation probably alters average plant (and, possibly, pollinator) abundance. For example, adaptation might increase seed production and recruitment into the plant population, or relax constraints on other life-history components. A subsequent section outlines an approach to coupled ecological and evolutionary dynamics.

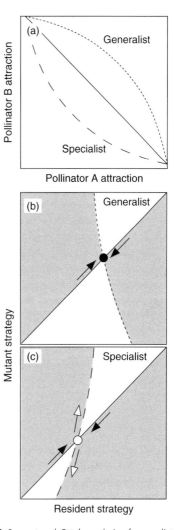

Figure 2.2 Sargent and Otto's analysis of generalist and specialist evolution. (a) Contrasting assumptions about joint attractiveness to different pollinators determine the model trade-offs for different reproductive strategies. In panels (b) and (c), a value on the x-axis indicates the reproductive strategy adopted by most individuals in the population, whereas a value on the y-axis indicates the strategy adopted by a rare 'mutant' individual. (b) When trade-offs favour the evolution of a generalist pollination system (dotted line, panel a), a resident strategy (value on the x-axis) can be invaded by a mutant adopting a strategy (value on the y-axis) such that the combination of resident and mutant strategies is in the grey area of the figure. Evolution by successive mutations of small effect will drive populations in the direction of the arrows, toward the stable equilibrium (large, solid circle). (c) Specific forms of trade-off result in initial evolution towards a generalist strategy, but as generalists become established (the population moves towards the open circle) selection favours evolution toward specialization. Speciation is one outcome.

2.3.2 Pollination as trade

In the models so far, the role of pollinators in plant evolution is only implied through their influence on gain curves. A more explicit, dynamic possibility recognizes pollinators as participants in coevolution, whereby pollinator foraging, morphology, and other features evolve in response to selection by the plant and other aspects of the pollinator's environment. Several approaches describe selection when plants and pollinators coevolve, including tightly coupled genetic interactions (e.g., Nuismer et al. 1999), or quantitative genetic interactions (e.g., Kiester et al. 1984). A phenotypic approach conceptually related to ESS analysis involves *optimal foraging*.

Optimal foraging models commonly assume that foraging strategies evolve to maximize a forager's average rate of net energy gain R. Suppose a pollinator of type j encounters and forages at plants of type i with probability λ_{ij} and that while visiting a plant of this type it extracts $g_{ij}(t_{ij})$ net energy during t_{ij} time units. A pollinator that includes n types of plants in its diet has an average rate of net energy gain

$$R_j = \frac{\sum_{i=1}^{n} \lambda_{ij} g_{ij}(t_{ij})}{1 + \sum_{i=1}^{n} \lambda_{ij} t_{ij}}, \qquad (5)$$

where time is scaled so that the average search time between encounters is 1 unit. The pollinator strategy involves two components: the plant types $i = 1, 2, \ldots, n$ included in the diet, and the optimal duration t_{ij}^* of each visit. An optimal pollinator includes plant type i if its maximum rate of net energy gain from that plant equals or exceeds the pollinator's current average rate of gain $R_j \geq g_{ij}(t_{ij}^*)/t_{ij}^*$, independent of encounter probability (e.g., Stephens and Krebs 1986, p. 33). Once included in the diet, a plant type continues receiving visits until the instantaneous rate of energy gain $dg_{ij}(t_{ij})/dt_{ij}|_{t_{ij} = t_{ij}^*}$ equals the pollinator's average rate of energy gain R_j.

How does optimal foraging relate to selection of plant reproductive characters? Optimal pollinator behaviour depends on encounter probability λ_{ij} and the rate $g_{ij}(t_{ij})/t_{ij}$ and amount $g_{ij}(t_{ij}^*)$ of energy extraction. For visually foraging pollinators, encounter probability depends primarily on floral display, including inflorescence size or flower size and pigmentation (Ohashi and Yahara 2001).

Energy extraction rate $g_{ij}(t_{ij})/t_{ij}$ depends on how readily floral morphology allows pollinators to consume nectar, and the amount of reward extracted $g_{ij}(t_{ij}^*)$ depends ultimately on nectar production itself. Thus essential parameters in the evolution of pollinator foraging map directly onto essential features of plant reproductive strategy.

The forgoing suggests that plants and pollinators engage in a form of trade (see Morgan 2000; more generally, Schwartz and Hoeksema 1998): plants exchange energy for reproductive opportunity. One insight from these considerations concerns the amount of reward that plants offer. Recall that a pollinator includes a plant species in its diet if that species offers an average rate of reward $g_{ij}(t_{ij}^*) / t_{ij}^*$ equal to or greater than the pollinator's average reward R_j. When pollinators include many individuals of one or several species in their diet, the reward of each plant contributes little to a pollinator's average reward, i.e., $dR_j / dg_{ij}(t_{ij}^*) \approx 0$. Therefore the individual plant must offer reward equal to or greater than the reward of other individual plants in the pollinator's repertoire: plant reward allocation is held hostage to community reward levels.

Now consider plants pollinated by pollinators that visit few species. The average rate of reward to such specialist pollinators is then sensitive to plant strategy, $dR_j / dg_{ij}(t_{ij}^*) \neq 0$. If reward production is costly to plants, specific relations between R_j and $g_{ij}(t_{ij}^*)$ lead to evolution of reduced reward (M. T. Morgan, unpublished analysis). This result provides an explanation for the evolution of rewardless flowers, and anticipates that rewardless pollination evolves in tightly coupled plant and pollinator systems. Such a prediction is broadly consistent with rewardless flowers in many species of specialized orchids, although a more balanced accounting would recognize that nectar is often a minor component of the overall physiological costs of flowers (Harder and Barrett 1992) and that orchids have additional unique features that may contribute to evolution of rewardless flowers (Harder 2000).

2.3.3 Ecological dynamics

Recognition that pollinators influence plant reproductive success leads to consideration of the ecological changes in plant and pollinator densities (e.g.,

Holland et al. 2002). For example, suppose that *per capita* recruitment is affected by the effects of plant density N on the fractions of cross-fertilization $T(N)$, self-fertilization $S(N)$ (including reduced fitness of selfed progeny w_s), and the probability that a seed recruits into the adult population $R(N)$ according to $\theta[S(N)w_s + T(N)]R(N)$, where θ represents ovule production. As illustrated by the thin curve in Fig. 2.3a, per capita recruitment peaks at an intermediate density. Also suppose that plant density does not affect adult mortality μ (Fig. 2.3a, thick line), so that changes in the biotic or physical environment (e.g., moisture availability, herbivore abundance, anthropogenic activity) increase or decrease the elevation of the mortality curve. Under these conditions, the instantaneous change in density per capita is

$$\frac{1}{N}\frac{dN}{dt} = \theta[S(N)w_s + T(N)]R(N) - \mu. \quad (6)$$

Equilibria occur when seed production and recruitment balance mortality ($\theta[S(N)w_s + T(N)] = \mu$: Fig. 2.3a, open and closed symbols). Forms of recruitment and regulation are open to considerable interpretation, and need investigation.

Figure 2.3a illustrates plausible aspects of recruitment in outcrossing populations (i.e., $S(N) = 0$). At low densities, plants have very low fertilization success $T(N)$, perhaps because intraspecific pollen transfer increases with mate availability, as assumed by Sargent and Otto (2006), or because of pollinator learning or other aspects of foraging not included in the basic optimal foraging model (eq. 5). Because population persistence requires that recruitment exceed mortality, there is an *extinction threshold* (Fig. 2.3a, open symbol) below which populations are driven deterministically to extinction, a condition known as the *Allee effect* (see Chapter 6). Fertilization and recruitment exceed mortality above the extinction threshold and densities increase until the population reaches its carrying capacity (Fig. 2.3a, closed symbol) at which some other factor, such as competition for pollinator visits $T(N)$ or limited seedling establishment sites $R(N)$, again reduces *per capita* fertilization and recruitment below mortality. The carrying capacity represents a stable equilibrium density.

Specialized interactions between plants and pollinators can result in a different scenario for

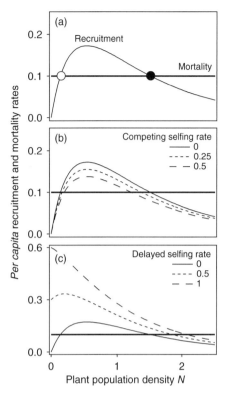

Figure 2.3 Population dynamics under self-fertilization. (a) Relations of *per capita* recruitment and mortality rates to population density in a random-mating population. The open circle is an extinction threshold, below which populations are driven deterministically to extinction. The positive relation of recruitment to density at low densities is an Allee effect. The solid circle is a stable equilibrium (carrying capacity). (b) Competing selfing decreases *per capita* recruitment, lowering the stable equilibrium and elevating the extinction threshold, and thereby eroding the ecological stability of the population. (c) Delayed (and prior) selfing enhances recruitment, and reduces or eliminates the extinction threshold and Allee effect.

ecological dynamics (e.g., Boucher 1985), because both partners are subject to carrying capacities, extinction thresholds, and Allee effects. The extinction thresholds and Allee effects are particularly important, because if either the plant or pollinator species falls below its extinction threshold both species will be driven deterministically to extinction. Alternative reproductive or energetic sources may eliminate the Allee effect for plants, pollinators, or both (Boucher 1985).

What kinds of evolution might occur in the ecological context represented by eq. 6? The relative fitness expression of eq. 1 is designed so that the average fitness equals 1, whereas for ecological models we often want to allow reproductive strategies to influence the number of offspring explicitly. Equation 1 can be rewritten, without altering its evolutionary interpretation, in terms of female reproductive (i.e., seed) output

$$\frac{1}{2}\left[W_f(z) + \overline{W}_f \frac{W_m(z)}{\overline{W}_m}\right].$$

This formula is the *per capita* recruitment of an individual with trait value z in a population at density N, so a reasonable density-dependent fitness measure is

$$W(z;N) = \frac{\theta}{2}\left[W_f(z;N) + \overline{W}_f(N)\frac{W_m(z;N)}{\overline{W}_m(N)}\right]R(N).$$

The phenotypic selection acting on trait z at density N and with predominant phenotype z^* is then approximately

$$\frac{1}{\overline{W}(N)} \frac{dW(z;N)}{dz}\bigg|_{z \approx z^*}. \qquad (7)$$

This expression reduces to the ESS equation (eq. 2) when population size remains constant during evolutionary time, or when male and female fertilities depend on trait value z, but not on population density N. The former condition might occur when individuals occupy a stable ecological niche, and the population is at its carrying capacity. Another, somewhat weaker, possibility is that evolutionary change is mutation limited. The possibility that evolutionary change is mutation limited separates "fast" ecological time from "slow" evolutionary time, and is an assumption of adaptive dynamics. Analysis of evolution under density dependence still considers a single dynamical equation (i.e., eq. 7), but with population size as an implicit function of current reproductive strategy. Interesting features are then possible, such as the evolution of generalist strategies until specialization becomes evolutionarily favourable (Sargent and Otto 2006).

Many real populations, especially those experiencing demographic changes imposed by anthropogenic activity (e.g., Davis *et al.* 2005), are likely to experience simultaneous change in population size N and trait value z. In these cases eqs 6 and 7 require simultaneous analysis. The tempo of change is likely to involve standing variation (rather than new mutation) for selection response, so that a quantitative genetic, rather than ESS, model may be appropriate.

2.4 Self-fertilization, phenotypic selection, and reproductive assurance

The evolution of self-fertilization is perhaps the most common evolutionary transition in plant reproductive biology (Stebbins 1974), and certainly the most thoroughly studied. A central focus contrasts the genetic transmission advantage of selfing with the fitness consequences of inbreeding depression (Lande and Schemske 1985), but here I focus instead on Lloyd's (1979; refined in 1992; Lloyd and Schoen 1992; Schoen and Lloyd 1992) recognition that the mechanism and timing of selfing have important consequences for the ecology and evolution of selfing (also see Chapters 1, 4 and 10).

Lloyd (1979) originally recognized "prior", "competing", and "delayed" modes of selfing, depending on when selfing occurs relative to outcrossing, and what role pollinators play in causing selfing. (Usage in this chapter follows Lloyd's 1979 terminology, but Lloyd [1992] changed his conception of selfing modes. A key area of confusion concerns "competing" selfing. Lloyd's 1979 characterization of competing selfing is most comparable to his 1992 "facilitated" selfing. Eckert uses Lloyd's 1992 concept of competing selfing in Chapter 10; Harder argues in Chapter 4 against the use of competing selfing entirely). The modes of self-pollination have consequences for ecological and evolutionary dynamics. Lloyd proposed that with competing selfing, lifetime inbreeding depression $\delta > 1/2$ is sufficient to maintain outcrossing. This scenario is central to Lande and Schemske's (1985) seminal contribution to understanding the genetics of selfing and the anticipated bimodal distribution of plant mating systems. However, the threshold inbreeding depression of 1/2 is restricted remarkably to models of competing selfing, as Lloyd's original (1979) analysis indicated that delayed selfing is *always* evolutionarily advantageous, whereas the advantage to prior selfing depends on the level of pollination.

Specifically, Lloyd noted that prior selfing is advantageous when the fitness of selfed progeny exceeds half the fraction of ovules fertilized by pollinators. The evolution of selfing in this case still depends on an inbreeding depression threshold, but the relevant threshold is always less than or equal to the threshold under competing selfing.

2.4.1 Ecological consequences of selfing mode

Recent work integrates Lloyd's original formulations of selfing into plant population dynamic models (e.g., Morgan *et al*. 2005; also Vallejo-Marin and Uyenoyama 2004; Porcher and Lande 2005a) and helps to develop Lloyd's insights about phenotypic selection and the evolution of selfing. Consider selfing caused by pollinators, either within or between flowers, which Lloyd (1979) referred to as competing selfing. This selfing mode does *not* provide reproductive assurance, because seed production is by assumption the same regardless of pollen receipt, and hence does not alleviate the Allee effect. In fact, in Lloyd's original (1979) analysis competing selfing provides a one-to-one trade-off of outcrossed seeds for selfed seeds (seed discounting), so that in the presence of inbreeding depression, selfed seeds have lower fitness than outcrossed seeds. In the ecological context of eq. 6, seed discounting and inbreeding depression decrease *per capita* population growth rates. Consequently, the extinction threshold increases and the carrying capacity decreases. Surprisingly, then, competing selfing makes the ecological dynamics of plant populations more tenuous.

Prior or delayed selfing (Fig. 2.3c) offer different possibilities than competing selfing. Both selfing modes occur in the absence of pollinators, hence providing reproductive assurance in the face of limited pollinator service (see Chapter 10). Ecologically, an increase in these modes of selfing reduces and perhaps eliminates the extinction threshold and increases the carrying capacity (Morgan *et al*. 2005). Though not a necessary outcome, greater prior or delayed selfing can also eliminate the Allee effect, so that increasing population density always decreases *per capita* recruitment. As a consequence of a lower extinction threshold and higher carrying capacity, populations that engage in prior or delayed selfing may be more resilient to fluctuations in ecological factors that influence plant recruitment or mortality.

Suppose that anthropogenic change decreases the habitat available for recruitment, or increases mortality due to edge effects (see Chapter 9). These changes alter the elevation of the recruitment or mortality curves in Fig. 2.3. For instance, increased mortality reduces the carrying capacity. Sufficient increases in mortality in the absence of reproductive assurance can lead to abrupt transitions from stable populations to extinction.

Changes in mortality or recruitment when selfing provides reproductive assurance can lead to novel consequences. For instance, increased mortality in the presence of substantial delayed selfing causes a more-or-less continuous shift from outcrossing to self-fertilization. Note that this change in mating system is *ecological*, rather than evolutionary: deviations from the single optimal reproductive strategy arise from environmental features influencing recruitment or mortality. In other formulations, increased prior selfing enhances recruitment into the population and hence the rate of outcross pollination (Cheptou 2004): although the *number* of selfed ovules increases, the *fraction* of selfed ovules decreases. Such ecologically driven changes in selfing rate might contribute significantly to the incidence of intermediate selfing rates (Vogler and Kalisz 2001; Goodwillie *et al*. 2005).

2.4.2 Evolutionary consequences of selfing mode

The importance of lifetime inbreeding depression $\delta > 1/2$ in models of competing selfing results from a kind of genetic accounting, because selfed progeny have twice as many copies of maternal alleles as outcrossed individuals. However, an ecological accounting is also at work, because selfed progeny that suffer inbreeding depression reduce *per capita* recruitment. Thus evolutionary (genetic transmission) and ecological (population dynamic) processes have conflicting influences on phenotypic selection. Resolution of this conflict can be disastrous. For example, suppose that $\delta < 1/2$ in an outcrossing population with recruitment only slightly greater than mortality. Analysis of eqs 5 and 7 shows that competing selfing variants increase in frequency, even as they reduce population growth rate below that required for persistence, so the population evolves into extinction (Morgan et al. 2005). Cheptou (2004) reached similar conclusions under different modelling scenarios.

Lloyd (1979) noted that greater prior selfing is evolutionarily advantageous when lifetime inbreeding depression is less than half the fraction of cross-fertilized ovules. Pollen-limitation studies (summarized in Ashman et al. 2004) suggest that this cross-fertilization fraction is typically ≈ 0.7, so that prior selfing evolves with a threshold lifetime inbreeding depression of $\delta \approx 0.35$. This threshold can be satisfied much more readily than the $\delta = 0.5$ threshold under competing selfing. This conclusion exposes a need to reassess the role of deleterious mutation in the maintenance of outcrossing, because substantial inbreeding depression occurs only with high genome-wide rates of deleterious mutation (e.g., $U > 1$) to strongly recessive alleles (e.g., Charlesworth et al. 1991). In contrast, genetic studies (e.g., Lynch et al. 1999; Schultz et al. 1999; Shaw et al. 2000; Shaw et al. 2003) do not observe such mutation rates, even though inbreeding depression is sometimes large (e.g., in long-lived species; Lande et al. 1994; Scofield and Schultz 2005). The observed incidences of outcrossing, mutation, and inbreeding depression remain paradoxical and present empirical and theoretical challenges. One possibility is that reproductive compensation (replacement of aborted embryos with viable embryos: Porcher and Lande 2005b) or some forms of density-dependent competition during establishment effectively weakens selection against deleterious mutations and hence allows genetic load to increase.

2.4.3 Variable pollination environments

The studies summarized so far model a constant pollination environment, albeit entertaining informally the possibility of directional change in recruitment or mortality. Many pollination biologists recognize spatial and temporal variation as essential components of plant reproduction (e.g., Barrett et al. 1989; Fausto et al. 2001; Thompson 2005; Chapters 8, 12, 15 and 16). Theoretical studies are beginning to include variable pollination environments explicitly (e.g., Pannell and Barrett 1998; Nuismer et al. 2003; Chapter 12).

A recent paper addresses how variable pollination influences the phenotypic selection of self-fertilization (Morgan and Wilson 2005). Lloyd's distinctions between modes of selfing again become important. For competing selfing, variation in pollinator service influences the number of ovules fertilized, but not the fraction of ovules fertilized that are selfed. As a consequence, stochasticity does not affect phenotypic selection of competing selfing, provided that the variable pollination environment allows population persistence. Likewise, delayed selfing is advantageous regardless of the number of ovules fertilized, so variable pollination environments influence the strength of phenotypic selection for delayed selfing, but not its unilateral evolutionary advantage.

Variable pollination environments have interesting consequences for the phenotypic selection of prior selfing (Morgan and Wilson 2005). For instance, variation in the pollination environment erodes the threshold required to maintain outcrossing, facilitating invasion of selfing variants, because the relevant fitness measure depends on variation in fitness (e.g., Gillespie 1976; Seger and Brockmann 1987). Compared with outcrossing individuals, reproduction through prior selfing reduces both the arithmetic mean (due to inbreeding depression) and variance (due to reproductive assurance) in fitness through time. The geometric mean is the appropriate fitness measure in a temporally variable environment, and it is less than or equal to the arithmetic mean, with the discrepancy increasing with the amount of variation. Thus greater variation in the pollination environment enhances the geometric mean fitness of prior selfing compared with outcrossing individuals, even while inbreeding depression reduces the arithmetic mean (see Fig. 2 of Morgan et al. 2005). In addition to reducing the threshold inbreeding depression required for the evolution of selfing, variable pollination allows intermediate stable prior selfing rates.

The preceding analyses involve ESS-style models with constant inbreeding depression. Complementary studies involve fluctuations in inbreeding depression due to changes in the competitive environment (e.g., Cheptou and Mathias 2001; Cheptou and Dieckmann 2002; Cheptou and Schoen 2002), for which geometric mean fitness is again important. Fluctuations in effective population size associated with variation in reproductive rates influence genetic load due to deleterious mutation (see Glemin 2003), so explicit models of realistic pollination modes and inbreeding depression (Porcher and Lande 2005a) in stochastic environments are an important direction for future development.

2.5 Discussion

Much current understanding of how phenotypic selection shapes plant reproductive character adaptation traces to the seminal works of Lloyd, D. Charlesworth and B. Charlesworth, and Charnov. These authors and the ESS and population genetic approaches they introduced help clarify the conceptual basis of selection, especially facilitating strategic and logical thinking about the action of selection on individuals. Whenever biological complexities overwhelm clear insight, the formulation in eq. 1 provides the investigator with a straight-forward mathematical framework for identifying and evaluating the logical consequences of relevant biological observations. A corollary is that model outcomes depend on underlying biological assumptions (Frank 1990), so a model's contribution should be judged by the fidelity with which it distills relevant biological detail.

Pollinators play an essential role in the phenotypic selection of reproductive characters. Reasoning about how pollinators influence gain curves or selection gradients can provide key insights into phenotypic selection. However, pollinators must sometimes be incorporated more explicitly, as in the selection of generalist versus specialist interactions (Fig. 2.1) and the consequences of variable pollination environments to the selection of self-fertilization (Fig. 2.3). Considerable work remains in formulating an appropriate theoretical context for these studies. The main challenges are to identify general and realistic ecological descriptions of plant and pollinator interaction and to develop a satisfactory approach to the analysis of simultaneous ecological and evolutionary change.

Plant reproductive structures are unique in their role as both targets of phenotypic selection and agents patterning genetic variation. Kelly (e.g., 1999a, b; Kelly and Williamson 2000) provided a fundamental illustration of this role, pointing out that non-random mating invalidates naive uses of *heritability* to describe between-generation evolutionary response when traits have a non-additive genetic basis. Furthermore, the genetic basis and rate of adaptive change probably differs between selfing and outcrossing species (Charlesworth 1992). This difference arises because selfing increases homozygosity and allows phenotypic selection to act on recessive and dominant variation. Other features are important in plant reproductive biology but not in other systems. Perenniality, for instance, influences the observed inbreeding depression (Morgan 2001; Scofield and Schultz 2005) and hence selection of such major features of plant reproduction as selfing.

Perhaps the most challenging aspect of selection is to develop non-tautological descriptions of floral diversity. The ESS and quantitative genetic approaches presented here describe clearly how phenotypic selection occurs. However, these descriptions consider phenotypic selection (the gain curves or fitness functions) as an externally prescribed feature of the environment: *if* gain curves accelerate, then selection favours the evolution of separate sexes, etc. Desire for a fuller accounting of dynamical evolutionary change motivates the inclusion of pollinators in the description of phenotypic selection. The success of this approach is incomplete: asking when generalist versus specialist pollination systems evolve probably results in an answer very similar to that for the evolution of separate sexes (*if* the relation between plant strategy and reproductive gain follows a particular pattern, then specialization results). Adaptive dynamic approaches sometimes anticipate evolutionary diversification (as in Sargent and Otto 2006), but again they are a realization of the underlying assumptions about the relations between reproductive traits and fitness.

Approaches like pollination as trade, and more complete inclusion of ecological interactions, may offer the best prospects for describing floral diversification non-tautologically.

Acknowledgments

I am grateful to Drs. Spencer Barrett and Lawrence Harder for the opportunity to write this chapter. Discussion with Josh Neely, Joe Rausch, and Jayson Osborne helped to clarify ideas. This research received support from NSF DEB-0128896.

References

Abrams PA, Harada Y, and Matsuda H (1993). On the relationship between quantitative genetic and ESS models. *Evolution*, **47**, 982–5.

Adams WT, Birkes DW, and Erickson VJ (1992). Using genetic markers to measure gene flow and pollen dispersal in forest tree seed orchards. In R Wyatt, ed. *Ecology and evolution of plant reproduction: new approaches*, pp. 37–61. Chapman and Hall, New York.

Aigner PA (2001). Optimality modeling and fitness trade-offs: when should plants become pollinator specialists? *Oikos*, **95**, 177–84.

Ashman T-L (2003). Constraints on the evolution of males and sexual dimorphism: field estimates of genetic architecture of reproductive traits in three populations of gynodioecious *Fragaria virginiana*. *Evolution*, **57**, 2012–25.

Ashman T-L (2005). The limits on sexual dimorphism in vegetative traits in a gynodioecious plant. *American Naturalist*, **166**, S5-16.

Ashman T-L, Knight TM, Steets J, et al. (2004). Pollen limitation of plant reproduction: Ecological and evolutionary causes and consequences. *Ecology*, **85**, 2408–21.

Barrett SCH, Morgan MT, and Husband BC (1989). The dissolution of a complex genetic polymorphism: the evolution of self-fertilization in tristylous *Eichhornia paniculata* (Pontederiaceae). *Evolution*, **43**, 1389–416.

Boucher DH, ed. (1985) *The biology of mutualism: ecology and evolution.*, Croom Helm Ltd, Beckenham, Kent.

Brown AHD (1990). Genetic characterization of plant mating systems. In AHD Brown, MT Clegg, AL Kahler, and BS Weir, eds. *Plant population genetics, breeding, and genetic resources*, pp. 145–62. Sinauer, Sunderland, MA.

Broyles SB and Wyatt R (1990). Paternity analysis in a natural population of *Asclepias exaltata*: multiple paternity, functional gender, and the 'pollen-donation' hypothesis. *Evolution*, **44**, 1454–68.

Burczyk J, Adams WT, Moran GF, et al. (2002). Complex patterns of mating revealed in a *Eucalyptus regnans* seed orchard using allozyme markers and the neighbourhood model. *Molecular Ecology*, **11**, 2379–91.

Charlesworth B (1980). *Evolution in age-structured populations*. Cambridge University Press, Cambridge.

Charlesworth B (1992). Evolutionary rates in partially self-fertilizing species. *American Naturalist*, **140**, 126–48.

Charlesworth B, Morgan MT and Charlesworth D (1991). Multilocus models of inbreeding depression with synergistic selection and partial self-fertilization. *Genetical Research (Cambridge)*, **57**, 177–94.

Charlesworth D and Charlesworth B (1979). The evolutionary genetics of sexual systems in flowering plants. *Proceedings of the Royal Society of London, Series B*, **205**, 513–30.

Charnov EL (1982). *The theory of sex allocation*. Princeton University Press, Princeton, NJ.

Charnov EL, Maynard Smith J, and Bull JJ (1976). Why be an hermaphrodite? *Nature*, **263**, 125–6.

Cheptou PO (2004). Allee effect and self-fertilization in hermaphrodites: reproductive assurance in demographically stable populations. *Evolution*, **58**, 2613–21.

Cheptou PO and Dieckmann U (2002). The evolution of self-fertilization in density-regulated populations. *Proceedings of the Royal Society of London, Series B*, **269**, 1177–86.

Cheptou PO and Mathias A (2001). Can varying inbreeding depression select for intermediary selfing rates? *American Naturalist*, **157**, 361–73.

Cheptou PO and Schoen DJ (2002). The cost of fluctuating inbreeding depression. *Evolution*, **56**, 1059–62.

Davis MB, Shaw RG, and Etterson JR (2005). Evolutionary responses to changing climate. *Ecology*, **86**, 1704–14.

Delph LF, Gehring JL, Frey FM, et al. (2004). Genetic constraints on floral evolution in a sexually dimorphic plant revealed by artificial selection. *Evolution*, **58**, 1936–46.

Fausto JA, Eckhart VM and Geber MA (2001). Reproductive assurance and the evolutionary ecology of self-pollination in *Clarkia xantiana* (Onagraceae). *American Journal of Botany*, **88**, 1794–800.

Frank SA (1990). Sex allocation theory for birds and mammals. *Annual Review of Ecology and Systematics*, **21**, 13–55.

Garant D and Kruuk LEB (2005). How to use molecular marker data to measure evolutionary parameters in wild populations. *Molecular Ecology*, **14**, 1843–59.

Geritz SAH, Kisdi E, Meszena G, *et al.* (1998). Evolutionary singular strategies and the adaptive growth and branching of the evolutionary tree. *Evolutionary Ecology*, **12**, 35–57.

Gillespie JH (1976). Natural selection for variance in offspring number: a new evolutionary principle. *American Naturalist*, **111**, 1010–4.

Glemin S (2003). How are deleterious mutations purged? Drift versus nonrandom mating. *Evolution*, **57**, 2678–87.

Goodwillie C, Kalisz S, and Eckert CG (2005). The evolutionary enigma of mixed mating systems in plants: occurrence, theoretical explanations, and empirical evidence. *Annual Review of Ecology, Evolution and Systematics*, **36**, 47–79.

Goulson D (2003). *Bumblebees: their behavior and ecology*. Oxford University Press, Oxford.

Haig D and Westoby M (1988). On limits to seed production. *American Naturalist*, **131**, 757–9.

Harder LD (2000). Pollen dispersal and the floral diversity of Monocotyledons. In K Wilson and D Morrison, eds. *Monocots: systematics and evolution*, pp. 243–57. CSIRO Publishing, Melbourne, Australia.

Harder LD and Barrett SCH (1992). The energy cost of bee pollination for *Pontederia cordata* (Pontederiaceae). *Functional Ecology*, **6**, 226–33.

Harder LD and Thomson JD (1989). Evolutionary options for maximizing pollen dispersal of animal-pollinated plants. *American Naturalist*, **133**, 323–44.

Hereford J, Hansen TF, and Houle D (2004). Comparing strengths of directional selection: How strong is strong? *Evolution*, **58**, 2133–43.

Holland JN, DeAngelis DL, and Bronstein JL (2002). Population dynamics and mutualism: Functional responses of benefits and costs. *American Naturalist*, **159**, 231–44.

Houle D (1991). Genetic covariance of fitness correlates: what genetic correlations are made of and why it matters. *Evolution*, **45**, 630–48.

Kelly JK (1999a). Response to selection in partially self-fertilizing populations. I. Selection on a single trait. *Evolution*, **53**, 336–49.

Kelly JK (1999b). Response to selection in partially self-fertilizing populations. II. Selection on multiple traits. *Evolution*, **53**, 350–7.

Kelly JK and Williamson S (2000). Predicting response to selection on a quantitative trait: A comparison between models for mixed-mating populations. *Journal of Theoretical Biology*, **207**, 37–56.

Kiester RA, Lande R, and Schemske DW (1984). Models of coevolution and speciation in plants and their pollinators. *American Naturalist*, **124**, 220–43.

Lande R (1976). Natural selection and random genetic drift in phenotypic evolution. *Evolution*, **30**, 314–34.

Lande R (1979). Quantitative genetic analysis of multivariate evolution, applied to brain:body size allometry. *Evolution*, **33**, 402–16.

Lande R and Arnold SJ (1983). The measurement of selection on correlated characters. *Evolution*, **36**, 1210–26.

Lande R and Schemske DW (1985). The evolution of self-fertilization and inbreeding depression in plants. I. Genetic models. *Evolution*, **39**, 24–40.

Lande R, Schemske DW, and Schultz ST (1994). High inbreeding depression, selective interference among loci, and the threshold selfing rate for purging recessive lethal mutations. *Evolution*, **48**, 965–78.

Lloyd DG (1977). Genetic and phenotypic models of natural selection. *Journal of Theoretical Biology*, **69**, 543–60.

Lloyd DG (1979). Some reproductive factors affecting the selection of self-fertilization in plants. *American Naturalist*, **113**, 67–79.

Lloyd DG (1982). Selection of combined versus separate sexes in seed plants. *American Naturalist*, **120**, 571–85.

Lloyd DG (1984). Gender allocation in outcrossing cosexual plants. In R Dirzo and J Sarukhán, eds. *Perspectives on plant population ecology*, pp. 277–300. Sinauer Associates, Sunderland, MA.

Lloyd DG (1985). Parallels between sexual strategies and other allocation strategies. *Experientia*, **41**, 1277–85.

Lloyd DG (1992). Self- and cross-fertilization in plants. II. The selection of self-fertilization. *International Journal of Plant Sciences*, **153**, 370–80.

Lloyd DG and Schoen DJ (1992). Self- and cross-fertilization in plants. I. Functional dimensions. *International Journal of Plant Sciences*, **153**, 358–69.

Lloyd DG and Webb CJ (1986). The avoidance of interference between the presentation of pollen and stigmas in angiosperms. I. Dichogamy. *New Zealand Journal of Botany*, **24**, 135–62.

Lloyd DG and Webb CJ (1992). The selection of heterostyly. In SCH Barrett, ed. *Evolution and function of heterostyly*, pp. 179–207. Springer-Verlag, Berlin.

Lynch M, Blanchard J, Houle D, *et al.* (1999). Perspective: spontaneous deleterious mutation. *Evolution*, **55**, 645–63.

Morgan MT (2000). Evolution of interactions between plants and their pollinators. *Plant Species Biology*, **15**, 249–59.

Morgan MT (2001). Consequences of life history for inbreeding depression and mating system evolution. *Proceedings of the Royal Society of London, Series B*, **268**, 1817–24.

Morgan MT and Conner JK (2001). Using genetic markers to directly estimate male selection gradients. *Evolution*, **55**, 272–81.

Morgan MT and Schoen DJ (1997). Selection on reproductive characters: floral morphology in *Asclepias syriaca*. *Heredity*, **79**, 433–41.

Morgan MT and Wilson WG (2005). Self fertilization and the escape from pollen limitation in variable pollination environments. *Evolution*, **59**, 1143–8.

Morgan MT, Wilson WG, and Knight TM (2005). Pollen limitation, population dynamics, and the persistence of plant populations: roles for pollinator preference and reproductive assurance. *American Naturalist*, **166**, 169–83.

Nuismer SL and Cunningham BM (2005). Selection for phenotypic divergence between diploid and autotetraploid *Heuchera grossularifolia*. *Evolution*, **59**, 1928–35.

Nuismer SL, Gomulkiewicz R, and Morgan MT (2003). Coevolution in temporally variable environments. *American Naturalist*, **162**, 195–204.

Nuismer SL, Thompson JN, and Gomulkiewicz R (1999). Gene flow and geographically structured coevolution. *Proceedings of the Royal Society of London, Series B*, **266**, 605–9.

Ohashi K, and Yahara T (2001). Behavioural responses of pollinators to variation in floral display size and their influences on the evolution of floral traits. In L Chittka and JD Thomson, eds. *Cognitive ecology of pollination*, pp. 274–96. Cambridge University Press, New York.

Pannell JR and Barrett SCH (1998). Baker's Law revisited: reproductive assurance in a metapopulation. *Evolution*, **52**, 657–88.

Porcher E and Lande R (2005a). The evolution of self-fertilization and inbreeding depression under pollen discounting and pollen limitation. *Journal of Evolutionary Biology*, **18**, 497–508.

Porcher E and Lande R (2005b). Reproductive compensation in the evolution of plant mating systems. *New Phytologist*, **166**, 673–84.

Sargent RD and Otto SP (2006). The role of local species abundance in the evolution of pollinator attraction in flowering plants. *American Naturalist*, **167**, 67–80.

Schoen DJ and Lloyd DG (1992). Self- and cross-fertilization in plants. III. Methods for studying modes and functional aspects of self-fertilization. *International Journal of Plant Sciences*, **153**, 381–93.

Schultz ST, Lynch M, and Willis JH (1999). Spontaneous deleterious mutation in *Arabidopsis thaliana*. *Proceedings of the National Academy of Sciences of the United States of America*, **96**, 11393–8.

Schwartz MW and Hoeksema JD (1998). Specialization and resource trade: biological markets as a model of mutualisms. *Ecology*, **79**, 1029–38.

Scofield D and Schultz S (2005). Mitosis, stature, and evolution of plant mating systems: low-Φ; and high-Φ; plants. *Proceedings of the Royal Society of London, Series B*, **273**, 275–82.

Seger J and Brockmann HJ (1987). What is bet-hedging? In PH Harvey and L Partridge, eds. *Oxford Surveys in Evolutionary Biology*, Vol. 4, pp. 182–211. Oxford University Press, Oxford.

Shaw RG, Byers DL, and Darmo E (2000). Spontaneous mutational effects on reproductive traits of *Arabidopsis thaliana*. *Genetics*, **155**, 369–78.

Shaw RG, Shaw FH, and Geyer C (2003). What fraction of mutations reduces fitness? A reply to Keightley and Lynch. *Evolution*, **57**, 686–9.

Smouse PE and Meagher TR (1994). Genetic analysis of male reproductive contributions in *Chamaelirium luteum* (L.) Gray (Liliaceae). *Genetics*, **136**, 313–22.

Stebbins GL (1970). Adaptive radiation of reproductive characteristics in angiosperms. I. Pollination mechanisms. *Annual Review of Ecology and Systematics*, **1**, 307–26.

Stebbins GL (1974). *Flowering plants: evolution above the species level*. Belknap Press, Cambridge, MA.

Stephens DW and Krebs JR (1986). *Foraging theory*. Princeton University Press, Princeton, NJ.

Thompson JN (2005). *The geographic mosaic of coevolution*. University of Chicago Press, Chicago.

Turelli M and Barton NH (1994). Genetic and statistical analyses of strong selection on polygenic traits: what, me normal? *Genetics* **138**, 913.

Vallejo-Marin M and Uyenoyama MK (2004). On the evolutionary costs of self-incompatibility: Incomplete reproductive compensation due to pollen limitation. *Evolution*, **58**, 1924–35.

Vogler DW and Kalisz S (2001). Sex among the flowers: the distribution of plant mating systems. *Evolution*, **55**, 202–4.

Waser NM, Chittka L, Price MV, et al. (1996). Generalization in pollination systems, and why it matters. *Ecology*, **77**, 1043–60.

Waxman D and Gavrilets S (2005). 20 Questions on adaptive dynamics. *Journal of Evolutionary Biology*, **18**, 1139–54.

Webb CJ and Lloyd DG (1986). The avoidance of interference between the presentation of pollen and stigmas in angiosperms. II. Herkogamy. *New Zealand Journal of Botany*, **24**, 163–78.

Zhang X-S and Hill WG (2005). Genetic variability under mutation selection balance. *Trends in Ecology and Evolution*, **20**, 468–70.

CHAPTER 3

Evolutionarily stable reproductive investment and sex allocation in plants

Da-Yong Zhang

MOE Key Laboratory for Biodiversity Science and Ecological Engineering, College of Life Sciences, Beijing Normal University, Beijing, P. R. China

Outline

Optimal allocation of limiting resources to various fitness-enhancing activities is central to both life-history and sex-allocation theory, providing the basis for different reproductive strategies. Allocation commonly varies with plant size, but this size dependence has not been investigated fully either theoretically or empirically. In this chapter, I first introduce a new approach to the classical model of sex allocation and apply it to quantify and explain within-population variation in sex allocation and reproductive effort of annuals and perennials. The model shows that the reproductive strategies of cosexual plants should equalize the marginal fitness returns through each fitness component (male and female function, and adult survival for perennials), unless constrained by insufficient available resources. In particular, if fitness gain through female function increases linearly with investment, then fixed amounts of resources should be allocated to male function and post-breeding survival. The model provides adaptive explanations for several features of resource allocation of cosexual plants, including the minimum size for reproduction in perennials, but not annuals; neutral or positive phenotypic correlations between male and female allocation and between reproduction and survival; and even why male function precedes female function in hermaphrodites. These models also identify the shapes of fitness gain curves as important determinants of optimal allocation and the evolution of sexual systems, including combined sexes and gender diphasy.

3.1 Introduction

Resource allocation is central to the life-history and sexual strategies of plants. According to the principle of allocation (Gadgil and Bossert 1970), individuals have limited resources to spend on the three basic functions that allow plants to succeed in their environments: growth, maintenance, and reproduction. As a consequence of this limitation, allocation to one function can increase only at the expense of other functions. Given these trade-offs, what is the optimal strategy? Theoretical modelling of reproductive effort and sex allocation seeks to answer this question (Schaffer 1974; Charlesworth 1980; Charnov 1982; Lloyd 1984; Bazzaz *et al.* 1987; Kozlowski 1992).

The expected success of a given strategy is often identified by analyzing Evolutionarily Stable Strategies (ESS: Maynard Smith 1982). In this approach, partial differential fitness equations are analyzed to determine whether a large, monomorphic population can be invaded by a rare variant with different allocation characteristics. An ESS can be viewed broadly as a strategy that is more successful than any other when adopted by all members of a population. Note that an ESS involves fitness maximization only when the mutant arrives in the population. Thus ESS is a weaker concept of optimization than the classical optimization approach.

Most evolutionary models for reproductive investment build upon Fisher's (1930) argument that equal average male and female fitness within populations imposes negative frequency-dependent selection on the sex ratio. Although Fisher considered dioecious organisms, his arguments extend readily to plants with varying degrees of cosexuality (Charnov 1982; Lloyd 1984). These models predict investment patterns that equalize marginal gains through investment in male and female function within the population (Lloyd 1988; Lloyd and Venable 1992; Morgan and Schoen 1997; Chapter 2), but the partitioning of sexual investment among individuals depends on the quantitative form of male and female fitness gains with increasing investment. If both male and female reproductive success increase linearly with their respective investments, then the ESS is equal male and female investment by each individual. Non-linearity in one or both gain relations can select for unequal optimal allocation.

Individuals differ extensively within and among populations in both reproductive investment (Willson 1983; Samson and Werk 1986; Weiner 1988; Bazzaz and Ackerly 1992; Reekie 1999) and sex allocation (Lloyd and Bawa 1984; Burd and Allen 1988; Bickel and Freeman 1993; Klinkhamer et al. 1997; Barrett et al. 1999; Wright and Barrett 1999; Sarkissian et al. 2001; Ishii 2004). Because of their modular growth, plants often differ considerably in size or resource status as a result of environmental variation (Harper 1977). In these circumstances, natural selection favours a genetically determined allocation rule specifying allocation in relation to size or environment, rather than a single genetically determined allocation (Lloyd and Bawa 1984). A few models address this issue (Charnov 1982; Willson 1983; Lloyd and Bawa 1984; Charnov and Bull 1985; Samson and Werk 1986; Frank 1987; Kakehashi and Harada 1987; Weiner 1988; Day and Aarssen 1997; Klinkhamer et al. 1997; Sakai and Sakai 2003; Cadet et al. 2004; Sato 2004), but most are too demanding mathematically to be grasped by most biologists, and they usually do not incorporate reproductive effort and sex allocation simultaneously (but see Zhang and Jiang 2002).

In this chapter, I investigate the theoretical effects of plant size for both annual and perennial plants through a new, graphical analysis of the ESS resource-allocation model and discuss their implications for the evolution of sexual systems. I implement a graphical approach, because it is more accessible to biologists who are not mathematically inclined and can be more powerful than analytical models, which must often rely on further simplifying assumptions or numerical simulation to obtain biologically meaningful results (e.g., Frank 1987; Cadet et al. 2004). To develop theory, I assume complete outcrossing. The effects of selfing on sex allocation are relatively well understood (Charlesworth and Charlesworth 1981; Charnov 1982; Lloyd 1987; Charlesworth and Morgan 1991; Brunet 1992; de Jong et al. 1999; Zhang 2000; Klinkhamer and de Jong 2002; Zhang and Jiang 2002).

3.2 The classic model of sex allocation for outcrossing species

Beginning with Fisher's (1930) famous argument for equal investment in male and female offspring, the study of sex allocation has been one of the most thoroughly studied areas of evolutionary biology. Classical sex-allocation models assume that the resources available for reproduction are fixed and can be devoted in varying amounts to male (pollen production) versus female functions (ovule and seed production). These conditions apply strictly only in models that assume an annual life cycle (Charlesworth and Morgan 1991; Brunet 1992). In perennials, unused resources from one reproductive episode can be used for growth and to increase survival to the next breeding season (Section 3.4).

Hermaphroditic plants propagate their genes through both male and female functions. Any parent's total fitness from the two sex roles is always additive (Lloyd 1984). Female fertility can be measured by the number of seeds that survive to adulthood, and male fertility by the number of sired offspring surviving to reproductive maturity. If f_i, m_i, and w_i are the female, male, and combined fitness contributions, respectively, of individual i, then

$$w_i = f_i + m_i. \quad (1)$$

If males compete to sire seeds in the population, m_i depends on the average mating success in the population. To characterize this dependence, suppose that a fraction, β, of all pollen produced in the population successfully enters the gene pool of the next generation (β is the product of mating probability and the proportional survival of sired offspring). For an outcrossing species, $\beta = \bar{f}/\bar{p}$, where $\bar{p}$ is the population average pollen production, and $\bar{f}$ is the population average female fertility (seed production). Therefore, the male fertility of a focal plant within this population that produces p_i pollen grains equals βp_i. Consequently, eq. 1 becomes

$$w_i = f_i + \beta p_i = f_i + \frac{\bar{f}}{\bar{p}} p_i. \quad (2)$$

An individual's male and female fitness contributions depend on the proportions of its available resources, R, that it allocates to male and female function, r_i and $1 - r_i$, respectively. In particular, male and female fertilities are assumed to depend on their respective resource inputs according to $m_i = \beta p(r_i R)$, the "male gain curve," and $f_i = f[(1 - r_i)R]$, the "female gain curve," respectively, where p and f convert resources into pollen and ovules (seeds). Note that the scaling factor β depends on overall sex allocation of the population and the quantity of available resources, although it is independent of an individual's allocation strategy (r_i). Also note that fitness via male or female function usually requires investment in attractive structures such as nectar or petals, which is ignored above. If resource availability limits seed production, this simplification is well justified under Bateman's (1948) principle that female reproductive success (seed production) is not affected by mate availability (the level of pollination), so that pollinator attraction serves mainly male function (Bell 1985).

The model that I derive considers absolute male and female fertilities (Lloyd 1984), rather than the more commonly used relative fertilities (Shaw and Mohler 1953; Charnov 1982; Morgan and Schoen 1997), for the following reasons. First, this approach allows comparison of the fitness returns for male and female investment on the same scale, unlike seed production f_i and pollen production p_i, which have arbitrary scales and are not generally comparable. This standardization to a common metric enables comparison of male and female gain curves on the same set of axes. Second, the formulation allows adult survival to be incorporated easily into the theoretical framework of sex allocation (Zhang and Jiang 2002; see also Section 3.4). Finally, population fitness under Bateman's principle usually depends on absolute female fertility, which may be useful when comparing the density-dependent competitive ability of species with different sexual systems (e.g., Wilson and Harder 2003).

3.2.1 Gain curves

Fitness gain curves are central in most sex-allocation models, as their shapes control the evolution of sexual systems and sex allocation. In general, fitness through either sex role is assumed to increase continuously with increasing investment. Some mechanisms may cause these relations to accelerate, such as increased fruit dispersal with fruit number and increased pollen dispersal with flower number, when pollinators are scarce; however, male and female fitness gain curves are typically expected to be either linear or decelerating (Charnov 1982; Brunet 1992; Campbell 2000; Cadet et al. 2004).

Female gain curves are often assumed to be linear (Charnov 1982; Brunet 1992), but if seeds compete with their siblings for limited establishment sites, local resource competition can cause female fitness to saturate. The importance of this competition can be tested by following the success of offspring from individual plants and determining how many survive to reproductive maturity, although this has rarely been done. The few studies of this kind, such as Rademaker and de Jong (1999), found no evidence for significant effects of local resource competition on female fitness (reviewed in Campbell 2000). Therefore, I will usually assume linear female gain relations (Fig. 3.1) for clarity and conceptual simplicity. Qualitative results often remain unchanged when considering saturating female curves.

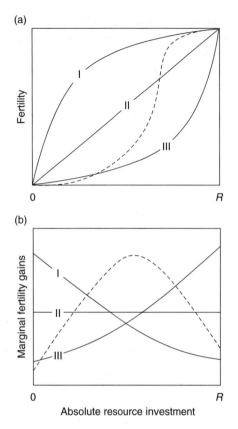

Figure 3.1 Three basic relations of fitness gain to increased resource investment: decelerating (I), linear (II), and accelerating (III). Despite having received scant attention, S-shaped curves (dashed lines) may also be relevant. Panel (a) illustrates changes in absolute fitness, whereas panel (b) depicts changes in marginal fitness (i.e., the first derivative of absolute fitness with respect to investment).

Male fitness curves probably decelerate as male investment increases in animal-pollinated plants (Fig. 3.1a) for several reasons. First, if pollen loss during transport increases as individual pollinators remove more pollen from flowers, because they can carry a limited amount of pollen or they groom more when they carry more pollen, the probability of pollen export per pollen grain decreases as investment in male function increases (Harder and Thomson 1989). Second, when a pollinator visits multiple flowers per plant, pollen picked up in one flower may be lost through grooming or deposited on stigmas or other flower parts of neighbouring flowers, reducing the pollen available for export (pollen discounting: Harder and Barrett 1995).

Third, if pollen disperses very locally or, if population density is extremely low, pollen grains can compete for a limited number of available ovules with sib pollen that lands on the same stigma (local pollen competition), again resulting in diminishing fitness returns (Lloyd 1984).

Unlike animal pollen vectors, wind and water cannot saturate with pollen, and pollen discounting is less likely to vary with the number of flowers produced (male investment). Therefore the male gain curve in outcrossed wind-pollinated plants is frequently suggested to be linear (Charlesworth and Charlesworth 1981; Charnov 1982; Lloyd 1984; Charlesworth and Morgan 1991), as long as the pollen dispersal distance is relatively large or equivalently population density is sufficiently high, so that local pollen competition is limited.

Some authors have proposed S-shaped fitness curves (Frank 1987). For example, a certain minimum investment in male function (i.e., flower production) may be required to attract any animal pollinators, whereas pollen export is often less efficient in plants with many flowers because insects visit more flowers on the same plant (Harder and Barrett 1995). However, the accelerating portion of an S-shaped curve is irrelevant to the evolution of sex allocation and the stability of hermaphroditism when each plant in a population has sufficient resources to invest in reproduction that its fitness gains level off (Fig. 3.2).

For this chapter, I assume linear female fitness returns on investment and nonlinear returns for male function. I restrict attention to this simple, yet plausible, form for the female fitness curve to make the exposition more transparent. Nonlinear fitness returns for both sexes can be handled in a straightforward, although more tedious, manner. Needless to say, given the importance of the shape of gain curves for sex-allocation theory, priority should be given to determining gain curves in natural populations, which has seldom been done.

3.2.2 ESS sex allocation

If the current sex allocation in a population, r, represents an ESS, any mutant strategy r_i must be less fit. This condition implies that eq. 2, as a function of r_i, is maximized when $r_i = r$, or

$$\left.\frac{\partial w_i}{\partial r_i}\right|_{r_i=r} = 0 \qquad (3)$$

(see also Chapter 2). Substituting eq. 2 into 3 yields

$$\left.\frac{\partial f_i}{\partial(1-r_i)}\right|_{r_i=r} = \left.\frac{\partial m_i}{\partial r_i}\right|_{r_i=r} = \frac{\bar{f}}{\bar{p}}\left.\frac{\partial p_i}{\partial r_i}\right|_{r_i=r},$$

or

$$\frac{\partial f}{\partial(1-r)} = \frac{\bar{f}}{\bar{p}}\frac{\partial p}{\partial r}. \qquad (4)$$

According to eq. 4 (and assuming that the male gain curve ultimately decelerates), the ESS occurs at the allocation for which the tangent to the male gain function on investment (right-hand side) has the same slope as the tangent to the female gain function (left-hand side), which in the cases examined here is the slope of the linear female curve (Fig. 3.2).

In other words, hermaphroditic plants should adopt reproductive strategies that equalize the magnitudes of marginal fertility gains through male and female function (Lloyd 1988). According to eq. 4, any gain in pollen production has to be scaled by mating success, $\beta = \bar{f}/\bar{p}$, to convert to a fertility gain that can be compared with the fertility gain of females. Also note that male and female fitness must be equal at an ESS; that is, $f = m$ (Fig. 3.2).

Sex-allocation models usually assume that male fertility is a power function of resources invested in male reproduction (Charnov 1979; Charnov 1982; Brunet 1992; Klinkhamer and de Jong 2002). Thus

$$f_i = k_f[(1-r_i)R]$$
$$m_i = p_i\frac{\bar{f}}{\bar{p}} = k_m(r_iR)^c\frac{\bar{f}}{\bar{p}}, \qquad (5)$$

where k_f and k_m are constants, and $0 < c < 1$ if the male gain curve decelerates. In this case, the ESS

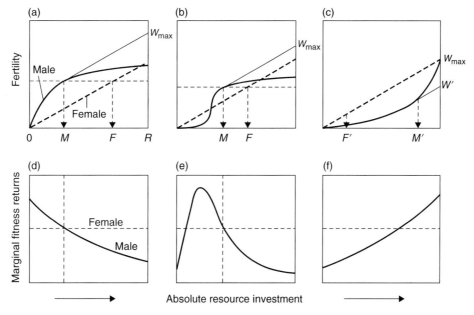

Figure 3.2 The ESS sex allocations with a linear female curve and (a) decelerating, (b) sigmoid, or (c) accelerating male gain curves. Panels (d–f) illustrate the corresponding marginal fitness returns. In case (a), the highest total fitness (w_{max}) is attained when $M (= rR)$ resources are allocated to male function, and $F = R - M = (1 - r) R$ are allocated to female function, with equal fitness returns (d). Furthermore, an ESS individual must gain fitness equally through investment in male function and in female function (dashed horizontal line at $w_{max}/2$). Hermaphroditism is evolutionarily stable to invasion by either pure male or pure female, because unisexual individuals' fitness must be lower than w_{max}. The ESS sex allocation is female biased because $F > M$. Case (b) leads to qualitatively similar results because the accelerating portion of the male gain curve is irrelevant to the evolution of sex allocation. However, in case (c), dioecy is stable, and a hermaphrodite with any investment in both male (M') and female (F') function must be less fit than unisexual individuals ($w' < w_{max}$). Note that male or female individuals must be equally fit within a dioecious population.

allocation to male function, r, is $c/(1+c)$, which is female biased ($<1/2$), and also independent of resource availability, R.

However, the independence of the sex allocation on R is an illusion caused largely by unique characteristics of the power function. Other expressions for saturating curves, especially those that rise to an asymptote, lead to different conclusions. For example, if male fitness is a Monod function of investment (Zhang and Wang 1994),

$$m_i = p_i \frac{\bar{f}}{\bar{p}} = \frac{k_m(r_i R)}{(r_i R) + D} \frac{\bar{f}}{\bar{p}},$$

the ESS sex allocation is

$$r = \frac{1}{1 + \sqrt{1 + (R/D)}}. \tag{6}$$

In this case, the ESS sex allocation (r) increases as resource availability, R, (or plant size) decreases. In such a situation, harsher environments (e.g., low soil fertility, dry soils, low light intensity) that reduce R increase the optimal allocation to male function, which is consistent with the frequent observations of increased maleness in stressful ecological conditions (Freeman et al. 1980). Although the Monod model is rather specific, it shows that sex allocation can shift environmentally, even without concomitant alterations in the shape of fitness gain curves. In other words, resource status or plant size can influence the overall sex allocation of the population.

Male gain curves need not be power functions of investment, although a linear female curve may often be a reasonable assumption. In fact, the power function was adopted originally merely for its flexibility in describing gain curves of different shapes and its tameness to mathematical analysis (Charnov 1979). Furthermore, the power function has the inherent unrealistic properties that, for the decelerating case, marginal fitness returns on investment approach infinity when investment is close to zero and fitness increases without bounds (rather than asymptotically) with increased investment.

3.2.3 Evolution of sexual systems

The shapes of gain curves determine the evolutionary stability of sexual systems. As illustrated in Fig. 3.2a, when the female gain curve (dashed curve) is linear and the male gain curve (thick solid curve) decelerates, investment of all the available R resources in one sex function or the other does not maximize fitness. Instead, the optimal investment devotes M resources to male function, taking advantage of the portion of the male gain curve that increases more rapidly than the female gain curve (i.e., higher marginal fitness through male function; Fig. 3.2d). Any investment in excess of M should be applied to female function (i.e., $F = R - M$), which then has the higher rate of return. All alternate allocation patterns result in lower total fitness than that realized from the optimal allocation (w_{max}). Therefore, the optimal allocation equalizes the marginal fitness returns through both sex roles (Fig. 3.2d), as noted by Lloyd (1988).

In contrast, when the female gain curve is linear and the male gain curve accelerates, individuals maximize their fitness by investing all of their R reproductive resources in one sex function or the other (i.e., $M = R$ or $F = R$; Fig. 3.2c). As the dotted line in Fig. 3.2c indicates, a hermaphrodite that invests M' and $F' = R - M'$ resources in male and female function, respectively, realizes lower total fitness (w') than either a pure female or a pure male. Note that females and males derive equal fitness, on average, because every offspring has a mother and a father (Fisher 1930). In addition, marginal fitnesses are not equalized by the optimal allocation pattern when at least one gain curve accelerates (Fig. 3.2f). If both male and female gain curves are linear, they must coincide; otherwise individuals of one sex will have a higher fertility than those of the other sex. Linear fitness gain curves for both sexes promote the evolution of a balanced optimal sex ratio (Charnov 1982).

S-shaped male curves do not lead to qualitatively new conclusions. They result in either hermaphroditism or dioecy, depending on how the relation of male marginal fitness to allocation intersects that of female function. With an S-shaped male gain curve, male and female marginal fitness may be equal at one (Fig. 3.2d and f) or two allocation patterns (Fig. 3.2e). In the latter case (Fig. 3.2e), only the higher allocation represents an ESS, with the lower allocation making incomplete and inefficient use of available resources. An S-shaped curve leads to a result similar to that of a saturating curve if all individuals have a sufficiently large quantity of resources to invest in reproduction.

Resource availability can alter the mating success parameter (β), but has no effect on whether the gain curve decelerates or accelerates. For S-shaped male gain curves, an initially stable hermaphroditism may become unstable as resource availability declines (Fig. 3.3), if the male gain curve accelerates within the restricted range of resource investment. Natural selection can therefore act differently among populations of the same species living in poor or rich environments. This result provides one explanation, among others (Delph 2003), for the empirical observation that gender dimorphic species are often found in more stressful ecological conditions than their sexually monomorphic relatives (Chapter 11). However, if the male gain curve only decelerates, environmental stress generally increases relative (but not absolute) investment in male function, although the sex allocation remains female biased (see, e.g., eq. 6).

Comparison of Fig. 3.2a and c reveals that hermaphroditism results in higher individual fitness (male and female fertility in total) than dioecy if both sexual systems have the same female gain function. Therefore, if dioecy evolves from hermaphroditism because the male gain curve changes from saturating to accelerating, the dioecious species will be less competitive than the hermaphroditic species, unless returns on female investment change simultaneously (e.g., lower inbreeding depression and/or lower self-fertilization rate). In this sense, dioecy is not a beneficial evolutionary strategy, which is consistent with Heilbuth's (2000) finding of lower species richness in dioecious clades. Models of relative allocation cannot bring out this conclusion as readily as the present absolute approach does.

3.3 Plant size and sex allocation

Above, I assumed that each individual in a population has equivalent resources (R) and that at the equilibrium every individual invests the same proportion of resources in male and female function,

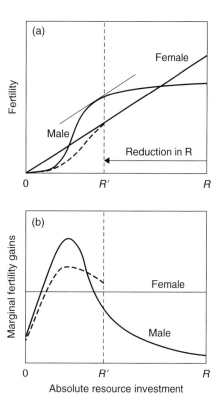

Figure 3.3 A decrease in resource availability from R in a resource-rich environment to R' in a stressful environment may shift the sexual system from hermaphroditism to dioecy, other things being equal. Heavy solid curves indicate the relations of absolute (a) or marginal fitness (b) to resource investment. The thin solid line in panel (a) depicts the tangent to the male gain curve that parallels the female gain curve. As indicated in Fig. 3.2a, the male investment associated with the tangent represents the optimal allocation to male function in the resource-rich environment. Dashed lines indicate male fertility (a) and marginal fertility returns (b) in a stressful environment with R' available resources. Female fitness returns are assumed to be unaffected by resource availability, whereas male fertility must be affected, because mating success, β, depends on both resource availability and the overall sex allocation of the population.

independent of individual or environmental conditions. For most plant species, these assumptions are clearly unrealistic, as a rapidly increasing number of studies report substantial variability in phenotypic gender in cosexual plants (reviewed in Klinkhamer et al. 1997; de Jong and Klinkhamer 2005).

Plant size, as represented by R, can have two main effects on sex allocation: "budget effects" or "direct effects" (Klinkhamer et al. 1997). With budget effects, larger plants have more resources to invest in reproduction and produce more flowers and seeds. Such effects may introduce a nonlinearity in fitness gain curves due to, for example, increased geitonogamy or local pollen competition. In contrast, with direct effects, plant size influences fitness returns directly. In the following sections, I apply the graphical approach to the problem of size-dependent sex allocation, a subject for which analytical models rapidly become abstruse or even intractable.

3.3.1 Budget effects of plant size on fitness gains

Situations in which individuals differ in the resources they have to invest, and each individual adjusts its allocation conditional on its resource status, have received the most theoretical analysis (e.g., Lloyd and Bawa 1984; Frank 1987). In this section, I focus on the budget effects of plant size in the absence of direct effects, as might be suitable for animal-pollinated plants. In this case, the same fitness gain curves apply to all size classes. Again I assume a linear female curve, and consider the three shapes for the male gain curve (Fig. 3.4), with qualitatively different outcomes. All the results can also be applied when the male fitness curve is linear, whereas the female curve is nonlinear.

If male fitness decelerates with increased resource investment, selection favours a hermaphrodite that invests in both sex functions, with gradually increased emphasis on female reproduction with increasing plant size (continuous gender adjustment; Fig. 3.4a and d). In Fig. 3.4a, individuals smaller than the threshold R_1, for which marginal fitness returns are equal for male and female investment (i.e., the tangent is parallel to the female gain curve), should be male, because investing in male function yields higher fitness returns than in female function. In contrast, individuals exceeding R_1 should maintain a fixed absolute investment in male function, because further investment in male function is less worthwhile (lower fitness returns) than in female function, with the remaining resources invested in female function. The threshold R_1 depends on both the average sex allocation of the population and the distribution of resources among individuals (cf. Frank 1987). In this case, the selected strategy never involves distinct gender phases, although this may result for accelerating or partially accelerating gain curves (Fig. 3.4b and c).

Several studies have presented results that are consistent with Fig. 3.4a. In monoecious populations of the perennial aquatic herb *Sagittaria latifolia*, female flower production increases steadily with ramet size, but male flower production remains constant (Sarkissian et al. 2001). The many herbivory studies that have found no influence of defoliation (which reduces the resources available to a plant) on male investment also provide indirect evidence for resource dependence of female, but not male effort. For example, simulated defoliation of a woodland orchid *Dactylorhiza maculata* reduced capsule production, but did not affect pollinium mass (Vallius and Salvonen 2000). In addition, small individuals in some animal-pollinated species act solely as males, and their breeding system may be mistakenly regarded as androdioecious (Charlesworth 1984). As Fig. 3.4a suggests, these plants may in fact benefit more from aborting seeds to enhance their paternal success.

Although increased femaleness with plant size is the rule in animal-pollinated plants (Klinkhamer et al. 1997), exceptions have been reported for *Asclepias syriaca* (Willson and Rathcke 1974) and *Narthecium asiaticum* (Ishii 2004). Increased maleness with size is predicted if the male fitness curve is more linear than the female curve, or if increased size directly enhances pollen dispersal, even in animal-pollinated plants. In either case, pollen production would be affected more by foliar herbivory than would ovule or seed production, a result that has been found in a few plants, such as wild radish *Raphanus raphanistrum* (Lehtila and Strauss 1999).

When male fitness varies sigmoidally with investment, as in Fig. 3.4b, the population may comprise female, male, and hermaphrodite individuals, a pattern similar to subdioecy. In this case, plants smaller than R_1 should be female, those between R_1 and R_2 should be male, and those larger than R_2 should be hermaphrodite. The frequency of three sex types depends primarily on

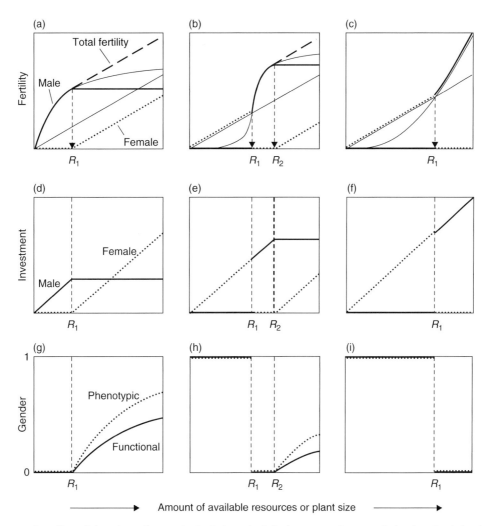

Figure 3.4 Budget effects of plant size on fitness gains (a–c), the optimal absolute resource investment in female and male function (d–f), and phenotypic and functional gender (g–i) with a linear female curve and decelerating (a, d, and g), sigmoidal (b, e, and h), or accelerating male gain curves (c, f, and i). In panels (a–c) thin lines indicate absolute fitness gain curves, the heavy solid and dotted lines depict the optimal male and female fertility, respectively, and the dashed line represents total fertility for individuals with sufficient resources to invest in both sex roles. With continuous gender adjustment (a), individuals smaller than R_1 for which the tangent is parallel to the female gain curve (i.e., male and female marginal fitness returns are equal) should be male, and those exceeding R_1 should maintain a constant absolute investment in male function and invest the remaining resources in female function (heavy solid line). Subdioecy (female, male, and hermaphrodite) is predicted for case (b), as a plant should be female when it is smaller than R_1, male when intermediate between R_1 and R_2, and hermaphrodite when larger than R_2. Gender diphasy is selected in case (c), for which small individuals ($< R_1$) are female and large individuals ($> R_1$) are male. The thresholds, R_1 and R_2, depend on the distribution of resources among individuals.

the population size structure, and the population may even include three distinct sexual morphs, namely, female, male, and hermaphrodite (subdioecy). If most individuals are large, the population would resemble that depicted in Fig. 3.4a with male and hermaphrodite morphs. However, if most individuals are relatively small the population will appear to be dioecious, although it actually exhibits gender diphasy. Only when both small and large individuals are well represented within the population is subdioecy possible. Barrett *et al.* (1999) examined patterns of resource allocation among the sex phenotypes in subdioecious populations of the diminutive geophyte *Wurmbea dioica*. They found that hermaphroditic individuals (their "fruiting males") were indeed significantly larger than unisexual plants, but males were not larger than females, thus providing partial support for the prediction of Fig. 3.4b.

When male fitness accelerates with increasing resource investment (Fig. 3.4c), the optimal strategy involves a choice between distinct sexes, rather than continuous adjustment of gender. In this case, small individuals ($<R_1$) are female and large individuals ($>R_1$) are male. Thus the shapes of male and female fitness curves are important for the type of gender modification selected, as it is for the selection of hermaphroditism (cosexuality) versus dioecy. Generally speaking, a low incidence of accelerating fitness curves is probably the major reason for the rarity of gender-choosing strategies among animal-pollinated plants. Without size variation among individuals, the selected sexual strategy is dioecy (Fig. 3.2c). In the presence of variation, male and female gain curves must intersect, and fitness is maximized at the resource threshold determined by the intersection of the two curves (see also Warner 1975).

Some unambiguous evidence for gender diphasy has been obtained in a few plant species, reviewed variously by Freeman *et al.* (1980), Lloyd and Bawa (1984), Schlessman (1988) and Korpelainen (1998). As Lloyd and Bawa (1984) discussed, diphasy is superior to genetic dimorphism (dioecy), because it allows plants to exploit the ability of whichever sex role is most advantageous in a particular subset of the circumstances that individuals encounter. However, diphasy will not be selected if plants cannot either evaluate which sex would promote fitness most, or assess their conditions accurately (Lloyd and Bawa 1984).

3.3.2 Direct effects of plant size on fitness gains

Plant size can affect fitness directly, so that returns from a given absolute amount of resources differ for small and large plants (Fig. 3.5). For instance, in a wind-pollinated plant, pollen released from a tall individual may disperse farther and be more successful than pollen from a short individual. Direct effects may also occur in some animal-pollinated plants, because pollinators may prefer taller plants among neighbours (Ishii 2004). With direct effects of plant size, fitness returns per unit of investment differ for small and large plants (Fig. 3.5), so that separate gain curves must be established for individuals in different size classes.

If both sexes have linear gain curves (Fig. 3.5a), direct fitness effects will most probably cause an abrupt shift from male to female at a certain size (Klinkhamer *et al.* 1997; Cadet *et al.* 2004). If the slope of the male gain curve switches from increasing less steeply than the female curve to more steeply at some threshold size, large individuals should be completely male and small individuals should be completely female (Fig. 3.5a and d). Gender diphasy is selected again through the direct effects of plant size. Although gender diphasy occurs in some species, it is relatively rare compared with the more gradually changing sex-allocation patterns found in simultaneous hermaphrodites.

Nonlinear fitness gain on investment is probably the rule, rather than the exception, even for wind-pollinated plants. Especially for very large plants, local pollen and/or local resource competition is very likely to cause saturating fitness gain curves. In this case, very large plants are likely to be hermaphrodite, no matter whether the male or female gain curve levels off at high investment. In Fig. 3.5b and c, the male gain curve decelerates and the female gain function is linear; the reverse is also possible, with a nearly linear male curve and a concave female curve (McKone *et al.* 1998).

In general, direct and budget effects of size can operate simultaneously, although they need not be equally important. When size affects fitness directly, abrupt sex change from one sex (female) to the other sex (male) is expected with increased size. When plants are small, fitness gain curves for both sexes will be nearly linear. Thus, small plants should be unisexual: male if the male gain curve is steeper than the female gain curve, and female if the reverse holds. Direct effects can cause sex change, if choosing one sex is advantageous for small plants but disadvantageous for moderately large plants, as illustrated by Fig. 3.5b. In this case,

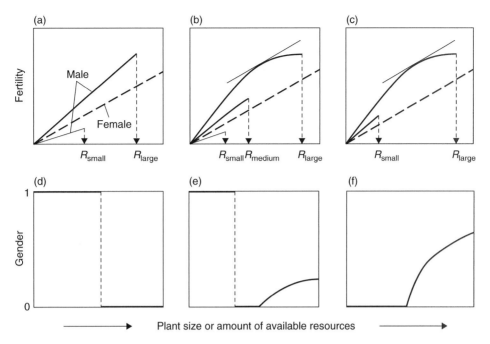

Figure 3.5 Effects of plant size on fitness returns (a–c) and gender (d–f) when a given investment causes differences in fitness returns between small and large plants (direct effect). Female fitness increases linearly with investment and is not influenced by plant size, whereas the male gain curve differs for different-sized individuals, and possibly levels off as investment becomes very large (b and c). R_{small}, R_{medium}, and R_{large} are the resource budget of small, middle, and large individuals, respectively. In panel (a), only budget effects exist, whereas in panels (b) and (c), both direct and budget effects operate simultaneously, with small plants benefiting from choosing one sex (female; b) or the other (male; c).

the optimal strategy involves a choice between distinct gender phases, rather than continuous gender adjustment, as plant size increases from small to moderately large (Fig. 3.5e). Abrupt sex change is selected even though budget effects may play an important role for very large plants. This conclusion differs from that of Cadet et al. (2004) who suggested that sex reversal is selected only in the absence of budget effects, that is, only when fitness gain curves are linear. Cadet et al. assumed linearity or nonlinearity of fitness gain curves to be independent of resource status, whereas I assume that male gain curves for small individuals are nearly linear and become more nonlinear for larger plants. If budget effects are weak for small and moderately large plants, abrupt sex change will occur before the budget effect of size comes into play.

If both small and moderately large plants benefit from choosing the same sex, as illustrated in Fig. 3.5c, a pattern of size-dependent sex allocation results, as if investment did not affect fitness directly (Fig. 3.5f). Despite the limited qualitative influence of direct effects in this case, they significantly affect the empirical measurement of fitness gains for different-sized individuals (for more discussion see Klinkhamer and de Jong 2002). The shape of the fitness gain curve is often estimated by regressing fitness against, say, flower number for a set of different-sized plants. This method may give misleading results if direct effects of plant size are important to fitness gains, even though they may have little qualitative effect on sex allocation patterns.

3.3.3 Why are male and female function often displaced temporally in cosexual plants?

Maternal provisioning of offspring (seed filling) necessarily occurs after fertilization, but many plants mature seeds and fruits for an extended period after male function has effectively terminated. Westoby and Rice (1982) proposed that deferment of maternal investment in offspring is selected to avoid wasting investment on seeds that will not be fertilized and/or to allow maternal choice among fertilized seeds. Alternatively, Day and Aarssen (1997) recognized that female function simply requires more time than male function, because of the time required for fruit production. Both hypotheses may apply for individual flowers/fruits, but they cannot explain why the flowering and fruiting stages are temporally displaced within plants in most species.

A closely related, but not identical, question is why some plants open all flowers simultaneously during a brief period, resulting in little overlap between sex functions, whereas others open flowers sequentially, so that the sex functions overlap extensively. Bawa (1983) suggested that the flower-production rate may be adjusted to match the resources available for fruit production, so that extended blooming allows better control over the relative investment in flowers and fruit than does mass blooming.

Using the sex-allocation model developed in Section 3.2.2, I now discuss a simple additional evolutionary explanation for the pattern of flowering and fruiting, which has largely been overlooked. I argue that temporal separation of sexual functions or flowering and fruiting may serve as an adaptive response to environmental stochasticity in resource availability. As Lloyd and Bawa (1984) clarified emphatically, most if not all size-dependent sex-allocation models assume that plants can assess their conditions accurately before the conditions affect their reproductive success. This assumption may hold reasonably well for monocarpic plants and species with large storage roots that form floral primordia during the preceding growth period, which remobilize stored assimilates for reproductive growth. However, many plants rely on current photosynthesis for their reproductive expenditure (Janzen 1976), so that the quantity of resources acquired during the breeding season must be unpredictable due to vagaries of environmental conditions or herbivory. If resource availability is uncertain, what is the best pattern of parental investment in male or female function? The optimal allocation rule is intuitively clear, namely, that at any time and any level of investment resources should be allocated to whichever sex function can yield higher fitness returns. This rule ensures that fitness is maximized for an individual in the face of unpredictable resources.

With a linear female gain curve and a decelerating male gain curve (Fig. 3.6a), an individual should not invest in female function when absolute male investment remains below the investment threshold (R_1), for which marginal fitness returns per unit of investment are equal through male and female function. As resource availability exceeds this threshold, the plant should switch completely to invest in female function. In this case, male and female expenditures will overlap temporally very little, although they use the same resource pool. Obviously, female investment cannot be postponed completely after male allocation, because ovules must be fertilized to develop into seeds, so that the optimal allocation pattern will also depend on the reproductive stages of other individuals in the population. Note that this mechanism could occur in concert with those proposed by Westoby and Rice (1982) and Day and Aarssen (1997).

When both female and male gain curves saturate, the transition from male to female investment should occur gradually, rather than abruptly (Fig. 3.6b and d). In Fig. 3.6b, R_1 and R_2 are the thresholds at which marginal male fitness equals the maximum and minimum marginal female fitnesses, respectively. Resources should be invested in both sex roles between these thresholds to maintain equal marginal fitness from investment in male and female function. The extent to which male and female investment overlap depends largely on the relative rate of saturation of absolute female fitness gains with investment: the closer to linearity (i.e., constant marginal returns), the smaller the range of resource availability between R_1 and R_2, and hence the briefer the overlap between sex functions as resource availability increases.

Almost all previous sex-allocation models treat the temporal displacement of male and female function in cosexual plants as an external constraint (Geber and Charnov 1986; Charlesworth and Charlesworth 1987; Burd and Head 1992; Seger and Eckhart 1996; Sato 2000, 2004), rather than as an

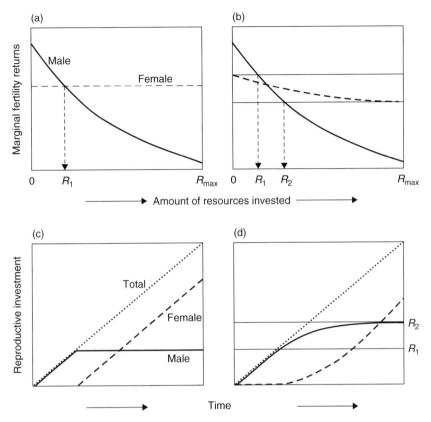

Figure 3.6 Temporal displacement of male and female function (largely fruit production) in cosexual plants as an adaptive strategy, with a decelerating male gain curve and either a linear (a and c) or decelerating female gain curve (b and d). Panels (a) and (b) illustrate the relations of marginal fitness to resource investment, whereas panels (c) and (d) depict the investment of resources in the sex roles and overall resource allocation during the reproductive season. Panels (c) and (d) indicate that total reproductive investment is assumed to increase proportionally with time, perhaps due to a constant production rate. In case (a) the optimal investment allocates all resources to male function and then switches completely to female function once investment exceeds the threshold (R_1) at which marginal fitness returns are equal for male and female function (c). In case (b), the plant should allocate resources only to male function until the threshold R_1 is reached, then invest simultaneously in both male and female function (d); when male investment approaches asymptote R_2 quickly, no further investment in male function is favoured.

adaptive strategy in itself. Indeed, it is not widely appreciated that an alternative adaptive explanation exists for the phenological pattern of flowering and fruiting. At present, few data are available to test the dynamic-allocation hypothesis thoroughly, so that this hypothesis warrants empirical study.

3.4 Joint evolution of reproductive effort and sex allocation in perennial plants

Most plants are perennial and face a trade-off between current and future reproduction. In contrast, sex-allocation models typically assume very simple life histories with non-overlapping generations and a fixed rate of reproduction (see Section 3.2). Complete understanding of sex allocation in perennial species requires analysis of sex allocation and reproductive effort within a single framework (Zhang and Wang 1994). In this section, I describe a model of size-dependent resource allocation by herbaceous perennial plants (Zhang and Jiang 2002), which incorporates budget effects of plant size. Direct effects of size could be analyzed by a straightforward extension of this approach.

Assume that a plant's size is determined environmentally and independently of its growth dynamics. For many temperate herbs that annually shed all vegetative parts, except for storage organs, plant size must have a substantial environmental component, because they must rebuild vegetative tissues every year. Pugliese (1987), Iwasa and Cohen (1989) and Pugliese and Kozlowski (1990) considered the optimal growth schedule of perennial plants and showed that herbaceous perennials without persisting vegetative parts should not increase their size after becoming reproductively active. I will also assume a trade-off between only reproduction and survival and will ignore other costs likely to be associated with reproduction, such as reductions in growth or future reproductive output (Zhang and Jiang 2002). Because only storage organs (rhizome, corm, bulb or tuber) persist during dormant periods, a plant's probability of surviving the rigors of the unfavourable season depends on its investment in supporting structures (or chemical substances) that maintain the viability of the storage organs.

Suppose that individual i with R_i units of resources allocates M_i resources to male function, F_i to female function, and the remainder $P_i = R_i - M_i - F_i$ to survival. Thus, the total reproductive allocation (E_i) is $E_i = (M_i + F_i)/R_i$ and sex allocation is $r_i = M_i/(M_i + F_i)$. The number of genes contributed by any mutant to the next reproductive season depends on its joint reproductive success as a pollen and seed parent and its post-breeding survival (S_i):

$$w_i = f_i + m_i + 2S_i.$$

Note that in perennial plants, a surviving adult contributes two copies of its genome to the next generation, thus S is multiplied by 2.

If each fitness component depends only on its resource input; or, $f_i = f(F_i)$, $m_i = m(M_i)$, and $S_i = S(P_i)$, subject to the constraint that $P_i + M_i + F_i = R_i$, natural selection favours an ESS that satisfies

$$\frac{\partial f}{\partial F} = \frac{\partial m}{\partial M} = \frac{\partial (2S)}{\partial P}. \quad (7)$$

This equation suggests that resource investment in pollen production, ovule and seed production, or survival to the next reproductive season must produce equal marginal fitness returns, if the ESS involves an intermediate allocation strategy (Lloyd 1988; Lloyd and Venable 1992; Morgan and Schoen 1997). At this allocation, the decrease in fitness caused by reduced investment in one activity is exactly compensated by the increase in fitness caused by increased investment in other activities. The population size structure affects the ESS only by influencing the population mean fertility of both sexes.

Assume that marginal fitness returns on initial resource investment in survival to the next reproductive season exceed those from male function, as shown in Fig. 3.7. In this case, the marginal fitness from investing in survival exceeds that of investing in either sex role for small individuals with less than S_1 resources (Fig. 3.7a and c), so they should invest all their resources in survival. Therefore, S_1 is the threshold size for reproduction. This explanation is not applicable to annuals, because they cannot survive to another reproductive season, and accordingly many annual plants have a very small minimum size for reproduction (Weiner 2004). If a plant's available resources exceed S_1, but it cannot invest more than S_2 resources on survival and S_3 resources on reproduction, it should reproduce only as a male, because the marginal fitnesses through both of these components exceed that of reproducing as a female (Fig. 3.7a and c). Only when an individual's resources exceed $(S_2 + S_3)$ should a plant invest in all three fitness components. In this case, investment in each function increases as plant size increases, so that large plants should produce both more flowers (pollen) and more seeds, and also survive better than small plants.

If the female gain curve is linear, as commonly expected, female fitness returns per unit of investment do not depend on the amount of resources already invested (Fig. 3.7a). Therefore, the absolute resource allocations to survival and male function will remain fixed once resource availability exceeds $S_2 + S_3$ (see Fig. 3.7b). In such cases, plants with more than $S_2 + S_3$ resources have fixed investments in survival, $P_i = S_2$, and male function, $M_i = S_3$, so the sex allocation $r_i = M_i/(M_i + F_i) = S_3/(R_i - S_2)$ must decrease and the total reproductive allocation ($E_i = 1 - S_2/R_i$) must increase, with plant size (R_i). Individuals smaller than the size threshold $S_2 + S_3$ should produce no

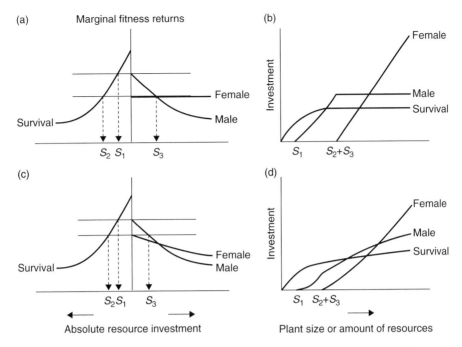

Figure 3.7 Size-dependent allocation to male function, female function, and post-breeding survival in perennial plants. Panels (a) and (c) illustrate the fitness returns per unit of investment through survival (to the left) and female or male function (to the right) in relation to absolute investment, with (a) linear or (c) decelerating female gain curves. Panels (b) and (d) illustrate absolute investment in each fitness component as a function of plant size or resource status, expected with linear (b) or decelerating (d) female fitness gain curves. Note that plants should invest in female function only when absolute investment in survival and male function exceeds S_2 and S_3, respectively, and thereafter they should invest in all three fitness components until the marginal fitness returns on each function are equal.

seeds, being either nonreproductive or functionally male only.

If the female gain curve also saturates, but less rapidly than the male gain curve, larger plants within a population will allocate disproportionately more resources to female function than smaller plants, even though plants generally allocate more resources to reproduction as they grow larger (eq. 7). As discussed in Section 3.3, increased femaleness with plant size is common in animal-pollinated plants, and in many plants herbivory induces a plastic shift towards maleness.

3.4.1 Size-dependent reproductive effort

Samson and Werk (1986) and Weiner (1988) proposed that size-dependent reproductive effort might arise allometrically. For example, if plants produce flowers singly in the axils of leaves, flower number cannot increase without increasing leaf number. These authors argued that such allometric constraints impose a linear relation between plant size and absolute reproductive output, and observations of such relations are often implicated as supporting this "allometric" hypothesis (e.g., Mendez and Obeso 1993; Schmid and Weiner 1993; Reekie 1998). Underlying allometry may control how reproductive effort varies, but as the present model predicts, an alternative adaptive explanation is also possible.

The model presented above both predicts a linear relation between reproductive output and plant size (Fig. 3.7a) and is consistent with several observations about which the allometric hypothesis is either silent or only partially compatible. The model predicts that perennial plants must achieve a minimum size (S_1) before reproducing, because plants smaller than S_1 realize highest fitness by investing all their resources in survival, rather than reproduction. In contrast, a minimum size for reproduction is assumed, rather than predicted, in the allometric-constraint model of Samson and Werk (1986) and Weiner (1988). According to Fig. 3.7, a minimum size for reproduction should not exist for annuals, and this is indeed often observed (Weiner 2004).

If the female gain curve is linear, adult survival should be constant for seed-producing plants and fairly independent of resource availability and investment in male function. In such cases, organisms may maintain high adult survival by varying their reproductive effort in response to environmental conditions. The association of high and relatively unvarying adult survival with highly variable reproductive rates is well documented for many perennial plants (Harper 1977).

Life-history theory commonly ignores male allocation, which is justified if the female gain curve is linear, even when resource availability varies extensively among individuals. Very small plants that do not set any seeds will suffer higher mortality, simply because they lack sufficient resources to maintain survival. However, if the gain curve is decelerating, rather than linear, adult survival will also increase with size, so that the omission of male allocation leads to an incomplete analysis.

3.5 Discussion and conclusion

Resource allocation commonly varies with plant size in many plant species, but the factors causing such variation and the significance of these patterns to plant life histories are not appreciated fully. Also, relatively few studies have attempted to quantify and explain the within-population variation among individuals. Using a simple graphical approach to the ESS model, I have shown that if fitness returns through allocation to female function, male function, and post-breeding survival vary differentially with plant size or resource availability, plants should modify their allocation to male versus female function and to reproduction versus survival according to their size. The theory developed in this chapter predicts that the reproductive strategies of hermaphroditic plants should equalize marginal fitness returns through male and female sexual function, and adult survival, unless constrained by insufficient available resources.

In developing this theory, I assumed obligate outcrossing. In contrast, hermaphrodites commonly self-fertilize, spanning the whole range from complete outcrossing to complete selfing (Barrett et al. 1997; Goodwillie et al. 2005). As Zhang (2000) showed, reproductive effort often correlates positively with increased selfing rate, because self-fertilization increases the genetic value of offspring from the standpoint of an invading allele that affects the selfing rate. Selfed progeny might contain two copies of the invading allele, whereas outcrossed progeny can contain only a single copy. Thus, selfing increases female fitness gains, which in turn select for both higher female allocation and higher total reproductive effort (Zhang 2000). However, the positive correlation between selfing rate and reproductive effort has yet to be demonstrated empirically. Furthermore, higher selfing in large plants with many flowers will also select for increased femaleness with plant size (de Jong et al. 1999).

Despite their paramount importance, fitness gain curves are exceedingly difficult to measure empirically (Brunet 1992; Klinkhamer et al. 1997; Campbell 2000; Klinkhamer and de Jong 2002), hindering the development and maturation of theory. Measurement of male or female fertility often requires parentage or paternity analysis with genetic markers, which is time consuming and feasible only in small populations with the precision necessary to describe a gain function. Another difficulty is that if direct effects of plant size are important, different-sized plants cannot be used to determine the gain curve, as in the very few attempts to measure gain curves in natural populations (Emms 1993; Klinkhamer et al. 1997; Klinkhamer and de Jong 2002). Furthermore, it is often unclear what resources should be measured to assess allocation. Most studies consider dry mass as a currency, but use of other currencies, such as nitrogen or phosphorus, can give different perspectives (e.g., McKone et al. 1998).

Life-history and sex-allocation theories typically assume trade-offs between reproduction and survival, or between male and female sexual functions. Phenotypic correlation between two life-history traits measured on a series of individuals has been used to quantify these trade-offs, but contrary to expectations, a number of studies have not found significant costs associated with reproduction (e.g., Galen 1993). Many authors argue that the absence of a negative correlation between life-history traits in natural populations is not evidence against trade-offs (van Noordwijk and de Jong 1986; de Jong 1993), but instead that variation in a third variable, such as plant size or resource availability, may mask any negative correlation between the traits under study. Indeed, the present model predicts that adaptive variation in allocation usually leads to a neutral (if the female gain curve is nearly linear) or positive (if the curve is decelerating) phenotypic correlation between reproductive allocation and survival, and between male and female investment (Fig. 3.7).

Charnov (1982) argued that a critical test for adaptive hypotheses about size-dependent resource allocation requires comparison between populations of a single species in different, isolated habitats. In contrast, evolutionary theory on size-related resource-allocation patterns developed in this chapter and elsewhere (Lloyd and Bawa 1984; Frank 1987; Klinkhamer *et al.* 1997) predicts that the size effect ought to be based on relative size in the breeding population: "large" depends on the size distribution in the entire breeding population, which may vary between habitats. Therefore, future studies of reproductive investment and sex allocation that consider variation among plants within populations, which avoid, or at least account for, complications arising from differences among populations are likely to be particularly fruitful.

Acknowledgements

I thank Spencer Barrett and Lawrence Harder for their invitation, guidance, and patience. They and two anonymous reviewers made very useful comments on the manuscript. My research on plant reproductive ecology has been jointly funded by the National Natural Science Foundation of China (30430160, 30125008), the Chinese Ministry of Education, and Beijing Normal University.

References

Barrett SCH, Harder LD, and Worley AC (1997). The comparative biology of pollination and mating in flowering plants. In J Silvertown, M Franco, and JL Harper, eds. *Plant life histories: ecology, phylogeny and evolution*, pp. 57–76. Cambridge University Press, Cambridge, UK

Barrett SCH, Case AL, and Peters GB (1999). Gender modification and resource allocation in subdioecious *Wurmbea dioica* (Colchicaceae). *Journal of Ecology*, **87**, 123–37.

Bateman AJ (1948). Intra-sexual selection in *Drosophila*. *Heredity*, **23**, 349–68.

Bawa KS (1983). Patterns of flowering in tropical plants. In GE Jones and RJ Little, eds. *Handbook of experimental pollination biology*, pp. 394–410. Scientific and Academic Editions, Van Nostrand Reinhold, New York, USA.

Bazzaz FA and Ackerly DD (1992). Reproductive allocation and reproductive effort in plants. In M Fenner, ed. *Seeds: the ecology of regeneration in plant communities*, pp. 1–26. CAB International, Wallingford, Oxon.

Bazzaz FA, Chiariello NR, Coley PD, and Pitelka LF (1987). Allocating resources to reproduction and defense. *Bioscience*, **37**, 58–67.

Bell G (1985). On the function of flowers. *Proceedings of the Royal Society of London, Series B*, **224**, 223–65.

Bickel AM and Freeman DC (1993). Effects of pollen vector and plant geometry on floral sex-ratio in monoecious plants. *American Midland Naturalist*, **130**, 239–47.

Brunet J (1992). Sex allocation in hermaphrodite plants. *Trends in Ecology and Evolution*, **7**, 79–84.

Burd M and Allen TFH (1988). Sexual allocation strategy in wind-pollinated plants. *Evolution*, **42**, 403–7.

Burd M and Head G (1992). Phenological aspects of male and female function in hermaphroditic plants. *American Naturalist*, **140**, 305–24.

Cadet C, Metz JAJ, and Klinkhamer PGL (2004). Size and the not-so-single sex: disentangling the effects of size and budget on sex allocation in hermaphrodites. *American Naturalist*, **164**, 779–92.

Campbell DR (2000). Experimental tests of sex-allocation theory in plants. *Trends in Ecology and Evolution*, **15**, 227–31.

Charlesworth B (1980). *Evolution in age-structured populations*. Cambridge University Press, Cambridge, UK.

Charlesworth D (1984). Androdioecy and the evolution of dioecy. *Biological Journal of the Linnean Society*, **22**, 333–48.

Charlesworth D and Charlesworth B (1981). Allocation of resources to male and female functions in hermaphrodites. *Biological Journal of the Linnean Society*, **15**, 57–74.

Charlesworth D and Charlesworth B (1987). The effect of investment in attractive structures on allocation to male and female functions in plants. *Evolution*, **41**, 948–68.

Charlesworth D and Morgan MT (1991). Allocation of resources to sex functions in flowering plants. *Philosophical Transactions of the Royal Society of London, Series B*, **332**, 91–102.

Charnov EL (1979). Simultaneous hermaphroditism and sexual selection. *Proceedings of the National Academy of Sciences of the United States of America*, **76**, 2480–4.

Charnov EL (1982). *The theory of sex allocation*. Princeton University Press, Princeton, NJ, USA.

Charnov EL and Bull JJ (1985). Sex allocation in a patchy environment: a marginal value theorem. *Journal of Theoretical Biology*, **115**, 619–24.

Day T and Aarssen LW (1997). A time commitment hypothesis for size-dependent gender allocation. *Evolution*, **51**, 988–93.

de Jong G (1993). Covariances between traits deriving from successive allocations of resources. *Functional Ecology*, **7**, 75–83.

de Jong TJ and Klinkhamer PGL (1989). Size dependency of sex allocation in plants. *Functional Ecology*, **3**, 201–6.

de Jong TJ and Klinkhamer PGL (2005). *Evolutionary ecology of plant reproductive strategies*. Cambridge University Press, Cambridge.

de Jong TJ, Klinkhamer PGL and Rademaker MCJ (1999). How geitonogamous selfing affects sex allocation in hermaphrodite plants. *Journal of Evolutionary Biology*, **12**, 166–76.

Delph LF (2003). Sexual dimorphism in gender plasticity and its consequences for breeding system evolution. *Evolution and Development*, **5**, 34–9.

Emms SK (1993). On measuring fitness gain curves in plants. *Ecology*, **74**, 1750–6.

Fisher RA (1930). *The genetical theory of natural selection*. Clarendon Press, Oxford, UK.

Frank SA (1987). Individual and population sex allocation patterns. *Theoretical Population Biology*, **31**, 47–74.

Freeman DC, Harper KT, and Charnov EL (1980). Sex change in plants: old and new observations and new hypotheses. *Oecologia*, **47**, 222–32.

Gadgil M and Bossert WH (1970). Life historical consequences of natural selection. *American Naturalist*, **104**, 1–24.

Galen C (1993). Cost of reproduction in *Polemonium viscosum*: phenotypic and genetic approaches. *Evolution*, **47**, 1073–9.

Geber MA and Charnov EL (1986). Sex allocation in hermaphrodites with partial overlap in male-female resource inputs. *Journal of Theoretical Biology*, **118**, 33–43.

Goodwillie C, Kalisz S, and Eckert CG (2005). The evolutionary enigma of mixed mating systems in plants: occurrence, theoretical expectation and empirical evidence. *Annual Review of Ecology, Evolution and Systematics*, **36**, 47–79.

Harder LD and Barrett SCH (1995). Mating costs of large floral displays in hermaphrodite plants. *Nature*, **373**, 512–5.

Harder LD and Thomson JD (1989). Evolutionary options for maximizing pollen dispersal of animal-pollinated plants. *American Naturalist*, **133**, 323–44.

Harper JL (1977). *The population biology of plants*. Academic Press, New York, USA.

Heilbuth JC (2000). Lower species richness in dioecious clades. *American Naturalist*, **156**, 221–21.

Ishii HS (2004). Increase of male reproductive components with size in an animal-pollinated hermaphrodite, *Narthecium asiaticum* (Liliaceae). *Functional Ecology*, **18**, 130–7.

Iwasa Y and Cohen D (1989). Optimal growth schedule of a perennial plant. *American Naturalist*, **133**, 480–505.

Janzen DH (1976). Effect of defoliation on fruit-bearing branches of the Kentucky coffee tree, *Gymnocladus dioicus* (Leguminosae). *American Midland Naturalist*, **95**, 474–8.

Kakehashi M and Harada Y (1987). A theory of reproductive allocation based on size-specific demography. *Plant Species Biology*, **2**, 1–13.

Klinkhamer PGL and de Jong TJ (2002). Sex allocation in hermaphrodite plants. In ICW Hardy, ed. *Sex ratios: concepts and research methods*, pp. 333–48. Cambridge University Press, Cambridge, UK

Klinkhamer PGL, deJong TJ and Metz H (1997). Sex and size in cosexual plants. *Trends in Ecology and Evolution*, **12**, 260–5.

Korpelainen H (1998). Labile sex expression in plants. *Biological Review*, **73**, 157–80.

Kozlowski J (1992). Optimal allocation of resources to growth and reproduction: implications for age and size at maturity. *Trends in Ecology and Evolution*, **7**, 15–19.

Lehtila K and Strauss SY (1999). Effects of foliar herbivory on male and female reproductive traits of wild radish, *Raphanus raphanistrum*. *Ecology*, **80**, 116–24.

Lloyd DG (1984). Gender allocations in outcrossing cosexual plants. In R Dirzo and J Sarukhán, eds. *Perspectives on plant population ecology*, pp. 277–300. Sinauer Associates, Sunderland, MA, USA.

Lloyd DG (1987). Allocations to pollen, seeds and pollination mechanisms in self-fertilizing plants. *Functional Ecology*, **1**, 83–9.

Lloyd DG (1988). A general principle for the allocation of limited resources. *Evolutionary Ecology*, **2**, 175–87.

Lloyd DG and Bawa KS (1984). Modification of the gender of seed plants in varying conditions. *Evolutionary Biology*, **17**, 255–338.

Lloyd DG and Venable DL (1992). Some properties of natural-selection with single and multiple constraints. *Theoretical Population Biology*, **41**, 90–110.

Maynard Smith J (1982). *Evolution and the theory of games*. Cambridge University Press, Cambridge, UK.

McKone MJ, Lund CP, and O'Brien JM (1998). Reproductive biology of two dominant prairie grasses (*Andropogon gerardii* and *Sorghastrum nutans*, Poaceae): Male-biased sex allocation in wind-pollinated plants? *American Journal of Botany*, **85**, 776–83.

Mendez M and Obeso JR (1993). Size-dependent reproductive and vegetative allocation in *Arum italicum* (Araceae). *Canadian Journal of Botany*, **71**, 309–14.

Morgan MT and Schoen DJ (1997). The role of theory in an emerging new plant reproductive biology. *Trends in Ecology and Evolution*, **12**, 231–4.

Pugliese A (1987). Optimal resource allocation and optimal size in perennial herbs. *Journal of Theoretical Biology*, **126**, 33–49.

Pugliese A and Kozlowski J (1990). Optimal patterns of growth and reproduction for perennial plants with persisting or not persisting vegetative parts. *Evolutionary Ecology*, **4**, 75–89.

Rademaker MCJ and de Jong TJ (1999). The shape of the female gain curve for *Cynoglossum officinale* and *Echium vulgare*: quantifying seed dispersal and seedling survival in the field. *Plant Biology*, **1**, 351–6.

Reekie EG (1998). An explanation for size-dependent reproductive allocation in *Plantago major*. *Canadian Journal of Botany*, **76**, 43–50.

Reekie EG (1999). Resource allocation, trade-offs, and reproductive effort in plants. In TO Vuorisalo and PK Mutikainen, eds. *Life history evolution in plants*, pp. 173–93. Kluwer Academic Publishers, London, UK.

Sakai A and Sakai S (2003). Size-dependent ESS sex allocation in wind-pollinated cosexual plants: fecundity vs. stature effects. *Journal of Theoretical Biology*, **222**, 283–95.

Samson DA and Werk KS (1986). Size-dependent effects in the analysis of reproductive effort in plants. *American Naturalist*, **127**, 667–80.

Sarkissian TS, Barrett SCH, and Harder LD (2001). Gender variation in *Sagittaria latifolia* (Alismataceae): is size all that matters? *Ecology*, **82**, 360–73.

Sato T (2000). Effects of phenological constraints on sex allocation in cosexual monocarpic plants. *Oikos*, **88**, 309–18.

Sato T (2004). Size-dependent sex allocation in hermaphroditic plants: the effects of resource pool and self-incompatibility. *Journal of Theoretical Biology*, **227**, 265–75.

Schaffer WM (1974). Selection for optimal life histories: the effects of age structure. *Ecology*, **55**, 291–303.

Schlessman MA (1988). Gender diphasy (sex choice). In J Lovett Doust and L Lovett Doust, eds. *Plant reproductive ecology: patterns and strategies*, pp. 139–53. Oxford University Press, New York.

Schmid B and Weiner J (1993). Plastic relationships between reproductive and vegetative mass in *Solidago altissima*. *Evolution*, **47**, 61–74.

Seger J and Eckhart VM (1996). Evolution of sexual systems and sex allocation in plants when growth and reproduction overlap. *Proceedings of the Royal Society of London, Series B*, **263**, 833–41.

Shaw RF and Mohler JD (1953). The selective significance of the sex ratio. *American Naturalist*, **87**, 337–42.

Vallius E and Salvonen V (2000). Effects of defoliation on male and female reproductive traits of a perennial orchid, *Dactylorhiza maculata*. *Functional Ecology*, **14**, 668–74.

van Noordwijk AJ and de Jong G (1986). Acquisition and allocation of resources: their influence on variation in life-history tactics. *American Naturalist*, **128**, 137–42.

Warner RR (1975). The adaptive significance of sequential hermaphroditism in animals. *American Naturalist*, **109**, 61–82.

Weiner J (1988). The influence of competition on plant reproduction. In J Lovett-Doust and L Lovett-Doust, eds. *Plant reproductive ecology: patterns and strategies*, pp. 228–45. Oxford University Press, Oxford, UK.

Weiner J (2004). Allocation, plasticity and allometry in plants. *Perspectives in Plant Ecology, Evolution and Systematics*, **6**, 207–15.

Westoby M and Rice B (1982). Evolution of seed plants and inclusive fitness of plant tissues. *Evolution*, **36**, 713–24.

Willson MF (1983). *Plant reproductive ecology*. John Wiley and Sons, New York.

Willson MF and Rathcke BJ (1974). Adaptive design of the floral display in *Asclepias syriaca* L. *American Midland Naturalist*, **92**, 47–57.

Wilson WG and Harder LD (2003). Reproductive uncertainty and the relative competitiveness of simultaneous hermaphroditism versus dioecy. *American Naturalist*, **162**, 220–41.

Wright SI and Barrett SCH (1999). Size-dependent gender modification in a hermaphroditic perennial herb. *Proceedings of the Royal Society of London, Series B*, **266**, 225–32.

Zhang DY (2000). Resource allocation and the evolution of self-fertilization in plants. *American Naturalist*, **155**, 187–99.

Zhang DY and Jiang XH (2002). Size-dependent resource allocation and sex allocation in herbaceous perennial plants. *Journal of Evolutionary Biology*, **15**, 74–83.

Zhang DY and Wang G (1994). Evolutionarily stable reproductive strategies in sexual organisms—an integrated approach to life-history evolution and sex allocation. *American Naturalist*, **144**, 65–75.

CHAPTER 4

Pollen and ovule fates and reproductive performance by flowering plants

Lawrence D. Harder and Matthew B. Routley

Department of Biological Sciences, University of Calgary, Alberta, Canada

Outline

Pollen and ovules experience diverse fates during pollination, pollen-tube growth, fertilization, and seed development that govern the male and female potential of flowering plants. This chapter identifies these fates and many of their interactions, and considers their theoretical implications for the evolution of pollen export and the production of selfed and outcrossed seeds. This analysis clarifies the importance of pollen quantity and quality for seed production. Our analysis emphasizes the asymmetry of pollen and ovule fates and considers its consequences for reproductive evolution. We also identify ovule limitation as a constraint on seed production, which has paradoxically not been recognized before, but is an implicit assumption of previous theoretical analysis of mating-system evolution. Ovule limitation increases the diversity of possible reproductive strategies. In addition to ovule limitation, we consider the implications of pollen and resource limitation for the evolution of self- and cross-fertilization. Resource limitation occurs only if plants produce more ovules than they can mature into seeds, which allows a mixture of selfing and outcrossing to be an optimal mating system in some circumstances. The chance of mixed mating being optimal is enhanced by trade-offs between self- and cross-pollination, and more diverse optimal combinations of selfing and outcrossing are possible should mixed mating be favoured. Our analysis illustrates the key role played by interactions between genetic and ecological influences on reproductive performance in the evolution of plant reproduction.

4.1 Introduction

Most plants produce millions of pollen grains and/or thousands of ovules during their lives, but only one of each will be represented in the next generation in a stable population, on average. Clearly, most pollen grains and ovules succumb to fates other than successful reproduction, which bear significant implications for the persistence and dynamics of plant populations (Morgan *et al.* 2005). These fates arise from diverse interactions with the abiotic environment, with other species, and with other pollen grains and seeds, as well as from a lack of such interactions. This dependence on variable abiotic and biotic environments renders plant mating highly stochastic and context dependent (e.g., Herrera 2002; Herrera 2004; Johnson *et al.* 2005). Nevertheless the diversity of pollen and ovule fates provides many opportunities for reproductive adaptation (see Lloyd 1979, 1992; Harder and Wilson 1998; Harder 2000).

Pollen grains and ovules experience very different environments, resulting in dissimilar fates. The independence of pollen grains from their parental sporophytes during much of their functional lives exposes them to the vagaries of weather, predation by pollen-consuming animals, misadventure during dispersal, competition with other pollen grains for access to ovules if they reach a stigma, and rejection by pistils. In contrast, angiosperm ovules are relatively protected within ovaries, so that their fates depend primarily on the quantity and quality of pollen entering the ovary and the availability of maternal resources for seed production, unless developing seeds are attacked by pre-dispersal seed predators. This asymmetry has many implications for the operation of sexual selection in plants (Skogsmyr and Lankinen 2002), the allocation of resources to the sex roles (Charnov 1982; Lloyd 1984; Chapter 3), and the evolution of floral and inflorescence characteristics (Lloyd 1984; Bell 1985; Bell and Cresswell 1998).

Despite the ecological and evolutionary importance of pollen and ovule fates, they have been subject to relatively little explicit integrated analysis. This neglect partly reflects the traditional separation of studies of the pollination, post-pollination, and seed-development phases of reproduction, so the dependence of later reproductive stages on earlier stages is often ignored. Lloyd (1979, 1992) initiated explicit studies of pollen and ovule fates after recognizing that when and how self-pollination occurs affects a plant's opportunities for cross-fertilization and pollen export, and the extent to which selfing provides reproductive assurance. This perspective has been incorporated increasingly in models of mating-system evolution (Schoen et al. 1996; Morgan et al. 1997; Harder and Wilson 1998; Chapter 2) and population dynamics (Morgan et al. 2005). Nevertheless, these studies have ignored ovule fates, other than whether they are fertilized or not. Lloyd's identification of alternative modes of selfing also motivated some empirical analysis of pollen fates (e.g., Eckert 2000; Harder 2000; Goodwillie et al. 2005; Johnson et al. 2005). However, these studies have been conducted largely independently of analyses of the relative incidence of pollen versus resource limitation of seed production (reviewed by Casper and Neisenbaum 1993; Ashman et al. 2004), or of studies of pollen-tube competition or seed development that were stimulated by interest in sexual selection during the 1980s and 1990s (reviewed by Korbecka et al. 2002; Skogsmyr and Lankinen 2002).

In this chapter, we characterize the diverse fates that await pollen and ovules and describe four theoretical examples of their implications for floral and mating-system evolution. We first identify pollen and ovule fates and many of their interactions and briefly review empirical estimates of the magnitudes of key fates. This overview clarifies the three limits on seed production: pollen receipt, ovule production, and resource availability. Given this foundation, we then consider the conditions under which a population could be invaded by a phenotype with novel floral traits that alter its pollen export or self-pollination. This analysis reveals that previous mating-system theory has largely ignored the consequences of resource limitation. Finally, we consider several implications of our results for current perspectives on angiosperm reproduction and its microevolution.

4.2 Pollen fates and ovule fates

The fates that await a plant's P pollen grains and O ovules are diverse and interact in complex ways (Fig. 4.1). These fates arise during five partially overlapping phases of the reproductive process: pollen dispersal, pollen-tube growth, ovule fertilization, seed development, and seedling establishment. Because of this sequential process, the opportunities for pollen and ovules to participate during each phase depend on the outcomes of all preceding phases. The conditional nature of pollen and ovule fates is particularly important because it determines the incidence and intensity of competition during two phases: competition among pollen tubes to fertilize ovules, and competition among developing seeds for maternal resources.

4.2.1 Pollen dispersal

Our characterization of pollen fates during dispersal (Table 4.1) integrates the perspectives of Lloyd (1979, 1992) and Harder (Harder and Wilson 1998; Harder 2000). Lloyd considered the pollen deposited on stigmas and emphasized the timing of self-pollination relative to cross-pollination and whether self-pollination occurred autonomously or with the aid of pollen vectors. In contrast, Harder considered all of the P pollen grains produced by a

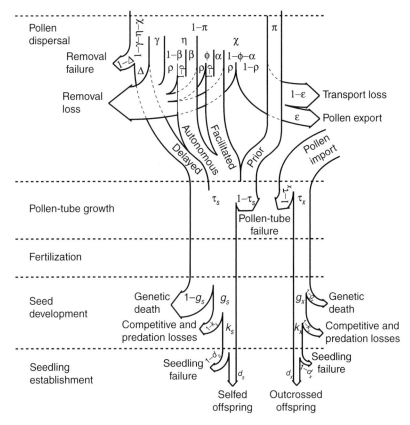

Figure 4.1 The general fates that await the P pollen grains and O ovules produced by a plant during pollen dispersal, pollen-tube growth, ovule fertilization, seed development, and seedling establishment. Each parameter indicates a proportion of the pollen grains, pollen tubes, zygotes, embryos, or seeds above it.

plant and explicitly divided self-pollination into components that reduce opportunities for pollen export (pollen discounting) or occur independently of export (non-discounting self-pollination). Similar to Lloyd, we recognize three phases of pollination (Table 4.1), although our terminology differs somewhat, as we explain below. Prior self-pollination occurs before flowers interact with pollen vectors and involves a fraction π of the P pollen grains produced by a plant. Processes that occur simultaneously with cross-pollination divide the remaining $(1-\pi)P$ pollen grains into four fractions: χ, pollen that could be exported to other plants; η, pollen that could be involved in self-pollination without affecting pollen export; γ, pollen that is displaced from flowers independently of self- or cross-pollination, such as by wind or rain; and any remainder, $1-\chi-\eta-\gamma$, which is involved in processes that occur once opportunities for pollen export cease. We now characterize the distribution of pollen that arises after prior self-pollination.

The $(1-\pi)\chi P$ pollen grains that could be exported experience the most diverse fates. A fraction α of exportable pollen may be involved in autonomous, intrafloral self-pollination. In addition, pollen vectors may deposit a fraction ϕ of the exportable pollen on the plant's own stigmas, resulting in facilitated self-pollination either within or between flowers (note that this usage differs from Lloyd's facilitated self-pollination, which involved only the intrafloral component). The remaining fraction, $1-\alpha-\phi$, is poised to leave the plant, except that pollen vectors could displace a fraction, ρ, which is lost from dispersal, adding to the removal loss associated with γ (ρ also

Table 4.1 Possible fates of the P pollen grains produced by a single plant and their effects on pollen export.

Timing relative to access to pollen vectors	Fate	Number of pollen grains involved	Effect on an individual's pollen export
Prior	Autonomous self-pollination	$P_p = \pi P$	Negative
Simultaneous[a]	Autonomous self-pollination	$P_a = (1-\pi)(\alpha\chi + \beta\eta)P$	Negative if $\alpha > 0$, strongly so if $\eta \cong 0$
	Facilitated self-pollination[b]	$P_f = (1-\pi)(\phi\chi + [1-\beta]\eta)(1-\rho)P$	Negative if $\phi > 0$, strongly so if $\eta \cong 0$
	Export	$P_e = (1-\pi)(1-\alpha-\phi)\chi(1-\rho)\varepsilon P$	
	Removal loss	$P_l = (1-\pi)[\{(1-\alpha)\chi + (1-\beta)\eta\}\rho + \gamma]P$	Negative, intensity increases with ρ
	Transport loss	$P_t = (1-\pi)(1-\alpha-\phi)\chi(1-\rho)(1-\varepsilon)P$	Negative, intensity increases with $1-\varepsilon$
Delayed[c]	Autonomous self-pollination	$P_d = (1-\pi)(1-\chi-\eta-\gamma)\Delta P$	None
	Removal failure	$P_r = (1-\pi)(1-\chi-\eta-\gamma)(1-\Delta)P$	None

Note that although pollen export cannot vary positively with simultaneous self-pollination, removal loss, or transport loss, for individual plants, such a relation is possible *among* plants if plants differ in their proportion of potentially exportable simultaneous pollen (χ). See Fig. 4.1 for parameter definitions.

[a] These fates require that $\pi < 1$.
[b] May include both intrafloral and interfloral (geitonogamous) components.
[c] These fates require that $\pi < 1$ and $\chi + \eta + \gamma < 1$.

reduces facilitated self-pollination). Of the remaining $(1-\pi)\chi(1-\alpha-\phi)(1-\rho)P$ pollen grains that leave the plant, a fraction ε is exported to conspecific stigmas and the remainder is lost during transport (including deposition on heterospecific stigmas).

The $(1-\pi)\eta P$ pollen grains that could be involved in self-pollination without affecting pollen export may participate in up to three fates. A fraction β of this pollen may be involved in autonomous, intrafloral self-pollination, and the remaining fraction $(1-\beta)$ could be displaced by pollinators and contribute to either facilitated self-pollination or removal loss with probability ρ (we assume the same ρ for discounting and non-discounting facilitated self-pollination for simplicity). Facilitated self-pollination could occur without pollen discounting if pollen that would otherwise have fallen from a flower during a pollinator visit instead lands on a stigma. Thus, facilitated self-pollination can include both discounting and non-discounting components, depending on the magnitudes of ϕ and $1-\beta$, respectively (Table 4.1).

Once cross-pollination ceases, a fraction Δ of the remaining $(1-\pi)(1-\chi-\eta-\gamma)P$ pollen grains could be involved in delayed self-pollination, with the remainder constituting pollen removal failure (Table 4.1). Delayed self-pollination can occur through various mechanisms, including the anthers collapsing onto the stigma, the stigma growing to contact the anthers, or the anthers brushing the stigma as the corolla falls (Chapter 10).

In summary, nine pollen fates arise during dispersal (Table 4.1). Self-pollination can occur autonomously before (*prior self-pollination*, P_p), simultaneously with (*simultaneous, autonomous self-pollination*, P_a), or after cross-pollination (*delayed self-pollination*, P_d). Self-pollination can also involve the action of pollen vectors (*facilitated self-pollination*, P_f) either within the flower that produced the pollen (*intrafloral, facilitated self-pollination*) or among flowers on the same plant (*geitonogamy*). In addition, *pollen export* (P_e) results when pollen reaches stigmas on other conspecific plants. Finally, pollination can fail owing to three fates: *removal loss* (P_l) while pollen vectors interact with the producing plant, *transport loss* as vectors move among plants (P_t), and *removal failure* (P_r). Pollen dispersal establishes the scope of potential reproductive output by determining the numbers of pollen grains involved in self-pollination ($P_s = P_p + P_a + P_f + P_d$), pollen export ($P_e$), and pollen import ($P_i$). Pollen export equals import within a closed population; however, this equality need not hold for individual plants.

According to our characterization of pollen fates, all modes of self-pollination, except for delayed self-pollination, can reduce a plant's ability to export pollen (Table 4.1). As Lloyd (1979, 1992) noted, this pollen discounting is a universal consequence of prior self-pollination. The extent to which simultaneous self-pollination discounts pollen export depends on how it occurs. Simultaneous self-pollination does not affect export if $\alpha = \phi = 0$. In contrast, every self-deposited grain reduces export opportunities when $\eta = 0$ and α or $\phi > 0$. Given the diversity of ways in which a plant's pollen can be deposited on its own stigmas, reality may typically lie between these extremes. Nevertheless, a negative relation between discounting, simultaneous self-pollination, and pollen export within individual plants may be difficult to detect from comparisons among plants, because both outcomes vary positively with the proportion of pollen remaining after prior self-pollination that could be exported to other plants, χ (see Table 4.1). For example, if all self-pollination is facilitated (i.e., $\pi = \alpha = \beta = \delta = 0$ and $0 < \phi < 1$), then the number of pollen grains involved in self-pollination and export equals $P_s = (\chi\phi + \eta)(1 - \rho)P$ and $P_e = \chi(1 - \phi)(1 - \rho)\varepsilon P$, respectively. If plants differ in their proportions of exportable pollen χ, self-pollination will vary positively and linearly with pollen export according to $P_s = \eta(1 - \rho)P + (\phi/[1 - \phi]\varepsilon)P_e$ (also see Harder 2000; Harder et al. 2000). A similar positive relation between pollen export and removal or transport loss may explain the paradox of species that use pollen-collecting bees to disperse their pollen (Harder and Wilson 1997).

The incidences of alternative pollen fates have not been measured completely for any plant species, but it is clear that most pollen does not reach stigmas (e.g., Fig. 4.2). Harder (2000) surveyed studies of monocots and found a strong dichotomy between species with granular pollen and orchids with pollen aggregated into pollinia. Species with granular pollen experienced relatively low removal failure (median = 7%), but only 1% of pollen removed from anthers reached conspecific stigmas (self- and cross-pollination), because of removal and transport losses (13 species). In contrast, orchids have much higher removal failure (median = 49%), but a median of 17% of the pollen removed from anthers reaches stigmas (11 species). Therefore, the evolution of pollinia seems to reduce transport loss considerably. Of the relatively small amount of pollen deposited on stigmas, self-pollination can contribute a variable fraction, as is illustrated by the extensive variation among species in the proportion of seeds that are self-fertilized (reviewed by Goodwillie et al. 2005).

Pollen fates also vary extensively within species. In the most complete study of pollen fates to date, Johnson et al. (2005) found that Disa cooperi (Orchidaceae) plants experienced large differences in pollination between two consecutive years, including 76 and 37% declines in self-pollination and pollen export, respectively. During both years, the percentage of pollen on stigmas attributed to geitonogamous self-pollination ranged from 0 to 100% among plants, with most plants exporting no pollen during two days of observation, but one plant exporting pollen to eight recipients. Similarly, several studies have shown that the incidence of geitonogamy varies positively among plants with the number of flowers that they display simultaneously (reviewed by Harder et al. 2004). These results illustrate that pollen fates during dispersal depend strongly on the characteristics of individual plants

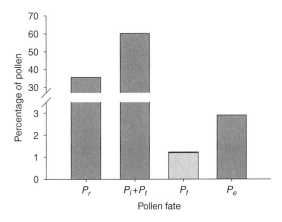

Figure 4.2 Fates of Disa cooperi (Orchidaceae) pollen during two nights' exposure to hawk-moth pollination, including pollen removal failure (P_r); pollen removal loss (P_l); pollen transport loss (P_t); facilitated self-pollination (P_f), including both geitonogamous (light greybar) and intrafloral deposition (dark grey bar); and pollen export (P_e). This species cannot self-pollinate autonomously. Based on data from Johnson et al. 2005. Note the different scaling on either side of the break along the ordinate.

and their pollination environment. However, some plants can adjust their characteristics to alter pollen fates adaptively in response to their recent pollination history. Specifically, if pollination shortens floral longevity, frequent pollinator visits reduce floral display size, limiting geitonogamy and its associated pollen discounting (Harder and Johnson 2005).

4.2.2 Pollen-tube growth

Once pollen reaches a stigma it must germinate and grow a pollen tube into the ovary. This process involves complex interactions between the pollen tube and pistil (Wheeler et al. 2001; Skogsmyr and Lankinen 2002; Stephenson et al. 2003), and competition among pollen tubes from the same or different pollen donors (Skogsmyr and Lankinen 2002; Armbruster and Rogers 2004; Bernasconi 2004). Pollen from some donors may fail completely in a particular pistil because of incompatibility reactions based on the genotype of the haploid pollen grain, or its diploid parent (homomorphic incompatibility), or, in heterostylous species, based on the anther level that produced a pollen grain (heteromorphic incompatibility: de Nettancourt 2001). However, self-incompatibility can weaken as flowers age (Good-Avila and Stephenson 2002; Goodwillie et al. 2004; Travers et al. 2004), allowing reproductive assurance if cross-fertilization is incomplete and adaptive implementation of mixed mating (Vallejo-Marín and Uyenoyama 2004). Nevertheless, the number of viable male gametophytes undergoes considerable attrition between pollination and fertilization. Indeed, the amount of pollen needed to maximize ovule fertilization typically exceeds the number of ovules in an ovary 5- to 10-fold (e.g., Mitchell 1997 and papers cited therein). Furthermore, relatively less self-pollen than cross-pollen survives during this post-pollination phase, even for species with weak self-incompatibility mechanisms (Bernasconi 2004). We represent the probability that a self- or cross-pollen grain on a stigma produces a pollen tube that reaches the ovary by τ_s and τ_x, respectively, where $\tau_s = 0$ for a self-incompatible species. τ_s and τ_x do not incorporate competition among pollen tubes, which we incorporate during the next phase of reproduction.

4.2.3 Ovule fertilization

The composition of a plant's zygotes depends on the numbers of prior, simultaneous, and delayed self-pollen tubes and cross-pollen tubes entering the ovary and the number of ovules that remain unfertilized (Table 4.2). We assume that all pollen grains germinate and grow pollen tubes at the same rate, so that fertilization occurs in the same order that pollen grains arrive on stigmas. Consequently, prior self-pollen (if present) fertilizes ovules first, followed by simultaneous self- and imported pollen, and then delayed self-pollen. As Table 4.2 summarizes, two or three quantitative outcomes are possible for each fertilization fate, depending on whether ovules remain unfertilized when pollen tubes enter the ovary and, if so, whether fewer or more pollen tubes enter the ovary than are needed to fertilize the remaining ovules.

Table 4.2 reveals several features of ovule fates. First, fertilization outcomes depend on pollen fates during pollen dispersal and pollen-tube growth. Second, the distinction between different modes of simultaneous self-pollination becomes irrelevant during fertilization, as the resulting self-pollen tubes are functionally identical. Third, the number of fertilized ovules can be limited by either pollen receipt, if $\tau_s(P_p + P_a + P_f + P_d) + \tau_x P_i < O$, or ovule production, if $\tau_s(P_p + P_a + P_f + P_d) + \tau_x P_i \geq O$. Note that pollen tubes compete only when fertilization is ovule-limited and that the likelihood of competition increases from prior to simultaneous through delayed phases. We assume that the probability that a pollen tube fertilizes an ovule equals 1 when fertilization is pollen-limited but equals the inverse of the number of competing pollen tubes when fertilization is ovule-limited (i.e., $1/[P_s + P_i]$: Table 4.2).

4.2.4 Seed development

After fertilization, zygotes become embryos and, together with associated tissues, consume maternal resources and develop into seeds. This process may fail for three reasons: death of zygotes or young embryos from the expression of lethal alleles (genetic death: Charlesworth and Charlesworth

Table 4.2 Possible fertilization fates of the O ovules produced by a single plant.

Timing relative to cross-fertilization	Fate	Number of ovules involved	Condition(s)
Prior	Self-fertilized	$F_p = \tau_s P_p$, or $F_p = O$	(1) $\tau_s P_p < O$ (2) $\tau_s P_p \geq O$
Simultaneous	Self-fertilized	$F_{af} = 0$ $F_{af} = \tau_s(P_a + P_f)$, or $F_{af} = \dfrac{\tau_s(P_a + P_f)(O - \tau_s P_p)}{\tau_s(P_a + P_f) + \tau_x P_i}$	(2) (3) $\tau_s (P_p + P_a + P_f) + \tau_x P_i < O$ (4) $\tau_s (P_p + P_a + P_f) + \tau_x P_i \geq O$
	Cross-fertilized	$F_x = 0$ $F_x = \tau_x P_i$, or $F_x = \dfrac{\tau_x P_i(O - \tau_s P_p)}{\tau_s(P_a + P_f) + \tau_x P_i}$	(2) (3) (4)
Delayed	Self-fertilized	$F_d = 0$ $F_d = \tau_s P_d$, or $F_d = O - \tau_s(P_p + P_a + P_f) - \tau_x P_i$	(4) (5) $\tau_s (P_p + P_a + P_f + P_d) + \tau_x P_i < O$ (6) $\tau_s (P_p + P_a + P_f + P_d) + \tau_x P_i \geq O$
	Unfertilized	$F_u = 0$ $F_u = O - \tau_s(P_p + P_a + P_f + P_d) - \tau_x P_i$	(6) (5)

All fates, except for prior self-fertilization, depend on ovules remaining unfertilized after all preceding fates are fulfilled. The second condition for a fate (first for prior) applies when fewer pollen tubes enter the ovary than are needed to fertilize the remaining ovules, and the third condition (second for prior) applies when pollen tubes compete for fertilizations. P_p, P_a, P_f, P_d, and P_i are defined in Table 4.1 (except P_i) and represent the numbers of a plant's P pollen grains that are involved in: prior self-pollination; autonomous, simultaneous self-pollination; facilitated self-pollination; delayed self-pollination; and pollen import, respectively. τ_s and τ_x are the proportions of self-pollen and outcrossed pollen grains on a stigma whose pollen tubes enter the ovary, respectively.

1987; Husband and Schemske 1996); competition among embryos for maternal resources (including preferential maternal allocation: Casper and Niesenbaum 1993; Korbecka et al. 2002); and consumption by pre-dispersal predators (Fenner and Thompson 2005). We propose that genetic death occurs before embryos consume appreciable maternal resources and that proportions g_s and g_x of the $F_s = F_p + F_{af} + F_d$ selfed and F_x outcrossed embryos, respectively, survive this phase. Because selfed offspring are homozygous at more loci than outcrossed offspring, they are more likely to express recessive lethal traits, so that $g_s < g_x$, unless the population bears negligible genetic load (Charlesworth and Charlesworth 1987; Husband and Schemske 1996). This characterization of early embryo mortality includes late-acting self-incompatibility, whereby interaction with a self-pollen tube disables an ovule, even if fertilization does not occur (de Nettancourt 2001). Embryos that survive genetic death then consume maternal resources. We assume that a maternal plant has sufficient resources to mature a fraction, m, of its ovules into seeds, so embryos compete for resources if the number of embryos surviving genetic death exceeds mO (i.e., $F_s g_s + F_x g_x > mO$). In general, the probabilities of selfed and outcrossed embryos becoming seeds after they survive genetic death are k_s and k_x, respectively, with $k_s = k_x = 1$ in the absence of resource competition. k_s and k_x can incorporate survival of pre-dispersal seed predation, but we ignore this process. Therefore, pre-dispersal, or early-acting, inbreeding depression equals $1 - (g_s k_s / g_x k_x)$.

For simplicity, we make two assumptions concerning resource competition. First, we assume that the timing of fertilization does not affect competitive outcomes. In contrast, embryos fertilized early compete more successfully than later embryos if competitive ability depends on the size of a developing seed (Ganeshaiah and Uma Shaanker 1994; Uma Shaanker et al. 1995). This priority necessarily alters mating outcomes, although whether it affects the optimal mating patterns is uncertain. Second, we assume that selfed and outcrossed embryos compete equally, surviving resource competition with a probability equal to the inverse of the total number of embryos that survive genetic death (i.e., $k_s = k_x = 1/[F_s g_s + F_x g_x]$). Although selfed embryos may compete less successfully than outcrossed embryos owing to greater expression of non-lethal deleterious alleles (Korbecka *et al.* 2002), this simplification has little effect on the optimal mating system (L. D. Harder, M. B. Routley, and S. A. Richards unpublished manuscript).

The concepts of seed and ovule discounting explicitly recognize the ability of selfing to reduce seed production. Lloyd (1992) defined seed discounting as reduced production of outcrossed seeds caused by self-fertilization. Barrett *et al.* (1996) distinguished this post-zygotic process from ovule discounting, a reduction in seed production caused when self-pollen tubes disable some ovules (see Chapter 13). However, neither ovule nor seed discounting is an inevitable outcome of late-acting self-incompatibility or self-fertilization in plants that produce more ovules than they can mature into seeds. Indeed, production of "excess" ovules may serve specifically to compensate for such losses with limited impact on female fertility (Porcher and Lande 2005; L. D. Harder, M. B. Routley, and S. A. Richards unpublished manuscript).

4.2.5 Seedling establishment and parental fitness

Once their development is complete, seeds disperse and some germinate and establish reproductive offspring. As with pre-dispersal survival, the probability of post-dispersal survival of outcrossed seeds (d_x) generally exceeds that of selfed seeds (d_s), which are more likely to express deleterious traits due to their higher homozygosity (Husband and Schemske 1996). Therefore, post-dispersal or late-acting inbreeding depression equals $1 - d_s/d_x$. In general, d_s and d_x are on the order of the inverse of a plant's lifetime seed production, because one successful seed is sufficient to replace a parental plant.

A plant's fitness (w) depends on its genetic contributions to the next generation through selfed seeds (S_s—two contributions per seed), its own outcrossed seeds ($S_♀$—one contribution per seed), and seeds sired on other plants ($S_♂$—one contribution per seed),

$$w = 2S_s d_s + S_♀ d_x + S_♂ d_x.$$

The ovule and pollen fates outlined above govern the details of these three fitness components.

4.3 Limits on seed production

Pollen and ovules contribute to the next generation only if they are represented in seeds from the producing plant and plants that import its pollen. Consequently, limits on seed production fundamentally govern population dynamics and reproductive evolution. To date, studies of seed-production constraints have considered two factors: pollen receipt, which affects fertilization success, and the availability of maternal resources during seed development, which determines maximum fecundity (reviewed by Casper and Niesenbaum 1993; Ashman *et al.* 2004). However, analysis of ovule fates exposes a third constraint, ovule limitation (L. D. Harder, M. B. Routley, and S. A. Richards unpublished manuscript), which bears important implications for the evolution of ovule production and mating-system evolution. Before considering these implications, we clarify the conditions that result in pollen, ovule, and resource limitation.

4.3.1 Pollen limitation

Pollen limitation occurs when (1) some of an individual's ovules remain unfertilized ($F_s + F_x < O$) and (2) too few embryos avoid genetic death and predation to compete for maternal resources ($F_s g_s + F_x g_x < mO$: Fig. 4.3). The first condition involves pollination quantity and indicates that plants do not compete for ovule fertilization, so an individual's fertilization success as a maternal and paternal parent depends only on the absolute numbers of pollen tubes entering its ovaries and those on other plants, respectively. This lack of competition is one reason why we refer to "simultaneous, autonomous self-pollination," rather than Lloyd's (1992) "competing self-pollination." This pollen-quantity aspect also

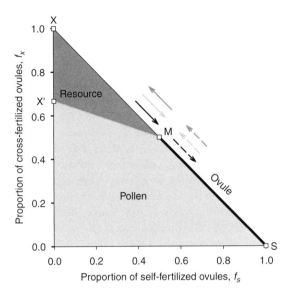

Figure 4.3 Relations of pollen limitation (light grey area), ovule limitation (heavy line), and resource limitation (dark grey area) of seed production to the proportions of self- and cross-fertilized ovules. In this example, the probability of a cross-fertilized zygote surviving genetic death exceeds the proportion of ovules that the plant can mature into seeds, given available resources (i.e., $m < g_x < 1$). The diagonal line (X – M – S) depicts combinations of self- and cross-fertilizations that result in fertilization of all ovules ($f_s + f_x = 1$). Along the transition from pollen to resource limitation (line X' – M) the proportion of zygotes surviving genetic death equals the proportion of ovules that can mature into seeds ($f_s g_s + f_x g_x = m$). The arrows indicate the direction of mating-system evolution given either resource limitation (solid arrows), or ovule limitation (dashed arrows). The directions of the arrows depend on the conditions that allow increased self-pollination, given these limits on seed production, as listed in Table 4.3 (black arrows, both conditions satisfied; dark grey arrows, neither condition satisfied; light grey arrow, resource-limited condition satisfied, but ovule-limited condition not satisfied).

clarifies that pollen limitation results from poor pollen dispersal (any of small χ, large ρ, or small ε) and/or limited self-pollination (any of small π, η, ϕ, or Δ) and/or poor pollen-tube performance (small τ_s and/or τ_x), and so is initiated by partial failure of male function, which precipitates subsequent partial failure of female function. Increased autonomous self-pollination can alleviate this aspect of pollen limitation from the female perspective (Chapter 10); however, it does little to relax the limitation of male performance. Therefore, reproductive assurance does not solve pollen limitation completely.

The second condition for pollen limitation involves aspects of pollination quality that determine the survival of young embryos (g_s and g_x), which has generally been overlooked (although see Ramsey 1995; Ramsey and Vaughton 2000; Ashman et al. 2004; Chapter 9). This aspect of pollen limitation has important consequences for the detection of pollen limitation, which is typically assessed by the addition of outcross pollen to flowers that are otherwise subject to natural pollination conditions. Consider a self-compatible plant with F_x cross-fertilized ovules and the remainder, $O - F_x$, self-fertilized, such that *fertilization is not pollen-limited*. In the absence of resource limitation, this plant would produce a total of $F_x g_x + (O - F_x) g_s = F_x (g_x - g_s) + O g_s$ seeds. Suppose that supplemental cross-pollen is applied to stigmas while flowers are young so that the number of cross-fertilized zygotes increases to $F_x + \Sigma$, decreasing the number of self-fertilized zygotes to $O - F_x - \Sigma$. Because outcrossed zygotes survive better than selfed zygotes ($g_x > g_s$), the replacement of self-fertilizations by cross-fertilizations caused by supplemental pollination increases seed production by $\Sigma(g_x - g_s)$ seeds. Such an elevation of seed production in response to supplemental cross-pollination would typically be interpreted as an indication of pollen limitation under natural conditions, even though fertilization was complete in both cases. This effect of pollen quality raises questions about whether pollen limitation is as common as supplementation experiments suggest (reviewed by Burd 1994; Ashman et al. 2004).

4.3.2 Ovule limitation

Ovule limitation occurs when all of a plant's ovules are fertilized ($F_s + F_x = O$) but too few zygotes avoid genetic death and predation to compete for maternal resources ($F_s g_s + F_x g_x \leq mO$: Fig. 4.3). This limit results when plants invest too few resources in ovule production during flower production, perhaps because resource availability improves between flower initiation and seed production. In this case, pollen tubes compete for fertilizations. In our model, pollen from different plants fertilizes ovules in proportion to its contribution to the pollen tubes that enter an ovary simultaneously (as in Holsinger's [1991] "mass-action" models: also see Lloyd 1992; Chapter 2), although biased competition is also possible. In addition, self- and cross-pollen tubes compete for access to ovules in a manner that depends on their relative timing, as outlined in Table 4.2. However, because of poor zygote survival, embryos develop without competition for maternal resources when ovule availability limits seed production, so $k_s = k_x = 1$. By not incorporating the possibility of differential survival of selfed and outcrossed embryos, all previous models of the consequences of different modes of self-pollination (e.g., Lloyd 1979, 1992; Schoen et al. 1996; Morgan and Wilson 2005) have implicitly considered this situation (see Section 4.4.2).

4.3.3 Resource limitation

Resource limitation occurs when more zygotes avoid genetic death than can mature into seeds, given the available maternal resources ($F_s g_s + F_x g_x > mO$: Fig. 4.3), whether or not fertilization is complete. Resource limitation imposes competition among developing embryos, so the probabilities that selfed and outcrossed embryos develop into seeds, k_s and k_x respectively, are < 1. In general, these probabilities depend on the number of competing embryos ($F_s g_s + F_x g_x$), the maximum number of seeds that can be produced (mO), and the relative competitive ability of selfed versus outcrossed embryos. Because resource limitation can occur without complete fertilization (Fig. 4.3), pollen tubes may fertilize ovules independently as described for pollen limitation, or they may compete for fertilizations as described for ovule limitation. Curiously, although resource limitation is a widely recognized constraint on seed production (e.g., Haig and Westoby 1988), its consequences for mating-system evolution have been considered only recently (Sakai and Ishii 1999; Porcher and Lande 2005; L. D. Harder, M. B. Routley, and S. A. Richards unpublished manuscript).

Resource competition occurs *only* if plants produce more ovules than they can mature into seeds (i.e., $m < 1$). Specifically, resource competition requires that the proportion of outcrossed zygotes surviving genetic death exceeds the maximum proportion of ovules that can mature into seeds ($g_x > m$). Such overproduction of ovules can occur for two reasons. The first cause is ecological and results from a mistake in resource allocation to ovule production versus seed development, which could occur if resource conditions decline between flower initiation and seed production. The second cause is adaptive and occurs, for example, if plants produce "extra" ovules to take advantage of unpredictably good pollination (Burd 1995), or resource availability, or to compensate for embryo losses during development to genetic death and/or predation (Porcher and Lande 2005; L. D. Harder, M. B. Routley, and S. A. Richards unpublished manuscript). The latter, evolutionary explanation may apply commonly, as a literature survey of 65 species found an average seed:ovule ratio of 0.6 for plants subject to excess hand cross-pollination, indicating that plants typically produce many more ovules than they mature into seeds (Fig. 4.4).

Reproductive compensation has important consequences for mating-system evolution (see Section 4.4) and the genetic load within populations (Porcher and Lande 2005; L. D. Harder, M. B. Routley, and S. A. Richards unpublished manuscript). Production of more ovules than can mature into seeds allows the genetic death of embryos soon after fertilization to have a limited impact on seed production by the maternal plant. By compensating for genetic deaths, extra ovules allow maternal plants to screen embryos *passively* for viable offspring, at the cost of producing the failed

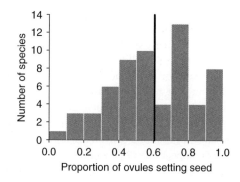

Figure 4.4 The proportion of ovules that develop into seeds for 65 species that were subjected to supplemental cross-pollination. The vertical line indicates the mean seed:ovule ratio.

ovules. This cost is probably relatively small, given that an ovule costs approximately 1% as much as a seed (C. A. Greenway and L. D. Harder unpublished data). Compensatory ovules would also facilitate *active* maternal choice among developing embryos with little cost (see Korbecka *et al.* 2002). Of particular relevance to mating-system evolution is the opportunity that compensatory ovules create for producing viable selfed offspring, which carry two haploid maternal genomes rather than one for outcrossed offspring, despite considerable genetic load in a population. Interestingly, this success of selfed offspring helps maintain genetic load, because parents produce a higher frequency of offspring that are heterozygous carriers for recessive lethal alleles than they would in the absence of reproductive compensation (Porcher and Lande 2005).

4.4 Examples of the roles of pollen and ovule fates in floral and mating-system evolution

To illustrate the evolutionary consequences of pollen and ovule fates, we consider the fitness differential between two phenotypes with different pollination and fertilization patterns,

$$w_2 - w_1 = 2(S_{s2} - S_{s1})d_s + (S_{\female 2} - S_{\female 1})d_x + (S_{\male 2} - S_{\male 1})d_x,$$

and identify circumstances in which phenotype 2 has an advantage (i.e., $w_2 - w_1 > 0$). The characteristics of this fitness differential depend on the details of pollination, whether pollen tubes compete for fertilization, whether embryos compete for maternal resources, and the proportion of the population comprised of phenotype 1 (z: phenotype 2 represents proportion $1 - z$). We now consider four of many possible evolutionary scenarios, showing the detailed derivation in the first case and simply summarizing the results for subsequent examples. Because the outcomes in several cases are frequency dependent, we focus on conditions that allow phenotype 2 to invade a population of phenotype 1 (i.e., $z \cong 1$) and consider equilibrium outcomes only when they are not frequency dependent. In all cases of resource competition, we assume that selfed and outcrossed embryos compete equally for maternal resources.

4.4.1 Improvements in pollen export

We illustrate the evolution of improved pollen export for plants that self-pollinate without affecting export, although the general conclusions also apply when self-pollination causes pollen discounting (results not shown). In the absence of pollen discounting, self-pollination equals $P_s = \eta P$, pollen export for the two phenotypes equals $P_{e1} = \chi_1 (1 - \rho_1)\varepsilon_1 P$ and $P_{e2} = \chi_2 (1 - \rho_2)\varepsilon_2 P$, and each plant of either phenotype imports $P_i = z P_{e1} + (1 - z) P_{e2}$ pollen grains. Thus, phenotype 2 might export pollen more successfully because its flowers attract more pollinators (increased χ) or they place pollen on pollinators' bodies where it is less susceptible to removal or transport losses (reduced ρ, increased ε).

In general, fitness depends on whether seed production is pollen-, ovule-, or resource-limited. With pollen limitation, the fitnesses of the two genotypes are

$$w_1 = 2\tau_s P_s g_s d_s + (P_i + P_{e1})\tau_x g_x d_x \text{ and}$$
$$w_2 = 2\tau_s P_s g_s d_s + (P_i + P_{e2})\tau_x g_x d_x,$$

resulting in a fitness differential of $w_2 - w_1 = (P_{e2} - P_{e1})\tau_x g_x d_x$. Consequently, phenotype 2 can invade a pollen-limited population of phenotype 1 (i.e., $w_2 > w_1$) as long as it exports more pollen (i.e., $P_{e2} > P_{e1}$: Table 4.3). In contrast, in an ovule-limited population, plants import enough pollen to fertilize all their O ovules, so the $\tau_s P_s + \tau_x P_i$ pollen tubes compete for fertilizations. If too few embryos survive genetic death to cause competition of maternal resources, the fitnesses of the two genotypes are

$$w_1 = \frac{2\tau_s P_s g_s d_s + \tau_x (P_i + P_{e1}) g_x d_x}{\tau_s P_s + \tau_x P_i} O \text{ and}$$
$$w_2 = \frac{2\tau_s P_s g_s d_s + \tau_x (P_i + P_{e2}) g_x d_x}{\tau_s P_s + \tau_x P_i} O,$$

resulting in a fitness differential of $w_2 - w_1 = \tau_x (P_{e2} - P_{e1}) g_x d_x O / (\tau_s P_s + \tau_x P_i)$. Increased pollen export is again favoured (Table 4.3);

in this case because cross-fertilizations by phenotype 2 displace self- and cross-fertilizations by phenotype 1. Finally, in a resource-limited population the $\tau_s P_s g_s + \tau_x P_i g_x$ embryos that survive genetic death exceed the number that can mature into seeds (mO), causing competition for maternal resources (i.e., $k_s = k_x = 1/[\tau_s P_s g_s + \tau_x P_i g_x]$). Now the fitnesses of the two genotypes are

$$w_1 = \frac{2\tau_s P_s g_s d_s + \tau_x (P_i + P_{e1}) g_x d_x}{\tau_s P_s g_s + \tau_x P_i g_x} mO \text{ and}$$
$$w_2 = \frac{2\tau_s P_s g_s d_s + \tau_x (P_i + P_{e2}) g_x d_x}{\tau_s P_s g_s + \tau_x P_i g_x} mO,$$

whether or not pollen tubes compete for fertilizations, and the fitness differential is $w_2 - w_1 = \tau_x(P_{e2} - P_{e1})g_x d_x mO/(\tau_s P_s g_s + \tau_x P_i g_x)$. Increased pollen export is again favoured (Table 4.3), in this case because an increase in the number of embryos sired by phenotype 2 displaces embryos sired by phenotype 1 during resource competition.

The preceding results demonstrate a general advantage to increased pollen export resulting from altered floral mechanisms or pollinator shifts that increase the proportion of exportable pollen (χ) or reduce removal or transport losses (decreased ρ or increased ε, respectively). Furthermore, because the fitness differential does not depend on the relative frequencies of the two phenotypes (z), a phenotype with higher pollen export should replace one with lesser export. This universal advantage arises in our model because increased pollen export alters the number and/or mixture of seeds produced on other plants, without affecting a plant's own seed production. In contrast, a change in a floral trait, such as increased dichogamy, that improves pollen export but reduces non-discounting self-pollination can be detrimental (results not shown). Nevertheless, the contrast between the expected general benefit of enhanced pollen export and the observation that plants with granular pollen export only 1% of their pollen, on average (see Section 4.2.1), suggests that the evolution of cross-pollination is subject to strong functional constraints.

4.4.2 Simultaneous, autonomous self-pollination without pollen discounting

Now consider a situation in which the two phenotypes differ in their ability to self-pollinate autonomously without pollen discounting, so $P_{s1} = \eta_1 P$ and $P_{s2} = \eta_2 P$, but they export equivalent amounts of pollen, $P_e = \chi(1 - \rho)\varepsilon P$. For example, suppose the petals of phenotype 2 retain more pollen that has fallen from anthers and would otherwise be lost, which is then transferred onto stigmas when flowers close at night. If seed

Table 4.3 Conditions for the invasion of a population of self-compatible plants (phenotype 1) by a phenotype (phenotype 2) that either exports more pollen, or self-pollinates more than the resident phenotype.

		Pollination contrast		
Limit on seed production	Pollen export[a]	Simultaneous, non-discounting self-pollination	Simultaneous, facilitated self-pollination	Delayed, autonomous self-pollination[b]
Pollen	$\frac{d_s}{d_x} > 0$	$\frac{d_s}{d_x} > 0$	$\frac{d_s}{d_x} > \frac{\tau_x \varepsilon}{2\tau_s} \cdot \frac{g_x}{g_s}$	$\frac{d_s}{d_x} > 0$
Resources	$\frac{d_s}{d_x} > 0$	$\frac{d_s}{d_x} > \frac{1}{2}$	$\frac{d_s}{d_x} > \frac{1}{2} + \frac{\tau_x \varepsilon}{2\tau_s} \cdot \frac{g_x}{g_s} + \frac{\eta + \phi_2 \chi}{2(1 - \phi_1)\chi}$	$\frac{d_s}{d_x} > \frac{1}{2}$
Ovules	$\frac{d_s}{d_x} > 0$	$\frac{d_s}{d_x} > \frac{1}{2} \cdot \frac{g_x}{g_s}$	$\frac{d_s}{d_x} > \left[\frac{1}{2} + \frac{\tau_x \varepsilon}{2\tau_s} + \frac{\eta + \phi_2 \chi}{2(1 - \phi_1)\chi}\right] \frac{g_x}{g_s}$	

See Fig. 4.3 for the implications of the conditions for resource and ovule limitation on mating-system evolution.
[a] Increased pollen export is favoured universally, because d_x always exceeds d_s.
[b] Increased delayed selfing cannot evolve if plants already experience enough self- and cross-pollination to fertilize all ovules.

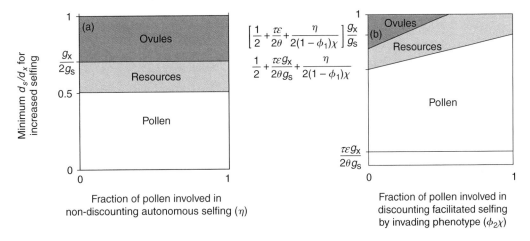

Figure 4.5 Relations of the minimum ratio of post-dispersal survival of selfed (d_s) and outcrossed seeds (d_x) that allows the invasion of a phenotype with increased self-pollination to aspects of self-pollination and the incidence of pollen limitation (white area and above), ovule limitation (light grey area and above), and resource limitation (dark grey area). Two contrasting situations are illustrated: (a) the novel phenotype has elevated autonomous, non-discounting self-pollination (η); (b) elevated facilitated self-pollination (both non-discounting and discounting: ϕ).

production is pollen-limited (i.e., no pollen-tube or embryo competition), phenotype 2 increases in frequency ($w_2 - w_1 > 0$) if it self-pollinates more than phenotype 1 ($\eta_2 > \eta_1$; Table 4.3, Fig. 4.5a). As demonstrated in Section 4.4.1, this situation also promotes increased pollen export. Therefore, any floral trait that alleviates pollen limitation is favoured, whether or not it increases self-pollination or pollen export, as long as the two pollination modes do not interact negatively.

Selection is more complex when pollen tubes compete for fertilizations and/or embryos compete for maternal resources. With competition for fertilization, but not for maternal resources (i.e., ovule limitation), increased non-discounting self-pollination is favoured if the survival of selfed *zygotes* relative to outcrossed *zygotes* exceeds 0.5 ($g_s d_s / g_x d_x > 0.5$: Table 4.3, Fig. 4.5a). In contrast, when resource availability limits seed production, phenotype 2 is favoured if the survival of selfed *seeds* relative to outcrossed *seeds* exceeds 0.5 ($d_s/d_x > 0.5$: Table 4.3, Fig. 4.5a).

Comparison of results for resource- and ovule-limited cases reveals a feature of classic mating-system theory that has not been recognized previously. Standard mating-system models (e.g., Lande and Schemske 1985) and Lloyd's (1979) model of competing selfing assumed that self-pollination did not affect pollen export and so are equivalent to the case modelled here. These models suggested that self-pollination is favoured if inbreeding depression (δ) after self-fertilization is < 0.5. In terms of our notation, $\delta = 1 - (g_s d_s / g_x d_x)$. Therefore, the classic result of mating-system theory is identical to our result for ovule limitation. In contrast, when seed production is resource-limited, we find a less stringent condition for increased selfing, which involves inbreeding depression only after seed production (post-dispersal inbreeding depression). Therefore, both classical mating-system models and Lloyd's more mechanistic model implicitly assume no resource limitation.

To appreciate the consequences of resource limitation for mating-system evolution, consider Fig. 4.3, which illustrates all possible combinations of self- and cross-fertilization. As we have demonstrated, any independent increase in self- or cross-pollination that lessens pollen limitation is always favoured, so that the optimal mating system will not be pollen-limited. Outcomes are more complex when enough ovules are fertilized to cause either ovule limitation (between points S and M in Fig. 4.3) or resource limitation (within triangle X–M–X' in Fig. 4.3). In these cases, the ultimate outcome of mating-system evolution depends on the survival of selfed zygotes relative to outcrossed zygotes ($g_s d_s / g_x d_x$) and of selfed seeds relative to outcrossed seeds (d_s / d_x). If both aspects of relative survival exceed 0.5, then increased selfing is favoured, whether or not seed production is ovule- or resource-limited (dashed and solid black arrows in Fig. 4.3), leading eventually to exclusive selfing (point S in Fig. 4.3). If neither aspect of relative survival exceeds 0.5, then increased outcrossing is always favoured (dark grey arrows in Fig. 4.3), resulting in exclusive outcrossing (any point between X and X′ in Fig. 4.3). Finally, if genetic death causes low survival of selfed zygotes compared with outcrossed zygotes ($g_s d_s / g_x d_x < 0.5$), but selfed seeds have high relative survival ($d_s / d_x > 0.5$), then mixed mating is favoured (light grey arrows in Fig. 4.3). In this case, the optimal mixture of self- and cross-fertilization lies at the transition from ovule to resource limitation (point M in Fig. 4.3), or

$$f_s^* = \frac{m - g_s}{g_x - g_s} \text{ and } f_x^* = \frac{g_x - m}{g_x - g_s},$$

and so depends only on the probabilities that selfed and outcrossed zygotes survive genetic death (g_s and g_x, respectively) and the proportion of ovules that can mature into seeds given the available maternal resources (m). Thus mixed mating can be an optimal mating system in the absence of pollen discounting if selfed zygotes are more susceptible to genetic death than outcrossed zygotes and seed production can be resource-limited. (The fourth combination of high relative survival of selfed zygotes and low relative survival of selfed seeds is mathematically impossible.) Note that none of the preceding results depends on the relative frequencies of competing phenotypes and so they apply to both invasion and equilibrium situations.

The solid black curve in Fig. 4.6 illustrates a specific case in which mixed mating is an evolutionary stable strategy (ESS), whereby no alternative mating pattern results in higher fitness. In addition to illustrating that the optimal mating system occurs at the transition between resource and ovule limitation, this figure reveals that the fitness differential between the resident and invading phenotypes declines more steeply away

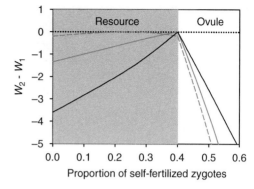

Figure 4.6 Examples of evolutionarily stable mixed mating when all ovules are fertilized and self-pollination either does not affect pollen export (black solid line) or causes pollen discounting (grey curves). The ESS occurs at the proportion of self-fertilization that equalizes the fitnesses of the resident and invading phenotypes (w_1 and w_2, respectively), so $w_2 - w_1 = 0$. For non-discounting self-pollination (black solid line) and some cases of discounting self-pollination (e.g., grey solid line), the ESS lies at the transition between resource limitation (grey area) and ovule limitation (white area). Other cases of discounting self-pollination result in other optimal mating systems (e.g., grey dashed line). For all examples, $P = 50\,000$ pollen grains, $O = 100$ ovules, $\chi(1 - \rho)\varepsilon\tau_x = 0.01$, $g_s = 0.3$, $g_x = 0.8$, $d_s / d_x = 0.71$, and $m = 0.6$. For the solid lines $\tau_x = 0.1$, whereas $\tau_x = 0.07$ for the dashed line.

from the ESS in the direction of ovule limitation, rather than resource limitation. Thus, ovule limitation imposes stronger selection toward the ESS than does resource limitation. Nevertheless the steep declines in fitness on either side of the ESS should result in strong stabilizing selection on the mating system.

4.4.3 Facilitated self-pollination

We now consider two phenotypes that rely on pollen vectors to facilitate both self-pollination ($P_s = [\phi\chi + \eta][1 - \rho]P$) and pollen export ($P_e = [1 - \phi]\chi[1 - \rho]\varepsilon P$). Specifically, we suppose that the phenotypes differ in the discounting component of self-pollination (ϕ_1 versus ϕ_2), which then alters pollen export. This situation could arise if phenotype 2 displays more flowers simultaneously than phenotype 1, so each pollinator visits more flowers, resulting in greater geitonogamy. Note that the $\phi\chi P$ pollen grains involved in pollen discounting cause a one-to-one reduction in pollen-export *opportunities*, as determined by the amount of pollen carried away from a plant ($[1 - \phi]\chi[1 - \rho]P$), but have a weaker effect on *realized* export, because of the lower probability of reaching a stigma owing to transport losses.

Not surprisingly, pollen discounting affects the outcomes of mating-system evolution. Two general outcomes are possible, compared with those observed when simultaneous self-pollination does not affect pollen export. One possibility involves the same ESS outcomes as observed for non-discounting self-pollination, namely, exclusive outcrossing, exclusive selfing, or mixed mating at the transition between resource and ovule limitation (Fig. 4.6, solid grey curve), although pollen discounting restricts the range of conditions that favour increased self-pollination, whether or not seed production is pollen-, ovule-, or resource-limited (Table 4.3, Fig. 4.5b). This situation leads to relatively straightforward explicit solutions, which enable comparison with the non-discounting case. We describe these solutions below. In contrast, the second possibility is more complex and we leave detailed analysis of it to a subsequent paper (L. D. Harder, M. B. Routley, and S. A. Richards unpublished manuscript). This case allows for a much greater range of mixed-mating systems, which are not constrained to lie at the transition between resource- and ovule-limitation (e.g., Fig. 4.6, dashed grey curve) and can even occur when resource limitation is not possible (i.e., $m = 1$).

In cases of discounting self-pollination that allow explicit solutions, the criterion for increased selfing observed in the absence of pollen discounting is incremented by at least $\tau_x \varepsilon g_x / 2\tau_s g_s$. This ratio includes all aspects of mating that are unique to outcrossing in the numerator and unique to selfing in the denominator, including the proportion of pollen grains that survive transport loss (ε), the proportions of self- and cross-pollen grains on stigmas that are represented by pollen tubes in the ovary (τ_s and τ_x, respectively), the proportions of selfed and outcrossed zygotes that survive genetic death (g_s and g_x, respectively), and the two-fold transmission advantage of selfed offspring. In general, we expect τ_x/τ_s and g_x/g_s to exceed 1, whereas ε is approximately 0.01 for species with granular pollen and 0.1 for species with pollinia (see Section 4.2.1). Therefore, for species with granular pollen $\tau_x \varepsilon g_x / 2\tau_s g_s$ probably lies between approximately 0.02 and 0.05, and so has a small effect on the threshold for increased self-fertilization. In contrast, for species with pollinia, $\tau_x \varepsilon g_x / 2\tau_s g_s$ probably lies between approximately 0.2 and 0.5, which greatly elevates the selfing threshold and could even exclude any possibility of increased discounting self-pollination. The common occurrence among orchids of features of the pollinarium that preclude self-pollen deposition until after a pollinator has left a plant (Darwin 1862; Peter and Johnson 2006) is consistent with this expectation

When pollen tubes compete for fertilization and/or embryos compete for maternal resources, the threshold for increased selfing is further increased by $(\eta + \phi_2\chi)/(2[1 - \phi_1\chi])$, which includes details of self-pollination (Fig. 4.5b). In general, this term renders increased selfing less likely to invade if the resident already self-pollinates extensively whether or not it involves pollen discounting (large η or ϕ_1), or if the floral or inflorescence traits of the variant greatly increase discounting above the prevailing level (large ϕ_2: Fig. 4.5b). Note that with pollen-tube competition this term is also multiplied by g_x/g_s, which is greater than 1, so that increased selfing is more likely to be favoured with resource limitation than with ovule limitation (Fig. 4.5b).

Our analysis of all outcomes of discounting, simultaneous self-pollination leads to a general conclusion about its effect on the evolution of mixed mating. This mating system is possible only with relatively strong inbreeding depression before seed dispersal ($2g_s d_s > g_x d_x$), but not later ($2d_s > d_x$). With pollen discounting, each pollen grain can be used in either self-pollination or potential export. If it is used in selfing, it has a high chance of reaching a stigma, but a low chance of becoming a viable offspring, because of pre-dispersal inbreeding depression. If it is used in potential export, the pollen grain has a low chance of reaching a stigma, because of losses during transport, but a high(er) change of becoming a viable offspring if it does reach a stigma. In this case, a plant could sire more offspring through partial selfing if the chance of export is lower than the losses due to inbreeding depression.

In contrast, pure outcrossing should be favoured if pollen transport losses are relatively low, whereas some selfing is favoured when transport losses are relatively high. This mechanism can occur regardless of the proportion of ovules that can mature into seeds (m).

4.4.4 Delayed self-pollination

That delayed self-pollination occurs after cross-pollination, such as when the stamens of wilting flowers collapse on stigmas, bears two immediate consequences for mating-system evolution. First, delayed self-pollination cannot affect a plant's outcross siring success, so the fitness differential between competing phenotypes depends only on their relative maternal contributions (also see Lloyd 1979). Second, delayed self-fertilization cannot alter the mixture of selfed and outcrossed zygotes if all ovules are fertilized, so it cannot mitigate the effects of poor pollen quality on seed production for ovule-limited plants. We now consider additional consequences of delayed self-pollination for pollen- and resource-limited plants.

Our analysis partially contradicts Lloyd's (1979) assertion that "delayed self-fertilization ⋯ is always advantageous whenever it is possible" (p. 71). Our results identify that delayed selfing is advantageous either (1) if it provides reproductive assurance when seed production is pollen-limited or (2) if it increases the number of selfed embryos competing for maternal resources when selfed seeds survive at least half as well as outcrossed seeds (i.e., $2d_s > d_x$: Table 4.3), or equivalently *post-dispersal* inbreeding depression is < 0.5. When selfed seeds survive poorly, delayed selfing is disadvantageous if it aggravates resource competition among developing seeds. Because predominantly outcrossing species suffer less post-dispersal inbreeding depression than selfing species (Husband and Schemske 1996), delayed selfing should be a more common mode of selfing for outcrossing species. The incompleteness of Lloyd's conclusions resulted because, as with most other analyses of mating-system evolution, he did not consider the direct consequences of resource competition, and so implicitly assumed that this mode of self-fertilization bears no resource consequences. Interestingly, Lloyd (1992) raised the possibility that resource consumption by seeds produced by delayed selfing could reduce a perennial parent's fitness by reducing its future survival and/or productivity during subsequent breeding seasons (also see Morgan *et al.* 1997).

4.5 Concluding discussion

4.5.1 The asymmetry of pollen and ovule fates

This review confirms the asymmetry of pollen and ovule fates that is obvious from the imbalance in pollen and ovule production, even though every sexually produced seed involves one pollen grain and one ovule. This asymmetry occurs for two reasons. Most important, outcrossing exposes pollen grains to many risks that are not experienced by ovules, particularly removal and transport losses. In addition, pollen fates depend on both the timing of alternative pollination modes and their dependence on pollen vectors, whereas ovule fates depend largely on the timing of fertilization. As a result, pollen is subject to a greater variety of fates, many of which do not result in direct genetic contributions to the next generation. We now briefly consider two implications of this asymmetry for the evolution of reproductive traits and its analysis.

The greater diversity and severity of risks confronted by pollen grains than by ovules requires plants to produce many more pollen grains than ovules. Following Lloyd (1965), Cruden (1977) used similar reasoning to explain the higher pollen:ovule ratios of outcrossing species than selfing species, although he focused on the delivery of pollen to fertilize seeds. Charnov (1982) criticized this argument, claiming that Cruden's explanation largely ignored paternal contributions to the fitness of hermaphrodites. Instead, Charnov proposed that elevated pollen production benefits outcrossing species by increasing competitive opportunities in pistils, whereas a similar increase in pollen production for a selfing species would simply aggravate competition among sibling grains (local mate competition). Although true, this argument ignores the contrasting risks involved in self-pollination versus pollen export. As a result, the relatively high pollen production of outcrossing species is undoubtedly necessary to compensate for pollen losses during transport. For example, Kjellberg *et al.* (2001) found that fig species pollinated by wasp species that collect pollen actively and carry it to receptive trees in pollen pockets on their thoraces, presumably limiting transport losses, produce significantly fewer anthers per ovule (and presumably a lower pollen:ovule ratio) than species pollinated by wasps that carry pollen passively and often groom it from their bodies.

Bateman's (1948) principle also considers the consequences of mating asymmetries between the sexes, but its applicability to the evolution of plant reproduction has been questioned (Burd 1994; Wilson *et al.* 1994; Ashman and Morgan 2004). Based on insightful experiments with *Drosophila*, Bateman proposed that resource availability typically limits female fertility, whereas mating opportunities limit male fertility, so sexual selection should favour female traits that promote mate quality and male traits that enhance mating frequency. In contrast to Bateman's expectation that female fertility is resource-limited, seed production often seems to be limited by pollen receipt (Knight *et al.* 2006; although see Section 4.3.1), stimulating the claim that Bateman's principle often may not apply to plants. Although this claim is strictly correct, pollen limitation need not imply that female and male traits therefore experience equal selection for increased mating opportunities. In particular, pollen limitation changes the nature, but not the existence, of asymmetries in the mating prospects of individual ovules and pollen grains. In particular, the greater diversity of mechanisms that can lead to pollen failure allow for greater variation in mating success through male function than through female function. Such variance differences underlie the role of sexual selection in the evolution of mating traits (Shuster and Wade 2003), including floral characteristics.

4.5.2 Limits on seed production and reproductive evolution

In addition to the widely recognized roles of pollen receipt and resource availability in limiting seed production, our model exposes the possibility of ovule limitation, which has several implications for reproductive evolution. Recognition of ovule limitation as an error in the allocation of reproductive resources to ovule versus seed production raises intriguing questions about the selection of ovule number that are beyond the scope of this chapter. In addition, identification of the possibility of ovule limitation broadens the variety of expectations for evolution of plant reproduction. The adaptive balance between pollen and resource limitation proposed by Haig and Westoby (1988; also see Ashman *et al.* 2004) is expected when selfed zygotes and seeds both have poor survival prospects compared with outcrossed zygotes and seeds (Fig. 4.3, point X′). In contrast, when selfed zygotes and seeds both survive relatively well, plants should be completely ovule-limited (Fig. 4.3, point S). Finally, if selfed zygotes survive relatively poorly, but selfed seeds are relatively successful, compared with outcrossed seeds, then the optimal reproductive policy can balance pollen, ovule, and resource limitation (Fig. 4.3, point M). In the latter case, plants benefit from producing extra ovules that allow them to identify viable selfed offspring that have survived genetic death (Porcher and Lande 2005; L. D. Harder, M. B. Routley, and S. A. Richards unpublished manuscript).

Our models of mating-system evolution differ from preceding analyses because they explicitly consider the consequences of resource limitation of seed production. Intriguingly, resource limitation imposes less stringent conditions on the evolution of self-fertilization than does the ovule limitation which is implicit, if unrecognized, in most mating-system models (Table 4.3). This contrast creates the opportunity for mixed mating to maximize parental fitness, especially if pollen discounting is limited. Given that most plants produce "extra" ovules (i.e., $m < 1$), which is a necessary condition for resource limitation, this mechanism may provide one of the few general, adaptive explanations for the common occurrence of mixed mating (see Goodwillie *et al.* 2005).

The theory presented in this chapter follows Lloyd's (1979, 1992) lead of expanding the analysis of mating-system evolution beyond the traditional genetic approach, which emphasized inbreeding depression (e.g., Lande and Schemske 1985; Charlesworth and Charlesworth 1987), to consider the influences of ecological factors (also see Uyenoyama *et al.* 1993; Goodwillie *et al.* 2005; Chapters 2, 6, 8, 10 and 12). A specific contribution of our models is the recognition that the post-dispersal performance of selfed seeds relative to outcrossed seeds (d_s/d_x) provides a more general criterion for mating system evolution than does lifetime inbreeding depression (see Table 4.3). In addition, our models link mating-system evolution directly to the fates of pollen and ovules and the alternative limits on seed production. This theory, and that of others, integrates the evolution of floral, fruit, and seed characteristics with that of the mating system, illustrating the interplay between reproductive ecology and evolution.

Acknowledgements

We thank Shane Richards for stimulating discussion about fate and Emmanuelle Porcher and Paul Wilson for comments on the manuscript. The Natural Sciences and Engineering Research Council of Canada funded this research through a Discovery Grant (LDH) and a post-doctoral fellowship (MBR).

References

Armbruster WS and Rogers DG (2004). Does pollen competition reduce the cost of inbreeding? *American Journal of Botany*, **91**, 1939–43.

Ashman TL and Morgan MT (2004). Explaining phenotypic selection on plant attractive characters: male function, gender balance or ecological context? *Proceedings of the Royal Society of London, Series B*, **271**, 553–9.

Ashman T-L, Knight TM, Steets JA, *et al.* (2004). Pollen limitation of plant reproduction: ecological and evolutionary causes and consequences. *Ecology*, **85**, 2408–21.

Barrett SCH, Lloyd DG, and Arroyo J (1996). Stylar polymorphisms and the evolution of heterostyly in *Narcissus* (Amaryllidaceae). In DG Lloyd and SCH Barrett, eds. *Floral biology: studies on floral evolution in animal-pollinated plants*, pp. 339–376. Chapman and Hall, New York, NY.

Bateman AJ (1948). Intra-sexual selection in *Drosophila*. *Heredity*, **23**, 349–68.

Bell G (1985). On the function of flowers. *Proceedings of the Royal Society of London, Series B*, **224**, 223–65.

Bell SA and Cresswell JE (1998). The phenology of gender in homogamous flowers: temporal change in the residual sex function of flowers of oil-seed rape (*Brassica napus*). *Functional Ecology*, **12**, 298–306.

Bernasconi G (2004). Seed paternity in flowering plants: an evolutionary perspective. *Perspectives in Plant Ecology, Evolution and Systematics*, **6**, 149–58.

Burd M (1994). Bateman's Principle and plant reproduction: the role of pollen limitation in fruit and seed set. *Botanical Review*, **60**, 83–139.

Burd M (1995). Ovule packaging in stochastic pollination and fertilization environments. *Evolution* **49**, 100–9.

Casper BB and Niesenbaum RA (1993). Pollen versus resource limitation of seed production: a reconsideration. *Current Science*, **65**, 210–3.

Charlesworth D and Charlesworth B (1987). Inbreeding depression and its evolutionary consequences. *Annual Review of Ecology and Systematics*, **18**, 237–68.

Charnov EL (1982). *The theory of sex allocation*. Princeton University Press, Princeton, NJ.

Cruden RW (1977). Pollen-ovule ratios: a conservative indicator of breeding systems in flowering plants. *Evolution*, **31**, 32–46.

Darwin CR (1862). *On the various contrivances by which British and foreign orchids are fertilised by insects*. John Murray, London.

de Nettancourt D (2001). *Incompatibility and incongruity in wild and cultivated plants*. 2nd edition. Springer, New York.

Eckert CG (2000). Contributions of autogamy and geitonogamy to self-fertilization in a mass-flowering, clonal plant. *Ecology*, **81**, 532–42.

Fenner M and Thompson K (2005). *The ecology of seeds*. Cambridge University Press, Cambridge.

Ganeshaiah KN and Uma Shaanker R (1994). Seed and fruit abortion as a process of self organization among developing sinks. *Physiologia Plantarum*, **91**, 81–9.

Good-Avila SV and Stephenson AG (2002). The inheritance of modifers conferring self–fertility in the partially self-incompatible perennial, *Campanula rapunculoides*. *Evolution*, **56**, 263–72.

Goodwillie C, Partis KL, and West JW (2004). Transient self-incompatibility confers delayed selfing in *Leptosiphon jepsonii* (Polemoniaceae). *International Journal of Plant Sciences*, **165**, 387–94.

Goodwillie C, Kalisz S, and Eckert CG (2005). The evolutionary enigma of mixed mating systems in plants: occurrence, theoretical expectation and empirical evidence. *Annual Review of Ecology, Evolution and Systematics*, **36**, 47–79.

Haig D and Westoby M (1988). On limits to seed production. *American Naturalist*, **131**, 757–9.

Harder LD (2000). Pollen dispersal and the floral diversity of Monocotyledons. In KL Wilson and D Morrison, eds. *Monocots: systematics and evolution*, pp. 243–57. CSIRO Publishing, Melbourne, Australia.

Harder LD and Johnson SD (2005). Adaptive plasticity of floral display size in animal-pollinated plants. *Proceedings of the Royal Society of London, Series B*, **272**, 2651–7.

Harder LD and Wilson WG (1997). Theoretical perspectives on pollination. *Acta Horticulturae*, **437**, 83–101.

Harder LD and Wilson WG (1998). A clarification of pollen discounting and its joint effects with inbreeding depression on mating-system evolution. *American Naturalist*, **152**, 684–95.

Harder LD, Barrett SCH, and Cole WW (2000). The mating consequences of sexual segregation within inflorescences of flowering plants. *Proceedings of the Royal Society of London, Series B* **267**:315–20.

Harder LD, Jordan CY, Gross WE, and Routley MB (2004). Beyond floricentrism: the pollination function of inflorescences. *Plant Species Biology*, **19**, 137–48.

Herrera CM (2002). Censusing natural microgametophyte populations: variable spatial mosaics and extreme fine-graininess in winter-flowering *Helleborus foetidus* (Ranunculaceae). *American Journal of Botany*, **89**, 1570–8.

Herrera CM (2004). Distribution ecology of pollen tubes: fine-grained, labile spatial mosaics in southern Spanish Lamiaceae. *New Phytologist*, **161**, 473–84.

Holsinger KE (1991). Mass-action models of plant mating systems: the evolutionary stability of mixed mating systems. *American Naturalist*, **138**, 606–22.

Husband BC and Schemske DW (1996). Evolution of the magnitude and timing of inbreeding depression in plants. *Evolution*, **50**, 54–70.

Johnson SD, Neal PR, and Harder LD (2005). Pollen fates and the limits on male reproductive success in an orchid population. *Biological Journal of the Linnean Society*, **86**, 175–90.

Kjellberg F, Jousselin E, Bronstein JL, Patel A, Yokoyama J, and Rasplus J-Y (2001). Pollination mode in fig wasps: the predictive power of correlated traits. *Proceedings of the Royal Society of London, Series B*, **268**, 1113–21.

Knight TM, Steets JA, and Ashman T-L (2006). A quantitative synthesis of pollen supplementation experiments highlights the contribution of resource reallocation to estimates of pollen limitation. *American Journal of Botany*, **93**, 271–7.

Korbecka G, Klinkhamer PGL, and Vrieling K (2002). Selective embryo abortion hypothesis revisited—a molecular approach. *Plant Biology*, **4**, 298–310.

Lande R and Schemske DW (1985). The evolution of self-fertilization and inbreeding depression in plants. I. Genetic models. *Evolution* **39**, 24–40.

Lloyd DG (1965). Evolution of self-compatibility and racial differentiation in *Leavenworthia* (Cruciferae). *Contributions from the Gray Herbarium of Harvard University*, **195**, 3–134.

Lloyd DG (1979). Some reproductive factors affecting self-fertilization in angiosperms. *American Naturalist*, **113**, 67–79.

Lloyd DG (1984). Gender allocations in outcrossing cosexual plants. In R Dirzo and J Sarukhán, eds. *Perspectives on plant population ecology*, pp. 277–300. Sinauer Associates, Sunderland, MA, USA.

Lloyd DG (1992). Self- and cross-fertilization in plants. II. The selection of self-fertilization. *International Journal of Plant Sciences*, **153**, 370–80.

Mitchell RJ (1997). Effects of pollination intensity on *Lesquerella fendleri* seed set: variation among plants. *Oecologia*, **109**:382–8.

Morgan MT, Schoen DJ, and Bataillon TM (1997). The evolution of self-fertilization in perennials. *American Naturalist*, **150**, 618–38.

Morgan MT, Wilson WG, and Knight TM. (2005). Plant population dynamics, pollinator foraging, and the selection of self-fertilization. *American Naturalist* **166**, 169–83.

Peter CI and Johnson SD (2006). Doing the twist: a test of Darwin's cross-pollination hypothesis for pollinarium reconfiguration. *Biology Letters*, 2, 65–8.

Porcher E and Lande R (2005). Reproductive compensation in the evolution of plant mating systems. *New Phytologist*, **166**, 673–84.

Ramsey M (1995). Ovule preemption and pollen limitation in a self-fertile perennial herb (*Blandfordia grandiflora*, Liliaceae). *Oecologia*, **103**, 101–8.

Ramsey M, and Vaughton G (2000). Pollen quality limits seed set in *Burchardia umbellata* (Colchicaceae). *American Journal of Botany*, **87**, 845–52.

Sakai S and Ishii HS (1999). Why be completely outcrossing? Evolutionary stable outcrossing strategies in an environment where outcross-pollen availability is unpredictable. *Evolutionary Ecology Research*, **1**, 211–22.

Schoen DJ, Morgan MT, and Bataillon T (1996). How does self-pollination evolve? Inferences from floral ecology and molecular genetic variation. *Philosophical Transactions of the Royal Society of London, Series B*, **351**, 1281–90.

Schuster SM and Wade MJ (2003). *Mating systems and strategies*. Princeton University Press, Princeton, NJ.

Skogsmyr I and Lankinen Å (2002). Sexual selection: an evolutionary force in plants? *Biological Reviews*, **77**, 537–62.

Stephenson AG, Travers SE, Mena-Ali JI, and Winsor JA (2003). Pollen performance before and during the autotrophic-heterotrophic transition of pollen tube growth. *Proceedings of the Royal Society, Series B*, **358**, 1009–18.

Travers SE, Mena-Ali J, and Stephenson AG (2004). Plasticity in the self-incompatibility system of *Solanum carolinense*. *Plant Species Biology*, **19**, 127–35.

Uma Shaanker R, Ganeshaiah KN, and Krishnamurthy KS (1995). Development of seeds as self-organizing units: testing the predictions. *International Journal of Plant Sciences*, **156**, 650–7.

Uyenoyama MK, Holsinger KE, and Waller DM (1993). Ecological and genetic factors directing the evolution of self-fertilization. *Oxford Surveys in Evolutionary Biology*, **9**, 327–81.

Vallejo-Marín M and Uyenoyama MK (2004). On the evolutionary costs of self-incompatibility: incomplete reproductive compensation due to pollen limitation. *Evolution*, **58**, 1924–35.

Wheeler MJ, Franklin-Tong VE, and Franklin FCH (2001). The molecular and genetic basis of pollen-pistil interactions. *New Phytologist*, **151**, 565–84.

Wilson P, Thomson JD, Stanton ML, and Rigney LP (1994). Beyond floral Batemania: gender biases in selection for pollination success. *American Naturalist*, **143**, 283–96.

PART 2

Ecological context of floral function and its evolution

The reliance of outcrossing plants on pollen vectors, whether animals, wind, or water, renders plant mating an ecological process. Of particular importance is the "pollination environment," which includes the abundance and proximity of potential mates, the abundance and efficacy of different pollen vectors, and the abundance and variety of other plant species that flower simultaneously. These aspects of the pollination environment govern the amount and quality of pollen exchanged within and among populations, which in turn influence the reproductive capacity of plants and their population dynamics.

In addition to ecological interactions that affect pollination directly, other aspects of a plant's environment can modify reproductive output. Obvious examples include the role of soil moisture availability on nectar production, which affects pollinator attraction, and herbivory of flowers, fruits, or seeds. Such influences on reproduction can act on longer time scales than pollination by affecting floral and fruit development.

The dependence of plant reproduction on both pollination and non-pollination aspects of the plant environment bears both evolutionary and practical implications. From an evolutionary perspective, environmental variation within and among populations and breeding seasons alters the nature of selection on reproductive traits. Such variation creates opportunities for contrasting adaptations to local pollination environments, promoting reproductive diversification. From a conservation, or agricultural perspective, environmental dependence makes plant reproduction susceptible to modification by human activity. Both perspectives emphasize the impossibility of understanding plant reproduction by considering plants in isolation from their environments. This theme pervades the chapters in Part 2, which explore the ecological context of plant reproduction.

Part 2 begins with James Cresswell's largely theoretical examination of gene dispersal via pollen in both natural and agricultural populations (Chapter 5). The dependence of pollen dispersal on the behaviour of pollen vectors involves two key components: the interactions between vectors and flowers, and vector movement within and among plant populations. Cresswell explores current understanding of each component for animal-pollinated species, identifying inadequate knowledge of pollinator movement as a particular constraint on advances in pollination biology. In addition, Cresswell considers the implications of pollen dispersal patterns for both the evolution of these patterns and the genetic isolation of agricultural crops, especially genetically modified varieties.

The perspective expands somewhat in Chapter 6, in which Monica Geber and David Moeller examine the consequences of plant species sharing pollinators for both sets of participants. By increasing the resource base for pollinators, simultaneous flowering by multiple plant species can provoke both short-term and long-term responses in the abundance and composition of pollinator faunas. In addition, pollinator sharing can involve co-flowering plant species in both positive (facilitation) and negative (competition) interactions with each other. The resulting interconnection of plant–pollinator communities creates complex environments for plant reproduction and its evolution, which can vary geographically with shifts in community composition. Geber and Moeller examine this complex interplay by systematically considering the range of responses by pollinator and plant populations to each other. Using specific floral traits and the incidence of selfing versus outcrossing as examples, they also explore the implications of community interactions for local adaptation of plant reproduction.

In contrast to ecological interactions that affect pollination directly, indirect effects associated with abiotic factors and biotic interactions other than pollination have received much less attention. Sharon Strauss and Justen Whittall address the effects of such non-pollinating agents on floral ecology and evolution in Chapter 7. To illustrate these effects, Strauss and Whittall examine a variety of traits, ranging from flower colour to the sexual system, which contribute fundamentally to pollination and mating, but which also play roles that implicate non-pollinating agents. As a consequence, selection on such traits can depend strongly on these less apparent roles. The evidence presented by Strauss and Whittall argues convincingly for an expanded perspective on floral function and serves as a clear reminder that adaptation involves compromise solutions to all of the functions in which flowers participate, directly and indirectly.

The timing of flowering importantly governs the reproductive ecology of individual plants and their populations by determining the exposure of flowers to abiotic conditions, pollinators, co-flowering plant species, herbivores, and seed dispersers. Chapter 8, by Gaku Kudo, considers the ecological effects of flowering phenologies, their consequences for phenological adaptation, and the extent to which flowering phenology affects the selection of other reproductive traits. Kudo specifically draws on his detailed studies in highly seasonal environments along alpine snow-melt gradients to illustrate the role of flowering phenology in structuring populations genetically, allowing for very local adaptation among temporally separated populations.

In the final chapter of Part 2, Marcelo Aizen and Diego Vázquez consider the effects of human modification of the environments for plant reproduction (Chapter 9). This chapter explicitly recognizes the ecological context of plant reproduction and explores the consequences of perturbation to this context. Because humans cause a variety of environmental effects, our modification of plant habitats can have diverse impacts on plant reproduction. To address this complexity of possible outcomes, Aizen and Vázquez provide a logical framework that recognizes the chain of specific perturbation effects on plant and/or pollinator characteristics through pollination to reproductive output. Despite contrasting effects of different perturbations, Aizen and Vázquez find considerable generality in the ultimate reproductive responses. Throughout their chapter, Aizen and Vázquez recommend valuable approaches to studying the conservation biology of plant reproduction.

Selected key references

Ashman T, Knight TM, Steets JA, et al. (2004). Pollen limitation of plant reproduction: ecological and evolutionary causes and consequences. *Ecology*, **85**, 2408–21.

Chittka L and Thomson JD (2001). *Cognitive ecology of pollination*. Cambridge University Press, Cambridge.

Kearns CA, Inouye DW, and Waser NM (1998). Endangered mutualisms: the conservation of plant-pollinator interactions. *Annual Review of Ecology and Systematics*, **29**, 83–112.

Lloyd DG and Barrett SCH (1996). *Floral biology: studies on floral evolution in animal-pollinated plants*. Chapman and Hall, New York.

Waser NM and Ollerton J (2006). *Plant-pollinator interactions: from specialization to generalization*. University of Chicago Press, Chicago.

CHAPTER 5

Models of pollinator-mediated gene dispersal in plants

James E. Cresswell

School of Biosciences, University of Exeter, UK

Outline

Gene dispersal by pollen is a critical process in the evolution and diversity of plants and their flowers. For many plants, animal pollination is an important mechanism of gene dispersal and therefore it has been widely studied. Three iconic patterns of animal-mediated gene dispersal emerge from these studies: the relation of gene dispersal to distance, the relation of gene dispersal to plant population size, and the variation in gene dispersal success among individual plants. I ask whether theoreticians have yet produced well-founded models that quantitatively explain and predict these patterns by tracing the historical development of potentially relevant models and their representations of pollinator movement and of flower-to-flower transfer of pollen or genes. I then focus on models that incorporate patterns of flower-to-flower gene transfer, or "paternity shadows," and introduce a further method for their empirical characterization. These models have not been exploited previously to address evolutionary questions associated with variation in gene dispersal among individual plants, and I therefore use them to begin to investigate the adaptation of floral form for specialization in pollinator use, and for maximum outcrossing. In conclusion, I show that important progress has been made towards explaining the three iconic patterns theoretically, particularly in regard to the spatial extent of gene dispersion, but two limitations are most obvious. First, the models falter when required to predict gene dispersal at the landscape scale, because of limited knowledge of long-distance movements by pollinators. Second, models based on paternity shadows are as yet inimical to representing individual-based floral variation and their use in addressing evolutionary questions presents an area for future development.

5.1 Introduction

Pollination is a key mechanism of gene dispersal in the flowering plants (Fenster 1991; Ghazoul 2005) that has the potential to exert a major influence on plant evolution. Pollen dispersal within and among populations maintains the cohesion of a species' gene pool, thereby diminishing the potential for local adaptation and speciation (Slatkin 1985). Among individuals, pollination influences reproductive success, thereby contributing to fitness differences and causing pollination-related traits to evolve by natural selection (Galen 1989). Overall, therefore, a full understanding of gene dispersal by pollen is critically important for explaining the evolution and diversity of plants and their flowers: but how is such understanding to be achieved?

The spatial extent of pollen-mediated gene dispersal had been appreciated by the nineteenth century, when Charles Darwin (1892, p. 379) wrote "With respect to the distance from which pollen is often brought, no one who has had any experience would expect to obtain pure cabbage seed, for instance, if a plant of another variety grew within two or three hundred yards. An accurate observer, the late Mr. Masters of Canterbury, assured me that he once had his whole stock of seeds "seriously affected with purple bastards" by some plants of purple kale which flowered in a cottager's garden at the distance of half a mile; no other plant of this variety growing any nearer." Ever since, pollen-mediated gene dispersal has been similarly quantified by tracing a marker gene from a point source (e.g., Crane and Mather 1943; Handel 1983; Rieger et al. 2002). Scientists now also analyse extant molecular variation to infer both the spatial extent of cross-pollination (Sork et al. 1999; Smouse et al. 2001) and patterns of mating among individual plants (Meagher 1986; Sork and Schemske 1992). For some species, persistent study has accumulated many empirical descriptions of gene dispersal via pollen (e.g., Damgaard and Kjellsson 2005), but theoretical principles are nevertheless required to explain the causal basis of these observations and to predict future events. For wind-pollinated plants, well-established models treat the airborne dispersal of pollen-like particles (Pasquill 1974; McCartney and Fitt 1985) and their impaction on a receptor (Perry et al. 1997). These generalized models serve as first approximations for making quantitative predictions and they usefully expose the key governing variables. Can similar tools be developed for animal-pollinated plants?

In this chapter, I consider the aims and achievements of theoretical models in explaining pollinator-mediated gene dispersal. I focus on models that incorporate knowledge of the fundamental causal mechanisms in pollination systems. Despite their evident utility (Gliddon 1999), I neglect other approaches that undertake predictive extrapolation from observed patterns of gene dispersal by fitting mathematical functions chosen primarily for the closeness of their fit to the observations. To illustrate the patterns that need explanation, I first introduce three iconic patterns of pollinator-mediated gene dispersal. I review the history of the development of models of pollinator-mediated gene dispersal based on the mechanisms of animal pollination and assemble a set of qualitative generalizations that arise from them. I also extend the models to a new question, namely the effect on gene dispersal of variation in the composition of the pollinator fauna. These models expose quantitative predictions and I show how they and the

understanding of pollination systems on which they are based can be tested. I also illustrate how the models can be applied to explore the evolutionary biology of pollination systems. Finally, I evaluate the extent to which the models explain the three iconic patterns of pollinator-mediated gene dispersal and consider their future prospects.

5.2 Three iconic patterns of pollinator-mediated gene dispersal

A general theory of gene dispersal by pollen should explain three iconic patterns. The first is the relation of gene dispersal to distance (Fig. 5.1), which has been measured many times and typically exhibits leptokurtic decay from a point source of marker genes over both metres (Crane and Mather 1943; Devlin and Ellstrand 1990; Cresswell 2005) and kilometres (Rieger et al. 2002; Austerlitz

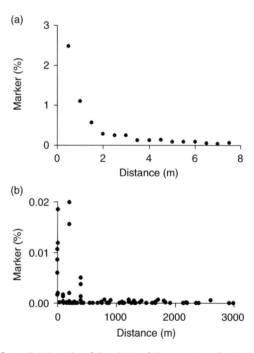

Figure 5.1 Examples of the relation of the percentage of marked seeds set by the unmarked plants to the distance of genetically unmarked plants from a patch of genetically marked plants. Panel (a) shows data from jute (*Corchorus olitorius*) pollinated by *Apis dorsata* collected by Datta, Maiti, and Basak (cited in Levin 1986). Panel (b) shows data from oilseed rape, or canola, *Brassica napus*, assembled from numerous studies (redrawn from Damgaard and Kjellsson 2005).

et al. 2004). The spatial extent of pollen dispersal is important ecologically because it determines the genetic isolation of population fragments and individual plants (Ellstrand 1992). Evolutionarily this relation reflects the amount and extent of gene flow, which tends to homogenize a plant species' gene pool and act against local adaptation (Slatkin 1985). Pollen dispersion is also critical in assessing the risk of genetic escape from genetically modified (GM) crops (Lutman 1999). Given these implications, a theory of pollen-mediated gene dispersal should expose the mechanisms responsible for the shape of the relation, explain differences in these relations among plant–pollinator systems, and produce quantitative predictions of gene dispersal in specified landscapes.

The second pattern (Fig. 5.2) is the decline in the proportion of seeds sired by foreign pollen with increasing size of a patch of flowers. This pattern has been predicted qualitatively from first principles (Handel 1983; Levin 1986) and it has emerged in empirical observations ranging in scope from the floral displays of individual plants (Harder and Barrett 1996) to the collective floral displays of a population (Klinger *et al.* 1992; Goodell *et al.* 1997; Richards *et al.* 1999), but with some exceptions (see Klinger *et al.* 1992; Richards *et al.* 1999). This density dependence is important because pollination by animals typically occurs among patches of flowers, with implications ranging from the genetic fate of small populations (Barrett and Kohn 1991) to the evolution of floral display size (Harder and Barrett 1996). An informative theory of gene dispersal should expose the mechanisms responsible for this density dependence and produce quantitative predictions.

The final pattern (Fig. 5.3) is the extensive variability in male reproductive success in plant populations, with some plants siring no seeds and others siring many (Devlin and Ellstrand 1990; Meagher 1991). Differential reproductive success is a fundamental requirement for Darwinian evolution, and variation in male success can be an important source of natural selection (Janzen 1977; Willson 1979). Thus a theory of gene dispersal should help pollination biologists to quantify variation in male success (Snow and Lewis 1993; Conner *et al.* 1996), identify its phenotypic correlates (Bell 1985; Stanton *et al.* 1986), and deduce the implications of these findings (Wilson *et al.* 1994; Skogsmyr and Lankinen 2002).

All three of the iconic patterns of pollen-mediated gene dispersal described above have been studied extensively by pollination biologists, but in a somewhat disparate fashion. Conceivably, a unifying theory of pollinator-mediated gene dispersal can address all three patterns, if it is based on fundamental mechanisms common to plant–pollinator interactions, because general theories are possible when some details do not matter (Battersby 2003). The search for these mechanisms and their incorporation into a synthetic theory represents a key challenge for the study of the ecology and evolution of flowers.

5.3 A historical perspective on the theory of pollinator-mediated gene dispersal

Bateman (1947) derived the first model of gene dispersal by animal pollination. His approach, which includes an attempt to "fit insect flight to a formula," appears inspired by the contemporary success of mathematical theory in population genetics (Dobzhansky 1937), although he did not explicitly consider the evolutionary consequences of gene dispersal. Instead, Bateman studied crop plants, either turnip

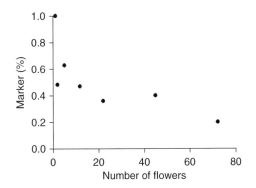

Figure 5.2 An example of the effect of the number of flowers in a population of genetically unmarked plants on the mean percentage of marked seeds set by the unmarked plants. The data are drawn from mixed sex, experimental populations of dioecious *Silene alba*, which was pollinated by moths, bees, and flies. Redrawn from Richards *et al.* (1999).

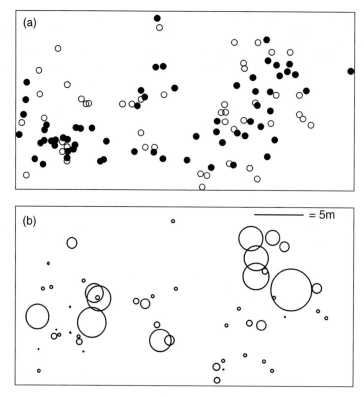

Figure 5.3 An example of the variation in paternal success within a plant population. Panel (a) illustrates the distributions of male and female plants (open and closed symbols, respectively) in a *Chamaelirium luteum* population (redrawn from Meagher 1986). Panel (b) depicts the number of progeny attributed to individual males by the diameter of each male's circle (redrawn from Meagher 1991).

(*Brassica rapa* L. subsp. *rapa*) or radish (*Raphanus sativus* L.), in artificially uniform arrays to identify a scientific basis for isolating agricultural plots of plants to ensure the production of varietally pure seed, although he actually studied gene dispersal through a contiguous population, rather than across a spatial disjunction.

Bateman considered the dispersal of a marker gene from a central patch of marked plants along a row of genetically unmarked plants by social bees. He deduced that bees deposited marked pollen in diminishing amounts at increasing distance from the marked plants, leading him to propose that a bee deposits a fixed proportion, $w < 1$, of the marked pollen on its body on the stigma of each unmarked flower that it visits. Bateman therefore imagined a process now known as "pollen carryover" (Thomson and Plowright 1980) and expressed the dispersal of pollen from a marked flower by assuming a geometric decay in the amount of marked pollen on a pollinator's body as it is deposited on unmarked flowers during successive visits. He also assumed that after the pollinator leaves the marked flower marked pollen constitutes a fraction v of the pollen it carries. Each visit to an unmarked flower reduces this fraction by proportion w, so when the pollinator visits the rth unmarked flower, the fraction vw^{r-1} of the pollen that it deposits on the stigma is marked pollen (see Table 5.1 for definitions of parameters and variables). To include pollinator movements in his model, Bateman focused on the pollen received by a particular unmarked flower. Specifically, he considered the proportion of pollinators that arrive at the flower on their rth flower visit after leaving the marked flowers, denoted N_r. The proportion of pollen on the flower's stigma deposited by pollinators who have travelled r flights between flowers from the marked flower, p_r, is then

Table 5.1 Definitions of the parameters and variables used in the chapter.

Parameter or variable	Definition
a	The number of genetically marked flowers visited by a pollinator before arriving at a patch of unmarked flowers
b	The number of flowers visited by a pollinator in a patch of unmarked flowers, also termed the "residence"
$B(x,t)$	The probability that a pollinator is at distance x from the marked patch of flowers t seconds after leaving it
c	The advection coefficient in a diffusion–advection model of pollinator movement
C	The characteristic number of flowers that a pollinator visits at a plant's display
d	A distance from the marked patch of flowers
D	The diffusion coefficient in a diffusion–advection model of pollinator movement
E	The probability that a pollinator arriving at the patch of unmarked flowers carries marked pollen
f_r	The rth component of the paternity shadow—the proportion of seed sired by a marked flower at the rth unmarked flower visited subsequently by a pollinator
F	The proportion of marked seed produced by an unmarked flower
F_r	The proportion of marked seed produced by the rth unmarked flower that a pollinator visits after leaving the patch of marked flowers
F	A matrix whose elements are F_r
M_d	The proportion of marked seed in flowers located at distance d from the patch of marked flowers
M	A matrix whose elements are M_d
N_r	The proportion of pollinators that arrive at an unmarked flower on their rth inter-flower flight after leaving the marked patch of flowers
$N_{d,r}$	The probability that a pollinator at distance d from the marked patch arrived to visit its rth unmarked flower after leaving the marked patch
N	A matrix whose elements are $N_{d,r}$
p_r	The proportion of pollen on an unmarked flower's stigma that is deposited by pollinators that arrive on their rth inter-flower flight after leaving the marked patch of flowers
$p(t)$	The probability that a grain of marked pollen remains on the pollinator's body t seconds after it has left the marked patch of flowers
q	The proportion of a flower's seeds that is not fertilized by animal pollination
r	The number of unmarked flowers that the pollinator has visited since leaving the marked patch of flowers
u	The probability that a pollen grain remains on a pollinator's body while in transit between flowers
v	The proportion of marked pollen on a pollinator's body immediately after it leaves a marked flower
w	The proportion of pollen on a pollinator's body removed by a stigma during a pollinator visit—controls geometric decay in Bateman's model of pollen dispersal
(*)e	A hypothesized value of any parameter, *
α	The proportion of the superior cross-pollinator in a pollinator fauna comprised of two pollinator species
δ	A small portion of a component of a paternity shadow
ρ	The fraction of a flower's ovules fertilized during a single pollinator visit
ξ	The proportion of marked seed produced by a patch of unmarked flowers
ψ	The number of marked fruits produced by a pollinator for every b unmarked flowers that it fertilizes fully

To illustrate historical continuity among the models, I have given chronological priority in notation. Thus, in this chapter some recent models have different notation from that presented in the original publications.

$$p_r = N_r v w^{(r-1)}. \quad (1)$$

Bateman tacitly assumed that marked pollen sires seeds in a fruit in proportion to its representation among the pollen on the flower's stigmatic surfaces, an assumption which is sometimes violated (Cresswell et al. 2001). He further proposed that a proportion q of seeds are fertilized by pollen delivered by means other than animal pollination, such as autonomous self-pollination (Lloyd and Schoen 1992) or wind pollination, so that the expected proportion of marked seed produced by a flower, F, is

$$F = (1-q) \sum_r p_r. \quad (2)$$

Equations 1 and 2 expose the governing influences of animal-mediated gene dispersal as follows. First, q represents the relative importance of animal pollinators. Second, N_r implicates pollinator movement. Third, v and w respectively implicate the amount of pollen transferred and its distribution among recipient plants.

In the absence of empirical data, Bateman necessarily assumed a simple process of pollen dispersal and he explored his model by adopting arbitrary values for its parameters, v and w. Even decades later, biologists relied on similarly arbitrary values when they began to adduce evidence for certain key aspects of theoretical population genetics, such as the spatially restricted gene flow that creates locally panmictic neighbourhoods (Levin et al. 1971). Several authors eventually demonstrated pollen carryover empirically almost simultaneously (Thomson and Plowright 1980; Waser and Price 1982; Lertzman and Gass 1983). These scientists established that the deposition of marked pollen decreased very rapidly over successively visited unmarked flowers, but that small amounts of marked pollen arrived many flowers later, resulting in a long-tailed dispersal curve. This tail frequently extended further than expected under Bateman's geometric decay model, indicating that the population of marked pollen on a pollinator's body was not decremented by a constant proportion during each flower visit (Morris et al. 1994; Harder and Wilson 1998). Nevertheless, the short range of most flower-to-flower pollinator movements (Levin and Kerster 1968; Schmitt 1980) indicated spatially restricted gene dispersal, despite pollen carryover (Levin et al. 1971). Consequently, further development of models for predicting the spatial extent of gene dispersal may have appeared unnecessary for population geneticists, and Bateman's theory remained the acme for over 40 years.

The characterization of pollen dispersal alerted ecologists to one of its consequences, self-pollination among a plant's flowers, or geitonogamy, which attracted substantial theoretical and experimental treatments (de Jong et al. 1993; Harder and Barrett 1996). Geitonogamy can occur when a pollinator visits multiple flowers on a plant's floral display. This process can reduce fitness by wasting a plant's pollen on its own flowers that would otherwise have a chance to fertilize seeds on other plants (Lloyd 1992; Harder and Wilson 1998) and, for self-compatible plants, by producing inbred seeds, which are susceptible to inbreeding depression (Charlesworth and Charlesworth 1987). Avoidance of inbreeding depression places a fitness premium on outcrossing and is a major influence shaping the evolution of many aspects of floral form and display (Barrett 2002). However, concerns originating in agriculture, rather than in evolutionary ecology, were particularly important in rekindling interest in predicting pollinator-mediated gene dispersal.

The first release of GM plants occurred in 1986 (Barber 1999). Ecologists advocated scientific study to inform risk assessments (Colwell et al. 1985), intensifying the imperative to quantify the potential spread of transgenes by pollen. Confinement strategies, such as the use of separation distances between GM and unmodified crops and border traps around fields of GM crops, were tested in field trials of insect-pollinated crops (Manasse 1992). These results exposed a need for theory that could estimate the permeability of plant populations to incoming transgenes. In this climate, Morris (1993) modelled pollen dispersal in plant arrays that were being used to investigate transgene confinement.

Morris (1993) predicted the spatial dispersion of marked pollen using partial differential equations to combine a diffusion–advection model of pollinator movements with an empirically determined, time-dependent function of pollen dispersal. Morris's general model for the amount of marked pollen arriving at an unmarked plant x metres away from the marked flowers is given by

$$P_{AD}(x) = \int_t B(x,t)p(t)dt, \qquad (3)$$

where $B(x,t)$ defines the probability that the pollinator has moved x metres from the marked flowers t seconds after leaving them and $p(t)$ is the probability density that a pollen grain remains on the pollinator's body for t seconds before being deposited onto a stigma. As in eq. 1, eq. 3 formulates pollen dispersal as the product of pollinator movements and associated pollen deposition, but whereas Bateman relied on an empirical description of pollinator movement, Morris attempted to represent the underlying process itself, which he introduced as a parametric diffusion model through $B(x,t)$ as follows:

$$B(x,t) = \frac{1}{2\sqrt{\pi Dt}} \exp\left(\frac{-(x-ct)^2}{4Dt}\right). \quad (4)$$

Equation 4 describes the probability density of possible pollinator locations as a normal distribution. The distribution's variance, $2Dt$, increases with time (diffusion), and its location, ct, moves away from the origin at a constant velocity c (advection). Parameters D and c are estimated from three attributes of pollinator movement: the mean inter-flower distance moved by a pollinator, the mean time between moves, and the probability that the pollinator moves away from the origin during each move. Morris derived an analytic solution for the probability density of pollen-dispersal distances which allows easy calculation of the effect on gene dispersal of changes in any of the movement attributes that affect D and c.

Morris's model illustrates how a mechanistic model of pollinator movements can be coupled with pollen deposition to predict the spatial extent of gene dispersal. As a diffusion-based model, it applies best to gene dispersal from a point source through a uniformly spaced plant population. In circumstances approaching this ideal, this model could be used in ecological or evolutionary studies to estimate neighbourhood sizes around individual plants and the spatial extent of local random mating. In the context of GM confinement, this model is best suited to predicting the width of field border required to isolate a conventional field, given that pollinators arrive at or near the field's edge. However, this model is less suited to the estimation of field-to-field cross-pollination in an arable landscape, because a diffusion process may be inappropriate for landscape-scale pollinator movements over kilometres.

Indeed, further progress on this question seemed stymied. Landscape-scale pollinator movements remained largely unknown, despite both the use of traditional mark-recapture techniques (Dramstad 1996) and advances in radar tracking (Osborne et al. 1999). Moreover, models of gene dispersal may have seemed unimportant to North American ecologists, who had recognized that transgenes could not be contained in the agricultural landscape following widespread commercialization (Kareiva et al. 1994). However, in Europe, where GM crops were not released commercially, a legislative threshold permitting at most 0.9% GM content in a "GM-free" product (Weekes et al. 2005) motivated assessment of whether GM gene dispersal into a conventional arable field could exceed this limit. Specifically, could the possibility of satisfying this requirement be determined without resolving the difficult problem of quantifying pollinator movement between fields? In addition, previous pollen-dispersal models required modification, because dealing with threshold GM content required quantification of the dispersion of paternity, not pollen.

These new demands were addressed as follows (Cresswell 2003). Assume that each pollinator arriving in a patch of unmarked flowers visits b flowers and fertilizes a fraction ρ of the ovules in each of them, so that the expected amount of seed produced by a pollinator's visit is ρb. Let E denote the proportion of pollinators carrying marked pollen that arrive in a patch of unmarked flowers and, following eq. 2, let ρF_r denote the expected proportion of marked seed produced by the rth-visited flower in the unmarked patch. If marked pollen arrives only by animal pollinators, the proportion of marked seed produced in the patch is

$$\xi = \frac{E \sum_{r=1}^{b} F_r}{b} \quad (5)$$

Note that ρ has cancelled from eq. 5, which therefore applies whether or not full fertilization requires more than one visit.

Equation 5 corresponds with the "portion-dilution" model (PDM) of Cresswell et al. (2002), where the parameter ψ replaces $\sum_{r=1}^{b} F_r$. Like eq. 5, the PDM considers pollinators moving from a marked patch of flowers into an unmarked patch. Similarly, once in the unmarked patch, each pollinator fertilizes ψ fruits with marked pollen for every b fruits that it fertilizes (Fig. 5.4). The model is of interest only when $\psi < b$, which restricts the model's application to patches of unmarked flowers that are also a source of compatible unmarked pollen. In these cases, the proportion of marked seed is

$$\xi = \frac{E\psi}{b}. \qquad (6)$$

Unlike eq. 5, which incorporates a geometric-decay model of pollen transfer, $\sum_{r=1}^{b} F_r$ (eq. 2), the PDM can draw on a wide range of alternative characterizations of flower-to-flower gene dispersal to characterize ψ. In effect, the PDM generalizes eq. 5.

Solution of eq. 6 based on measured gene dispersal in a GM crop, *Brassica napus* L., pollinated by bumble bees (*Bombus* spp.) yielded $\psi = 1.2$ and $b \approx 500$ in an arable field. Thus, the PDM predicts that bumble bee-mediated gene flow into arable fields (i.e., $E = 1$) introduces foreign genes into a maximum of 0.2% of seeds (Cresswell *et al.* 2002), which is substantially below current regulatory thresholds in Europe. By quantifying gene dispersal into a patch of flowers in relation to the relative rates of arrival of pollen of intrinsic versus extrinsic origin, the PDM extends some of the fundamental results from the study of outcrossing versus geitonogamy in self-compatible plants (Harder and Barrett 1996) to patches of flowers of any size, including the collective floral displays of populations. Consequently, the PDM begins to apply an understanding of common pollination mechanisms across scales and therefore suggests itself as the basis of a unifying theory.

5.4 Qualitative generalizations from the portion-dilution model

Theoretical models expose the factors governing a biological system. I now inspect the PDM (eq. 6) in search of general, qualitative implications and elaborate it to encompass some further complexities of natural pollination systems. Specifically, the PDM models gene dispersal by a single kind of pollinator, whereas flowers commonly receive visits from multiple pollinator species (Waser and Ollerton 2006). I therefore extend the PDM to apply to a pollinator assemblage.

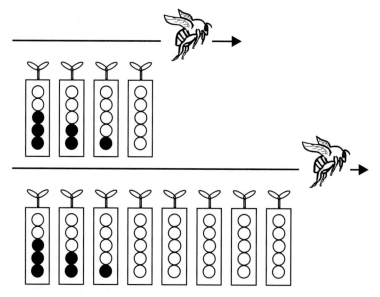

Figure 5.4 Schematic of the "portion-dilution" model, which describes the influence of the number of flowers pollinated during a bout in an unmarked patch, b (residence), on the overall proportion of genetically marked seeds, ξ. Two residence lengths are indicated, with the shorter above. Each upright rectangle represents the fruit of an unmarked flower with its constituent seeds represented by circles. Filled circles represent marked seeds and open circles represent unmarked seeds resulting from transfer of pollen within the unmarked patch. The amount of marked seed produced when a bee fertilizes b flowers (i.e., the total number of fruits full of filled circles) is denoted by ψ.

According to the PDM, the proportion of seed with marked paternity, ξ, varies directly with the probability that a pollinator arrives at the patch of unmarked flowers carrying marked pollen, E. This probability is likely to decline with increasing separation between the marked and unmarked patches for two reasons. First, marked pollen may be lost from a pollinator's body as it travels, either because the pollinator grooms (Thomson 1986) or because pollen is dislodged by airstreams or pollinator movements (Chapter 4). Second, with increasing separation a pollinator is increasingly likely to encounter alternative patches of flowers, thereby increasing the likelihood that intervening visits exhaust the marked pollen that it carries before it arrives at the distant unmarked patch. The decline with separation distance in the frequency of marked seed (Fig. 5.1) is therefore expected; however, limited knowledge of pollinator movement precludes prediction of the precise nature of this relation at the landscape scale. When the relevant patches of flowers are the floral displays of individual plants, E represents the probability that a pollinator carries the pollen of a particular individual. E is influenced by the relative attractiveness to pollinators of an individual's display (Mitchell 1993), but the implications of this relation for reproductive success remain to be quantified. Presumably, variation in E among individuals causes some of the extensive variation in male reproductive success within populations (Fig. 5.3).

According to the PDM, ξ varies inversely with the number of flowers that a pollinator visits in the unmarked patch, b, or the pollinator's "residence" in the unmarked patch (Fig. 5.4). This consequence arises because the incoming marked pollen carried by pollinators is increasingly diluted as unmarked pollen accumulates on pollinators' bodies during successive flower visits. Generally, residence increases with the number of available flowers in a patch when patches are comprised of either individual flowers in a plant's display (Geber 1985; Klinkhamer et al. 1989; Robertson 1992; Goulson et al. 1998; Ohashi and Yahara 2002) or the collective display of multiple plants (Sih and Baltus 1987; Cresswell and Osborne 2004). Thus, small patches of flowers are relatively most susceptible to incoming gene dispersal (Fig. 5.2).

In the PDM, ξ depends directly on the total number of fruits that each pollinator fertilizes with marked pollen during its residence in the unmarked patch, ψ. If the pollinator's residence is short, ψ may increase with residence, b, because of the number of successive flower visits required to deplete the pollinator of marked pollen. However, residences that deplete marked pollen maximize ψ. Apart from the possible effect of residence, the governing influences on ψ are poorly understood, but they probably include certain key properties of the pollinator, such as its capacity to carry pollen, its tendency to remove pollen from body surfaces by grooming, and the match between the pollinator's body and the sexual architecture of flowers. Nevertheless, generalizations about the likely magnitude of ψ are currently elusive.

Given so many potential sources of variation, ψ probably differs among a plant's pollinator species, so the overall gene dispersal into a patch of flowers probably depends on the composition of the pollinator fauna. Consider a patch of unmarked flowers whose pollinator fauna includes species A and B, of which A comprises proportion α. Each individual of species A (or B) arriving at the patch visits b_A (or b_B) flowers before leaving. Individuals of A (or B) carry marked pollen with probability E_A (or E_B), and the pollen they deliver sires marked seeds in the equivalent of ψ_A (or ψ_B) fruits during a single bout in the unmarked patch. For simplicity, assume that both pollinators remain in the unmarked patch long enough to exhaust the marked pollen they carry, so that ψ_A and ψ_B are constants. As a result, the proportion of marked seed in the patch equals

$$\xi_{A,B} = \frac{\alpha E_A \psi_A + (1-\alpha) E_B \psi_B}{\alpha b_A + (1-\alpha) b_B}. \qquad (7)$$

Unless the pollinator species are functionally equivalent ($\xi_A = \xi_B$), three generic relations between the proportion of marked seed, $\xi_{A,B}$, and the relative abundance of species A, α, are possible, denoted types I, II, and III (Fig. 5.5). Note, that these generic relations arise exactly only when ψ_A and ψ_B are constants. If instead the delivery of marked pollen (ψ_A and ψ_B) increases with pollinator residence in the unmarked patch (b_A

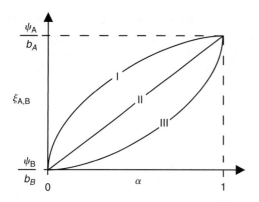

Figure 5.5 Three possible relations (types I, II, and III) between the proportion of the pollinators belonging to species A (α) in a fauna of two species and the proportion of a patch's seeds resulting from incoming gene dispersal, $\xi_{A,B}$ (see eq. 7). In this example, all pollinators arrive at a patch with marked pollen ($E_A = E_B = 1$) and pollinator species A produces the greatest proportion of marked seeds when it is the sole pollinator. Type I occurs when $b_A > b_B$, type II when $b_A = b_B$, and type III when $b_A < b_B$.

and b_B), $\xi_{A,B}$ will vary with α in a more complex manner than suggested in Fig. 5.5.

Suppose that species A is the superior cross-pollinator (i.e., $E_A\psi_A > E_B\psi_B$), so replacement of A by B (i.e., decreased α) invariably reduces dispersal of marker genes into the unmarked patch, $\xi_{A,B}$. However, $\xi_{A,B}$ is proportional to α only when both species visit the same number of flowers in the unmarked patch (i.e., $b_B = b_A$; Fig. 5.5, type II). If, instead, species B has a shorter residence than A (i.e., $b_B < b_A$), replacement of the superior cross-pollinator, species A, by a "weaker diluter" offsets the decline in $\xi_{A,B}$ (Fig. 5.5, type I). In contrast, if species B has a longer residence than A (i.e., $b_B > b_A$), replacement of A by a "stronger diluter" accentuates the decline in $\xi_{A,B}$ (Fig. 5.5, type III). These results have implications for the effects of changes in pollinator composition on gene dispersal. First, the relative influence of different pollinators on gene dispersal is complicated, as it depends on three attributes of a pollinator species (E, ψ, and b). Second, changes in the pollinator fauna of a patch of flowers need not cause proportionate changes in gene dispersal into the patch. Instead, the sensitivity of gene dispersal to faunal change depends on the relative residence of each species in the unmarked patch.

If the patch of unmarked flowers is the display of an individual, self-compatible plant, eq. 7 also can be used to model outcrossing rates, potentially exposing insights into the evolution of flowers and floral display when avoidance of geitonogamy is important (Barrett 2002). Generally, selection is expected to result in plants specializing on the "most effective" pollinator (Stebbins 1970; although see Chapter 2). Effectiveness has previously been characterized as the product of the frequency of a pollinator's visits and the amount of pollen that it delivers per visit (Primack and Silander 1975; Mayfield et al. 2001; but see Vázquez et al. 2005), which views effectiveness only in terms of reproductive assurance. If eq. 7 is construed as the proportion of outcrossed seeds, plant attributes that modify the composition of the pollinator fauna, such as accessibility of floral rewards and flower colour (Faegri and van der Pijl 1971), can be targets of selection for outcrossing. Under such selection, the proportion of the pollinator fauna represented by species A, α, adapts to maximize $\xi_{A,B}$, and the relation of $\xi_{A,B}$ to α describes a fitness surface or adaptive landscape. Hence, the gradient of the curves in Fig. 5.5 indicates the strength of selection on traits that affect α. Figure 5.5 therefore indicates that selection for increased outcrossing, $\xi_{A,B}$, invariably favours plants whose flowers specialize to eliminate an inferior cross-pollinator (i.e., species B when $E_A\psi_A > E_B\psi_B$), but selection becomes weaker as complete specialization approaches when the inferior cross-pollinator is also the weaker diluter (i.e., $b_B < b_A$; Fig. 5.5, type I). Therefore, selection for outcrossing cannot strongly oppose the presence of an inferior cross-pollinator with relatively short residence, which may persist at low frequency in a plant's pollinator fauna. This result may explain, in part, the relative rarity of complete specialization for pollinator use among plants (Fenster et al. 2004).

Overall, explanation of pollinator specialization (or the lack thereof) in self-compatible plants is complicated by the duality of plants' fitness objectives, namely reproductive assurance and the avoidance of inbreeding depression (Morgan and Wilson 2005; Chapter 2). However, the above analysis suggests that pollinator residence may be an important feature in distinguishing the importance of these objectives because, all else being equal (i.e., $E_A\psi_A \approx E_B\psi_B$), only selection for reproductive assurance can favour specialization on a pollinator with a longer residence on a plant's floral display.

5.5 In pursuit of quantitative predictions from the portion-dilution model

In addition to their consistency with existing knowledge, models are tested by the correspondence between their predictions and observations. Tests of the PDM require estimates of its parameters. Pollinator movements can be observed directly, so estimates of pollinator residence, b, and the probability that a pollinator arrives at unmarked flowers carrying marked pollen, E, are feasible in principle, although the difficulty of observing long-distance pollinator movements poses technical problems for evaluating E in all but the smallest landscapes, where initial tests are therefore best conducted. A simple method for estimating the number of fruits fertilized with marked pollen during a pollinator's residence, ψ, requires two steps: (1) compel a pollinator both to visit genetically marked flowers until it becomes fully charged with marked pollen and then to visit a long series of unmarked flowers, and (2) assay the progeny of the visited unmarked flowers for the marker (Cresswell 1994). Alternatively, any pattern in pollinator-mediated gene dispersal among two or more flowers, such as ψ, can be derived, in principle, from the pattern in flower-to-flower gene dispersal that follows a pollinator's visit to a single marked flower, or its "paternity shadow" (Cresswell et al. 2002). Reference to "shadow" invites comparison with the concepts of "seed shadow" and "pollen shadow" (Janzen 1983), which denote the spatial dispersion of seeds and pollen export around a producing plant, whereas a paternity shadow has no spatial reference, being instead "cast" by a single flower over a series of recipient flowers visited subsequently by a pollinator. The paternity shadow is transformed into a spatial dispersion only once the location of the recipient flowers is additionally identified. Therefore, once determined, the paternity shadow affords great scope for predicting spatial gene dispersal, because it remains only to fix the position of the recipient flowers by reference to pollinator movements.

The paternity shadow for a species with monomorphic and mutually compatible flowers is formalized as follows. A pollinator visits a marked flower, $r = 0$, and then disperses marked pollen to the next $r = 1, 2, \ldots, m$ conspecific flowers that the pollinator visits, after which the marked pollen is depleted. Assume that a single pollinator visit delivers enough pollen to fertilize all of the ovules in each recipient flower. Because of deposition and loss during transport, marked pollen fertilizes a diminishing proportion of the seed in each successive unmarked flower, so that $f_1 \geq f_2 \geq \ldots \geq f_m > 0$ and $f_r = 0$ for $r > m$. The set $\{f_r\}$ for $r = 1$ to m comprises the paternity shadow.

When pollinators visit a marked flowers followed by b unmarked flowers, the proportion of marked seed in the rth-visited unmarked flower is

$$F_r = \sum_{r}^{a} f_r \qquad (8)$$

(Fig. 5.6). When pollinators carry only marked pollen after leaving marked plants and subsequent visits to unmarked plants deplete this pollen (i.e., $b \geq a \geq m$), ψ equals

$$\psi = \sum_{r=1}^{m} f_r + \sum_{r=2}^{m} f_r + \ldots + f_m = \sum_{r=1}^{m} r f_r. \qquad (9)$$

If pollinators are not fully charged with marked pollen after leaving the marked flowers (i.e., $a < m$), then (Cresswell 2005)

$$\psi_a = \sum_{r=1}^{a} r f_r + a \sum_{r=a+1}^{m} f_r. \qquad (10)$$

A paternity shadow can be estimated in at least two ways. In the first, a pollinator is required to visit first a single marked flower and then a long series of previously unvisited, unmarked flowers. The proportion of marked seed in the rth-visited unmarked flower yields the rth component of the paternity shadow, f_r (Cresswell et al. 2002). This procedure is difficult to implement under field conditions and must usually be undertaken in a laboratory. However, a laboratory procedure

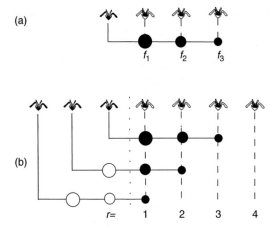

Figure 5.6 Examples of paternity shadows. (a) The paternity shadow of a genetically marked flower (shaded) extending across $m=3$ unmarked flowers. The relative size of the circle for the rth component of the paternity shadow, f_r, indicates the proportion of that flower's seed that is fertilized by marked pollen. (b) The paternity shadows of three marked flowers extending across a collection of unmarked flowers. The proportion of marked seed at the rth-visited flower, F_r, equals the sum of f_r associated with the shaded circles directly below each flower (see eq. 8). The total marked paternity, ψ, is the sum of the f_r associated with all the shaded circles (see eq. 9).

cannot characterize an ecologically realistic paternity shadow if pollen transfer under field conditions is influenced by variation in nectar and pollen availability (Galen and Plowright 1985; Cresswell 1999) that is not represented in the laboratory. To remedy this situation, Cresswell and Hoyle (2006) developed a second method to estimate paternity shadows, which requires study of pollinator movements and gene dispersal in a row of unmarked plants with a centrally located patch of one or more marked plants, namely, the simple field array used by Bateman (1947).

Consider that the marked plants occupy the centre (origin) in the row of unmarked plants. Pollinator movements are recorded to estimate $N_{d,r}$, the probability that a pollinator is located on a plant at distance d from the origin after leaving the marked plants and making r inter-flower flights (eq. 1). When pollination is complete, the proportion of marked seed at each plant location, M_d, is determined. If the row is symmetrical about the origin, then each M_d is the mean of the two plants located at $+d$ and $-d$. Consider pollinators that each leave the central marked plant and carry marked pollen among the unmarked flowers of the row. Hypothesize that the proportion of marked paternity that each pollinator produces at the rth unmarked flower that it visits in the row is $(F_r)_e$. Drawing on eq. 1, the expected spatial distribution of marked paternity, $(M_d)_e$, given the observed pollinator movements, $N_{d,r}$, is

$$(M_d)_e = \sum_{r=1}^{m} N_{d,r}(F_r)_e. \tag{11}$$

Now estimate $(F_r)_e$ by adjusting its hypothesized values until the expected distribution of marked paternity, $(M_d)_e$, matches the observed distribution of marked seed, M_d, as closely as possible. This can be achieved, for example, by using the method of least squares, which minimizes the residual sum of squares (SSR) between the observed and expected values,

$$\text{SSR} = \sum_d [M_d - (M_d)_e]^2, \tag{12}$$

subject to the constraints $F_r \geq 0$ and $F_r > F_{r+1}$. Note that maximum-likelihood methods are better suited when a non-normal sampling distribution around M_d must be specified. Given estimates of all F_r, the individual components of the paternity shadow can be calculated from eq. 8 as $f_r = F_r - F_{r+1}$.

This procedure is illustrated in the following hypothetical matrix representation of eq. 11.

$$\begin{matrix} \mathbf{N} & \mathbf{F} & \mathbf{M} \end{matrix}$$
$$\begin{bmatrix} 0.71 & 0.25 & 0.04 \\ 0.19 & 0.50 & 0.31 \\ 0.10 & 0.25 & 0.65 \end{bmatrix} \begin{bmatrix} (F_1)_e \\ (F_2)_e \\ (F_3)_e \end{bmatrix} \begin{bmatrix} 0.50 \\ 0.20 \\ 0.10 \end{bmatrix} \tag{13}$$

The elements of $\mathbf{N}$ are the probabilities of pollinator arrivals at location d (row) after move r (column). For example, 19% of pollinators arriving at a plant at $d=2$ from the marked plants arrived after their first inter-flower flight beyond the marked plants. Note that the rows of $\mathbf{N}$ need not sum to 1. The elements of $\mathbf{M}$ are the observed proportions of marked seed at each plant location, M_d. For example, 20% of the seed progeny collected at $d=2$ were marked. Solution of the matrix equation for $\mathbf{F}$ under the constraints listed above (e.g., with the *Solver* routine in Microsoft® Excel) yields $F_1 = 0.65$, $F_2 = 0.15$, and $F_3 = 0.00$. In this contrived case, the solution is virtually exact and SSR ≈ 0. The components of the paternity shadow are therefore estimated as $f_1 = 0.50$, $f_2 = 0.15$, and $f_3 = 0.00$. This paternity shadow can now be used to estimate gene dispersal from any set of observed pollinator movements; that is, an unknown **M** can be solved in eq. 13 for any newly observed **N** by using the previously estimated **F**. Estimates of **M** obtained in this way may be valid under a wide range of ecological conditions, if the paternity shadow is a conservative attribute of the plant–pollinator interaction, as seems to be the case for *B. napus* (Cresswell and Hoyle 2006). In addition, the sum of the elements of **F** estimates ψ in the PDM (eq. 6), which can then be used to estimate gene flow into patches of flowers. In practice, the dimensions of **N** and **M** must be set before the experiment to encompass the anticipated extent of pollen carryover (Cresswell and Hoyle 2006).

This matrix framework also allows direct fitting of the components of the paternity shadow (Cresswell and Hoyle in press) for one of the highly leptokurtic, parametric curves describing flower-to-flower pollen transfer (Morris *et al*. 1995), such as the exponential power function (Cresswell 2005). In the latter case, the elements of **F** are determined through eq. 8 by

$$(f_v)_e = \exp(\phi v^\beta). \tag{14}$$

In a least-squares approach, regression is implemented by varying ϕ and β to minimize SSR (eq. 12). Using this approach, Cresswell and Hoyle (2006) identified a paternity shadow that, when combined with a description of bee movements, explained virtually all of the spatial variation in the dispersion of marker genes from centrally located marked plants in a row of unmarked *B. napus* ($r^2 = 90\%$).

5.6 Evolutionary biology of the paternity shadow

Floral variation among flowering plants has been studied extensively (Lloyd and Barrett 1996; Barrett 2002). Traditionally, many floral traits in hermaphroditic plants have been interpreted as anti-selfing mechanisms that safeguard maternal function, but more recent work recognizes their importance in also promoting pollen dispersal and paternal function (Barrett 2002). The paternity shadow amalgamates these perspectives, because it determines both maternity on a plant's floral display and also the paternity achieved by the plant on other plants' displays (Cresswell *et al*. 2002). Thus, insights into floral adaptation and evolution should follow the identification of an ideal paternity shadow that maximizes overall fitness through male and female functions, particularly if the form of the paternity shadow can be linked with particular floral traits. In this section, I consider the relation of plant fitness to the form of the paternity shadow.

The paternity shadow is a meta-trait that emerges from the mechanics of pollination and, when cross-pollination occurs, it involves the flowers of more than one plant. Before speculating about adaptation of the paternity shadow, I suggest a mechanism by which it could evolve, which involves the contact between the areas of the pollinator's body that carry pollen and anthers and stigmas (Berg 1960). Suppose that anthers deposit pollen on a localized area of pollinators' bodies, so that the pollen accumulates in a bivariate normal distribution centred at position $x,y = 0,0$ (Fig. 5.7). Also, suppose that the frequency with which stigmas contact the pollinator's body follows an identical distribution that is also located at $x,y = 0,0$. All else being equal, pollen located far from $x,y = 0,0$ on pollinators will remain on the pollinator longer, and so travel further than centrally located pollen. Therefore, a variant plant with stamens that place pollen consistently away from $x,y = 0,0$ on the pollinator will also produce a variant paternity shadow. If heritable variation in the paternity shadow has implications for

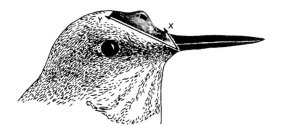

Figure 5.7 Hypothetical bivariate distribution of contact between a pollinator (rufous hummingbird) and anthers and stigmas. *Source*: Lertzman and Gass, 1983, Fig. 24–8.

individual fitness, the shape of the paternity shadow will evolve under selection.

Equation 9 indicates how the paternity shadow might evolve in response to selection for increased outcrossing. Suppose that pollinators visit a characteristic number of flowers, C, at a plant's display and consider a modification of floral architecture that delays deposition of pollen in such a way that a small fraction of the paternity shadow, denoted δ, is realized at the $(r+1)$th-visited unmarked flower instead of at the rth unmarked flower. The variant paternity shadow improves the plant's outcrossing, denoted ψ_C, by δ, but only provided that $(r+1) \leq C$, because otherwise δ is not originally devoted to self-fertilization in the unmodified paternity shadow. It is biologically reasonable to require that pollen carryover declines monotonically and therefore that f_i cannot exceed f_{i+1}, on average. Under this constraint, ψ_C is maximized by a step-shaped paternity shadow with $f_r = 1/C$ for $r \leq C$ and $f_r = 0$ for $r > C$. Note that an identical value for ψ_C can be obtained by any paternity shadow where $\sum_{r=C}^{r=\infty} f_r = 1/C$, because the plant can benefit at most C times from any particular component of its paternity shadow. Therefore, although selection for outcrossing can flatten a paternity shadow for f_r within the range $r < C$, it alone cannot adjust its shape outside this range. Consequently, pollen carryover to flower $r > C$ cannot be explained by this model as an adaptation for outcrossing. The preceding model is formulated around the paternity shadow and a currency of outcrossed seeds, but note that a model of identical form applies to pollen carryover in a self-incompatible plant, where the optimized currency is outcrossed pollen, which generalizes the result among plants with hermaphrodite flowers.

Pollen travelling on a pollinator is at risk of being lost from the pollination process during transport (Chapter 4) either because pollinators groom during inter-plant flights (Thomson 1986; Harder 1990; Rademaker *et al.* 1997; Harder and Wilson 1998), or because it may otherwise become dislodged from the pollinator's body (Thomson and Thomson 1989; Johnson *et al.* 2005). The risks of transport loss can offset the potential paternity gained from extending pollen dispersal.

Assume that any pollen grain survives the trip between recipient flower r and flower $r+1$ with probability u. Equation 9 implies that delaying deposition of some of a small fraction of f_r, denoted δ, to the next flower improves ψ_C when $r\delta < (r+1)\delta u$, or

$$u > \frac{r}{r+1}, \qquad (15)$$

provided that $(r+1) \leq C$. As before, the optimal form of the paternity shadow under this model is a step-function, but the exact location of the inflection depends on u (eq. 15). Thus, the optimally adaptive paternity shadow will be of limited extent and have an abrupt diminution when pollinators groom frequently and thoroughly.

In reality, pollen-dispersal curves (and, by inference, paternity shadows) are neither step-functions nor particularly limited in extent (Morris *et al.* 1994), even in species that typically display only one flower (Thomson 1986). Possibly, the discrepancy between the preceding optimality analysis and these observations arises because plants cannot control the pattern of pollen dispersal finely, which is, for plants without pollinia, merely a simple physical consequence of spreading friable, granular pollen to receptive surfaces via animal pollinators. More likely, the optimality analysis is incomplete. Indeed, two further adaptive mechanisms have yet to be considered. First, a long-tailed pattern of pollen dispersal can be adaptive if the risks of inter-flower travel are offset by the benefits of avoiding inbreeding depression caused by localized biparental inbreeding. If the neighbours of individual plants are relatives, there is a premium on long-distance pollen dispersal (but see Waser and Price 1983). Extensive dispersal may be achieved by a low probability of pollen deposition during any flower visit, so that most pollen will not be delivered to stigmas until the pollinator has moved away from the donor plant's close relatives. Second, the optimum shape of the paternity shadow may be affected by variation in the characteristic number of flowers visited by pollinators per plant, C, which may arise through pollinator behaviour. Thus, the observed patterns of pollen dispersal, and consequent paternity shadows, may be adaptive, although this remains a speculation.

Such speculations are usually resolved by experiment, but it is problematic to test empirically whether a paternity shadow, or a pattern of pollen carryover, is adaptive. For example, selection should constrict paternity shadows when pollinators groom frequently, but plant species served by intensely grooming pollinators necessarily exhibit restricted pollen carryover, because of the resulting loss during transport. Therefore, simply observing a short paternity shadow associated with intense pollinator grooming is not sufficient evidence for inferring adaptation in pollen-dispersal patterns. A more incisive approach would determine the extent to which a plant's paternity shadow could respond to selection for outcrossing. Thus, it is necessary to demonstrate that phenotypic variation in floral architecture or pollen properties affects the paternity shadow. Demonstrations of this kind are currently rare (but see Waser and Price 1984). In conclusion, pollen carryover and the possession of a paternity shadow are, of course, adaptive for avoiding pollen discounting and inbreeding depression. However, observed patterns of pollen carryover do not conform to that predicted by the preceding simple models of maximum outcrossing. Furthermore, the extent to which floral traits control the paternity shadow remains to be determined empirically. Therefore, whether observed pollen carryover and paternity shadows represent adaptations is an open question.

5.7 Prospects for the theory of pollinator-mediated gene dispersal

A general theory of gene dispersal should explain the three iconic patterns identified in Section 5.2. The first pattern is the relation of gene dispersal to distance (M_d versus d), which typically exhibits leptokurtic decay from a point source of marker genes (Crane and Mather 1943; Rieger *et al.* 2002; Austerlitz *et al.* 2004). The key determinants of this pattern, restricted pollinator movements and rapid attenuation of pollen carryover, are well known, and the theory reviewed in this chapter allows us to predict this relation well over small spatial scales, such as the rows of plants studied by Bateman (Cresswell 2005). In contrast, the spatial pattern of gene dispersal is neither explained fully nor predicted easily at the landscape scale, because of the difficulty of describing long-distance pollinator movement and quantifying the probability that a pollinator arrives with marked pollen (E). Theory about pollinator movements deriving from either diffusion–advection models (Morris 1993) or optimal foraging theory (Cresswell *et al.* 2000) has made little impact on landscape-scale questions about pollinator-mediated gene dispersal, despite their critical importance in the conservation of rare and fragmented plant populations (Young *et al.* 1996) and to the management of GM confinement. Furthermore, emerging generalizations about the relative permeability of different landscapes to gene flow (Ghazoul 2005) are based largely on inference from collected empirical results, rather than theoretical derivation. Therefore, further development of the theory of pollinator movements poses an immediate challenge.

The second iconic pattern is the decline in the proportion of seeds sired by foreign pollen with increasing population size (Handel 1983; Klinger *et al.* 1992; Goodell *et al.* 1997). As the number of flowers visited by pollinator in a patch typically increases with the number of flowers available in a patch, the PDM explains this pattern qualitatively (eq. 6) and can predict it quantitatively if its parameters are evaluated (Cresswell *et al.* 2004). These predictions remain untested. Formulating a well-founded theory for the causes of this iconic pattern presents no immediate challenge, but experimental tests of the theory are required, as are further confirmatory empirical demonstrations of the occurrence of the pattern at the population scale.

The final iconic pattern considered here is differential paternal success in plant populations (Fig. 5.3). The models reviewed in this chapter offer a framework in which the likely governing mechanisms can be discussed. For example, the PDM (eq. 6) indicates that the representation of a marked gene among the seed progeny of a patch of unmarked flowers is proportional to the probability that a pollinator arrives with marked pollen (E). Pollen from floral displays of individual plants that are more attractive to pollinators is therefore better represented on pollinators, which will improve the paternal success of those plants. Consequently, traits that affect the attractiveness of an individual's floral display could evolve as adaptations by natural selection. Consideration of the PDM thus again indicates that progress awaits further investigation of the ecology of E. Reflection on another parameter of the PDM, the number of marked fruits produced by a pollinator for every b unmarked flowers that it fertilizes fully, ψ, also exposes paths towards new understanding. In Section 5.6, I showed how male success depends on the form of an individual plant's paternity, but this offered no insight into the floral traits or pollinator attributes that control the form of the paternity shadow. ψ could be measured using the method associated with eqs 11–14 so that ψ could be compared among floral variants to identify its phenotypic determinants.

Overall, continued study has brought us closer to Bateman's (1947) goal of producing a quantitative theory of pollinator-mediated gene dispersal, and more is now known about certain components necessary to the theory, such as flower-to-flower gene dispersal and the patterns of pollinator movements. However, important limitations remain. For example, the minimum separation distance required to isolate two patches of flowers genetically cannot be predicted from first principles, although sound generalizations have emerged from amassed observations. In effect, the relative permeability of different landscapes to gene flow cannot be discerned easily, a limitation that arises from lack of understanding of pollinator movements. In addition, models based on paternity shadows have yet to be harnessed fully to address evolutionary questions, such as the adaptation of floral traits and display. The models must be expanded to incorporate the stochastic nature of pollen carryover and individual variation among paternity shadows. As the iconic patterns of pollinator-mediated gene dispersal remain only partially explained, much science remains to be done.

Acknowledgements

My thanks to W. F. Morris, M. Hoyle, an anonymous reviewer, and this volume's editors for greatly improving the quality of this chapter.

References

Austerlitz F, Dick CW, Dutech C, *et al.* (2004). Using genetic markers to estimate the pollen dispersal curve. *Molecular Ecology*, **13**, 937–54.

Barber S (1999). Transgenic plants: field testing and commercialisation including a consideration of novel herbicide resistant rape (*Brassica napus* L.). In PJ Lutman, ed. *Gene flow and agriculture*, pp. 3–12. British Crop Protection Council, Farnham, Surrey.

Barrett SCH (2002). The evolution of plant sexual diversity. *Nature Reviews Genetics*, **3**, 274–84.

Barrett SCH and Kohn JR (1991). Genetic and evolutionary consequences of small population size in plants: implications for conservation. In DA Falk and KE Holsinger, eds. *Genetics and conservation of rare plants*, pp. 3–30. Oxford University Press, New York.

Bateman AJ (1947). Contamination of seed crops. III Relation with isolation distance. *Heredity*, **1**, 303–36.

Battersby S (2003). The left-handed universe. In P Tallack, ed. *The science book*, pp. 380. Weidenfeld and Nicolson, London.

Bell G (1985). On the function of flowers. *Proceedings of the Royal Society of London, Series B*, **224**, 223–65.

Berg RL (1960). The ecological significance of correlation pleiades. *Evolution*, **14**, 171–80.

Charlesworth D and Charlesworth B (1987). Inbreeding depression and its evolutionary consequences. *Annual Review of Ecology and Systematics*, **18**, 237–68.

Colwell RK, Norse EA, Pimental D, *et al.* (1985). Genetic engineering in agriculture. *Science*, **229**, 111–12.

Conner JK, Rush S, and Jennetten P (1996). Measurements of natural selection on floral traits in wild radish (*Raphanus raphanistrum*). II. Selection through lifetime male and total fitness. *Evolution*, **50**, 1137–46.

Crane MB and Mather K (1943). The natural cross-pollination of crop plants with particular reference to radish. *Annals of Applied Biology*, **30**, 301–8.

Cresswell JE (1994). A method for quantifying the gene flow that results from a single bumblebee visit using transgenic oilseed rape, *Brassica napus* L. cv. Westar. *Transgenic Research*, **3**, 134–7.

Cresswell JE (1999). The influence of nectar and pollen availability on pollen transfer by individual flowers of oil-seed rape (*Brassica napus*) when pollinated by bumblebees (*Bombus lapidarius*). *Journal of Ecology*, **87**, 670–7.

Cresswell JE (2003). Towards the theory of pollinator-mediated gene flow. *Philosophical Transactions of the Royal Society of London, Series B*, **358**, 1005–8.

Cresswell JE (2005). Accurate theoretical prediction of pollinator-mediated gene dispersal. *Ecology*, **86**, 574–8.

Cresswell JE and Hoyle M (2006). A mathematical method for estimating patterns of flower-to-flower gene dispersal from a simple field experiment. *Functional Ecology*, **20**, 245–51.

Cresswell JE and Osborne JL (2004). The effect of patch size and separateness on bumblebee foraging in oilseed rape (*Brassica napus*): implications for gene flow. *Journal of Applied Ecology*, **41**, 539–46.

Cresswell JE, Osborne JL, and Goulson D (2000). An economic model of the limits to foraging range in central place foragers with numerical solutions for bumblebees. *Ecological Entomology*, **25**, 249–55.

Cresswell JE, Hagen C, and Woolnough JM (2001). Attributes of individual flowers of Brassica napus L. are affected by defoliation but not by intraspecific competition. *Annals of Botany*, **88**, 111–17.

Cresswell JE, Osborne JL, and Bell SA (2002). A model of pollinator-mediated gene flow between plant populations with numerical solutions for bumblebees pollinating oilseed rape. *Oikos*, **98**, 375–84.

Damgaard C and Kjellsson G (2005). Gene flow of oilseed rape (*Brassica napus*) according to isolation distance and buffer zone. *Agriculture Ecosystems and Environment*, **108**, 291–301.

Darwin CR (1892). *The effects of cross and self-fertilisation in the vegetable kingdom*. D. Appleton and Co., New York.

de Jong TJ, Waser NM, and Klinkhamer PGL (1993). Geitonogamy: the neglected side of selfing. *Trends in Ecology and Evolution*, **8**, 321–5.

Devlin B and Ellstrand NC (1990). Male and female fertility variation in wild radish, a hermaphrodite. *American Naturalist*, **136**, 87–107.

Dobzhansky T (1937). *Genetics and the origin of species*. Columbia University Press, New York.

Dramstad WE (1996). Do bumblebees (Hymenoptera: Apidae) really forage close to their nests? *Journal of Insect Behavior*, **9**, 163–82.

Ellstrand NC (1992). Gene flow by pollen—implications for plant conservation genetics. *Oikos*, **63**, 77–86.

Faegri K and van der Pijl L (1971). *The principles of pollination ecology*, 2nd edition. Pergamon Press, Oxford.

Fenster CB (1991). Gene flow in *Chamaecrista fasciculata* (Leguminosae) I. Gene dispersal. *Evolution*, **45**, 398–409.

Fenster CB, Armbruster WS, Wilson P, et al. (2004). Pollination syndromes and floral specialization. *Annual Review of Ecology, Evolution, and Systematics*, **35**, 375–403.

Galen C (1989). Measuring pollinator-mediated selection on morphometric floral traits: bumblebees and the alpine sky pilot, *Polemonium viscosum*. *Evolution*, **43**, 882–90.

Galen C and Plowright RC (1985). The effects of nectar level and flower development on pollen-carryover in inflorescences of fireweed (*Epilobium angustifolium*) (Onagraceae). *Canadian Journal of Botany*, **63**, 488–91.

Geber MA (1985). The relationship of plant size to self-pollination in *Mertensia ciliata*. *Ecology*, **66**, 762–72.

Ghazoul J (2005). Pollen and seed dispersal among dispersed plants. *Biological Reviews*, **80**, 413–43.

Gliddon CJ (1999). Gene flow and risk assessment. In PJW Lutman, ed. *Gene flow and agriculture: relevance for transgenic crops*, pp. 49–56. British Crop Protection Council, Farnham, UK.

Goodell K, Elam DR, Nason JD, and Ellstrand NC (1997). Gene flow among small population of a self-incompatible plant: An interaction between demography and genetics. *American Journal of Botany*, **84**, 1362–71.

Goulson D, Stout JC, Hawson SA, and Allen JA (1998). Floral display size in comfrey, *Symphytum officinale* L. (Boraginaceae): relationships with visitation by three bumblebee species and subsequent seed set. *Oecologia*, **113**, 502–8.

Handel SN (1983). Pollination ecology, plant population structure, and gene flow. In LA Real, ed. *Pollination biology*, pp. 163–211. Academic Press, Orlando, FL.

Harder LD (1990). Pollen removal by bumble bees and its implications for pollen dispersal. *Ecology*, **71**, 1110–25.

Harder LD and Barrett SCH (1996). Pollen dispersal and mating patterns in animal-pollinated plants. In DG Lloyd and SCH Barrett, eds. *Floral biology: studies on floral evolution in animal-pollinated plants*, pp. 140–90. Chapman and Hall, New York.

Harder LD and Wilson WG (1998). A clarification of pollen discounting and its joint effects with inbreeding depression on mating system evolution. *American Naturalist*, **152**, 684–95.

Janzen DH (1977). A note on optimal mate selection by plants. *American Naturalist*, **111**, 365–71.

Janzen DH (1983). Seed and pollen dispersal by animals: convergence in the ecology of contamination and sloppy harvest. *Biological Journal of the Linnean Society*, **20**, 103–13.

Johnson SD, Neal PR, and Harder LD (2005). Pollen fates and the limits on male reproductive success in an orchid population. *Biological Journal of the Linnean Society*, **86**, 175–90.

Kareiva P, Morris W, and Jacobi CM (1994). Studying and managing the risk of cross-fertilization between transgenic crops and wild relatives. *Molecular Ecology*, **3**, 15–21.

Klinger T, Arriola PE, and Ellstrand NC (1992). Crop-weed hybridization in radish (*Raphanus sativus*): effects of distance and population size. *American Journal of Botany*, **79**, 1431–5.

Klinkhamer PGL, de Jong TJ, and Debruyn GJ (1989). Plant size and pollinator visitation in *Cyanoglossum officinale*. *Oikos*, **54**, 201–4.

Lertzman KP and Gass CL (1983). Alternative models of pollen transfer. In CE Jones and RJ Little, eds. *Handbook of experimental pollination biology*, pp. 474–89. Scientific and Academic Editions, New York.

Levin DA (1986). Breeding structure and genetic variation. In MJ Crawley, ed. *Plant ecology*, pp. 217–52. Blackwell Scientific Publications, Oxford.

Levin DA and Kerster HW (1968). Local gene dispersal in *Phlox*. *Evolution*, **22**, 130–9.

Levin DA, Kerster HW, and Niedzlek M (1971). Pollinator flight directionality and its effect on pollen flow. *Evolution*, **25**, 113–18.

Lloyd DG (1992). Self-fertilization and cross-fertilization in plants. II. The selection of self-fertilization. *International Journal of Plant Sciences*, **153**, 370–80.

Lloyd DG and Barrett SCH (1996). *Floral biology: studies on floral evolution in animal-pollinated plants*. Chapman and Hall, New York.

Lloyd DG and Schoen DJ (1992). Self- and cross-fertilization in plants I. Functional dimensions. *International Journal of Plant Sciences*, **153**, 358–69.

Lutman PJW (1999). *Gene flow and agriculture*. British Crop Research Council, Farnham, Surrey.

Manasse RS (1992). Ecological risks of transgenic plants: effects of spatial dispersion on gene flow. *Ecological Applications*, **2**, 431–8.

Mayfield MM, Waser NM, and Price MV (2001). Exploring the "most effective pollinator principle" with complex flowers: bumblebees and *Ipomopsis aggregata*. *Annals of Botany*, **88**, 591–6.

McCartney HA and Fitt BDL (1985). Mathematical modelling of crop disease. In CA Gilligan ed. *Advances in plant pathology*, Vol. 3, pp. 107–43. Academic Press, London.

Meagher TR (1986). Analysis of paternity within a natural population of *Chamaelirium luteum*. 1. Identification of most-likely parents. *American Naturalist*, **128**, 199–215.

Meagher TR (1991). Analysis of paternity within a natural population of *Chamaelirium luteum*. II. Patterns of male reproductive success. *American Naturalist*, **137**, 738–52.

Mitchell RJ (1993). Adaptive significance of *Ipomopsis aggregata* nectar production: observation and experiment in the field. *Evolution*, **47**, 25–35.

Morgan MT and Wilson WG (2005). Self-fertilization and the escape from pollen limitation in variable pollination environments. *Evolution*, **59**, 1143–8.

Morris WF (1993). Predicting the consequences of plant spacing and biased movement for pollen dispersal by honeybees. *Ecology*, **74**, 493–500.

Morris WF, Mangel M, and Adler FR (1995). Mechanisms of pollen deposition by insect pollinators. *Evolutionary Ecology*, **9**, 304–317.

Morris WF, Price MV, Waser NM, et al. (1994). Systematic increase in pollen carryover and its consequences for geitonogamy in plant populations. *Oikos*, **71**, 431–40.

Ohashi K and Yahara T (2002). Visit larger displays but probe proportionally fewer flowers: counterintuitive behaviour of nectar-collecting bumble bees achieves an ideal free distribution. *Functional Ecology*, **16**, 492–503.

Osborne JL, Clark SJ, Morris RJ, et al. (1999). A landscape-scale study of bumble bee foraging range and constancy, using harmonic radar. *Journal of Applied Ecology*, **36**, 519–33.

Pasquill F (1974). *Atmospheric diffusion*. Ellis Horwood Ltd., Chichester, UK.

Perry RH, Green DW, and O'Hara Maloney J, eds. (1997). *Perry's chemical engineer's handbook*, 7th edition. McGraw-Hill, New York, NY.

Primack RB and Silander JA (1975). Measuring the relative importance of different pollinators to plants. *Nature*, **255**, 143–4.

Rademaker MCJ, de Jong TJ, and Klinkhamer PGL (1997). Pollen dynamics of bumble-bee visitation on *Echium vulgare*. *Functional Ecology*, **11**, 554–63.

Richards CM, Church S, and McCauley DE (1999). Influence of population size and isolation on gene flow by pollen in *Silene alba*. *Evolution*, **53**, 63–73.

Rieger MA, Lamond M, Preston C, et al. (2002). Pollen-mediated movement of herbicide resistance between commercial canola fields. *Science*, **296**, 2386–8.

Robertson AW (1992). The relationship between floral display size, pollen carryover and geitonogamy in *Myosotis colensoi* (Kirk) Macbride (Boraginaceae). *Biological Journal of the Linnean Society*, **46**, 333–49.

Schmitt J (1980). Pollinator foraging behaviour and gene dispersal in *Senecio* (Compositae). *Evolution*, **34**, 934–43.

Sih A and Baltus MS (1987). Patch size, pollinator behavior, and pollinator limitation in catnip. *Ecology*, **68**, 1679–90.

Skogsmyr I and Lankinen Å (2002). Sexual selection: an evolutionary force in plants. *Biological Reviews*, **77**, 537–62.

Slatkin M (1985). Gene flow in natural populations. *Annual Review of Ecology and Systematics*, **16**, 393–430.

Smouse PE, Dyer RJ, Westfall RD, and Sork VL (2001). Two-generation analysis of pollen flow across a landscape. I. Male gamete heterogeneity among females. *Evolution*, **55**, 260–71.

Snow AA and Lewis PO (1993). Reproductive traits and male-fertility in plants—empirical approaches. *Annual Review of Ecology and Systematics*, **24**, 331–51.

Sork VL and Schemske DW (1992). Fitness consequences of mixed-donor pollen loads in the annual legume *Chamaecrista fasciculata*. *American Journal of Botany*, **79**, 508–15.

Sork VL, Nason J, Campbell DR, and Fernandez JF (1999). Landscape approaches to historical and contemporary gene flow in plants. *Trends in Ecology and Evolution*, **14**, 219–24.

Stanton ML, Snow AA, and Handel SN (1986). Floral evolution: attractiveness to pollinators increases male fitness. *Science*, **232**, 1625–7.

Stebbins GL (1970). Adaptive radiation of reproductive characteristics in angiosperms, I: Pollination mechanisms. *Annual Review of Ecology and Systematics*, **1**, 307–26.

Thomson JD (1986). Pollen transport and deposition by bumble bees in *Erythronium*: influences of floral nectar and bee grooming. *Journal of Ecology*, **74**, 329–41.

Thomson JD and Plowright RC (1980). Pollen carryover, nectar rewards and pollinator behaviour with special reference to *Diervilla lonicera*. *Oecologia*, **46**, 68–74.

Thomson JD and Thomson BA (1989). Dispersal of *Erythronium grandiflorum* pollen by bumblebees: implications for gene flow and reproductive success. *Evolution*, **43**, 657–61.

Vázquez DP, Morris WF, and Jordano P (2005). Interaction frequency as a surrogate for the total effect of animal mutualists on plants. *Ecology Letters*, **8**, 1088–94.

Waser NM and Ollerton J, eds. (2006) *Plant-pollinator interactions: From specialization to generalization*. University of Chicago Press.

Waser NM and Price MV (1982). A comparison of pollen and fluorescent dye carry-over by natural pollinators of *Ipomopsis aggregata* (Polemoniaceae). *Ecology*, **63**, 1168–72.

Waser NM and Price MV (1983). Optimal and actual outcrossing in plants, and the nature of the plant-pollinator interaction. In CE Jones and RJ Little, eds. *Handbook of experimental pollination biology*, pp. 341–59. Academic and Scientific Editions, New York.

Waser NM and Price MV (1984). Experimental studies of pollen carryover: effects of floral variability in *Ipomopsis aggregata*. *Oecologia*, **62**, 262–8.

Weekes R, Deppe C, Allnutt T, *et al.* (2005). Crop-to-crop gene flow using farm scale sites of oilseed rape (*Brassica napus*) in the UK. *Transgenic Research*, **14**, 749–59.

Willson MF (1979). Sexual selection in plants. *American Naturalist*, **113**, 777–90.

Wilson P, Thomson JD, Stanton ML, and Rigney LP (1994). Beyond floral Batemania: gender biases in selection for pollination success. *American Naturalist*, **143**, 283–96.

Young A, Boyle T, and Brown T (1996). The population genetic consequences of habitat fragmentation for plants. *Trends in Ecology and Evolution*, **11**, 413–8.

CHAPTER 6

Pollinator responses to plant communities and implications for reproductive character evolution

Monica A. Geber[1] and David A. Moeller[2]

[1] Department of Ecology and Evolutionary Biology, Cornell University, Ithaca, NY, USA
[2] Department of Plant Biology, University of Minnesota, St. Paul, MN, USA

Outline

Reproduction by most plant species occurs in a community of other flowering species and a diverse fauna of potential pollinators. This community context shapes both pollinator behaviour and ecology and the outcome and evolution of plant reproduction. We consider the effects of communities of pollinator-sharing plant species on three pollinator responses, behaviour (functional response, preference and constancy), demography (numerical and aggregative responses), and community structure (diversity and relative abundance of pollinator taxa). Pollinator responses may differ between multi-species plant communities and single-species populations and these differences can alter patterns of selection on, and evolution of, plant reproductive traits. Plant-pollinator studies that ignore community context can therefore misrepresent the causes of selection on reproductive traits. Pollinator responses to plant communities also influence the extent of heterospecific pollen transfer, a form of interference competition, and affect whether plant species compete (exploitatively) or facilitate each other for pollinator visits. We describe how interference competition, exploitative competition, and facilitation can have opposing selective effects on important reproductive characters, such as floral attractive features, flowering time, and mating-system traits. As an example of the implications of community interactions, we review our work on *Clarkia*, a genus of annual plants that shows facilitative interactions among pollinator-sharing congeners. This facilitation affects pollinator availability and reproductive success, and appears to shape patterns of selection on key floral traits (herkogamy and protandry) affecting plant mating systems (outcrossing versus selfing). Last, we explore potential broad-scale consequences of community interactions for the biogeography of floral variation using data on regional species richness and mating systems of *Clarkia* taxa.

6.1 Introduction

Plants rarely grow solely with conspecifics (single-species population), but instead co-occur and flower with other plant species in a community. Furthermore, studies of community patterns of plant-pollinator interactions clearly show that plant species are typically visited by multiple species of potential pollinators and that most flower visitors visit multiple plant species (Waser and Ollerton 2006). Thus, co-flowering plant species often share flower visitors, and most pollinators are embedded in a community of visitors with whom they share floral resources. In 1983, Rathcke and Waser published seminal papers on the nature of plant species interactions for pollinator services in communities. Surprisingly, their broad-ranging treatment of the subject appears not to have stimulated much empirical study of the community context of plant-pollinator interactions. Studies of plants and their pollinators often emphasize interactions between a pair of plant-pollinator species or at most between one plant and a few of its pollinators. This approach is particularly common for evolutionary studies of pollinators as agents of selection on floral, inflorescence and whole-plant reproductive traits (Fenster et al. 2004). Although the individual plant species considered in each such study is embedded in a plant community, the effects of other plant species on the ecology and evolution of the target plant are often ignored (but see Armbruster 1985; Fishman and Wyatt 1999; Caruso 2000; Hansen et al. 2000).

In this chapter, we ask how the presence of pollinator-sharing plants alters interactions between plants and pollinators relative to interactions in single-species plant populations. Altered interactions result from changes in pollinator "responses" to multiple versus single species. We ask how differences in response can lead in turn to different trajectories of reproductive-character evolution between the two ecological settings. We argue that pollinator-mediated selection in communities cannot always be predicted from studies in single-species populations.

Our discussion begins by considering the properties of plant communities that affect interactions with pollinators and that differ from single-species populations (e.g., interspecific variation in the quantity, quality, and type of reward: Fig. 6.1, Box A, Section 6.2). Next we discuss the direct effects of these plant-community attributes on three pollinator responses: pollinator behaviour, demography, and community structure (i.e., identity, diversity, and relative abundance of pollinator species: Fig. 6.1, Box B, solid arrow (1); Section 6.3). We follow with a discussion of the direct effects of these responses on patterns of selection on plant reproductive traits (Fig. 6.1, Box C, solid arrow (2); Section 6.4). The sum of these "direct" effects causes an indirect effect of plant communities on reproductive character evolution (Fig. 6.1, broken

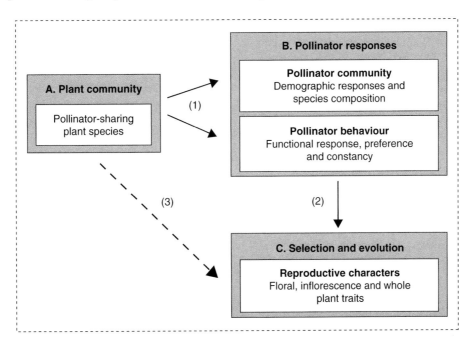

Figure 6.1 Framework for the evolution of plant reproductive traits as mediated by pollinator responses to a community of pollinator-sharing plants. The plant community affects pollinator community structure and foraging behavior (solid arrows, (1)). In turn, these responses influence patterns of selection and evolution in plant reproductive characters (solid arrow, (2)). The combined effect of (1) and (2) produces an indirect effect of plant communities on reproductive character evolution in member species (broken arrow, (3)).

arrow (3)). We then describe an example of the indirect effect of plant communities on selection on mating systems from our work on *Clarkia xantiana*, a plant species that co-occurs with congeners and shares with them a suite of specialist solitary-bee visitors (Section 6.5).

Before proceeding, we define the meaning of several terms that are used throughout the chapter. We refer to pollinators and flower visitors interchangeably, although we realize that not all flower visitors pollinate. In addition, by plant community we refer to a subset of species that share a common set of flower visitors. Among the shared visitors, consideration can generally be restricted to the plants' most abundant and/or effective pollinators. Descriptions of plant-pollinator networks in communities have sometimes identified "nested" subsets of plants and pollinators that interact strongly (Dicks *et al.* 2002; Bascompte *et al.* 2003). We focus on these subsets as they are good places to start in studies of plant-community effects on reproductive character evolution. Finally, we emphasize the role of pollinator responses to communities for *in situ* evolution of plant reproductive characters and the consequent distribution of adaptive character states among species. However, similar patterns of reproductive character distribution can arise by ecological sorting among a regional pool of species, where a subset of species co-exist because they already possess the reproductive attributes favoured by pollinator-mediated selection (see Armbruster *et al.* 1994).

6.2 Properties of plants and communities

Plant communities differ from single-species populations for the simple reason that species vary in relative abundance and in floral, inflorescence, and whole-plant traits that influence the quantity (e.g., nectar volume), quality (e.g., nectar chemistry), and type (e.g., nectar, resin, oil) of rewards per flower or per plant. If species offer rewards of the same quality and type, a pollinator can readily use one species' rewards in the place of another's, (i.e., species provide substitutable resources: Tilman 1982). More often, however, rewards will not be fully substitutable across species. A pollinator may then avoid species that offer rewards of the wrong type (e.g., nectarless flowers for butterflies) or of low quality, especially if the species is rare. On the other hand, a pollinator may visit multiple species that provide non-substitutable, but essential or complementary rewards.

Species differ in traits that affect a pollinator's search time (e.g., flower colour, scent, number) and handling time (e.g., flower and inflorescence shape, flower scent, nectar guides), which determine a pollinator's rate of resource extraction or *proficiency* on a plant. The quantity, quality, and type of reward, along with a pollinator's proficiency, all influence the *profitability* of flowers and plants to a pollinator. Finally, interspecific variation in profitability per plant, combined with variation in species' relative abundances, determines the potential *resource value* of different species and the combined resource value of the entire community to a pollinator.

Species also differ in the seasonal and diurnal schedule of flowering, and these differences affect resource availability over time. In the presence of a single plant species, a pollinator's foraging may be limited by its flowering duration during the day or season. In contrast, interspecific segregation of flowering time in a community can extend the period of resource availability (Chapter 8). For example, co-flowering African acacias release pollen at different times of the day and pollinators make coordinated transitions from one species to another as their pollen becomes available (Stone *et al.* 1998).

Interspecific variation in flower morphology, the location of rewards, and the placement of sexual organs affect where on a pollinator's body pollen is picked up and how it is deposited on stigmas. The placement and timing of maturation of sexual organs within flowers, as well as in the sexual phase of flowers within and among inflorescences, also influence the likelihood of pollen transfer within or between flowers of the same plant and between plants of the same or different species. For example, in self-compatible plants, outcrossing from pollen transfer between conspecifics is facilitated by separation in the placement (herkogamy) and the timing of maturation (dichogamy) of sexual organs.

These floral, inflorescence, and whole-plant attributes also vary among individuals of a species, but intraspecific variation is smaller than interspecific variation. Nevertheless, this intraspecific variation supplies the "raw material" for pollinator-mediated selection and evolution. This chapter focuses on how pollinator responses to variation in abundance and traits *among* species in communities shape patterns of selection on these same traits *within* a species.

6.3 Plant-community effects on pollinator responses

Pollinator responses to interspecific variation in plant attributes are manifold. We focus on three responses—behavioural responses, demographic responses, and pollinator-community composition (pollinator species identity, abundance, and diversity)—and consider differences in responses to communities and to single-species populations. Note that although we contrast plant communities with single-species populations, these two ecological contexts represent extremes of a continuum; similar comparisons apply to species-rich versus species-poor communities.

6.3.1 Pollinator behavioural responses

Once a pollinator begins foraging in a community, three behavioural responses affect the reproductive success of a community member: the pollinator's functional response (i.e., the relation of consumption rate to resource density: Holling 1959), its preference for one or more species, and its fidelity (constancy) to a species while foraging.

Functional responses have been described in relation to the density of a single resource or of multiple resources. In general, a pollinator's consumption rate increases with the density of suitable flowers, as search time declines. At high floral density, the foraging rate reaches an asymptote because a pollinator's consumption becomes limited by handling time on flowers or it becomes satiated. Functional responses can increase linearly (type I response), in a saturating manner (type II), or in a sigmoidal fashion (type III) to the asymptote, with the latter two responses being more common (Holling 1959). Type II responses are typical of oligolectic consumers that use only a few plant species, whereas type III responses are typical of generalist consumers that switch between food hosts (Schenk and Bacher 2002). Type III functional responses also characterize consumers that move shorter distances in high-density patches (area-restricted foraging; Murdoch 1969; Keasar *et al.* 1996) or learn to find and handle hosts better once resources are encountered. In Section 6.4.2, we consider how a pollinator's functional response to the presence of multiple species may differ from its response to single species.

Pollinator preference and flower constancy are uniquely applicable to plant communities and irrelevant in single-species population. Preference occurs when a pollinator visits one species more frequently than expected based on its resource density and is probably shaped by interspecific variation in the resource value of a species to a pollinator. Preference clearly influences the relative visitation rate to flowers of different species. Constancy is a measure of the degree to which a pollinator restricts its visits to one plant species during a foraging bout (or several foraging bouts). The limited ability of pollinators to remember floral characters of more than one species at a time is thought to be a major contributing factor to constancy (Chittka *et al.* 1999). Constancy can also be economically beneficial when pollinators bypass species with lower rewards for ones with higher rewards (Gegear and Thomson 2004). Constant pollinators fly more frequently between plants of the same species and thus are better at transferring conspecific pollen than inconstant ones. Preference and constancy have been discussed extensively elsewhere (Waser 1986; Chittka *et al.* 1999; Chittka and Thomson 2001) and are largely beyond the scope of this chapter.

6.3.2 Pollinator demographic responses

Interspecific variation in reward quantity, quality, and type can affect the size of pollinator populations through numerical or aggregative responses. Numerical responses are changes in the per capita reproductive rate of consumers (e.g., pollinators) associated with resource availability (Holling 1965). Aggregative responses, on the other hand, involve the redistribution (i.e., movement) of consumers among patches (e.g., communities) of varying resource density (Murdoch 1977, Bosch and Waser 1999). As with functional responses, the form of numerical and aggregative responses can be of type I, II or III.

Surveys of pollinator populations indicate that pollinator abundance can track the availability of floral resources. For example, during a long-term study of euglossine bees in tropical forests, the abundance of all bees varied 4-fold and the abundance of individual species varied up to 14-fold among years (Roubik 2001). These dramatic fluctuations in population size were largely explained by the effect of climatic variation on flower production.

Whether pollinator abundance results from a numerical or aggregative response may be difficult to determine in any given situation, and indeed both responses may operate simultaneously. Perhaps, the most convincing evidence of a numerical response comes from changes in pollinator abundance between years of differing flower abundance in pollinator species that nest at or near foraging sites (Strickler et al. 1996; Gathmann and Tscharntke 2002) or return to the same foraging site year after year. For example, Minckley et al. (1994) found that the reproductive success of a specialist solitary bee increased three-fold during a year of abundant flowering in their sunflower host compared to years of limited flowering. The elevated reproductive success caused a three-fold increase in bee abundance the year after copious flowering. Potts et al. (2003) also found that variation in bee abundance among communities in Israel correlated positively with nectar abundance, though the relation was weak. Furthermore, variation in bee abundance during one year depended more strongly on floral abundance during the previous year than during the same year, suggesting a numerical response.

Aggregative responses should be most characteristic of long-distance foragers that can assess patch variation in resource pools. Studies of single plant species indicate that pollinators often select high-density or large patches of plants over low-density or small patches (e.g., Kunin 1997; Grindeland et al. 2005) and there is no reason to believe that pollinators cannot assess site variation in the combined resource value of multiple plant species. For example, social pollinators, such as the generalist honey bee (*Apis mellifera*), survey floral resources over broad areas (>10 km) and often focus their foraging on a small subset of highly profitable patches (Visscher and Seeley 1982; Beekman and Ratnieks 2000). Furthermore, Steffan-Dewenter et al. (2002) found that honey bees responded to variation in resource patchiness in a fragmented landscape (percentage of semi-natural habitat) only at large spatial scales (up to a 3000 m radius) whereas solitary bees responded only at small spatial scales (<750 m radius). Social animals that share information are particularly likely to exhibit aggregative responses, because few individuals incur the search cost of scouting for high-quality resource patches that all solitary foragers would sustain. Nevertheless, solitary foragers, such as trap-lining hummingbirds and some euglossine bees that travel long distances in search of resources, may also exhibit aggregative responses (Thomson et al. 1987).

The population size of a pollinator can be larger or smaller in communities than in single-species population, depending on whether the resource value is higher or lower in the former than in the latter. If floral rewards are not substitutable across species and some are non-essential or unprofitable, the combined resource value and hence pollinator population size may be lower in a plant community than in a population consisting entirely of a plant species with essential and/or profitable rewards. However, we argue that a community often offers more predictable resources than single-species populations. First, many pollinators are generalists and even specialist pollinators typically consume resources from several, usually closely-related species. Thus, multiple plant species may typically provide partially substitutable or complementary resources to pollinators and therefore offer a combined resource value that exceeds that of a single species. For example, Williams and Tepedino (2003) found that, despite added travel costs, the solitary bee, *Osmia lignaria*, regularly foraged on two plant species, even when one species grew further from nests or was rarer than the other, apparently because one species was more profitable as a source of pollen and the other as a source of nectar. Second, as noted above, interspecific displacement in diurnal and seasonal flowering extends the period of resource availability, which can then support a larger population of a pollinator that forages throughout the day or the season (Waser and Real 1979; Rathcke and Lacey 1985). For example, six mass-flowering dipterocarps share thrips pollinators, which persist at very low abundance between mass-flowering events and then build up rapidly in number as successive species flower (Ashton *et al.* 1988).

Last, plant species generally differ in the environmental conditions best suited for growth and reproduction (e.g., soil moisture, shade, nutrients)—conditions that vary both spatially and temporally within sites. Communities may therefore support higher plant densities and more rewards than single-species populations, because spatial environmental variation within sites allows for greater plant occupancy. Floral rewards may also be more predictable in communities if optimal environmental conditions for one species are followed during the next year by optimal conditions for another species (Chesson and Huntley 1989). High resource predictability should dampen fluctuations in pollinator population size among years.

6.3.3 Pollinator-community structure

Many plant species are visited by a diversity of pollinators, whether or not they grow alone or as members of a community. However, the assemblages of pollinators visiting different species are not identical. Therefore, communities probably present more abundant and diverse pollinator faunas than single-species populations, exposing plant species to altered patterns of visitation in the two settings. Indeed, Potts *et al.* (2003, 2004) found that bee species richness correlated strongly with floral species richness among communities and also varied with nectar resource diversity, an index of the diversity of nectar quantity and sugar concentration among plant species.

In concluding this review of pollinator responses, we emphasize that community effects on pollinator behaviour, demography, and community structure occur at multiple spatial and temporal scales. Behavioural and numerical responses are sensitive to the resource value of species of one community, aggregative responses vary with differences in resource value among communities, and pollinator communities reflect resource distribution within and among plant communities. From a temporal perspective, community context influences pollinator behaviour only on species that flower simultaneously, whereas community context can influence demography and pollinator-community structure even when species flower sequentially. In the case of sequential flowering, the effects of pollinator demographic responses and community structure on plant reproduction may differ between early- and late-flowering species.

6.4 Consequences of pollinator responses for selection on plant reproductive traits

A plant's reproductive success depends primarily on the quantity and quality of seeds it produces and sires. These fitness components are, at least partially, a function of pollen receipt and export and the latter, in turn, are determined by floral visitation rate and the effectiveness of pollinators at transferring pollen between compatible mates (Stebbins 1970; Herrera 1987; Chapters 2, 4, and 5). Floral visitation rate depends on pollinator abundance (demographic responses), consumption rate (functional response), and, in communities, the partitioning of pollinator visits among species (preference). The per-visit effectiveness of pollen transfer depends partially on pollinator constancy during a foraging bout, although preference increases pollinator fidelity to a preferred species.

The presence of multiple plant species can affect pollinator-mediated interspecific interactions in three ways. First, plants are subject to the exchange of heterospecific pollen, which can be viewed as a form of interference competition. The export of an individual's pollen to another species is always detrimental, because pollen is lost to incompatible mates (Waser 1978; Campbell and Motten 1985; Murcia and Feinsinger 1996). The receipt of heterospecific pollen can also interfere with fertilization by conspecific pollen (e.g., Waser 1978; Bell et al. 2005) or result in inferior hybrids (Chapter 18).

In addition, plant species may interact competitively or facilitatively for pollinator visits (Rathcke 1983). Facilitation has traditionally received less emphasis than competitive interactions, but has nevertheless been documented in a few instances (Thomson 1981; Johnson et al. 2003; Moeller 2004). Competition for pollinator visits is a form of exploitative competition for a shared resource (Waser 1983). For the purposes of this discussion, we define facilitation as the enhancement of visitation rate in the presence of multiple species (i.e., the opposite of exploitative competition). Our definition of facilitation differs from that of other researchers. For example, Rathcke defined facilitation as the net effect of beneficial increases in visitation rate and detrimental interspecific pollen exchange. We separate the effects of heterospecific pollen exchange from facilitation for pollinator visits, because they can have different selective consequences for plant reproductive traits. In a model of interspecific interactions that included enhanced visitation, heterospecific pollen transfer, and interspecific competition for seedling establishment, Feldman et al. (2004) defined facilitation as the demographic rescue from extinction of a rare species by a more common one when the latter boosts the former's population growth rate. We have elected to exclude competition for non-pollinator resources to distinguish it from pollinator-mediated selection.

Interspecific pollen transfer can occur simultaneously with either competition or facilitation for pollinator visits. The net effects of these interspecific interactions on reproductive success can obviously reduce or increase population growth rates of community members. However, even when growth rates remain the same, selection within species in response to interactions can occur if variation in reproductive success is linked to variation in phenotypes that influence the interaction.

Finally, we note that the net effect of interactions need not be reciprocal among species. For example, one species may suffer from the presence of others, but the latter may experience either no effect or benefit from the presence of the first (Ratchke 1983). Non-reciprocal effects may be common when species differ in relative density or rank order of pollinator preference.

6.4.1 Heterospecific pollen transfer

Selection to minimize heterospecific pollen transfer will always favour traits that promote pollinator constancy and may also favour interspecific divergence in the timing of pollen release during the day (Armbruster 1985; Stone et al. 1998), as both changes minimize temporal overlap in visits by the same pollinator and the chance that a pollinator carries a mixed pollen load. Selection may also favour species to forgo pollinator-dependent outcrossing to diminish the detrimental effects of heterospecific pollen. For example, self-pollination has evolved in *Arenaria uniflora* where it co-occurs with *Arenaria glabra*. This evolution probably reflects selection against interference from *A. glabra* pollen, because pollen receipt from *A. glabra* results in non-viable seed and reduces overall seed set (Fishman and Wyatt 1999).

Selection should also favour divergence among species in floral characters that affect where a pollinator contacts anthers and stigma. These traits include the position of sexual organs and features, such as flower shape and nectar guides, that manipulate how pollinators handle flowers (Grant 1950; Waser 1983).

In short, the detrimental effects of heterospecific pollen transfer favour interspecific differentiation in the spatial and temporal mechanics of pollen transfer, at the same time that selection favours convergence in these same traits among individuals within a species.

6.4.2 Pollinator visitation rate

A pollinator's functional and demographic responses to community resource density may or may not translate into higher visitation rates per flower to a member species relative to that which it experiences in isolation (Bosch and Waser 1999). A simple graphical model illustrates this point and

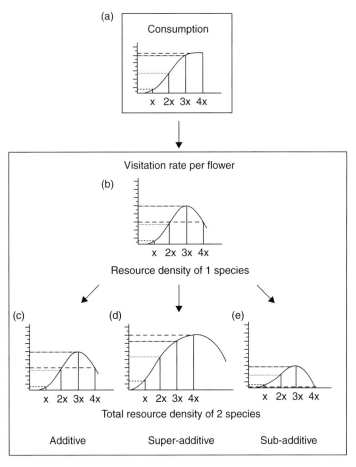

Figure 6.2 Relations between the pollinator functional response and per-flower visitation rate in a single-species population and the per-flower visitation rates of co-flowering species in a community. (a) Functional response and (b) visitation rate per flower by a pollinator to resource density in a single-species plant population. The visitation rate per flower is positively density dependent at low resource levels and negatively density dependent at high resource levels. In a two-species community, the visitation rate per-flower to both species can be (c) an additive, (d) super-additive, or (e) sub-additive function of the visitation to each species alone.

shows the kind of data that would elucidate response differences between communities and single-species populations. Consider, for example, a pollinator's functional response to the resource density of a single species growing alone (Fig. 6.2a). The translation of this response into visitation rate per flower is shown in Fig. 6.2b. The visitation rate first increases (i.e., is positively density-dependent at low resource density), but then decreases as the functional response saturates (i.e., negatively density dependent at high resource density). This transition from positive to negative density dependence is expected regardless of the shape of the functional response. Now consider a community of both species. If the functional responses and visitation rate per flower for both species in isolation are identical, then visitation rate on both species combined may remain the same (Fig. 6.2c; additive model). This result could arise if the two species provide perfectly substitutable rewards and pollinators do not discriminate against either species. In this case, the visitation rate per flower on both species at density x (total density $2x$) will be higher than the rate in isolated populations of either species at density x, but the same as the visitation rate in isolated populations of either species at density $2x$. If the two species have resource densities of x and $3x$, respectively, in a community, visitation rate to the rare species (x) is higher in the presence of the common one, but the latter now suffers from the presence of the first. Still, each species would do equally well in isolated populations at resource densities of $4x$.

The visitation rate per flower in two-species communities need not combine additively. It might be super-additive (Fig. 6.2d), if, for example, the two species provide complementary resources that are required by a pollinator. Alternatively, the visitation rate could be sub-additive, relative to the rate on each species alone (Fig. 6.2e), if a pollinator must adopt a different search image or handling method on each species and is therefore less efficient at obtaining resources from both. In the super-additive case, the rare (density x) and common ($3x$) species both benefit in the presence of the other (facilitation; i.e., visitation rate at the combined density, $4x$, is higher than at the density of each species alone, x and $3x$). In the sub-additive case, both species suffer in the presence of the other (competition).

The same logic can be applied if the functional responses and per-flower visitation rates in isolation differ between the two species (e.g., one species produces lower-quality rewards). In this case, the more profitable species is likely to suffer in the presence of the other. In contrast, the less profitable one may benefit from the first, unless a pollinator can discriminate between the two species and prefers the profitable species. In the latter case, the cost to the profitable species from the presence of the less profitable one will diminish or disappear, as will the benefit to the less profitable species.

The graphical model can also be applied to numerical and aggregative responses to co-flowering plants. Because functional and demographic responses can operate simultaneously, their net effect on visitation rate per flower may differ from each response taken alone. For example, if a demographic response to the combined resource value of a community is very strong, the number of pollinators in a patch will also be high and the resource density required to satiate an individual pollinator will be higher than if fewer pollinators were present. In other words, pollinators compete for plant resources. In turn, the visitation rate per flower will peak at a higher resource density (super-additive) and both species may be facilitated by the presence of the other.

Functional and demographic responses differ in an important way. As noted previously, the functional response is sensitive to only the combined resources of co-flowering species, whereas demographic responses can extend across sequentially flowering species, and, in the case of numerical responses, across years. Because demographic responses extend through time, the shape of the visitation function for a species should not differ between isolated populations and communities, except that, in communities, increases in pollinator abundance earlier during the season will elevate the visitation rate for a late-flowering species at low density (larger y-intercept, Fig. 6.2). Thus, facilitation should occur most often in species that succeed others in flowering (Waser and Real 1979), whereas early-season species should experience benefits only if pollinators or their offspring stay or return to the same site between years.

Although pollinator responses to resource density of single species have been demonstrated often, very little is known about the actual shape of the responses and even less is known about pollinator responses to the combined resources of multi-species communities. Only comparative studies of a focal species at varying density, both in isolation and in combination with other species, can provide the relevant data. If species compete for pollinator visits (e.g., sub-additive response), selection may result in an "arms race", whereby each species evolves to become more profitable and preferred by a pollinator. Selection could also favour interspecific divergence in flowering time to minimize direct competition for pollinator visits. When species interact facilitatively, functional and demographic responses can also generate varying forms of selection. For species that flower simultaneously, strong pollinator responses may result in selection for convergence in attractive characters (Brown and Kodric-Brown 1979; Johnson et al. 2003a; Chapter 8). On the other hand, for sequentially flowering species demographic responses need not generate selection for either convergence or divergence in reproductive traits.

Patterns of selection are likely to differ between co-flowering species when the effects of competition or facilitation are asymmetric. For example, when species' rewards are substitutable and pollinators do not discriminate between them, a rare species can experience facilitation from the strong functional response engendered by a common species (Feldman et al. 2004; see Fig. 6.2c). Selection may then favour convergence in floral characters and flowering phenology by the rare species on the common one. When species differ in profitability, less rewarding species may be selected to avoid direct competition with profitable ones through shifts in seasonal phenology, to mimic the outward appearance of profitable species (Gumbert et al. 2001; Johnson et al. 2003a), or selection may favour floral mechanisms that enhance self-pollination such that less rewarding species forgo entirely their reliance on pollinators (Lloyd 1965, 1992; Chapter 10). Mimicry can cause pollinators to mistake less rewarding species for rewarding ones. For example, rewardless species receive more visits when growing near rewarding species that they resemble in flower colour (Johnson et al. 2003b). Asymmetry in benefit between rewardless and rewarding plant species is very much like the asymmetry experienced by palatable versus unpalatable prey that are attacked by the same predators. Both interactions select for mimicry of the beneficiary on the benefactor (Johnson 1994), and the benefit to the "mimic" is expected to diminish as its abundance relative to the "model" increases.

6.4.3 Pollinator-community composition

If pollinator assemblages differ between plant communities and single-species populations, a plant species may encounter a new flower visitor in the presence of other plant hosts (Ginsberg 1983). For example, in agricultural landscapes, pollinators attracted to native plants are likely to spread to neighbouring crops and vice versa (Kremen et al. 2002; Ricketts et al. 2004; see also Memmott and Waser 2002; Chapter 9). Similarly, in polyculture systems, pollinators attracted to one crop can visit interplanted crops (Jones and Gillett 2005).

Just as different pathogen and herbivore species are deterred by, and act as selective agents for, different plant defences, different pollinators can favour different trait optima in attractive characters and rewards (Galen et al. 1987; Campbell et al. 1997). Thus, intra- and interspecific variation in floral phenotype can correlate with changes in the dominant pollinator at small and large spatial scales (e.g., Galen 1989; Schemske and Bradshaw 1999). However, the role of spatial variation in plant (and pollinator) community structure in the evolution of plant reproductive characters is largely unknown (Chapter 15).

Higher pollinator diversity may also expose plant species to more variable selection in communities than in single-species populations. Highly variable selection generally slows adaptation to specific biotic or abiotic factors and favours a generalized phenotype (Levins 1968). On the other hand, greater pollinator diversity might expose large inequalities in the effectiveness of different pollinator species and so strengthen selection for adaptations that increase preferential visitation by the best pollinators and exclude the worst pollinators (but see Aigner 2004). For example, does a plant species pollinated most effectively by long-tongued pollinators evolve greater barriers to short-tongued pollinators (e.g., longer or thicker corollas) when it co-flowers with species that attract short-tongued pollinators?

6.5. Community context and mating-system evolution in *Clarkia*

Studies of the ecological consequences of pollinator sharing have focused mainly on the negative effects of competition for pollinator visits and interspecific pollen transfer. In contrast, only a few studies have investigated the ecological and evolutionary consequences of positive interactions. In this section, we relay results from our own research on positive interactions for pollination between the annual plant, *C. xantiana*, and its congeners. First, we address the effects of pollinator sharing by *Clarkia* species on visitation rate and pollen limitation in *C. xantiana* and assess the possible influences of pollinator demography, behaviour, and community structure on its reproductive success. We then describe experimental studies on the evolutionary consequences of pollinator responses to *Clarkia* communities for mating-system evolution in *C. xantiana*. Although our work is certainly incomplete, we use it to illustrate a research approach that merges ecological and evolutionary perspectives on floral biology in plant communities. We end with a discussion on intriguing biogeographic patterns of mating-system distribution that suggest facilitation throughout the genus.

6.5.1 Study system

Clarkia comprises 44 annual species endemic to western North America and reaches its highest diversity in California (Lewis and Lewis 1955). Three attributes of the genus make it especially suitable for community-level studies of pollination. First, there is considerable sympatry among taxa and co-existence is common over small spatial scales (Lewis and Lewis 1995). Second, virtually all *Clarkia* are bee pollinated, and, in most species, the most common visitors are a small set of specialized solitary bees (MacSwain et al. 1973). Thus, the limited size of the plant-pollinator network makes the system tractable for observational and manipulative studies. Finally, most *Clarkia* exhibit a great deal of intraspecific variation in floral traits, which makes them ideal for microevolutionary studies. Our work has concentrated on a focal species, *C. xantiana*, which includes a predominantly outcrossing subspecies (ssp. *xantiana*) and a parapatric, predominantly selfing subspecies (ssp. *parviflora*) (Eckhart and Geber 1999).

6.5.2 Pollinator responses to *Clarkia* communities

Across most of the range of the outcrossing subspecies, 60% of *C. xantiana* populations co-occur with congeners (*Clarkia cylindrica*, *Clarkia unguiculata*, *Clarkia speciosa*). Congeners overlap in flowering time, but flowering modes tend to be staggered, with *C. xantiana* being the last to flower (Moeller 2004). The flight period of generalist and specialist bees extends through the flowering of several *Clarkia* species, so that pollinators can benefit from the prolonged period of resource availability in multi-species communities (MacSwain et al. 1973).

Comparative studies and experimental manipulations of community composition during three years have shown that *C. xantiana* receives more frequent pollinator visits and suffers less pollen limitation of seed set in the presence of congeners than in their absence (Fig. 6.3; Moeller 2004, 2005; Moeller and Geber 2005). The difference in pollinator availability stems principally from a greater abundance of specialists, rather than a change in bee species richness, in communities (Moeller 2005). In turn, higher pollinator abundance enhances pollen receipt, as hand-pollination experiments conducted during three years indicate less pollen limitation of seed production in *C. xantiana* in the presence versus the absence of congeners (Fig. 6.3; Moeller 2004).

Numerical and/or aggregative pollinator responses to *Clarkia* communities are likely to be the causes of higher visitation rates and seed production in *C. xantiana* populations coexisting with

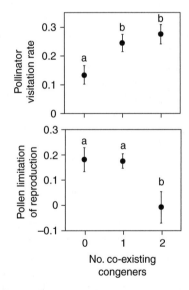

Figure 6.3 The effect of co-existing *Clarkia* congeners on pollinator availability (number of bee visits per plant per site census) and pollen limitation of reproduction in *Clarkia xantiana*. Pollen limitation was measured as the difference in seed set between open-pollinated flowers with or without supplemental hand-pollination, standardized by the seed set of flowers with supplemental pollen. Low values indicate little pollen limitation. Plots show least-square means (± SE) from an ANOVA that accounted for the effects of plant population size and population density. Different lower-case letters indicate a significant difference between factor levels based on the Tukey-Kramer test (from Moeller 2004).

congeners. Numerical responses may be important because the specialist bees nest within *Clarkia* communities and tend to forage locally (e.g., Burdick and Torchio 1959). Consequently, the resource value of the entire community probably affects the per capita reproductive rate of bees. Pollinator abundance may also be influenced by aggregative responses over hundreds of metres to a few kilometres, a scale which typically encompasses multiple, distinct *Clarkia* communities. Finally, the possibility that stronger functional responses contribute to higher visitation rates to *C. xantiana* in diverse communities cannot be excluded. However, the staggered flowering periods of *Clarkia* species should limit the strength of these responses.

Higher visitation rates do not appear to reflect a preference by bees for *C. xantiana*. The two most common visitors, *Lasioglossum pullilabre* and *Hesperapis regularis*, visited *C. xantiana* 26 and 16% less often than expected given its frequency (51%) in mixed patches with *C. speciosa* (L. Evanhoe and M.A. Geber unpublished data). Undervisitation of *C. xantiana* may be balanced by relatively low interspecific pollen transfer for two reasons. First, the staggered flowering of *Clarkia* species should, by itself, minimize heterospecific pollen transfer during the late portion of *C. xantiana*'s flowering. Second, even when species overlap in flowering time and grow in mixed patches, a large percentage (83%) of foraging transitions between plants by specialist bees were between conspecifics.

Pollen limitation of seed set also tends to be lower in large *C. xantiana* populations, particularly those of high density, suggesting strong Allee effects where plants are scarce (Moeller 2004). Based on these and the previous results, we predicted that facilitative effects of congeners mitigate Allee effects and reduce extinction risk in small populations. A survey of 85 populations showed that small populations of *C. xantiana* occur more often with congeners whereas populations isolated from congeners tend to be large (Moeller 2004).

6.5.3 Consequences of community context for selection on the mating system

Mating-system variation among *C. xantiana* populations correlates geographically with variation in population size and *Clarkia* species diversity: selfing populations (ssp. *parviflora*) tend to be smaller, of lower density, and occur largely outside the range of other outcrossing *Clarkia* species (Moeller and Geber 2005). We tested for the effects of population size and plant-community context on patterns of selection on two mating-system traits: herkogamy and protandry. We introduced large and small experimental populations of *C. xantiana* into sites where congeners were present or absent. Experimental populations consisted of plants derived from crosses within and between the two subspecies and exhibited a wide range of floral phenotypes. We found that selection strongly favoured traits that promote self-pollination (reduced herkogamy and protandry) in small, but not large, populations (Moeller and Geber 2005). In small populations, mating-system traits experienced weaker selection when congeners were present (Fig. 6.4a). The strength of selection on herkogamy depended largely on pollinator availability, which was influenced by community context (Fig. 6.4b; Moeller and Geber 2005).

These results are consistent with the hypothesis that reproductive assurance is an important factor shaping mating-system evolution in this system. This causal mechanism for the evolution of selfing is further supported by geographic variation in pollinator abundance and floral traits in *C. xantiana* (see Chapter 10). Pollinator abundance (on a per flower basis) is 4.4 times higher in ssp. *xantiana* than in ssp. *parviflora* populations and specialist solitary bees are absent from the exclusive range of ssp. *parviflora* (Fausto et al. 2001; Moeller 2006). Furthermore, common-garden studies of subspecies *xantiana* show that genetic differences among populations in herkogamy are correlated with pollinator abundance, particularly of specialists, and with the composition of pollinator communities (Fig. 6.4c; Moeller 2006).

6.5.4 Biogeographic patterns of mating-system variation and plant-community diversity

Patterns of reproductive character variation across a species' range may extend to larger biogeographic patterns, depending on whether the nature of interactions between a set of pollinator-sharing

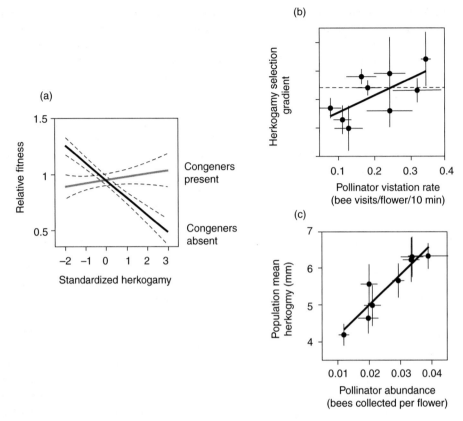

Figure 6.4 Evolutionary consequences of community context for herkogamy (anther-stigma separation). (a) Directional selection gradients (± SE) for herkogamy in small experimental populations introduced into communities where *Clarkia* congeners (*C. cylindrica* and *C. unguiculata*) were present or absent (from Moeller and Geber 2005). (b) The strength of selection on herkogamy among experimental populations varies significantly with bee pollinator visitation rates (from Moeller and Geber 2005). (c) Genetic differences in herkogamy among populations of subspecies *xantiana* correlate significantly with mean pollinator abundance (Moeller, 2006).

plant species is largely consistent across geographic regions. For example, if facilitative interactions are pervasive in *Clarkia*, outcrossing taxa (species or subspecies) should be most common in regions of high species diversity, and vice versa for self-pollinating taxa. To test this hypothesis we classified the mating system of *Clarkia* taxa based on petal size and herkogamy. Although taxa are neither exclusively outcrossing nor selfing (all taxa are self-compatible), there is a strong bimodal distribution in petal size that correlates with herkogamy (Fig. 6.5A), and this bimodality is consistent with the view that taxa are modally outcrossing or selfing (Wyatt 1988). We then used data on the geographic distribution of *Clarkia* taxa in each of 30 floristic provinces in the western United States: 24 provinces in California and 6 neighbouring provinces in the Pacific Northwest and Baja California, Mexico (Plate 1; Lewis and Lewis 1955; Hickman 1993) to examine the relation between mating-system frequency and species diversity. The geographic data do not indicate whether *Clarkia* species co-exist on a local scale, but co-occurrence is common in the genus (Lewis and Lewis 1955). As predicted, the frequency of selfing taxa is lowest in the centre of the genus' range where species diversity is highest (e.g., Sierra Nevada foothills, Outer Coast Range), and highest at the periphery of the range where diversity is lowest (e.g., Great Basin, Pacific Northwest, Baja California) (Plate 1).

The number of selfing taxa varies linearly with the number of outcrossing taxa across biogeographic provinces, with a slope significantly <1 (Fig. 6.5B). Thus, selfing taxa are overrepresented in regions

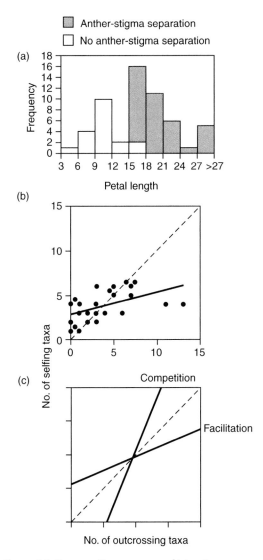

Figure 6.5 Biogeographic consequences of interactions among pollinator-sharing *Clarkia* species. (a) Petal size and herkogamy vary bimodally among *Clarkia* taxa, with taxa at the lower end of the distribution being primarily selfing and those at the upper end primarily outcrossing. (b) The relation between the number of selfing and outcrossing *Clarkia* taxa in 30 floristic provinces of the western United States (middle panel; see Plate 1 for all provinces, except a portion of the Rocky Mountains). *Clarkia* species or subspecies that contain both outcrossing and selfing forms, were assigned a value of 0.5 for each form. The slope is significantly less than one ($y = 0.25x + 2.84$; 95% confidence interval for slope: 0.08 – 0.42). (c) Expected consequences of competition and facilitation for biogeographic patterns of mating-system distribution. If competition among *Clarkia* species predominates across communities, selfing taxa should be overrepresented in regions with high *Clarkia* diversity (slope > 1), whereas the reverse should hold if species interact facilitatively (slope < 1). A mixture of competition and facilitation across communities would result in a slope near 1.

where outcrossing taxa are uncommon and vice versa. If instead co-occurring *Clarkia* species compete consistently for pollination, the slope of the relation should exceed 1 (Fig. 6.5c). If interactions vary from competitive to facilitative, the relation would not differ significantly from the 1:1 line. Although only correlative, the biogeographic pattern of mating-system distribution in *Clarkia* suggests that community context may have important and wide-ranging effects on patterns of floral diversity. The predictions arising from this exercise may be profitably extended to other taxonomic groups for which co-existence and pollinator sharing are common.

6.6 Conclusions and future directions

Understanding of the community context of pollination and floral evolution is in its infancy. Ecological studies of pollinator-mediated plant interactions most often focus on pollinator behaviour (e.g., preferences, flower constancy) at small spatial scales, through experimental manipulation of the abundance and dispersion of a target species. In contrast, functional and demographic responses have received little attention. Functional responses can also be evaluated through manipulations within communities, but will most likely require experimental or observational comparisons among communities. Demographic responses will continue to be overlooked unless the spatial and temporal scales of study are expanded.

Published work on pollinator-sharing plant species suggests that interactions among plant species are more often competitive (exploitative or interference) or neutral than facilitative. Whether this apparent bias in results is truly representative of the real world or is due to an absence of data is unclear. For example, co-flowering plant species may exhibit evidence of interference competition at a local scale (over metres) but floral visitation may be considerably higher in the presence of pollinator-sharing species, so that reproductive success is still greater in communities than in single-species populations.

Ecological studies on interactions among pollinator-sharing species far outnumber evolutionary studies on the selective consequences of these interactions. Even though most studies of natural selection on reproductive characters are, in fact, conducted in plant communities, the role of co-occurring species in shaping patterns of selection in a focal species is not isolated from other selective causes. The few evolutionary studies on the community context of pollination have revealed fascinating results. There is evidence of both competitive and facilitative interactions among pollinator-sharing taxa and these interactions appear to affect selection on traits involved in pollinator attraction and the functional fit between pollinator and flower (Armbruster 1985; Caruso 2000; Hansen *et al.* 2000) and traits that influence the mating system (Fishman and Wyatt 1999; Moeller and Geber 2005).

Evolutionary studies of the community context of pollinator-mediated selection will be most useful when combined with ecological studies of the mechanisms underlying positive or negative interspecific interactions. For example, negative interactions among pollinator-sharing species can be caused by exploitative competition for pollinator visits or interference competition via heterospecific pollen transfer. These two mechanisms cause selection on different traits (e.g., traits affecting pollinator preferences versus constancy), which can be tested only through evolutionary studies. Systems in which plant-community context varies geographically among populations of a focal plant species provide a particularly useful arena for evolutionary studies, because they allow for the combination of process-oriented studies of pollination and natural selection with pattern-oriented studies of population differentiation and character evolution.

Regardless of general patterns, it is important to recognize that the nature of reproductive interactions between species may vary over time. The long-term studies that are necessary to evaluate variability in the nature of pollinator-mediated interactions and in patterns of selection on reproductive traits, as a function of community context, have never been conducted.

In addition to their relevance to understanding plant reproductive ecology and floral evolution, the effects of community context on pollinators have important implications for conservation biology. Food-web analyses have indicated strong linkages among plant and pollinator species, but the implications of these results for the stability of plant-pollinator networks remain less clear (Chapter 9). Long-term monitoring of pollinator populations in natural or manipulated communities can provide important information on whether pollinator populations are in decline and on the importance of multi-species communities to the maintenance of pollinator populations and species diversity (Kearns *et al.* 1998; Chapter 15).

Acknowledgements

We wish to acknowledge the inspiration provided by B. Rathcke's and N. Waser's early work on plant-pollinator interactions in plant communities. We also thank L. D. Harder, S. C. H. Barrett, D. P. Vázquez, M. V. Price, N. M. Waser, and an anonymous reviewer for very helpful comments on an earlier draft. This work was supported by NSF funding to M. Geber (DEB-0515428) and to P. Tiffin and D. Moeller (DEB-0515466).

References

Aigner PA (2004). Floral specialization without trade-offs: optimal corolla flare in contrasting pollination environments. *Ecology*, **85**, 2560–9.

Armbruster WS (1985). Patterns of character divergence and the evolution of reproductive ecotypes of *Dalechampia scandens* (Euphorbiaceae). *Evolution* **39**, 733–52.

Armbruster WS, Edwards ME, and Debevic EM (1994). Floral character displacement generates assemblage structure of western Australian triggerplants (*Stylidium*). *Ecology*, **75**, 315–29.

Ashton PS, Givnish TJ, and Appanah S (1988). Staggered flowering in the Dipterocarpaceae: new insights into floral induction and the evolution of mast fruiting in the aseasonal tropics. *American Naturalist*, **132**, 44–66.

Bascompte J, Jordano P, Melián CJ, and Oleson JM (2003). The nested assembly of plant-animal mutualistic networks. *Proceedings of the National Academy of Sciences, the United States of America*, **100**, 9383–7.

Beekman M and Ratnieks FLW (2000). Long-range foraging by the honey-bee, *Apis mellifera* L. *Functional Ecology*, **14**, 490–6.

Bell JM, Karron JD, and Mitchell RJ (2005). Interspecific competition lowers seed production and outcrossing in *Mimulus ringens*. *Ecology*, **86**, 762–71.

Bosch M and Waser NM (1999). Effects of local density on pollination and reproduction in *Delphinium nuttallianum* and *Aconitum columbianum*. *American Journal of Botany*, **86**, 871–9.

Brown BJ, Mitchell RJ, and Graham SA (2002). Competition for pollination between an invasive species (purple loosestrife) and a native congener. *Ecology*, **83**, 2328–36.

Brown JH and Kodric-Brown A (1979). Convergence, competition, and mimicry in a temperate community of hummingbird-pollinated flowers. *Ecology*, **60**, 1022–35.

Burdick DJ and Torchio PF (1959). Notes on the biology of *Hesperapis regularis* (Cresson). *Journal of the Kansas Entomological Society*, **32**, 83–7.

Campbell DR and Motten AF (1985). The mechanism of competition for pollination between two forest herbs. *Ecology*, **66**, 554–63.

Campbell DR, Waser NM, and Melendez-Ackerman EJ (1997). Analyzing pollinator-mediated selection in a plant hybrid zone: Hummingbird visitation patterns on three spatial scales. *American Naturalist*, **149**, 295–315.

Caruso CM (2000). Competition for pollination influences selection on floral traits of *Ipomopsis aggregata*. *Evolution*, **54**, 1546–57.

Chesson P and Huntley N (1989). Short-term instabilities and long-term community dynamics. *Trends in Ecology and Evolution*, **4**, 293–8.

Chittka L and Thomson JD, eds. (2001). *Cognitive ecology of pollination: animal behaviour and floral evolution*. Cambridge University Press, Cambridge, U.K.

Chittka L, Thomson JD, and Waser NM (1999). Flower constancy, insect psychology, and plant evolution. *Naturewissenschaften*, **86**, 361–77.

Dicks LV, Corbet SA, and Pywell RF (2002). Compartmentalization in plant-insect flower visitor webs. *Journal of Animal Ecology*, **71**, 32–43.

Eckhart VM and Geber MA (1999). Character variation and geographic distribution of *Clarkia xantiana* A. Gray (Onagraceae): flowers and phenology distinguish two subspecies. *Madroño*, **46**, 117–25.

Fausto JA, Eckhart VM, and Geber MA (2001). Reproductive assurance and the evolutionary ecology of self-pollination in *Clarkia xantiana* (Onagraceae). *American Journal of Botany*, **88**, 1794–800.

Feldman TS, Morris WF, and Wilson WG (2004). When can two plant species facilitate each other's pollination? *Oikos*, **105**, 197–207.

Fenster CB, Armbruster WS, Wilson P, Dudash MR, and Thomson JD (2004). Pollination syndromes and floral specialization. *Annual Review of Ecology, Evolution and Systematics*, **35**, 375–403.

Fishman L and Wyatt R (1999). Pollinator-mediated competition, reproductive character displacement, and the evolution of selfing in *Arenaria uniflora* (Caryophyllaceae). *Evolution*, **53**, 1723–33.

Galen C (1989). Measuring pollinator-mediated selection on morphometric floral traits: bumble bees and the alpine sky pilot, *Polemonium viscosum*. *Evolution*, **43**, 882–90.

Galen C, Zimmer KA, and Newport MA (1987). Pollination in floral scent morphs of *Polemonium viscosum*: A mechanism for disruptive selection on flower size. *Evolution*, **41**, 599–606.

Gathmann A, and Tscharntke T (2002). Foraging ranges of solitary bees. *Journal of Animal Ecology*, **71**, 757–64.

Gegear RJ and Thomson JD (2004). Does the flower constancy of bumble bees reflect foraging economics? *Ethology*, **110**, 793–805.

Ginsberg HS (1983). Foraging ecology of bees in an old field. *Ecology*, **64**, 165–75.

Grant V (1950). The flower constancy of bees. *Botanical Review*, **16**, 379–98.

Grindeland JM, Sletvold N, and Ims RA (2005). Effects of floral display and plant density on pollinator visitation rate in a natural population of *Digitalis purpurea*. *Functional Ecology*, **19**, 383–90.

Gumbert A, Kunze J, and Chittka L (2001). Floral colour diversity in plant communities, bee colour space and a null model. *Proceedings of the Royal Society of London, Series B*, **266**, 1711–16.

Hanson TF, Armbruster WS, and Antonsen L (2000). Comparative analysis of character displacement and spatial adaptations as illustrated by the evolution of *Dalechampia* blossoms. *American Naturalist*, **156**, S17-34.

Herrera CM (1987). Components of pollination "quality": comparative analysis of a diverse insect assemblage. *Oikos*, **50**, 79–90.

Hickman JC, ed. (1993). *The Jepson manual: Higher plants of California*. University of California Press, Berkeley, CA.

Holling MP (1959). Some characteristics of simple types of predation and parasitism. *Canadian Entomologist*, **91**, 385–98.

Holling MP (1965). The functional response of predators to prey density and its role in mimicry and population regulation. *Memoirs of Entomological Society of Canada*, No. 45, 3–60.

Johnson S D. (1994). Evidence for Batesian mimicry in a butterfly-pollinated orchid. *Biological Journal of the Linnaen Society*, **24**, 225–235.

Johnson SD, Alexandersson R, and Linder HP (2003a). Experimental and phylogenetic evidence for floral mimicry in a guild of fly-pollinated plants. *Biological Journal of the Linnean Society*, **80**, 289–304.

Johnson SD, Peter CI, Nilsson LA, and Ågren J (2003b). Pollination success in a deceptive orchid is enhanced by co-occurring rewarding magnet plants. *Ecology*, **84**, 2919–27.

Jones GA, and Gillett JL (2005). Intercropping with sunflowers to attract beneficial insects in organic agriculture. *Florida Entomologist*, **88**, 91–6.

Kearns CA, Inouye DW, and Waser NM (1998). Endangered mutualisms: the conservation of plant-pollinator interactions. *Annual Review of Ecology and Systematics*, **29**, 83–112.

Keasar T, Shmida A, and Motro U (1996). Innate movement rules in foraging bees: Flight distances are affected by recent rewards and are correlated with choice of flower type. *Behavioral Ecology and Sociobiology*, **39**, 381–8.

Kremen C, Williams NM, and Thorp RW (2002). Crop pollination from native bees at risk from agricultural intensification. *Proceedings of the National Academy of Sciences of the United States of America*, **99**, 16812–6.

Kunin WE (1997). Population size and density effects in pollination: pollinator foraging and plant reproductive success in experimental arrays of *Brassica kaber*. *Journal of Ecology*, **85**, 225–34.

Laverty TM (1992). Plant interactions for pollinator visits: a test for magnet species effect. *Oecologia*, **89**, 502–8.

Levins R (1968). *Evolution in changing environments: some theoretical explorations*. Princeton University Press, Princeton, NJ.

Lewis H and Lewis ME (1955). *The genus Clarkia*. University of California Press, Berkeley, CA.

Lloyd DG (1965). Evolution of self-compatibility and racial differentiation in *Leavenworthia* (Cruciferae). *Contributions from the Gray Herbarium of Harvard University*, **195**, 3–134.

Lloyd DG (1992). Self- and cross-fertilization in plants. II. The selection of self-fertilization. *International Journal of Plant Sciences*, **153**, 370–80.

MacSwain JW, Raven PH, and Thorp RW (1973). Comparative behavior of bees and Onagraceae. IV. *Clarkia* bees of the western United States. *University of California Publications in Entomology*, **70**, 1–80.

Memmott J and Waser NM (2002). Integration of alien plants into a native flower-pollinator visitation web. *Proceedings of the Royal Society of London, Series B*, **269**, 2395–9.

Minckley RL, Wcislo WT, Yanega D, and Buchmann SL (1994). Behavior and phenology of a specialist bee (*Dieunomia*) and sunflower (*Helianthus*) pollen availability. *Ecology*, **75**, 1406–19.

Moeller DA (2004). Facilitative interactions among plants via shared pollinators. *Ecology*, **85**, 3289–301.

Moeller DA (2005). Pollinator community structure and sources of spatial variation in plant-pollinator interactions in *Clarkia xantiana* ssp. *xantiana*. *Oecologia*, **142**, 28–37.

Moeller DA (2006). Geographic structure of pollinator communities, reproductive assurance, and the evolution of self-pollination. *Ecology*, **87**, 1510–22.

Moeller DA and Geber MA (2005). Ecological context of the evolution of self-pollination in *Clarkia xantiana*: population size, plant communities, and reproductive assurance. *Evolution*, **59**, 786–99.

Murcia C and Feinsinger P (1996). Interspecific pollen loss by hummingbirds visiting flower mixtures: effects of floral architecture. *Ecology*, **77**, 550–60.

Murdoch WW (1969). Switching in general predators. Experiments on predator specificity and stability of prey populations. *Ecological Monographs*, **39**, 335–54.

Murdoch WW (1977). Stabilizing effects of spatial heterogeneity in predator-prey systems. *Theoretical Population Biology*, **11**, 252–73.

Potts SG, Vulliamy B, Roberts S, et al. (2004). Nectar resource diversity organizes flower-visitor community structure. *Entomologia Experimentalis et Applicata*, **113**, 103–7.

Rathcke B (1983). Competition and facilitation among plants for pollination. In L Real, ed. *Pollination biology*, pp. 305–25. Academic Press, Orlando, FL.

Rathcke B and Lacey EP (1985). Phenological patterns of terrestrial plants. *Annual Review of Ecology and Systematics*, **16**, 179–214.

Ricketts TH, Daily GC, Ehrlich PR, and Michener CD (2004). Economic value of tropical forest to coffee production. *Proceedings of the National Academy of Sciences of the United States of America*, **101**, 12579–81.

Roubik DW (2001). Ups and downs in pollinator populations: when is there a decline? *Conservation Ecology*, **5**, 2. [online] http://www.consecol.org/vol5/iss1/art2.

Schemske DW, and Bradshaw HD Jr (1999). Pollinator preference and the evolution of floral traits in monkeyflower (*Mimulus*). *Proceedings of the National Academy of Sciences of the United States America*, **96**, 11910–5.

Schenk D, and Bacher S (2002). Functional response of a generalist insect predator to one of its prey species in the field. *Journal of Animal Ecology*, **71**, 524–31.

Stebbins GL (1970). Adaptive radiation of reproductive characteristics in Angiosperms I: Pollination mechanisms. *Annual Review of Ecology and Systematics*, **1**, 307–26.

Steffan-Dewenter I, Münzenberg U, Bürger C, Thies C, and Tscharntke T (2002). Scale-dependent effects of landscape context on three pollinator guilds. *Ecology*, **83**, 1421–1432.

Stone GN, Willmer P, and Rowe JA (1998). Partitioning of pollinators during flowering in an African *Acacia* community. *Ecology*, **79**, 2808–27.

Strickler K, Scott VL, and Fischer RL (1996). Comparative nesting ecology of two sympatric leafcutting bees that differ in body size (Hymenoptera: Megachilidae). *Journal of the Kansas Entomological Society*, **69**, 26–44.

Tilman D (1982). *Resource competition and community structure*. Princeton University Press, Princeton, NJ.

Thomson JD (1981). Spatial and temporal components of resource assessment by flower-feeding insects. *Journal of Animal Ecology*, **50**, 49–59.

Thomson JD, Peterson SC, and Harder LD (1987). Response of traplining bumble bees to competition experiments—Shifts in feeding location and efficiency. *Oecologia*, **71**, 295–300.

Visscher PK and Seeley TD (1982). Foraging strategy of honey bee colonies in a temperate deciduous forest. *Ecology*, **63**, 1790–801.

Waser NM (1978). Interspecific pollen transfer and competition between co-occurring plant species. *Oecologia*, **36**, 223–36.

Waser NM (1983). Competition for pollination and floral character differences among sympatric plant species: a review of evidence. In CE Jones and RJ Little, eds. *Handbook of experimental pollination biology*, pp. 277–93. Van Nostrand Reinhold, New York, NY, USA.

Waser NM (1986). Flower constancy: definition, cause, and measurement. *American Naturalist*. **127**, 593–603.

Waser NM and Ollerton, eds. (2006). *Plant-pollinator interactions: From specialization to generalization*. University of Chicago Press. Chicago, IL, USA.

Waser NM and Real L (1979). Effective mutualism between sequentially flowering plant species. *Nature*, **281**, 670–2.

Williams NM and Tepedino VJ (2003). Consistent mixing of near and distant resources in foraging bouts by the solitary mason bee, *Osmia lignaria*. *Behavioral Ecology*, **14**, 141–9.

Wyatt R (1988). Phylogenetic aspects of the evolution of self-pollination. In LD Gottlieb and SK Jain, eds. *Plant evolutionary biology*, pp. 109–31. Chapman and Hall, London, UK.

CHAPTER 7

Non-pollinator agents of selection on floral traits

Sharon Y. Strauss and Justen B. Whittall

Section of Evolution and Ecology, University of California, Davis, CA, USA

Outline

Despite the dominating role of pollinators in floral evolution, mounting evidence reveals significant additional, often antagonistic, influences of abiotic and biotic non-pollinator agents. Even when pollinators and other agents impose selection on floral traits in the same direction, the role of other agents is frequently overlooked. Maintenance of genetic variation in floral traits and divergence from trait optima for pollination can result from both indirect selection on correlated traits and direct selection on floral traits. For example, in numerous species, periods of heat or drought favour pink- or purple-flowered individuals over white-flowered ones, because associated anthocyanins in vegetative tissues enhance stress tolerance. Conflicting selection on floral traits may also occur directly when floral antagonists and mutualists share the same preferences. We review the evidence for influences of abiotic and biotic non–pollinator agents of selection on several floral traits: petal colour, flower and display size, flower shape, nectar composition, flowering phenology, and breeding system. Despite growing evidence of the importance of non-pollinator selection, few studies have explored the relative strength of selection from pollinators versus other sources. In several cases, pollinators are not the strongest current source of selection on floral traits, despite perhaps being the driving factor shaping floral traits historically. Future studies will benefit from a synthetic approach that recognizes the entire ecological context of floral adaptation and combines field experiments with genetic studies to determine the relative roles of pollinators and non-pollinator agents in floral evolution. The study of floral evolution will be enhanced by approaches that incorporate a broader context that includes both abiotic and biotic agents of selection.

7.1 Introduction

Lloyd and Barrett's (1996) edited volume *Floral Biology: Studies on Floral Evolution in Animal-Pollinated Plants* began with two contrasting chapters: an English translation of Sprengel's "The secret of nature in the form and fertilization of flowers discovered," a pioneering treatise published in 1793 on the relation between flowers and their pollinators; and Herrera's chapter "Floral traits and plant adaptation to insect pollinators: a devil's advocate approach," which questioned the universality of these observations (Herrera 1996). Sprengel's interpretation of floral function from his direct observations in the wild provided insights into the intimate interactions between flowers and their pollinators. These insights were controversial at the time, because Sprengel assigned practical functions to features long thought to be divinely created. Herrera questioned some of the dogma that developed from the Sprengel-inspired field of pollination ecology, and revealed a broader ecological context underlying floral diversity. In particular, Herrera suggested that several factors, especially the diversity of the pollinator community, impose ecological and genetic constraints on floral adaptation. In this review, we expand on Herrera's critical perspective by highlighting the importance of multi-species interactions and abiotic agents of selection in shaping floral diversity.

Countless floral adaptations have undoubtedly arisen in response to selection from pollinators. The widespread convergent evolution in suites of floral traits across distinct plant families provides some of the most compelling evidence for the predominant role of pollinators in shaping floral adaptations (reviewed in Fenster *et al.* 2004). Furthermore, pollinators, as major determinants of mating patterns in many plants, are one of the primary drivers of plant diversification (Dodd *et al.* 1999; Chapter 17). For example, in comparative studies across angiosperms, animal-pollinated lineages have significantly more species than abiotically pollinated lineages (Dodd *et al.* 1999). More specifically, Bradshaw and Schemske (2003) showed that the shift from bee to hummingbird pollination in *Mimulus* section *Erythranthe* involved both floral pigmentation and flower shape caused by a few genes of major effect. The divergent preferences of different pollinators open the door to diversifying selection on floral design and display, and further reproductive isolation among floral morphs, thus paving the way to speciation (Grant 1949; Sargent 2004). The importance of pollinators as selective agents in these situations is indisputable.

Despite the primacy of pollinators as selective agents in many systems, several lines of evidence suggest that they are not the sole agents of selection on flowers, nor are they necessarily the most important selective agents in specific cases. Multi-species interactions have been incorporated broadly in the study of the evolution of plant defences; yet such approaches are much rarer in pollination studies (but for exceptions see Chapters 6, 8, and 15, and several studies cited herein). In many cases, floral traits may have evolved initially in response to selection from pollinators, but are now under stronger current selection from other community members (Herrera 1993). That is, once a plant has "locked in" to a particular suite of pollinators, other selective agents may drive subsequent modifications of floral traits. For example, the interplay between selection from enemies and pollinators has been well documented in comparative studies of *Dalechampia* (Armbruster 1997; Armbruster *et al.* 1997). Large, showy involucral bracts surrounding inconspicuous flowers were presumably favoured in ancestral species of this clade as a trait that attracted pollinators. In some species, these usually immobile bracts close over flowers at night. Experimental manipulations showed that bract closure prevents 90% of nocturnal herbivory on flowers. Nocturnal bract closure, and probably also bract size, appear to be under strong current selection from herbivores. Other floral modifications in *Dalechampia* that provide a defensive function against antagonists include enlarged sepals with long trichomes on pistillate flowers that cover developing fruits and resin secreting glands. Acquisition of some of these traits is associated with subsequent diversification within this clade (Armbruster 1997). Thus, traits shaped originally by pollinators were later modified by selection from floral and fruit antagonists. Both pollinators and non-pollinators appear to have played an important role in the morphological and taxonomic diversification of *Dalechampia*.

In many cases, floral traits may represent an adaptive compromise to selection caused by both pollinator and non-pollinating agents (Table 7.1). For example, heat stress and drought typically favour anthocyanin-producing petal morphs over white-flowered morphs (Section 7.3.1; Table 7.2). Given such effects, our understanding of floral evolution will be best served by a pluralistic approach that recognizes the range of selective influences on flowers and identifies biotic and abiotic factors that may shape floral traits in addition to pollinators.

When multiple agents influence selection on a floral trait, their effects may be either reinforcing or antagonistic, relative to the direction of selection imposed by pollinators. When floral traits exhibit close or coincident optima with respect to interactions with both pollinator and non-pollinator agents, the contributions of non-pollinator agents are often overlooked, because most investigators cease searching for other agents of selection when trait characteristics are consistent with selection from pollinators alone (Fig. 7.1). Irwin (2006) provided an exceptional example by demonstrating coincident selection on many floral traits of *Ipomopsis aggregata* from both fitness-reducing nectar

Table 7.1 Examples of antagonistic selection on floral traits by pollinators and non-pollinator agents of selection.

Species	Floral trait	Pollinator-mediated selection	Non-pollinator-mediated selection	References
Raphanus sativus	Flower colour	Pollinators prefer anthocyanin-less flowers	Herbivores prefer anthocyanin-less flowers	Strauss et al. 2004
Ipomoea purpurea	Flower colour	Self-pollination leads to higher fitness in anthocyanin-less flowers	Heat stress decreases flower production and fertilization success in anthocyanin-less flowers	Coberly and Rausher 2003
Polemonium viscosum	Floral shape	Bumble bees prefer open, flared corolla	Ants damage open, flared corollas more	Galen and Butchart 2003
		Bumble bees prefer larger corollas	Drought stress at high altitudes favours smaller corollas	Galen 2000
Phlox drummondii	Flower colour	Unknown	White-flowered individuals are competitively inferior	Levin and Brack 1995
Erysimum mediohispanicum	Stalk height, flower number, petal length, flower shape	Pollinators prefer taller plants with more flowers, longer petals	Browsing ungulates prefer taller plants; correlations between shape and height traits result in indirect selection against long petals	Gomez 2003
Castilleja linariaefolia	Calyx length, flower number, plant height	Pollinators prefer shorter calyces	Seed predators prefer shorter calyces	Cariveau et al. 2004
Fragaria virginica	Flower size and number of flowers per plant	Pollinators prefer larger flowers and more flowers per plant	Weevils are attracted to (and destroy) larger flowers and more flowers per plant	Ashman et al. 2004
Calyptrogyne ghiesbreghtiana	Number of flowers per inflorescence	Bats prefer inflorescences with numerous flowers	Katydids damage more flowers on taller inflorescences	Cunningham 1995
Silene dioica	Flower size, style length	Pollinators prefer larger flowers with longer styles	Smut spores are differentially deposited on larger flowers due to pollinator preferences	Elmqvist et al. 1993
Datura stramonium	Nectar volume	Hawk moths prefer flowers with higher nectar volumes	Increased visitation also increases oviposition by pollinator/herbivore	Adler and Bronstein 2004
Geranium sylvaticum	Gender	Pollinators prefer hermaphrodites	Floral herbivores also prefer hermaphrodites	Asikainen and Mutikainen 2005
Clarkia xantiana ssp. xantiana	Flower colour	No pollinator preference	Grasshoppers prefer fruits from plants without red-spotted petals	V. M. Eckhart unpublished data

Table 7.2 Differential fitness effects of non-pollinator agents for taxa that are polymorphic for floral anthocyanins.

Taxon	Non-pollinator agent	Effect	Difference between morphs (all significant differences)	Morphs with greater fitness listed first	Reference
Cirsium palustris	Drought stress	Reduced seed set and biomass	47%	Pink–purple/white	Warren and Mackenzie 2001
Digitalis purpurea	Drought stress	Reduced seed set and biomass	17%	Pink–purple/white	Warren and Mackenzie 2001
Echium plantagineum	Competition	Reduced biomass	45%[a]	Blue–purple/white	Burdon et al. 1983
Holcus lanatus	Drought stress	Reduced seed set and biomass	51%	Pink–purple/white	Warren and Mackenzie 2001
Ipomoea purpurea	Heat stress	Reduced flowers/plant, reduced fertility	12%	Purple/white	Coberly and Rausher 2003
Linanthus parryae	Spring rainfall	Population frequency fluctuations	N/A	Blue/white	Schemske and Bierzychudek 2001
Phlox drummondii	Moisture availability	Reduced survivorship and flower production	38%	Red/white	Levin and Brack 1995
Polygonum persicaria	Drought stress	Reduced seed set and biomass	17%	Pink–purple/white	Warren and Mackenzie 2001
Vicia sepium	Drought stress	Reduced seed set and biomass	27%	Pink–purple/white	Warren and Mackenzie 2001
Clarkia xantiana ssp. xantiana	Grasshoppers	Damage to fruits	N/A	Red petal spot/no petal spot	V. M. Eckhart unpublished data
Raphanus sativus	Herbivores (various)	Performance and damage	N/A	Pink/bronze versus white/yellow	Irwin et al. 2003
Claytonia virginica	Herbivores and pathogens	Herbivore damage	700% more damage	Whiter/redder	Frey 2004
		Infection by rust	Infection rates	Redder/whiter	

[a] Seed-set data not available. Fitness reduction estimated from decrease in mean dry weight during field experiments.

robbers and mutualist pollinators, even though these traits are typically thought to reflect just the actions of pollinators. In contrast, conflicting selection from non-pollinator agents may cause floral traits to deviate from optima favoured by pollinators, or may maintain polymorphisms in discrete traits (Fig. 7.1). Consequently, the impacts of non-pollinator agents on floral evolution are more likely to be detected (and reported) when selection from non-pollinator agents conflicts with that of pollinators.

In this review, we focus on the interplay between selection by pollinators and non-pollinator agents on floral traits. We begin by briefly reviewing the evolutionary roles of pollinators and non-pollinator agents in shaping several floral traits, recognizing that the vast majority of these examples show opposing selection imposed by these different agents. We then synthesize these examples in a discussion of the relative strengths of pollinator and non-pollinator agents during floral evolution. To illustrate this interaction, we focus on two case studies, *Raphanus sativus* and *Ipomoea purpurea*, for which multiple agents of selection on petal colour have been explored in detail. Last, we outline a framework for future studies to take a more pluralistic approach to understanding the evolution of floral form and function.

7.2 Selection on reproductive traits by non-pollinator agents

In the following sections, we review the evidence that non-pollinator agents of selection can play a major

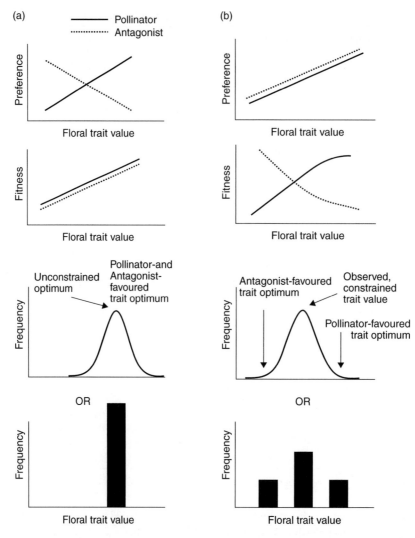

Figure 7.1 Floral trait evolution as a function of selection from multiple agents. When fitness-reducing antagonists and fitness-enhancing pollinators have opposing preferences (column a), they exert coincident selection favouring the same trait optimum, or favouring monomorphic populations when floral traits are discrete. When pollinators and antagonists share the same floral preferences (column b), plants may exhibit traits that reflect a compromise between values that maximize fitness through interactions with antagonists (left arrow) and pollinators (right arrow). Conflicting selection on discrete traits should maintain a balanced polymorphism (in this example, alleles are co-dominant). Other optima are possible and depend on preference and fitness functions.

role in shaping reproductive traits. We appreciate the numerous mechanisms through which floral adaptations may be constrained (developmental, biochemical, molecular, for example; Chapter 14), but have focused our attention on the role of ecological agents, abiotic and biotic. We organize our discussion by floral characters that have typically been considered solely for their role in pollinator attraction, efficacy, and reward: flower colour, size, shape, number, nectar, breeding system, and phenology. This review illustrates that viewing these traits exclusively as adaptive responses to pollinator-mediated selection often leads to an incomplete perspective on both their function and evolution.

7.2.1 Petal colour

Petal colour provides a visual cue that stimulates pollinator sensory systems and that selectively attracts certain types of pollinators (Grant 1949; Stebbins 1974; Melendez-Ackerman and Campbell 1998; Hodges *et al.* 2002). As a corollary, shifts in petal colour can promote speciation through reduced gene flow between colour morphs in association with concurrent changes in pollinator identity (Schemske and Bradshaw 1999; Hodges *et al.* 2002; Bradshaw and Schemske 2003). Nevertheless, anthocyanins, the pigments responsible for most flower colours, also have numerous non-pollinator functions and are often correlated with abiotic and biotic non-pollinator roles.

Anthocyanins are the most common floral pigments in angiosperms and are also often associated with tolerance to abiotic stresses (Table 7.2). An early review of anthocyanin pigmentation in plants listed 23 genera in which floral pigmentation correlated with vegetative tissue pigmentation (Onslow 1925). Although this list has grown substantially during the following decades, some of the ecological and evolutionary implications of correlations between floral and vegetative pigmentation have been revealed only recently (Warren and Mackenzie 2001; Coberly and Rausher 2003). For species with anthocyanin polymorphisms in both floral and vegetative tissues, pigmented individuals often tolerate stressful conditions like drought and heat better than anthocyanin-less morphs (Grace and Logan 2000; Warren and Mackenzie 2001; Steyn *et al.* 2002: Table 7.1). In a study of colour polymorphism in the British flora, anthocyanin-based flower colour polymorphisms (species with pink, blue, or purple flowers that also have white-flowered forms) occurred most commonly (Warren and Mackenzie 2001). Furthermore, experimental investigation of five of these taxa from different plant families demonstrated higher fitness for pigmented individuals than for unpigmented individuals under artificially imposed drought conditions (Warren and Mackenzie 2001; see Table 7.2 for additional examples). The maintenance of anthocyanin polymorphisms in the spikelets of wind-pollinated grasses (e.g., *Holcus lanatus* and *Poa trivialis*) is further testament to the non-pollinator mediated role of these pigments (Hubbard 1984).

Correlations in anthocyanin expression in different plant tissues may lead to indirect selection on flower colour or even predispose plants to a particular evolutionary trajectory. For example, petal colour in *Clarkia* correlates with anthocyanin content in seedlings. Anthocyanins in vegetative parts may make seedlings more robust to abiotic stresses (Bowman 1987) and may thus maintain floral polymorphisms through indirect selection on seedling traits. In a similar example, anthocyanin-containing floral bracts in *Dalechampia* may have originated as a result of indirect selection on stem and leaf pigments (Armbruster 2002). Alternatively, the original trait values under selection from pollinators may depend on prior vegetative character states under selection from other agents. For example, in *Acer*, the evolution of red or purple flowers evolved in lineages with anthocyanins in leaves, whereas pale-green or yellow flowers evolved in lineages without anthocyanins in vegetative structures (Armbruster 2002).

The evolutionary fate of anthocyanin-based flower colour polymorphisms can be partly determined by correlated changes in the expression of anthocyanins in vegetative tissues. The role of anthocyanins in vegetative tissues was addressed elegantly in a series of experiments that demonstrated selection on flower colour through correlations between petal colour and heat tolerance. Using *Ipomoea purpurea*, Coberly and Rausher (2003) identified white-flowered mutants caused by a deficient chalcone synthase enzyme; this mutation occurs at the *A* locus, which is the first dedicated step in the anthocyanin biosynthetic pathway. White-flowered individuals are particularly susceptible to heat stress: pigment-less mutants have 12% lower fitness than pigmented individuals, because of decreased flower production and lower fertilization success at higher temperatures. By modelling the dynamics of a polymorphic population mating under heat stress, Coberly and Rausher predicted morph ratios that were consistent with the low frequency of chalcone synthase mutants found in natural conditions. These results are particularly interesting with respect to another white-flowered *I. purpurea* morph, which occurs at much higher frequencies in nature and is caused by a mutation at the *W*, rather than the *A*, locus. This morph produces anthocyanins in vegetative tissues, but not flowers, due to a mutation in a tissue-specific regulatory gene. The ability of these white-flowered individuals to make anthocyanins in the leaves may confer heat tolerance, unlike the chalcone synthase mutants, which lack anthocyanins completely. These results suggest that pleiotropic effects of anthocyanins in vegetative tissues may constrain the types of mutations leading to white-flowered species (Durbin *et al.* 2003).

Flower colour polymorphisms that reflect differential tolerance to abiotic stresses may be maintained by fluctuating environmental conditions, even in the absence of pollinator preferences. In *Linanthus parryae*, blue-flowered morphs are more fit than white-flowered morphs during years of drought (Schemske and Bierzychudek 2001), whereas white morphs are more fit during years of high spring precipitation. No pollinator preferences were detected between morphs, nor did morphs differ in water-use efficiency (measured with carbon isotopes), although samples for the latter trait were taken during only a single, wet year. In this case, biotic agents that might also respond to altered precipitation patterns, such as herbivores, cannot be ruled out as agents of selection.

Petal colour may also be correlated with traits involved in biotic interactions, and may thus be subject to indirect selection from non-pollinator biotic agents. Anthocyanin-based petal colour differences have also been associated with competitive ability. In a transplant experiment with *Phlox drummondii*, white-flowered *Phlox* had 38% lower fitness (survivorship and fecundity) than pink-flowered plants when these morphs were grown together in competition (Levin and Brack 1995). Petal colour may also be associated with differences in vegetative or fruit defence traits, and thus may respond to indirect selection from antagonists. For example, herbivory induced higher glucosinolate concentrations in the leaves of *Raphanus sativus* morphs that produce petals with anthocyanins (pink and bronze) than in non-anthocyanin producing morphs (yellow and white) (Plate 2;

Strauss *et al.* 2004). In herbivore trials, anthocyanin-containing morphs decreased herbivore performance compared with anthocyanin-less morphs (Irwin *et al.* 2003). Correlations between petal colour and herbivore defence, or other vegetative traits, have also been observed for other species. Beetle larvae performed better on leaves of *I. purpurea* plants with white petals versus blue/purple petals (Simms and Bucher 1996). Artificial selection for higher concentrations of morphine alkaloids in opium poppy shifted petal colour frequency, suggesting a genetic correlation between flower colour and alkaloid production (Gyulane *et al.* 1980). In *Clarkia xantiana* ssp. *xantiana*, pollinators visit morphs with a wine-red spot on the petals and those lacking spots with equal frequency (Geber and Eckhart 2005), but grasshoppers regularly damage more fruits of unspotted morphs. Choice tests showed that this preference by grasshoppers persisted in the absence of other plant cues, so that petal spots appear to correlate genetically with other traits that make fruits less palatable to seed predators, and to be under selection from grasshoppers (V.M. Eckhart unpublished data and personal communication). Together these examples illustrate that several biotic factors, besides pollinators, favour anthocyanin-producing morphs over unpigmented individuals.

7.2.2 Flower shape

Floral shape has traditionally been considered an adaptation for pollinator attraction and manipulation (e.g., Darwin 1859; Bradshaw *et al.* 1998; Gomez 2003) and may be a driving force in angiosperm diversification (Sargent 2004). Yet, flower shape can also be under selection from agents other than pollinators. In one of the best-documented cases, nectar-robbing ants altered pollinator-mediated selection on corolla shape in *Polemonium viscosum* (Galen and Cuba 2001; Galen and Butchart 2003). Specifically, bumble bees preferred plants with more open, flared corollas and pollinated them more effectively, but ants selected against these individuals by causing more damage during nectar robbing. The result of this antagonistic selection is a sub-optimal flower (from the pollination perspective) with narrower, more tubular flowers that protect the styles. Differences in the relative abundance of ants and bumble bees at different elevations moderate selection on flower shape in *P. viscosum* and maintain genetic variation in corolla traits. Similarly, in *Erysimum mediohispanicum*, pollinator behaviour selected for longer petals and wider flowers in the absence of browsing ungulates, but not when ungulates were present (the natural condition) (Gomez 2003).

Other opportunities for non-pollinator agents to affect selection on floral shape may arise from constraints on fruit shape or size. For example, selection on fruit size and shape from seed predators could, in turn, select on ovary shape and may also influence other aspects of floral morphology (as in *Dalechampia*). To our knowledge, this possibility has received relatively little attention.

7.2.3 Flower size and display size

Pollinators often prefer large flowers, but bigger flowers may be costly in some environments (reviewed in Galen 1999). Flower development requires considerable water, because most change in petal size from bud to flower involves hydraulic cell expansion (Galen 1999, 2000). In *P. viscosum*, the amount of water taken up by flower buds accounts for 66% of the variation in petal size. In this case, pollinator attraction and increased drought tolerance appear to be at odds, because the diversion of water to developing flowers also compromises photosynthetic rates under drought conditions (Galen 2000). In *Rosmarinus officinalis*, a similar pattern exists across an elevational range, with smaller flowers in more stressful environments of dry coastal Mediterranean regions and larger flowers in moist, rich-soiled mountainous regions (Herrera 2005). Even though pollinators probably prefer larger rosemary flowers, smaller flowers may be favoured by the resource-cost compromise of the arid coastal environment. In this example, Herrera did not exclude phenotypic plasticity as the source of variation in floral traits. Nevertheless, these studies indicate how flower size may reflect a compromise between environmental stress (Clausen *et al.* 1940) and pollinator preferences (favouring large-flowered individuals).

Perhaps the greatest opportunity for conflicting selection from pollinators and other agents on floral traits occurs when pollinators and plant antagonists both use the same cues to locate and manipulate plants (Brody and Mitchell 1997). Often, larger flowers or more flowers per stem enhance attraction of both pollinators and floral antagonists. In wild strawberry (*Fragaria virginica*), herbivorous weevils prefer larger flowers and more flowers per plant, as do pollinators (Ashman *et al.* 2004). Similarly, katydids damage more flowers on taller inflorescences of tropical *Calyptrogyne ghiesbreghtiana*, and bat pollinators visit relatively more flowers on inflorescences with many flowers, although the incidence of bat visitation correlates negatively with katydid damage (Cunningham 1995). These patterns of use suggest that katydids and bats exert opposing selection on floral display.

Florivores that consume petals and prefer large-flowered plants can also reduce plant fitness through indirect effects on pollination. In *Nemophila menziesii*, many flowers experience floral herbivory and floral herbivores discriminate among flowers by colour, size, and gender in this gynodioecious species (McCall 2006). In this case, damaged flowers attract fewer pollinators, import less pollen, and are more pollinator limited than undamaged flowers. Similarly, experimental exclusion of florivores allowed *Isomeris arborea* flowers to produce three times more nectar than damaged flowers and twice as many anthers as those on exposed plants (Krupnick *et al.* 1999). In response, pollinators discriminated against damaged *Isomeris* flowers and visited patches of damaged plants less often than protected patches (Krupnick *et al.* 1999). Despite these clear effects of florivores on plant fitness and pollinator limitation (especially through male function), the intensity of selection imposed by florivores relative to that imposed by pollinators remains unknown.

Other fitness-reducing interactions promoted by large flowers result when pollinators transmit floral disease. Anther-smut sterilizes *Silene dioica* flowers and thus strongly reduces the fitness of diseased plants (Elmqvist *et al.* 1993). In an elegant comparison of floral morphology in populations exhibiting different disease rates, Elmqvist *et al.* (1993)

showed that plants from a non-diseased population produced larger flowers with longer styles than plants from highly diseased populations. In a common garden in the most diseased population, plants from a large-flowered healthy population received approximately four times more pollen and nine times more spores per flower than plants from the resident diseased population, indicating that larger flowers promote pollination, either through enhanced attraction or pollen exchange with individual pollinators. However, 20% of plants from the healthy population subsequently became diseased, whereas no plants from the small-flowered, local diseased population were infected (Elmqvist et al. 1993). In such cases, the direction and strength of selection on flower size will probably differ among populations and fluctuate through time as plant, pollinator, and pathogen densities vary among years, thereby maintaining genetic diversity for flower size (Elmqvist et al. 1993).

7.2.4 Nectar

Abundant nectar generally attracts more pollinators (reviewed in Mitchell 2004), but production of copious nectar bears costs through both pollination (e.g., Harder and Thomson 1989; Harder et al. 2001) and the action of abiotic and non-pollinating biotic agents. Although few studies have addressed the costs of nectar production in stressful environments, presumably the same selective conflicts exist between pollinator preference and drought tolerance as those described in the preceding section for flower size. The positive effects of water availability and temperature on nectar production are well documented (e.g., Zimmerman and Pyke 1988; Wyatt et al. 1992; Mitchell 2004). In fact, experimentally induced water stress reduced nectar volume of Chamerion (Epilobium) angustifolium more dramatically than flower size, sugar concentration, or plant height (Carroll et al. 2001). The high variation in nectar volume caused by environmental conditions complicates the estimation of the heritability and selection on nectar traits (Mitchell 2004).

Nectar characteristics are assumed to be optimized for pollinator reward and manipulation, but the use of nectar by non-pollinator species may also shape selection on nectar composition, quantity, or presentation. Adler and Bronstein (2004) supplemented nectar in *Datura stramonium* and found increased oviposition by *Manduca sexta*, a sphingid moth that pollinates flowers and lays eggs on plants. When herbivores also function as pollinators, their contrasting roles and their fitness effects may be linked inextricably.

Nectar robbers may also influence selection on nectar production, as their activity generally reduces plant fitness (reviewed in Irwin et al. 2001). The frequency of nectar robbing often varies with sugar concentration and other nectar components (Irwin et al. 2004), and these preferences can parallel those exhibited by pollinators (Gardener and Gillman 2002). Aside from the effects of nectar-robbing ants studied by Galen and colleagues, the impacts of robbers on the selection of nectar traits have received little attention.

Floral and extra-floral nectar also reward other plant mutualists, such as the predators of herbivores (like wasps and ants: Patt et al. 1999). The correlation between the composition of floral and extra-floral nectar and the association between the nectar preferences of pollinators and predators may also allow non-pollinator agents to influence selection on nectar traits. Again, our general lack of knowledge on the heritability of nectar traits (Mitchell 2004) precludes clear conclusions about correlated selection and constraints on nectar traits.

Herbivory can also affect nectar rewards when herbivores induce defensive chemicals in leaves or flowers, which are also incorporated in nectar. This side-effect of herbivore defence may incur costs in pollinator service and visit duration (Strauss et al. 1999), and may alter selection from pollinators. Euler and Baldwin (1996) showed that nicotine induced in foliage in response to damage also increased in concentration in the corollas of wild tobacco and the surrounding air; however, nicotine emissions reduced greatly at night, when pollinators forage on nectar. Secondary compounds can also occur constitutively in some floral nectar. "Toxic" nectar may deter floral antagonists, selectively eliminate unwanted pollinators, or simply be an unavoidable consequence of producing toxic compounds in other plant tissues (reviewed in Adler 2000). Alkaloid levels in nectar, flowers, and leaves correlate strongly across approximately 30 *Nicotiana* species (L.S. Adler, M. Gittinger, G.E. Morse, and M. Wink unpublished data). The presence of these compounds in nectar may be exaptations derived from plant defence and may be under selection from both herbivores and pollinators.

As an important aside, we note that even when pollinators are the primary selective agents on floral traits, the direction and strength of selection may be mediated by co-occurring community members. For example, several species in *Mimulus* section *Erythranthe* exhibit the hummingbird-pollination syndrome (red tubular flowers); however, some of these species offer less nectar than other, co-occurring hummingbird-pollinated species (Beardsley *et al.* 2003). If this difference in nectar rewards represents evolutionary divergence, the low-reward *Mimulus* species may be Batesian mimics of other hummingbird-pollinated species that offer large rewards (Brown and Kodrick-Brown 1979). Thus, although pollinators act as important, and perhaps the primary, selective agents, the trajectory of traits can also be influenced by concomitant selection from co-occurring species (i.e., *Mimulus* growing alone could be under selection to offer larger nectar rewards).

7.2.5 Sexual systems

In addition to floral traits, non-pollinating agents can affect the selection of sexual systems, including the relative incidence of selfing and outcrossing, and the diversity of mating types within populations. The evolution of self-pollination is commonly ascribed to the inconsistency or complete absence of pollinators (Chapter 10); however, the mating system also depends on other environmental aspects that affect reproduction. Self-pollination by annuals is typically favoured in environments with extremely short growing seasons, where rapid life cycles and time limitation are characteristic (Runions and Geber 2000; Mazer *et al.* 2004). For example, selfing has evolved from outcrossing at least 12 times in *Clarkia* in association with a reduction in flower size (Mazer *et al.* 2004). These small-flowered taxa often occur only at the range margins of the outcrossing parental species where environmental conditions are extreme (Runions and Geber 2000). Stressful abiotic conditions that favour small flowers as an epiphenomenon of the effects of stress on flower size may also drive increased selfing rates by increasing the proximity of anthers and stigmas (Snell and Aarssen 2005). Similar associations of reduced flower size, reduced resource availability, and increased selfing occur in numerous annual plant genera (Guerrant 1989), probably as an aggregate response to a variety of selective influences. Which selective agent has primacy in such adaptations is almost impossible to ascertain.

Breeding-system evolution may also be subject to the action of non-pollinating agents, especially herbivores. Ashman (2002; Chapter 11) presented considerable evidence that gynodioecy and dioecy may be selected because they reduce the impacts of floral or pre-dispersal seed predators on seed production. In both dioecious and gynodioecious species, male and hermaphroditic plants typically experience more herbivore damage than female plants (Ågren *et al.* 1999), sometimes in parallel with pollinator preference for hermaphrodites over females (Asikainen and Mutikainen 2005). In gynodioecious *Geranium sylvaticum*, patterns of pollination and herbivory in several populations during multiple years suggested that benefits through pollinator preferences did not outweigh the substantial detrimental effects of floral herbivory experienced by hermaphrodites (Asikainen and Mutikainen 2005). Therefore, floral herbivores may be the selective agent maintaining females in *G. sylvaticum*, although the relative importance of pollination and herbivory appears to fluctuate annually.

A phylogenetic perspective on this problem could be informative. An analysis of whether the evolution of dicliny is associated with the presence of flower-feeding herbivores (and their preferences) could test the role of non-pollinator agents. For example, *Anthonomus* weevils are notorious, injurious, specialized flower feeders and may be associated with clades with high frequencies of dicliny. Analysis of such broad-scale patterns may enhance understanding of the relations between herbivores and floral traits in the evolution of plant breeding systems.

7.2.6 Flowering phenology

The timing of flowering is commonly triggered by reliable environmental cues, such as day length and temperature (Stinchcombe et al. 2004; Amasino 2005). Similarly, pollinators must use reliable cues for their development to ensure that pollen and nectar are available when they emerge (e.g., Danforth 1999). Therefore, both plants and their pollinators could respond directly to weather conditions, using weather as a proxy for the presence of one another.

Flowering time in many plants will be constrained by factors other than pollinators (e.g., to avoid early frost and to allow sufficient time for fruit maturation), and the role of pollinator availability as a contributing factor in the evolution of flowering time remains to be determined (see Chapter 8). Shifts in flowering time have been observed in response to recent climate change (Primack et al. 2004; Molau et al. 2005). In some cases, flowering order within a community remains relatively stable, but the date of first flowering correlates strongly with climate. For example, flowering by species in 37 genera in the vicinity of Boston, Massachusetts, currently begins an average of eight days earlier than it did 80 years ago (Primack et al. 2004). In other systems, flowering time reflects character displacement among co-flowering species in response to competition for pollinators (see Chapter 8). Understanding the relative roles of selection associated with pollinators and abiotic factors in determining phenology is increasingly important in terms of appreciation of the impacts of climate change on plant reproductive success.

In addition to abiotic factors, interactions with non-pollinating animals, such as florivores, pre-dispersal seed predators, and seed dispersers, clearly affect flowering phenology. Florivory can affect flowering time directly, by damaging reproductive organs, or indirectly by decreasing attractiveness to pollinators (Krupnick and Weis 1999), leading to pollinator limitation (Krupnick et al. 1999; McCall 2006). For florivory to affect selection on flowering time, damage must be distributed unevenly during the current flowering phenology. For example, Kliber and Eckert (2004) found that the proportion of *Aquilegia canadensis* flowers eaten by ungulates increased as the season progressed (across 12 populations ungulates consumed 22% of primary flowers, 33% of secondary flowers, and 44% of tertiary flowers), perhaps because flowers on early-flowering inflorescences are less conspicuous. Such effects have also been documented at the community level in a survey of herbivore damage to the petals of 41 herbaceous species in a limestone grassland in central England (Breadmore and Kirk 1998). Slugs caused most petal damage early and late during the flowering season, perhaps owing to cooler temperatures. Although florivores can clearly affect flowering time, the relative role of florivores and pollinators in the selection of flowering time remains understudied.

In one of the few studies to examine the relative importance of pre-dispersal seed predators versus pollinators as selective agents on flowering phenology, Pilson (2000) showed that flowerhead-feeding pyralid and tortricid moths are the primary agents of selection on flowering phenology of *Helianthus annuus*. Late-flowering plants experienced much less damage than early-flowering plants, and selection analyses that excluded moth damage detected no other factors favouring late-flowering individuals. This study is one of few to demonstrate that selection on flowering phenology as an escape from insect attack can take primacy over mate availability or pollinator abundance.

Delayed flowering time in response to herbivory is supported by the theoretical models of Winterer and Weis (2004), which integrate stress-imposed delays in flowering time, assortative mating, and stress-resistance evolution. Specifically, they showed that stress-imposed delays in phenology can lead to assortative mating among resistant genotypes and eventually fixation of alleles for late flowering under certain conditions. Therefore, selection on phenology from environmental stressors (abiotic and biotic) could result in flowering times different from that predicted by peak pollinator availability.

Flowering phenologies different from the peak availability of pollinators can also result from selection imposed by dispersal timing. In an interesting study of a parasitic mistletoe (*Tristerix corymbosus*) the contrasting schedules of hummingbird pollinators and marsupial dispersers affected flowering time (Aizen 2003). Flowers opening during winter and late spring received fewer hummingbird visits and had reduced pollination and fruit set than those that opened during autumn or early spring. However, fruits produced during winter benefited from high removal rates and dispersal during summer when the primary disperser, the marsupial *Dromiciops australis*, was raising offspring. In this case, optimization of fruit dispersal with the timing of marsupial activity may be as important as, if not more important than, pollination success in determining flowering time (Aizen 2003). These results suggest that the activity period of seed dispersers can shape the evolution of flowering phenology, even though dispersal agents do not interact with flowers. Of the flowering traits discussed, phenology is among the most likely to reflect strong influences of non-pollinator agents of selection, because of its genetic reliance on abiotic cues and as a mechanism to escape environmental stressors (Winterer and Weis 2004).

7.3 Relative strengths of pollinator and non-pollinator agents of selection

Because plants occur in multi-species communities and are also constrained by abiotic conditions, we advocate a pluralistic approach in which multiple sources of selection are considered in studies of floral evolution. Understanding not just whether an agent exerts selection, but the relative importance of that agent compared with others, is of particular value.

Several studies have documented the selective impact of various agents on floral traits, but only a few have compared the relative strengths of selection from pollinators and other agents simultaneously. Notably, Cariveau *et al*. (2004) used structural equation modelling and path analysis to show that seed predators currently exert stronger selection on calyx length, flower production, and plant height in *Castilleja linariaefolia* than do pollinators, with calyx length experiencing opposing selection from pollinators and seed predators. Irwin (2006) also demonstrated that nectar robbers and pollinators impose weak, conflicting selection on floral traits of *Ipomopsis aggregata*. In this case, weak linkage between pollination and seed set attenuated selection from both pollinators and robbers, as did marked yearly variation in selection. In contrast, Galen and Cuba (2001) showed that conflicting selection between nectar robbers and pollinators of *Polemonium viscosum* favoured a different optimal corolla flare than expected from selection by bumble bee pollinators alone. Finally, browsing ungulates eat so many fruits and flowers of *Erysimum mediohispanicum* that pollinators enhance fitness only in the absence of herbivores (Gomez 2005). Exclusion of ungulates from plants for seven years resulted in divergence in flower shape and stalk height from that of exposed plants in a direction consistent with pollinator-mediated selection. These results demonstrate that the importance of pollinators as contemporary selective agents on floral traits can depend on the selective effects of other community members. Indeed, conflicting selection is not limited to the effects of biotic interactors. For example, drought stress selects for, and maintains, smaller flowers in *P. viscosum*, even though pollinators prefer larger-flowered individuals (Galen 2000).

Escape from herbivory may be the most important current cause of higher fitness for female plants than hermaphrodite plants in some gynodioecious species (Delph *et al*. 2004). Pollen-bearing flowers in dioecious and gynodioecious species tend to be larger and receive more visits from pollinators; however, larger flowers and bigger displays can also attract more floral herbivores (e.g., Ashman *et al*. 2004). As mentioned above, the relative importance of seed predators and pollinators may explain the maintenance of gynodioecy in *Geranium sylvaticum*. Thus, the detriments of herbivory on hermaphrodites can outweigh the benefits of attracting more pollinators (Asikainen and Mutikainen 2005). In this case, the maintenance of gynodioecy or dioecy may reflect balanced selection from herbivores and from pollinators (Chapter 11).

A recent study of 24 Japanese populations of the endangered *Primula sieboldii* during three years indicates that the relative strengths of selection from pollinators versus other agents varied seasonally and spatially (Matsumura and Washitani 2000). This variation depended on the degree to which antagonists, such as herbivores, disease, and abiotic factors, affected fitness and discriminated among floral phenotypes through direct or indirect selection. During one year, differences in seed set among populations correlated strongly with pollinator visitation. In contrast, during the other two years seed set varied among populations in relation to the abundance of antagonistic seed predators and fungi, not pollinator visitation. Selection during years with abundant seed predators probably reflects predator preferences more than pollinator preferences, as in the *Erysimum* example (Gomez 2003). However, this prediction must be qualified by the recognition that selection from pollinators can occur through effects on both seed set (female fitness) and pollen export and male fitness, which Matsumura and Washitani (2000) did not measure.

Self-pollinating populations/species/morphs may be particularly responsive to selection from non-pollinating agents, because they are less subject to pollinator-mediated selection. Species or genotypes that primarily self-fertilize typically produce small flowers with few pollen grains per ovule. For example, in the *Primula* populations described above, the long-styled, self-compatible morph had greater fitness than the short-styled morph when pollinators were limiting (Matsumura and Washitani 2000). Because of their greater susceptibility to pollen limitation, short-styled plants may be more responsive to selection from pollinators. Conversely, long-styled plants may have greater opportunity to adapt to seed predators, if fitness is less affected by conflicting selection from pollinators. To our knowledge, the hypothesis that selfing species or morphs may be more responsive to selection on floral or fruit traits from non-pollinator agents has not been tested.

7.3.1 A case study of *Raphanus sativus*

Raphanus sativus is another case in which the roles of multiple agents in the maintenance of flower colour variation in naturalized populations are being addressed experimentally. The four colour morphs of *R. sativus* (Plate 2) are determined by two alleles at each of two loci, with Mendelian inheritance (Panetsos 1964; Irwin and Strauss 2005). Yellow-flowered plants express carotenoid pigments and are recessive at both loci ($ppww$), whereas pink-petalled, anthocyanin-containing forms have dominant alleles at both loci ($P\_W\_$). White- and bronze-flowered plants (the latter expressing both carotenoids and anthocyanins) have at least one dominant allele at one locus and are homozygous recessive at the other ($ppW\_$ and P_ww, respectively). Frequencies of petal morphs vary among sites in California (Panetsos 1964; S. Y. Strauss and R. E. Irwin unpublished data).

The expected effects of pollinator preferences on morph ratios have been examined at a site at Bodega Bay, California, where the radish population is predominantly yellow-flowered ($\approx 1/2$ the population), with white flowers also fairly common ($\approx 1/3$) and pink and bronze relatively rare ($\approx 1/12$ each). Irwin and Strauss (2005) tested for any advantages of pollen from different colour morphs by pollinating stigmas of plants of known genotype with mixtures of equal amounts of pollen from each colour morph ("equal-pollinated"). The progeny ratios did not deviate from Mendelian expectations, indicating neither a siring advantage to any colour morph as a result of pollen competition, nor incompatibilities between morphs. Experimental hand-pollinations were then used to compare progeny ratios of open-pollinated flowers and equal-pollinated flowers with those of flowers pollinated with pollen mixtures that reflected the morph frequencies in the field ("null" pollinations). Null pollinations simulated pollinators foraging randomly with respect to morph colour. Experimental pollinations of adjacent flowers on 200 plants continued throughout the flowering season. Based on 8000 progeny, Irwin and Strauss found an over-representation of morphs with the yellow allele (yellow and bronze flowers) in open-pollinated seeds compared with progeny from "null" crosses and the parental generation (Fig. 7.2). They also found that the vast majority of white, pink, and bronze plants were heterozygous at the field site (e.g., pink plants were much more

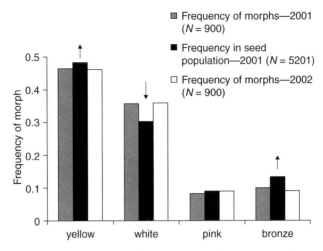

Figure 7.2 Selection from pollinators and unknown agents on flower colour in *Raphanus sativus*. Pollinators of *R. sativus* favour yellow flowers and alleles, resulting in a seed population with significantly more yellow alleles than the 2001 adult population that produced it (arrows denote change in frequency between 2001 parental and seed populations as a result of pollinator preferences). During the following year (2002), morph frequencies in flowering adults were identical to those during 2001. These results suggest that selection by non-pollinator agents against the yellow allele occurs during the intervening life stages.

likely to be $PpWw$ than $PPWW$, $PPWw$, or $PpWW$), a result consistent with preferences of pollinators and the large number of yellow flowers in the population. These results demonstrate that bee pollinators strongly prefer yellow morphs, as has also been observed in closely related *Raphanus* species (Kay 1976; Stanton 1987). Consequently, the yellow- and bronze-flowered morphs should increase in frequency if pollinators impose the primary selection on morph ratios. In contrast to this expectation, the ratios of petal morphs during the next flowering season did not differ significantly from those during the previous year (Fig. 7.2). Soil cores do not indicate extensive, long-lived seed banks in this system (S.Y. Strauss unpublished data), so that this contradiction between observed and expected morph ratios suggests that non-pollinator agents may select against yellow morphs at different life stages.

This hypothesis is supported by the observation that morphs differ in the glucosinolate concentrations in their leaf tissue (Strauss *et al.* 2004), so that that herbivores may play a role in maintaining petal variants. Strauss *et al.* (2004) tested the herbivory hypothesis with a glasshouse experiment using the third-generation progeny of controlled crosses with yellow mothers and pigmented sires. These crosses controlled for differences in genetic background (all progeny had yellow mothers with associated background) and also controlled for maternal effects. Half of the siblings from each family experienced experimental damage from *Pieris rapae* larvae, whereas the other half were left undamaged. Anthocyanin-containing morphs induced greater concentrations of indole glucosinolate in response to herbivore damage than did yellow morphs (Fig. 7.3). Herbivore preference and performance were also assessed on similarly created plant siblings. Whereas no herbivores exhibited preferences when allowed access to undamaged rosettes, which do not differ in glucosinolates, most performed differentially on different morphs once plants were damaged: yellow- and white-flowered plants supported two-fold faster growth of aphid colonies and better slug performance than anthocyanin-containing pink and bronze morphs. In addition, all herbivores with access to flowering plants preferred yellow morphs. These results support the hypothesis that plant antagonists also exert selection on flower colour, and that the direction of this selection is opposite to that imposed by pollinators.

Current studies on this system are using two approaches to address different aspects of the

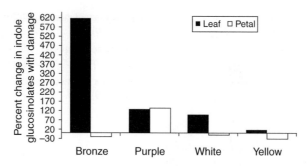

Figure 7.3 Percentage change in indole glucosinolate concentrations in leaves and petals of different colour morphs of *Raphanus sativus* after damage by *Pieris rapae* butterflies. Plants were grown in a glasshouse to control for maternal effects and were all progeny of yellow-flowered mothers to control for genetic background (see Strauss *et al.* 2004 for additional experimental details). Morphs did not differ in constitutive (undamaged) levels of glucosinolates.

herbivory hypothesis. First, to determine whether the presence/absence of herbivores changes the relative fitness of colour morphs in the field, herbivore densities have been reduced with insecticides and molluscicides for seven years. To date, seed production by yellow-flowered plants in herbivore-removal plots has increased relative to control plots in some years; however, in the presence of herbivores, the fitness of yellows has typically equalled that of the other morphs. Thus, in terms of fertility selection, selection from herbivores does not appear strong enough to prevent the spread of yellow alleles in the population, given strong pollinator preferences for yellow flowers. However, herbivores may have large impacts on seedling viability. Thus, to assess viability selection on flower colour during different life stages, seedlings at the cotyledon stage are being "rescued" in the field with virtually no mortality, brought into the greenhouse, and grown to flowering. Frequencies of the different floral morphs at the cotyledon and flowering stages can then be compared. Selection from herbivores or other antagonists during early life stages may limit the spread of yellow alleles. Alternatively, the bronze morph may act as a sink for yellow alleles, as these plants typically produce fewer seeds than the other morphs. Understanding the important selective agents on this colour polymorphism is continuing to require knowledge of the genetic basis of the trait, an understanding of correlations between vegetative and floral traits, consideration of selection on flower colour at several life-history stages, and consideration of multiple sources of selection.

7.4 Synthesis of ecological and genetic observations

The effects of non-pollinating agents of selection have been investigated from ecological and genetic perspectives. Numerous experiments in controlled and field conditions have documented significant correlations between a floral trait, like flower colour, and abiotic/biotic factors, such as heat stress and herbivory. At the same time, studies of developmental and molecular genetics have begun describing the multiple functions of associated genes and where they are expressed in the plant. Ideally, future studies will integrate analysis of the genetic basis of floral traits and the relative strength of selection from pollinators versus ecological non-pollinator agents. Knowledge of the genetic basis of these correlations is needed to determine whether they are caused by the same genes, physically linked genes, or genes co-segregating due to selection against recombinants (see Chapter 14). How traits are linked affects the constraints on the response to conflicting selection. Such detail will require genetic and molecular mapping of the floral trait and the associated correlation (herbivory, stress tolerance, etc.). Assessment of the roles of pollinator and non-pollinator agents will require additional studies that simultaneously explore (between-generation) selection responses, not just their (within-generation) effects on plant fitness

(Strauss et al. 2004). The synthesis of ecological and genetic approaches could be approached both empirically and with appropriate theory.

As described in Section 7.2.1, Coberly and Rausher's (2003) study of flower colour in *Ipomoea* exemplifies the synthesis of ecological and genetic approaches. By using a floral trait with a known genetic basis, they identified a correlated effect between the floral trait and an abiotic adaptation (tolerance to heat stress). These data were then incorporated as parameters in a model that predicted the relative fitness consequences from pollinators and non-pollinator agents of selection. Such analysis of the linkages between traits will be key to understanding how direct and indirect selection from multiple agents shape trait evolution.

7.5 Community context of trait evolution

As organisms evolve in the context of communities and multi-species interactions, the relative importance of various interactors as selective agents may shift. Because all species in a community can evolve, the strength of selection, and perhaps even the direction, may change through time, even in response to the same kind of interactions. Such dynamic selective landscapes may be the underpinnings of biological diversity (Schemske 2002).

Imagine a plant that is resistant to herbivory and that experiences strong selection from pollinators for large flowers. If a herbivore overcomes the plant's resistance and imposes strong effects on plant fitness, then selectivity of herbivores among plant genotypes (large flowered or not) may be more important than the preferences of pollinators. If floral and defensive traits are linked, then this condition can translate into a shift away from the optimal floral trait for pollination, perhaps opening the door for new, more effective pollinators to invade. The reciprocal interplay between antagonists and mutualists described in this example may also apply to other aspects of plant evolution.

The primacy of pollinators as agents of selection on floral traits is superseded when the fitness consequences of other agents surpass those of the pollinators, and the traits affecting interactions with pollinators and other agents are genetically linked, or are one and the same. The preceding review reveals that the non-pollinator agents of selection on floral traits are many and diverse. Consequently, selection on floral traits may seldom arise from the action of pollinators alone, given that these traits function in complex natural communities and environments.

Acknowledgements

Lawrence Harder and Spencer Barrett provided helpful comments on an earlier draft. The authors thank Vince Eckhart for contributing his unpublished results on fruit herbivory in *Clarkia xantiana* ssp. *xantiana* and Andrew McCall for sharing his unpublished results on florivory in *Nemophila menziesii*. JBW gratefully acknowledges support from a Comparative Biology Postdoctoral Fellowship in the Section of Evolution and Ecology; SYS acknowledges support from the NSF DEB-9807083.

References

Adler LS (2000). The ecological significance of toxic nectar. *Oikos*, **91**, 409–20.

Adler LS and Bronstein JL (2004). Attracting antagonists: does floral nectar increase leaf herbivory? *Ecology*, **85**, 1519–26.

Ågren JA, Danell K, Elmqvist T, Ericson L, and Hjalten J (1999). Sexual dimorphism and biotic interactions. In MA Geber, TE Dawson, and LF Delph, eds. *Gender and sexual dimorphism in flowering plants*, pp. 217–46. Springer-Verlag, Berlin.

Aizen MA (2003). Influences of animal pollination and seed dispersal on winter flowering in a temperate mistletoe. *Ecology*, **84**, 2613–27.

Amasino RM (2005). Vernalization and flowering time. *Current Opinions in Biotechnology*, **16**, 154–8.

Armbruster WS (1997). Exaptations link evolution of plant-herbivore and plant-pollinator interactions: a phylogenetic inquiry. *Ecology*, **78**, 1661–72.

Armbruster WS (2002). Can indirect selection and genetic context contribute to trait diversification? A transition-probability study of blossom-colour evolution in two genera. *Journal of Evolutionary Biology*, **15**, 468–86.

Armbruster WS, Howard JJ, Clausen TP, *et al*. (1997). Do biochemical exaptations link evolution of plant defense and pollination systems? Historical hypotheses and experimental tests with *Dalechampia* vines. *American Naturalist*, **149**, 461–84.

Ashman TL (2002). The role of herbivores in the evolution of separate sexes from hermaphroditism. *Ecology*, **83**, 1175–84.

Ashman TL, Cole DH, and Bradburn M (2004). Sex-differential resistance and tolerance to herbivory in a gynodioecious wild strawberry. *Ecology*, **85**, 2550–9.

Asikainen E and Mutikainen P (2005). Preferences of pollinators and herbivores in gynodioecious *Geranium sylvaticum*. *Annals of Botany*, **95**, 879–86.

Beardsley PM, Yen A, and Olmstead RG (2003). AFLP phylogeny of *Mimulus* section *Erythranthe* and the evolution of hummingbird pollination. *Evolution*, **57**, 1397–410.

Bowman RN (1987). Cryptic self-incompatibility and the breeding system of *Clarkia unguiculata* (Onagraceae). *American Journal of Botany*, **74**, 471–6.

Bradshaw HD Jr and Schemske DW (2003). Allele substitution at a flower colour locus produces a pollinator shift in monkeyflowers. *Nature*, **426**, 176–8.

Bradshaw HD Jr, Otto KG, Frewen BE, McKay JK, and Schemske DW (1998). Quantitative trait loci affecting differences in floral morphology between two species of monkeyflower (*Mimulus*). *Genetics*, **149**, 367–82.

Breadmore KN and Kirk WDJ (1998). Factors affecting floral herbivory in a limestone grassland. *Acta Oecologica*, **19**, 501–6.

Brody AK and Mitchell RJ (1997). Effects of experimental manipulation of inflorescence size on pollination and pre-dispersal seed predation in the hummingbird-pollinated plant *Ipomopsis aggregata*. *Oecologia*, **110**, 86–93.

Brown JH and Kodrick-Brown A (1979). Convergence, competition, and mimicry in a temperate community of hummingbird-pollinated flowers. *Ecology*, **60**, 1022–35.

Burdon JJ, Marshall DR, and Brown AHD (1983). Demographic and genetic changes in populations of *Echium plantagineum*. *Journal of Ecology*, **71**, 667–79.

Cariveau D, Irwin RE, Brody AK, Garcia-Mayeya LS, and von der Ohe A (2004). Direct and indirect effects of pollinators and seed predators to selection on plant and floral traits. *Oikos*, **104**, 15–26.

Carroll AB, Pallardy SG, and Galen C (2001). Drought stress, plant water status, and floral trait expression in fireweed, *Epilobium angustifolium* (Onagraceae). *American Journal of Botany*, **88**, 438–46.

Clausen J, Keck DD, and Hiesey WM (1940). *Experimental studies on the nature of species. I. Effect of varied environments on western North American plants*. Carnegie Institute of Washington, Washington, DC.

Coberly LC and Rausher MD (2003). Analysis of a chalcone synthase mutant in *Ipomoea purpurea* reveals a novel function for flavonoids: amelioration of heat stress. *Molecular Ecology*, **12**, 1113–24.

Cunningham SA (1995). Ecological constraints on fruit initiation by *Calyptrogyne ghiesbreghtiana* (Arecaceae): Floral herbivory, pollen availability, and visitation by pollinating bats. *American Journal of Botany*, **82**, 1527–36.

Danforth BN (1999). Emergence dynamics and bet hedging in a desert bee, *Perdita portalis*. *Proceedings of the Royal Society of London, Series B*, **266**, 1985–94.

Darwin CR (1859). *On the origin of species by means of natural selection or the preservation of favoured races in the struggle for life*. John Murray, London.

Delph LF, Gehring JL, Frey FM, Arntz AM, and Levri M (2004). Genetic constraints on floral evolution in a sexually dimorphic plant revealed by artificial selection. *Evolution*, **58**, 1936–46.

Dodd ME, Silvertown J, and Chase MW (1999). Phylogenetic analysis of trait evolution and species diversity variation among angiosperm families. *Evolution*, **53**, 732–44.

Durbin ML, Lundy KE, Morrell PL, Torres-Martinez CL, and Clegg MT (2003). Genes that determine flower color: the role of regulatory changes in the evolution of phenotypic adaptations. *Molecular Phylogenetics and Evolution*, **29**, 507–18.

Elmqvist T, Liu D, Carlsson U, and Giles BE (1993). Anther-smut infection in *Silene dioica*—variation in floral morphology and patterns of spore deposition. *Oikos*, **68**, 207–16.

Euler M and Baldwin IT (1996). The chemistry of defense and apparency in the corollas of *Nicotiana attenuata*. *Oecologia*, **107**, 102–12.

Fenster CB, Armbruster WS, Wilson P, Dudash MR, and Thompson JD (2004). Pollination syndromes and floral specialization. *Annual Review of Ecology, Evolution and Systematics*, **35**, 375–403.

Galen C (1999). Why do flowers vary? The functional ecology of variation in flower size and form within natural plant populations. *BioScience*, **49**, 631–40.

Galen C (2000). High and dry: drought stress, sex-allocation trade-offs, and selection on flower size in the alpine wildflower *Polemonium viscosum*. *American Naturalist*, **156**, 72–83.

Galen C and Butchart B (2003). Ants in your plants: effects of nectar-thieves on pollen fertility and seed-siring capacity in the alpine wildflower, *Polemonium viscosum*. *Oikos*, **101**, 521–8.

Galen C and Cuba J (2001). Down the tube: pollinators, predators, and the evolution of flower shape in the alpine skypilot, *Polemonium viscosum*. *Evolution*, **55**, 1963–71.

Gardener MC and Gillman MP (2002). The taste of nectar—a neglected area of pollination ecology. *Oikos*, **98**, 552–7.

Geber MA and Eckhart VM (2005). Experimental studies of adaptation in *Clarkia xantiana*. II. Fitness variation across a subspecies border. *Evolution*, **59**, 521–31.

Gomez JM (2003). Herbivory reduces the strength of pollinator-mediated selection in the Mediterranean herb *Erysimum mediohispanicum*: consequences for plant specialization. *American Naturalist*, **162**, 242–56.

Gomez JM (2005). Long-term effects of ungulates on performance, abundance, and spatial distribution of two montane herbs. *Ecological Monographs*, **75**, 231–58.

Grace S and Logan B (2000). Energy dissipation and radical scavenging by the plant phenylpropanoid pathway. *Philosophical Transactions of the Royal Society of London, B*, **355**, 1499–510.

Grant V (1949). Pollination systems as isolating mechanisms in angiosperms. *Evolution*, **3**, 82–97.

Guerrant EO (1989). Early maturity, small flowers and autogamy: a developmental connection? In JH Bock and YB Linhart, eds. *The evolutionary ecology of plants*, pp. 61–84. Westview Press, Boulder, CO.

Gyulane L, Peter T, and Gyulane V (1980). Some results in poppy *Papaver somniferum* breeding 1. Breeding of winter poppy. *Herba Hungarica*, **19**, 45–54.

Harder LD and Thomson JD (1989). Evolutionary options for maximizing pollen dispersal of animal-pollinated plants. *American Naturalist*, **133**, 323–44.

Harder LD, Williams NM, Jordan CY, and Nelson WA (2001). The effects of floral design and display on pollinator economics and pollen dispersal. In L Chittka and JD Thomson, eds. *Cognitive ecology of pollination*, pp. 297–317. Cambridge University Press, Cambridge.

Herrera CM (1993). Selection on floral morphology and environmental determinants of fecundity in a hawkmoth-pollinated violet. *Ecological Monographs*, **63**, 251–75.

Herrera CM (1996). Floral traits and plant adaptation to insect pollinators: A devil's advocate approach. In DG Lloyd and SCH Barrett, eds. *Floral biology: studies on floral evolution in animal-pollinated plants*, pp. 65–87. Chapman and Hall, New York.

Herrera CM (2005). Flower size variation in *Rosmarinus officinalis*: individuals, populations and habitats. *Annals of Botany*, **95**, 431–7.

Hodges SA, Whittall JB, Fulton M, and Yang JY (2002). Genetics of floral traits influencing reproductive isolation between *Aquilegia formosa* and *Aquilegia pubescens*. *American Naturalist*, **159**, S51–60.

Hubbard CE (1984). *Grasses: A guide to their structure, identification, uses, and distribution in the British Isles*. Penguin, London.

Irwin RE (2006). The consequences of direct versus indirect species interactions to selection on traits: pollination and nectar robbing in *Ipomopsis aggregata*. *American Naturalist*, **167**, 315–28.

Irwin RE and Strauss SY (2005). Flower color microevolution in wild radish: Evolutionary response to pollinator-mediated selection. *American Naturalist*, **165**, 225–37.

Irwin RE, Brody AK, and Waser NM (2001). The impact of floral larceny on individuals, populations, and communities. *Oecologia*, **129**, 161–8.

Irwin RE, Strauss SY, Storz S, Emerson A, and Guibert G (2003). The role of herbivores in the maintenance of a flower color polymorphism in wild radish. *Ecology*, **84**, 1733–43.

Irwin RE, Adler LS, and Agrawal AA (2004). Community and evolutionary ecology of nectar. *Ecology*, **85**, 1477–8.

Kay QQN (1976). Preferential pollination of yellow-flowered morphs of *Raphanus raphanistrum* by *Pieris* and *Eristalis* spp. *Nature*, **261**, 230–2.

Kliber A and Eckert CG (2004). Sequential decline in allocation among flowers within inflorescences: proximate mechanisms and adaptive significance. *Ecology*, **85**, 1675–87.

Krupnick GA and Weis AE (1999). The effect of floral herbivory on male and female reproductive success in *Isomeris arborea*. *Ecology*, **80**, 135–49.

Krupnick GA, Weis AE, and Campbell DR (1999). The consequences of floral herbivory for pollinator service to *Isomeris arborea*. *Ecology*, **80**, 125–34.

Levin DA and Brack ET (1995). Natural selection against white petals in *Phlox*. *Evolution*, **49**, 1017–22.

Lloyd DG and Barrett SCH, eds. (1996). *Floral biology: Studies on floral evolution in animal-pollinated plants*. Chapman and Hall, New York.

Matsumura C and Washitani I (2000). Effects of population size and pollinator limitation on seed-set of *Primula sieboldii* populations in a fragmented landscape. *Ecological Research*, **15**, 307–22.

Mazer SJ, Paz H, and Bell MD (2004). Life history, floral development and mating system in *Clarkia xantiana* (Onagraceae): Do floral and whole-plant rates of development evolve independently. *American Journal of Botany*, **91**, 2041–50.

McCall AC (2006). Natural and artificial floral damage induces resistance in *Nemophila menziesii* (Hydrophyllaceae) flowers. *Oikos*, **112**, 660–6.

Melendez-Ackerman E and Campbell DR (1998). Adaptive significance of flower color and inter-trait correlations in an *Ipomopsis* hybrid zone. *Evolution*, **52**, 1293–303.

Mitchell RJ (2004). Heritability of nectar traits: Why do we know so little? *Ecology*, **85**, 1527–33.

Molau U, Nordenhall U, and Eriksen B (2005). Onset of flowering and climate variability in an alpine landscape: a 10-year study from Swedish Lapland. *American Journal of Botany*, **92**, 422–31.

Onslow MW (1925). *The anthocyanin pigments of plants*. Cambridge University Press, London.

Panetsos CP (1964). Sources of variation in wild populations of *Raphanus* (Cruciferae). University of California, Berkeley, Ph.D. Thesis.

Patt JM, Hamilton GC, and Lashomb JH (1999). Responses of two parasitoid wasps to nectar odors as a function of experience. *Entomologia Experimentalis et Applicata*, **90**, 1–8.

Pilson D (2000). Herbivory and natural selection on flowering phenology in wild sunflower, *Helianthus annuus*. *Oecologia*, **122**, 72–82.

Primack D, Imbres C, Primack R, Miller-Rushing A, and Tredici P (2004). Herbarium specimens demonstrate earlier flowering times in response to warming in Boston. *American Journal of Botany*, **91**, 1260–4.

Runions CJ and Geber MA (2000). Evolution of the self-pollinating flower in *Clarkia xantiana* (Onagraceae). I. Size and development of floral organs. *American Journal of Botany*, **87**, 1439–51.

Sargent RD (2004). Floral symmetry affects speciation rates in angiosperms. *Proceedings of the Royal Society of London, Series B*, **271**, 603–8.

Schemske DW (2002). Ecological and evolutionary perspectives on the origin of tropical diversity. In R Chazdon and T Whitmore, eds. *Foundations of tropical biology: key papers and commentaries*, pp. 163–73. University of Chicago Press, Chicago.

Schemske DW and Bierzychudek P (2001). Perspective: evolution of flower color in the desert annual *Linanthus parryae*: Wright revisited. *Evolution*, **55**, 1269–82.

Schemske DW and Bradshaw HD Jr (1999). Pollinator preference and the evolution of floral traits in monkeyflowers (*Mimulus*). *Proceedings of the National Academy of Sciences of the United States of America*, **96**, 11910–5.

Simms EL and Bucher MA (1996). Pleiotropic effects of flower-color intensity on herbivore performance in *Ipomoea purpurea*. *Evolution*, **50**, 957–63.

Snell R and Aarssen L (2005). Life history traits in selfing versus outcrossing annuals: exploring the "time limitation" hypothesis for the fitness benefit of self-pollination. *BMC Ecology*, **5**, 1–14.

Stanton ML (1987). Reproductive biology of petal color variants of wild populations of *Raphanus sativus* L.: I. Pollinator response to color morphs. *American Journal of Botany*, **74**, 176–85.

Stebbins GL (1974). *Flowering plants, evolution above the species level*. Belknap Press, Cambridge, MA.

Steyn WJ, Wand SJ, Holcroft M, and Jacobs G (2002). Anthocyanins in vegetative tissues: a proposed unified function in photoprotection. *New Phytologist*, **155**, 349–61.

Stinchcombe JR, Weinig C, Ungerer M, *et al.* (2004). A latitudinal cline in flowering time in *Arabidopsis thaliana* modulated by the flowering time gene *FRIGIDA*. *Proceeding of the National Academy of Sciences of the United States of America*, **101**, 4712–7.

Strauss SY, Siemens DH, Decher MB, and Mitchell-Olds T (1999). Ecological costs of plant resistance to herbivores in the currency of pollination. *Evolution*, **53**, 1105–13.

Strauss SY, Irwin RE, and Lambrix VM (2004). Optimal defense theory and flower petal colour predict variation in the secondary chemistry of wild radish. *Journal of Ecology*, **92**, 132–41.

Warren J and Mackenzie S (2001). Why are all colour combinations not equally represented as flower-colour polymorphisms? *New Phytologist*, **151**, 237–41.

Winterer J and Weis AE (2004). Stress-induced assortative mating and the evolution of stress resistance. *Ecology Letters*, **7**, 785–93.

Wright JW, Stanton ML, and Scherson R (2006). Local adaptation to serpentine and non-serpentine soils in *Collinsia sparsiflora*. *Evolutionary Ecology Research*, **8**, 1–21.

Wyatt R, Broyles SB, and Derda G (1992). Environmental influences on nectar production in milkweeds (*Asclepias syriaca*) and *A. exaltata*. *American Journal of Botany*, **79**, 636–42.

Zimmerman M and Pyke GH (1988). Experimental manipulations of *Polemonium foliosissimum*: effects on subsequent nectar production, seed production and growth. *Journal of Ecology*, **76**, 777–89.

CHAPTER 8

Flowering phenologies of animal-pollinated plants: reproductive strategies and agents of selection

Gaku Kudo

Graduate School of Environmental Earth Science, Hokkaido University, Sapporo, Japan

Outline

The time and pattern of flowering strongly influence the reproductive success of animal-pollinated plants by controlling the overlap of flowering with temporally variable abiotic and biotic conditions that affect mating and seed production. Various factors influence the ecology and evolution of flowering phenologies, including features of the mating environment during flowering, herbivory on flowers and developing seeds, the period available for seed development, and seed dispersal conditions. This diversity of influences as well as spatial and temporal variation in flowering conditions complicate selection on flowering phenologies. Nevertheless, many examples demonstrate that flowering phenologies serve as adaptive reproductive strategies. In contrast, environments with brief growing periods may not allow evolutionary adjustment of flowering phenologies. In such cases, restrictions on mating imposed by the flowering phenology influence selection on other reproductive traits. Furthermore, variation in flowering time within and among populations caused by differences in abiotic environments imposes assortative mating, which can create genetic structure among environments. Based on studies of populations along snowmelt gradients, I demonstrate that site-specific flowering phenologies influence the quality and quantity of seed production, mating-system evolution, and spatial genetic structure of alpine plants. Finally, I discuss outstanding issues concerning the ecology and evolution of flowering phenologies, including the need to clarify the biological interactions on which selection acts, adequate evaluation of fitness, and experimental approaches that incorporate genetically determined phenological variation.

8.1 Introduction

The reproductive schedule is a key life-history trait, because it controls an individual's exposure to variable biotic and abiotic conditions that influence reproductive success. For animal-pollinated plants, relevant environmental variables affected by flowering time and pattern include pollinator and mate availability, predation intensity, temperature, and water availability. Flowering phenology refers to the seasonal occurrence of mating within plants, populations, and communities, and thus represents the reproductive behaviour of flowering plants at various biological scales. As a life-history trait, flowering phenology should be subject to strong selection and so serve as a reproductive strategy that promotes fitness. In addition, because a plant's flowering phenology influences its mating environment, it can affect selection on other reproductive traits and spatial genetic structure (i.e., spatial pattern of genetic similarity among plants) within and between populations.

Flowering schedule can serve as a reproductive strategy, given that some variation in phenological traits is genetically determined (e.g., Weis and Kossler 2004). However, the importance of flowering phenologies as reproductive strategies compared with other reproductive traits has been subject to considerable debate (Rathcke and Lacey 1985; Kochmer and Handel 1986; Ollerton and Lack 1992; Fox and Kelly 1993). Much of this disagreement reflects inconsistent trends in phenological patterns that result from four aspects of the timing of plant reproduction (also see Section 8.3.6). First, the flowering phenology is highly susceptible to environmental conditions, which obscures the identification of adaptive features. Second, multiple competing factors determine selection on the flowering phenology, complicating consistent and predictable responses of flowering behaviour. Third, phenological responses to selection may be limited by phylogenetic constraints and physiological capacity. Finally, changes in flowering phenology must be compatible with other time-dependent processes, such as growth, seed development, and seed dispersal.

In contrast to its strategic aspects, the role of flowering phenology in the selection of other reproductive traits has been scarcely considered in previous phenological studies. This limited attention results largely because of the need to study consistent patterns of phenological variation within and between populations to test the role of flowering phenology as an agent of selection. Comparative studies along environmental gradients that affect the seasonality of reproduction allow such replicated analysis. In particular, snowmelt gradients in alpine ecosystems offer an ideal natural experimental system, where the reproductive schedules of plants vary extensively, depending on the local time of snowmelt (Kudo 1991).

In this chapter, I consider factors that influence the evolution of the flowering phenology, determine its effectiveness as a reproductive strategy, and consider the extent to which environmentally induced variation in flowering phenology affects the evolution of other reproductive traits, including mating systems and dispersal traits. First, I identify the fundamental components of the flowering phenologies of individual plants, their populations, and communities. I then consider the abiotic and biotic factors that determine the benefits and costs of alternate phenologies for animal-pollinated plants, thereby influencing phenological adaptation. These factors act either directly on floral function or indirectly through reproductive processes that follow flowering. Next, I illustrate the role of environmentally limited phenologies in selection on other aspects of plant reproduction, using studies of alpine plants conducted by my research group as a case study. Finally, I discuss unresolved phenological problems that await analysis. Although flowering phenologies can be considered from perspectives that range from individual flowers through entire plant communities, I focus primarily on variation in flowering time within populations and communities.

8.2 Components of flowering phenologies

The flowering phenology is not a single trait of an individual, population, or community, but instead represents the aggregate outcomes of several time-dependent processes. I now briefly review current theoretical and empirical understanding of these components.

8.2.1 Individual phenology

The flowering phenology of individual plants depends on their time of first flowering, the rate at which flowers open, and floral longevity. The time of first flowering determines the seasonal schedule of reproduction, especially its overlap with peak periods of the availability of reproductive resources, including nutrients, water, light, and pollinators. Flowering rate and floral longevity simultaneously determine a plant's display size (number of open flowers per inflorescence and number of flowering inflorescences) and the duration of its flowering period (Meagher and Delph 2001; Harder and Johnson 2005). Indeed, drawing an analogy from resource-based population growth, Meagher and Delph (2001) recognized flowering rate and floral longevity as demographic parameters that govern the dynamics of floral display. These dynamics can influence reproductive success by affecting both pollinator attraction and pollination efficiency, especially the incidence of geitonogamy and pollen discounting, through their effects on pollinator behaviour (reviewed in Harder et al. 2001; Harder and Johnson 2005).

Of the three components of individual flowering phenology, only floral longevity has received special attention (reviewed by Ashman 2004). In general, longevity is assumed to be optimized to maximize pollen import and export within constraints imposed by the resource costs of maintaining individual flowers. An intriguing feature of floral longevity is its sensitivity to pollination in many plant species (van Doorn 1997), which allows extension of a plant's flowering period in response to infrequent pollinator visits. However, floral longevity is usually studied on its own, so its role in a plant's total flowering phenology remains largely unexplored.

8.2.2 Population phenology

At the population level, flowering phenology emerges from the aggregate phenology of individual plants, which determine flowering time, synchrony among plants, and the duration and skewness of population flowering. Consequently, a population's flowering phenology affects frequency- and density-dependent aspects of plant reproduction, such as pollinator attraction, detection by herbivores, and the availability of potential mates, thereby governing both reproductive success and mating patterns. Through these effects, flowering phenology influences the formation of genetic structure within populations, which can accelerate local evolution. In addition, phenological differences among populations caused by genetic differences and local environmental conditions, including site-specific biological interactions, determine the opportunity for gene flow among local populations and population differentiation.

Studies of variation in flowering phenology within populations have primarily addressed three topics: the extent to which plants flower synchronously, the mode of selection on flowering time (directional, stabilizing, or disruptive), and the optimal duration of the flowering period. These subjects are related, because directional selection concentrates flowering early or late during the season, stabilizing selection synchronizes flowering during a brief period, and disruptive selection favours staggered flowering or a mixture of synchronized and extended flowering.

Synchronous and asynchronous flowering have contrasting benefits. Synchronous flowering helps to attract pollinators and seed dispersers by the mass-display effect, to satiate flower and seed predators, and to promote outcrossing by maximizing the number of potential mates (Rathcke and Lacey 1985; Marquis 1988; O'Neil 1997; Ollerton and Diaz 1999). In contrast, asynchronous flowering encourages the movement of pollinators and seed-dispersing animals within the population and reduces the risk of exposure to poor conditions during pollination, seed dispersal, and/or seedling survival (Rathcke and Lacey 1985; Melampy 1987; Marquis 1988). Such contrasting benefits of simultaneous versus staggered flowering complicate the selection of flowering synchrony. In a pioneering study, Augspurger (1981) demonstrated experimentally that the synchronous flowering of *Hybanthus prunifolius* enhanced pollinator attraction (social bees) and helped satiate pre-dispersal seed predators (microlepidopteran and dipteran larvae), causing stabilizing selection on flowering phenology. Similar results were reported for *Astragalus scaphoides*, which predominately flowers during alternate years, as inflorescence herbivory and seed predation decreased and fruit set increased during years of synchronous flowering (Crone and Lesica 2004). In contrast, Marquis' (1988) study of *Piper arieianum* showed higher seed set during peak flowering, but fruits produced by off-peak flowers escaped seed predation, causing disruptive selection on synchrony.

Flowering pattern within a population influences pollinator attraction. A positively skewed phenology increases attraction of pollinators to a plant species during its initial flowering (O'Neil 1997). For example, early-flowering plants of *Phlox drummondii* are more conspicuous and attract more pollinators, so that fruit set declines with flowering time, causing directional selection for early flowering (Kelly and Levin 2000). In contrast, some studies have reported that pollinator visitation correlates negatively with flowering density due to intraspecific competition for pollinators (e.g., Melampy 1987; Gómez 1993; Kelly and Levin 2000). Asynchronous flowering would be beneficial in this case.

Synchronized flowering can be selected when the availability of suitable mates limits reproductive success, which is particularly important for species with self-incompatibility and/or polymorphic sexual systems. O'Neil (1997) demonstrated such stabilizing selection in *Lythrum salicaria*, a tristylous herb in which individuals of a particular morph can mate only with individuals of another morph. Similarly, Ollerton and Diaz (1999) observed strong stabilizing selection for flowering synchrony in *Arum maculatum*, a fly-pollinated monoecious herb with very brief stigma receptivity (12–18 h).

Differences in flowering period among plants within a population or between closely adjacent populations limit pollen dispersal to simultaneously flowering individuals. This positive assortative mating has several genetic and evolutionary consequences (Fox 2003). Obviously mating isolation from individuals that flower at different times promotes genetic differentiation in accordance with the extent of flowering separation. If flowering time varies in response to environmental heterogeneity, then spatial genetic structure can arise at a regional scale (see Section 8.4.3). If the relative abundance of plants flowering at different times varies, the resulting inequality in mating opportunities can result in stabilizing selection for increased flowering during the period of peak flowering (see Kirkpatrick and Nuismer 2004). Alternatively, when environmental conditions favour a shift in flowering time, the restriction of gene flow via pollination among plants exposed to contrasting selection may accelerate local adaptation (McNeilly and Antonovics 1968).

Selection on flowering phenology can act extremely locally within populations, if the spatial gradient in selection is sufficiently strong and pollen dispersal is limited. Schemske's (1984) study of *Impatiens pallida*, a self-pollinating annual, clearly illustrates this possibility. In his study area, beetles killed many plants in the forest interior during the middle of the flowering period. Presumably as a consequence of the resulting selection, plants in the forest interior flowered 12 days earlier than those at the forest edge, only 50 m away.

8.2.3 Community phenology

The community flowering pattern represents the combined phenologies of its constituent species. Analyses of this pattern have focused primarily on whether biological interactions among species produce regular, random, or aggregated sequences of flowering within communities. These patterns have been considered primarily in the context of interspecific competition through pollination (also see Section 8.3.2), based on the hypothesis that natural selection minimizes competition for pollinators and the risk of interspecific pollination among species sharing pollinators, resulting in sequential flowering. Therefore, most studies have compared the distribution of peak flowering time or flowering overlap among species to distinguish regularity (divergence) from aggregation (convergence), based on the null hypothesis of a random distribution (Kochmer and Handel 1986; Ollerton and Lack 1992; Fox and Kelly 1993).

Many studies of community flowering have incorporated comparisons with patterns expected in the absence of competitive interactions. However, studies of hummingbird-pollinated species in Costa Rican tropical forest have reached inconsistent conclusions, ranging from claims of regular to aggregated distributions, depending on the null hypotheses considered (reviewed by Rathcke and Lacey 1985). This confusion indicates the importance of identifying appropriate biological parameters and relevant periods for randomization. For example, Morales *et al.* (2005) hypothesized that in regions with restricted flowering seasons fewer species may flower early and late, rather than during the middle of the flowering season. This "mid-domain" effect results simply from geometric constraints on the arrangement of flowering periods, because flowering must be preceded by floral-bud development and followed by fruit maturation. Morales *et al.*'s results from two alpine sites during six years were largely consistent with this geometric null hypothesis, although flowering occurred somewhat less often than expected during early spring, when the risk of frost damage to flowers was high.

Several analyses of community flowering phenologies have found that the observed patterns did not deviate from random expectations (e.g., Wieder et al. 1984; Kochmer and Handel 1986); however, these studies may not involve relevant comparisons. In particular, if competition for pollination influences the temporal distribution of phenologies, it should affect only species visited by the same suite of pollinators. Indeed, examples of divergence in flowering times within communities involve species that share a specific pollinator guild, including bees, hummingbirds, thrips, or bats (reviewed by Rathcke and Lacey 1985; Sakai et al. 1999; Lobo et al. 2003). In contrast, species visited by diverse pollinators or generalist pollinators may compete less with each other and so are more likely to exhibit random flowering sequences. These studies suggest that the following factors should be considered when evaluating the ecological and evolutionary significance of flowering pattern within communities: the frequency and nature of competition among individual plant species, the fidelity of pollinators to specific plant species, and the pattern of pollinator availability and pollination efficiency during the flowering season.

Recent phenological studies within plant communities have considered the advantages of flowering convergence in high-diversity tropical forests (Sakai et al. 1999; Lobo et al. 2003). In aseasonal, tropical rain forests in Southeast Asia, various taxonomic groups flower *en masse* at irregular intervals of 3–10 years—a process called "general flowering." For example, 57% of reproductive events among 305 species in a Borneo rain forest occurred during general-flowering periods (Sakai et al. 1999). Such aggregated flowering of multiple species is thought to activate pollinators, resulting in higher pollination success than if species flowered asynchronously.

8.3 Flowering phenologies as reproductive strategies

Selection should act on flowering phenologies to enhance reproductive output in a particular abiotic and biotic environment, within constraints imposed by other plant functions and phylogenetically inherited development patterns. Nevertheless, several studies report no evidence of selection on flowering time (Murray et al. 1987; Gómez 1993; Ollerton and Lack 1998). I now consider whether studies of various abiotic and biotic aspects of flowering and post-flowering events (Fig. 8.1) are consistent with flowering phenologies serving as reproductive strategies.

8.3.1 Abiotic influences on phenology

Aspects of the physical environment, such as temperature (including risk of frost), light availability, precipitation (including burial by snow), and day length, strongly determine the favourable period for flowering on both ecological and

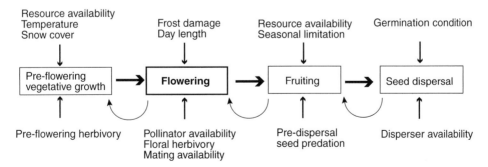

Figure 8.1 Factors that affect the reproductive phenologies of plants during pre-flowering growth, flowering, fruiting, and seed dispersal. Major abiotic and biotic influences are shown above and below each phase, respectively. Curved arrows indicate feedbacks between successive phases. For example, pre-dispersal seed predation can affect direct selection on fruiting phenology and indirect selection on flowering time.

evolutionary time scales. Indeed, environmental conditions that provide reliable information about forthcoming conditions, especially photoperiod and soil temperature or moisture, have commonly been co-opted as physiological cues for the initiation of flowering (e.g., Pavón and Briones 2001; Keller and Körner 2003). In seasonal tropics, most herbs and shrubs flower during rainy seasons, reflecting seasonal resource availability, whereas trees flower during both wet and dry seasons (Rathcke and Lacey 1985). In temperate deciduous forests the initiation of flowering by understory plants is strongly limited by low temperature during early spring, whereas flowering conditions degrade severely during late spring due to low light availability caused by canopy closure (e.g., Motten 1986). Alpine plants provide obvious examples of the environmental control of phenology, as spring snowmelt initiates their growing season and flowering patterns within communities change along a snowmelt gradient (Kudo 1991; Section 8.4). Furthermore, plants in seasonal environments require sufficient time to develop fruit, which additionally limits the potential flowering period (Primack 1987). For example, plants that produce large fruit tend to bloom early to allow for fruit maturation. Clearly, such seasonal variation in the availability of suitable conditions for flowering and fruiting can impose strong selection on phenological traits, particularly flowering time.

8.3.2 Pollination

Given that flowers function primarily to produce pollen and ovules and to facilitate pollination and fertilization, temporal variation in the availability of suitable pollination conditions should often dominate selection on flowering phenologies. A clear example of this influence can be found in the spring flowering by wind-pollinated deciduous trees before leaves expand and reduce air flow within canopies (Rathcke and Lacey 1985). However, a plant's pollination environment also includes other species that rely on the same pollen vectors, so that indirect interactions between plant species mediated by their pollen vectors can influence the selection of flowering phenologies.

By sharing pollinators, sympatric species can engage in three main types of interactions: competition, facilitation, and resource parasitism (Fig. 8.2). Competition for pollination can occur by

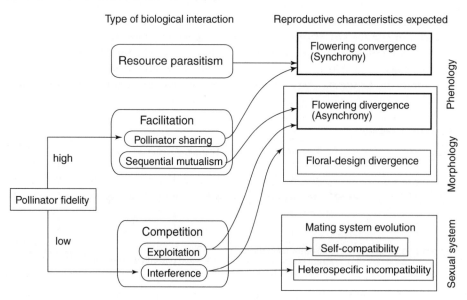

Figure 8.2 Types of biological interactions caused by plant species sharing pollinators (resource parasitism, facilitation, competition) and their influences on reproductive characteristics (flowering phenology, floral morphology, breeding system).

competition for pollinators (resource exploitation) and/or interference via interspecific pollination (reviewed by Rathcke 1983). Competition for pollinators results from differential pollinator preference for particular plant species that offer abundant floral resources and/or low foraging costs, causing pollinator limitation of seed production for competitively inferior species (e.g., Brown et al. 2002). In contrast, improper pollen transfer results from low pollinator fidelity and causes pollen wastage, decreased seed production because of reduced pollen import, stigma closing, and/or pollen allelopathy (reviewed by Brown et al. 2002), and formation of hybrids (Chapter 18). Bell et al. (2005) also demonstrated that interspecific pollination can reduce the proportion of cross-fertilized seeds, which may increase losses to inbreeding depression. Exploitation competition for pollinators may select for flowering divergence or increased selfing for competitively inferior species. Interference competition should also select for flowering divergence, but additionally favours the evolution of floral mechanisms that reduce heterospecific pollination, and/or heterospecific incompatibility that prevents hybridization (although see Chapter 16).

Strong interspecific competition can select for a phenological shift of flowering by competitively inferior species, even into a period of lower pollinator activity. For example, de Jong and Klinkhamer (1991) demonstrated selection for early-season flowering in *Cynoglossum officinale*, before the period of peak bumble bee activity, which reduced flowering overlap with *Echium vulgare*, a more attractive species. Similarly, *Blandfordia grandiflora*, a bird-pollinated geophyte, begins flowering before most migratory birds arrive, resulting in pollen limitation of seed production (Ramsey 1995), apparently to reduce competition for pollinators with *Banksia serrata*, a very attractive and rewarding shrub.

Facilitation of pollination can result from "pollinator sharing" or "sequential mutualism" (Rathcke 1983). Pollinator sharing occurs when species attract more pollinators by flowering simultaneously than they would by flowering separately. The effectiveness of pollinator sharing depends on either pollinator fidelity or pollen from different species being transported on different parts of pollinators' bodies to prevent interspecific pollination. Real (1983) demonstrated theoretically that flowering convergence can be adaptive when: pollinators respond quickly to local resource conditions; foraging efficiency of pollinators increases with floral densities due to reduced flight distance, learning, or development of a search image; and simultaneous flowering reduces foraging uncertainty within a habitat. In contrast, sequential mutualism occurs when a species realizes higher pollination by flowering immediately after an attractive species and inheriting the first species' pollinators. "General flowering" by dipterocarp trees in aseasonal forests of southeast Asia (Section 8.2.3) may reflect facilitation in which staggered flowering during the general-flowering period allows pollinator sharing.

For co-flowering plant species visited by the same pollinators, the extent of pollinator fidelity is a crucial factor in determining whether their interaction will be facilitative (pollinator sharing) or competitive (interference), as illustrated in Fig. 8.2. The extent of fidelity depends on both floral traits of co-flowering species and the pollinator taxa (e.g., Jones 2001), so that features of interactions between co-flowering species may vary with the local fauna and flora. Little is known about such variation.

Resource parasitism through pollination is evident in "floral mimicry," whereby a species realizes increased pollination because its flowers resemble those of an unrelated species with respect to morphology, colour, spectral reflectance, inflorescence architecture, and flowering time (Rathcke 1983; Johnson et al. 2003a). In Batesian mimicry, a rare unrewarding species receives pollinator visits as a result of its resemblance to a more common, rewarding species (Johnson 1994; Johnson et al. 2003a). Such mimicry is rare and confined mainly to orchids (reviewed by Johnson et al. 2003a). In contrast, with Müllerian mimicry, character convergence among two or more rewarding species increases the effective density of reward for pollinators, improving pollination (Rathcke 1983). This process would involve cooperative pollinator sharing, rather than resource parasitism in the strict sense, but no unequivocal examples of floral

Müllerian mimicry are known. However, highly attractive species with rich rewards serve as magnet species, creating a localized abundance of pollinators that benefits pollination of species with very different flowers (Johnson *et al.* 2003b). This magnet-species effect is a form of parasitic facilitation.

8.3.3 Herbivory

Herbivory of reproductive organs occurs throughout the reproductive process, from flower production to seed dispersal (Fig. 8.1). Thus, temporal variation in herbivory can impose selection on the reproductive phenology of plants (Ollerton and Lack 1998; Pilson 2000; Chapter 7). Indeed, loss of reproductive organs to herbivores can alter the pattern of resource allocation for reproduction, thereby influencing flowering phenology (Wright and Meagher 2003). Because many plants invest resources disproportionately in seed development, pre-dispersal seed predation has more severe resource and fitness consequences than herbivory during flower production or flowering (Ollerton and Lack 1998). For example, a recent study of *Castilleja linariaefolia* detected stronger phenotypic selection on floral traits by pre-dispersal seed predators than by pollinators (Cariveau *et al.* 2004). As described above (Section 8.2.2), the most common phenological response to herbivory involves phenological escape (Schemske 1984; Marquis 1988), although synchronized mass flowering can also counteract herbivory losses by satiating seed predators (Augspurger 1981).

Herbivory of flowers or fruits by multiple species that differ in their intensity and timing of damage complicates selection on phenology. To identify significant factors affecting reproductive phenology, the relative importance of individual herbivores on plant fitness should be quantified. For example, several seed-feeding insects attack developing seeds of wild sunflower, *Helianthus annuus*, during different portions of the flowering season. Phenotypic selection analyses revealed that pre-dispersal seed predation by two insect species favoured later flowering (Pilson 2000).

The response to selection for phenological escape from herbivorous damage can be retarded if pollinators are less abundant during a contiguous period. For example, early flowering of *Vaccinium hirtum*, when bumble bees were uncommon, reduced seed set, whereas later flowering increased seed predation by weevils and fly larvae (Mahoro 2002). In contrast, early flowering of *Polemonium foliosissimum* enhanced pollination, but directional selection for early flowering was counteracted by a higher risk of seed predation (Zimmerman 1980). These results illustrate that the response to selection on flowering phenologies can reflect a balance between the pollination and herbivory consequences of alternative flowering times.

Herbivory can also cause apparent reproductive competition among plant species when plants share the same floral or seed herbivores (Huntly 1991). For example, the same seed predators (anthomyiid fly larvae) attack hummingbird-pollinated *Ipomopsis aggregata* and bumble bee-pollinated *P. foliosissimum*. Because the seed predators prefer *P. foliosissimum*, *I. aggregata* individuals that flower simultaneously with *P. foliosissimum* suffer less seed loss, which could induce selection in *I. aggregata* for increased flowering overlap. However, such interspecific interactions vary between years and among sites (Brody 1997), emphasizing the importance of relevant fitness assessment based on long-term observations to evaluate herbivory effects on phenological adaptation.

8.3.4 Seed dispersal and germination

Despite occurring well after flowering, aspects of seed biology can influence the evolution of flowering phenologies, particularly for plants in seasonal environments. If the period of fruit production depends directly on flowering time, it can influence selection on the flowering phenology (Primack 1987). *Tristerix corymbosus*, a hummingbird-pollinated mistletoe, flowers during winter in the Argentine Andes, even though hummingbirds are least abundant during this season, because fruit maturation requires several months and its marsupial seed disperser is most active during summer (Aizen 2003). Similar influences may determine the flowering periods of many fleshy-fruited species that rely on seasonally available vertebrate seed dispersers. In temperate regions of the northern hemisphere, fleshy-fruited plant species commonly bloom during spring and mature their fruits during autumn and winter when frugivorous birds migrate southward (reviewed by Kimura *et al.* 2001). In overwintering regions for these birds, fleshy fruits ripen as migratory birds arrive and continuous fruiting lasts throughout the overwintering period. This temporal overlap between fruit production and bird migration is advantageous for both plants and birds, and the timing of fruiting may determine the flowering phenology of bird-dispersed plants (Thompson and Willson 1979). In contrast, bird-dispersed plant species that occupy habitats with few frugivorous birds tend to produce fruits more continuously throughout the year, in association with asynchronous flowering (Thompson and Willson 1979; Kimura *et al.* 2001).

Unlike seed dispersal, seasonal variation in opportunities for seedling establishment seems to have limited effect on flowering phenology. Such limited association is understandable for species in which seed germination is preceded by, and often requires, a period of seed dormancy. However, if seed dormancy is lacking or less effective, the timing of flowering and fruiting may significantly affect post-dispersal reproductive success. *Ochradenus baccatus*, a desert shrub, flowers primarily in association with winter rainfall, but large individuals flower continuously, even during hot, dry summers (Wolfe and Burns 2001). Winter flowers produce heavier seeds with higher germination rate than summer flowers, reflecting a diminishing resource pool through the flowering season. However, seedlings from winter flowers are exposed to dry spring conditions, whereas seedlings from the summer flowers can use winter precipitation and have higher survival. Thus *O. baccatus* illustrates disruptive selection on flowering time caused by selection on seedling establishment.

8.3.5 Sexual system

Various sexual systems and patterns of gender expression involve or influence the flowering pattern within individuals and populations. Differences in flowering phenology between sex morphs of polymorphic species have been interpreted in the context of sexual selection, which predicts more attractive and prolonged flowering of male flowers than female flowers. In dioecious species, male plants commonly bloom earlier and longer and produce larger inflorescences than female plants (Charlesworth *et al.* 1987). In *Panax trifolium*, a sex-changing ginseng, male plants present pollen longer than the protandrous hermaphrodites, completely overlapping the presentation of receptive stigmas by hermaphrodites (Schlessman *et al.* 1996). Similarly, in gynodioecious *Thymus vulgaris*, female plants have a shorter flowering period, which peaks later than that of hermaphrodites (Ehlers and Thompson 2004). These differences alter the mating environment during the flowering seasons of individuals, which seems to have selected for temporal variation in sex allocation within hermaphrodites. In particular, pollen production varied negatively with seed production in hermaphrodites, with maximal pollen production during peak flowering of female plants (Ehlers and Thompson 2004). A strong correlation between flowering phenology and sex allocation of hermaphrodite plants was also revealed by an experiment using late-flowering mutant lineages of *Arabidopsis* (Baker *et al.* 2005). In particular, the late-flowering genotype produced larger floral parts and its sex allocation was highly female biased. Together, these results suggest that the mating environment can influence the joint evolution of flowering phenology and sex-allocation patterns.

Dichogamy is a widespread floral trait (Bertin and Newman 1993) involving contrasting temporal patterns of sex function within individual flowers. This intrafloral separation for female and male function reduces interference between the sex roles of hermaphrodite flowers, increasing mating success of both functions (Lloyd and Webb 1986). Several patterns of dichogamy are evident among angiosperms, with protandry being most prevalent (Bertin and Newman 1993). Heterodichogamy, which involves a mixture of protogynous and protandrous genotypes within a population, is particularly interesting as it promotes cross-fertilization without causing temporal variation in sex ratio (e.g., Kimura *et al.* 2003).

8.3.6 Limits on phenological evolution

Despite the many factors that promote specific phenological characteristics, the evolution of flowering phenologies may be limited in some circumstances by either intrinsic constraints or features of selection.

Adaptive responses of flowering phenology can be restricted by the need for sufficient time for vegetative growth before flowering. In plants with determinate growth, vegetative growth occurs separately from reproduction, because terminal meristems differentiate into inflorescences after the production of vegetative tissue. In contrast, plants with indeterminate growth can grow and flower simultaneously, because growth involves terminal meristems, whereas lateral shoots produce inflorescences. If the onset of flowering affects vegetative development, then selection acting on aspects of vegetative morphology, including shoot architecture, height, and leaf number can limit the evolution of flowering phenology (Diggle 1999).

The occurrence of invariant developmental patterns within a clade can preclude adaptive responses of plant phenologies (Kochmer and Handel 1986; Ollerton and Lack 1992; Ollerton and Diaz 1999). Phylogenetically related species tend to have more similar flowering patterns than unrelated species (Ollerton and Lack 1992). Consequently, even intense interspecific competition for pollination may fail to modify the flowering phenologies of co-flowering related species (but note that even small adjustments of flowering overlap may successfully decrease the intensity of competition in some cases). In contrast, phylogenetic constraints are not ubiquitous, as the phenologies of shrub communities in the eastern Mediterranean illustrate. These communities include many related shrub species, but their flowering patterns seem unrelated to phylogenetic relationships (Petanidou et al. 1995).

Flowering phenology may evolve slowly if the intensity and direction of selection vary temporally or spatially. Pollinator availability and the intensity of herbivore attack can vary within and between years in response to weather conditions and influences on the dynamics of pollinators and herbivores that act independently of the focal plant species (e.g., Herrera 1995; Aizen 2001). Furthermore, plants that occupy diverse habitats may interact with different pollinator and herbivore assemblages, which cause spatial variation in the intensity and direction of selection (Johnston 1991). The relative flowering abundance of other plant species that compete for, or facilitate, pollination can also vary spatially, depending on climatic conditions and resource availability, or temporally because of vegetation dynamics. If selection acts sporadically on flowering time, competing species may coexist without shifts in their phenologies. For example, unrelated hummingbird-pollinated species in an aseasonal tropical cloud forest flower randomly, with no evidence of displacement in flowering periods (Murray et al. 1987). This pattern may exist because density-dependent competition for pollination occurs intermittently within years and the relative flowering intensity of individual species varies from year to year, causing inconsistent yearly fluctuations in interspecific interactions.

Alternatively, co-flowering species may experience limited competition for pollination if they share a diverse assemblage of generalist pollinators. Such pollinator sharing is probably more common than divergence in flowering time or specialization for pollinators (Rathcke 1988). In such cases, contrasting selection by different pollinators may result in little net selection on floral traits, including flowering phenologies (Johnson and Steiner 2000; Thompson 2001).

The examples presented in Sections 8.3.1–8.3.5 provide ample evidence that flowering phenologies enhance reproductive performance and so function as reproductive strategies. However, these sections also illustrate that the success of a particular phenology depends on interactions between diverse abiotic and biotic factors that act before, during, and after flowering. To the extent that these influences cause opposing or variable selection, adaptive phenological response may be limited (Gómez 1993; Brody 1997; Ollerton and Lack 1998). Nevertheless, many features of flowering phenologies seem to promote reproductive function of flowering plants.

8.4 Flowering phenology and selection on plant reproduction: case studies of alpine plants

Species in alpine ecosystems exhibit extensive phenological variation, because asynchronous snowmelt among snow patches creates mosaics of local environments that differ in the timing of conditions suitable for flowering and reproduction (Kudo 1991; Stanton et al. 1997; Yamagishi et al. 2005; Kudo and Hirao 2006; Fig. 8.3). Thus, local snowmelt gradients in alpine ecosystems provide replicated opportunities to evaluate the ecological significance of phenological variation on reproductive outcomes and the formation of genetic structure. In this section, I describe three aspects of the reproductive ecology of alpine plants associated with phenological variation, based on studies in the Taisetsu Mountains of Hokkaido, Japan, conducted by my research group, namely: pollination quantity and quality, interspecific competition for pollination and mating-system evolution, and metapopulation structure.

8.4.1 Flowering-time effects on pollination success and seed quality

Pollinator availability and pollination efficiency can vary during the growing season (Herrera 1995; Aizen 2001), so that the timing of flowering can affect pollination success and seed quality. The influence of seasonal changes in pollinator availability on fruit-set success of alpine plants is evident from both interspecific (Kudo and Suzuki 2002) and intraspecific comparisons (Kudo 1993; Kudo and Hirao 2006). In wind-blown fellfields where little snow accumulates during winter, flowering starts during early June and lasts until early August. In contrast, in snowbeds that accumulate thick snow cover during winter, the onset of flowering is usually delayed until mid-summer, depending on snowmelt timing, and lasts until late September (Fig. 8.3). Major pollinators during the early season include flies and over-wintered, queen bumble bees (*Bombus* spp.), but their activity is generally low due to cool weather and they visit infrequently and sporadically. In contrast, many bumble bee workers and flies visit flowers during the warm mid-season.

Reflecting the seasonal pattern in pollinator availability, species that flower early (early June to early July) within a community tend to receive low pollinator service, resulting in low relative fruit set, whereas species that flower later (late July to mid-August) experience good pollination and high fruit set (Fig. 8.4). Similar trends have been reported within individual species distributed along snowmelt gradients (e.g., Kudo 1993; Kudo and Hirao 2006). These results suggest an advantage of later flowering for successful seed production, although flowering too late often leads to reproductive failure, because of insufficient time for seed maturation.

My colleagues and I have also assessed the genetic composition of *Rhododendron aureum* seeds and found that seasonal differences in pollinator behaviour affect patterns of pollen movement and seed quality (Hirao et al. 2006). In *R. aureum* populations that flowered during mid-June, nectar-feeding bumble bee queens visited flowers opportunistically and flew long distances between visits. In contrast, in late-snowmelt populations that flowered during late July, pollen-collecting bumble bee workers flew short distances, often visiting multiple flowers per plant. These differences in pollinator behaviour resulted in production of more seeds per fruit, with seeds being sired by fewer pollen donors at early-flowering sites than at late-flowering sites (Fig. 8.5). The reduced seed set at late-season sites might result from higher self-pollination and seed abortion by inbreeding depression. This conclusion was based on the result of a hand-pollination experiment in which outcrossed flowers produced 10 times more seeds than self-pollinated flowers (Hirao et al. 2006). The increased number of sires at late-season sites might be due to frequent visits of worker bees during anthesis, resulting in the deposition of diverse outcrossing pollen even when the proportion of self-pollen dominates.

8.4.2 Interspecific competition for pollination along snowmelt gradients

When a brief growing season limits opportunities for phenological divergence to avoid competition for pollination between co-flowering species,

150 ECOLOGY AND EVOLUTION OF FLOWERS

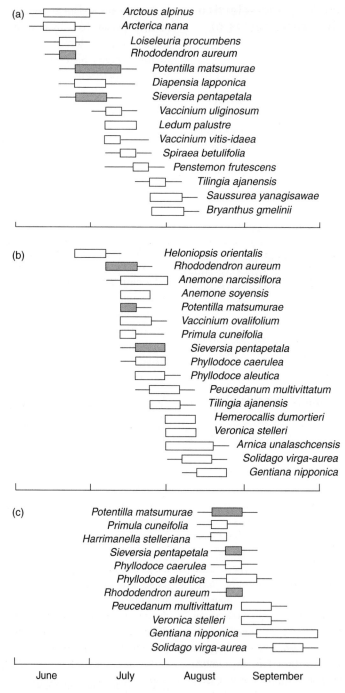

Figure 8.3 Flowering sequences of major insect-pollinated alpine plant species in (a) a fellfield, (b) an early-snowmelt snowbed, and (c) a late-snowmelt snowbed in the Taisetsu Mountains, Hokkaido, Japan. During 2005, snow disappeared from these sites in mid-May, on June 25, and on August 5, respectively. These sites were separated by a few hundred metres on a slope. For each species, the line illustrates its total flowering period within a 20 × 20 m plot and the bar depicts the period during which >30% of inflorescences displayed flowers. Closed bars identify three species that occupied all three plots.

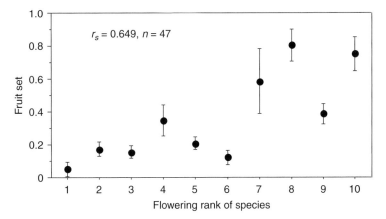

Figure 8.4 Proportion of flowers setting fruit (mean ± SE) for major ericaceous species growing in fellfield communities in the Taisetsu Mountains ranked in order of their flowering periods during the season: 1, *Arctous alpinus*; 2, *Arcterica nana*; 3, *Loiseleuria procumbens*; 4, *Rhododendron aureum*; 5, *Diapensia lapponica*; 6, *Vaccinium uliginosum*; 7, *Ledum palustore*; 8, *Rhododendron camtschaticum*; 9, *Vaccinium vitis-idaea*; 10, *Bryanthus gmelinii*. Fruit set correlates significantly with seasonal rank ($P < 0.001$, Spearman rank correlation). Revised from Kudo and Suzuki (2002). Refer to Fig. 8.3a for the flowering periods of most of these species.

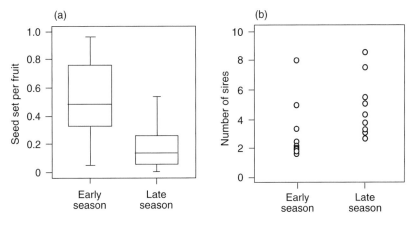

Figure 8.5 Differences in (a) seed production per ovule and (b) the number of pollen donors siring seeds in *Rhododendron aureum* fruits from populations that flowered during mid-June (early season) and late July (late season). Seed paternity was determined with microsatellite markers. Box plots in (a) show the median, interquartile range (box), and 10th and 90th percentiles. Analysis of a generalized linear model detected a significant difference in relative seed set between populations ($P < 0.001$). The number of sires was estimated for 10 outcrossed seeds per fruit and was significantly greater in the late-flowering population ($P < 0.05$: Mann–Whitney U-test). Revised from Hirao et al. (2006).

selection may favour mating-system changes for the competitively inferior species, such as the evolution of autogamous selfing in response to exploitation competition and/or heterospecific incompatibility in response to interference competition (Fig. 8.2). Comparative studies among sites separated by a few hundred metres along a snowmelt gradient have demonstrated such responses for *Phyllodoce aleutica*, which competes with *P. caerulea* (Kudo and Kasagi 2005). The flowering periods of these species overlap extensively at local sites along the snowmelt gradient (see Fig. 8.3) and they share bumble bee workers as pollinators. At sites with early or intermediate snowmelt, bumble bees prefer *P. caerulea*, an obligate outcrosser, because it produces more nectar than *P. aleutica*. At these sites, *P. aleutica* receives relatively fewer pollinator visits, which commonly deliver pollen from *P. caerulea*, so that seed production by *P. aleutica* is severely pollen limited (Fig. 8.6a). *Phyllodoce aleutica* at sites with early and intermediate snowmelt exhibit moderate self-compatibility after hand self-pollination (Fig. 8.6b) and an ability to self-pollinate autonomously (Fig. 8.6c). In addition, *P. aleutica* plants at these sites exhibit strong heterospecific incompatibility (Fig. 8.6d). In contrast, *P. aleutica* plants from sites that flower late, because of delayed snowmelt, possess traits that promote outcrossing. At these sites, the competitor, *P. caerulea*, is at its phenological limit and it produces fewer flowers and less nectar per flower than at sites that allow earlier flowering. Because of this change in relative attractiveness, bumble bees shift their preference from *P. caerulea* to *P. aleutica* during late season, increasing conspecific pollination of *P. aleutica* and alleviating pollen limitation (Fig. 8.6a). Correspondingly, *P. aleutica* from late-snowmelt sites obligately outcross, as they exhibit strong self-incompatibility (Fig. 8.6b). Interestingly, these late-flowering plants set as much seed after pollination with *P. caerulea* pollen as with conspecific pollination (Fig. 8.6d). Although the physiological relation between self-compatibility and heterospecific compatibility is unclear in this species, these studies strongly implicate adaptive responses to contrasting local competitive regimes caused by environmentally induced flowering phenology.

8.4.3 Metapopulation structure and phenological separation

Pollen dispersal is a primary mechanism of gene flow over long distances for most insect-pollinated plants (Levin 1987). Because gene flow via pollination requires that mating partners flower simultaneously, differences in flowering time

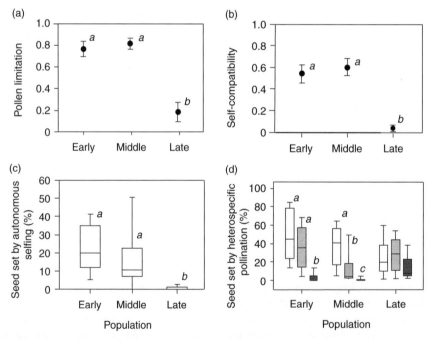

Figure 8.6 Variation in pollen limitation and mating ability of *Phyllodoce aleutica* among populations along a snowmelt gradient. (a) Mean (±SE) pollen limitation, as measured by the proportional reduction in seed production under natural pollination compared with hand cross-pollination. (b) The mean (±SE) extent of self-compatibility at each plot expressed as the ratio of seed set after hand self-pollination to cross-pollination. (c) Autonomous seed production by bagged flowers [median, interquartile range (box), and 10th and 90th percentiles]. (d) Seed set after hand-pollination with conspecific pollination (open), *Phyllodoce caerulea* pollen (dark grey), and mixture of conspecific and *P. caerulea* pollen (light grey) at each plot [median, interquartile range (box), and 10th and 90th percentiles]. Contrasting letters adjacent to results for different sites indicate significant differences ($P < 0.05$). Refer to Kudo and Kasagi (2005) for details.

caused by a heterogeneous snowmelt pattern isolate plants at neighbouring locations genetically. In an analysis of genetic differentiation of snowbed plants among snow patches, we considered isolation by both spatial distance and phenological difference along snowmelt gradients (Hirao and Kudo 2004). We detected genetic structure associated with phenological differences in *Veronica stelleri* (Fig. 8.7c and f) and *Gentiana nipponoica* (Fig. 8.7d and g), whereas genetic structure reflected geographic distance more strongly in *Peucedanum multivittatum* (Fig. 8.7b and e). The results for *V. stelleri* and *G. nipponoica* illustrate that directional gene flow induced by phenological isolation can affect the genetic structure of alpine plant populations at the landscape scale (also see Yamagishi *et al.* 2005).

The results for *P. multivittatum* and similar findings of stronger spatial than temporal isolation for *Ranunculus adoneus* in the Rocky Mountains of Colorado, USA (Stanton *et al.* 1997), reveal that phenological isolation does not affect the genetic structure of snowbed plants consistently. Species-specific patterns of genetic structure may result from the details of pollinator behaviour. The temporal, but not spatial, differentiation observed for *Gentiana nipponoica* probably reflects the ability of their highly mobile pollinators, nectar-foraging bumble bees, to disperse pollen long distances. In contrast, the significant spatial, but not temporal, genetic isolation of *P. multivittatum* may reflect local pollen flow by its dipteran pollinators. Although seed dispersal also contributes to spatial genetic structure, all species in our study seemed to disperse seeds locally, so that pollen flow probably influences genetic structure more strongly.

The existence of spatial genetic variation caused by phenological segregation could alter the nature of natural selection, even among neighbouring sites (Linhart and Grant 1996). Snowmelt time determines flowering time, a plant's access to the pollinator fauna, the diversity of competing plants, and season length for growth and reproduction of alpine plants, all of which could affect the nature of selection at sites with different flowering times. Our studies of alpine plants demonstrate that such differences can affect genotypic and phenotypic variation in reproductive traits at the landscape scale. Consequently, flowering phenology is a significant landscape feature in the ecology and evolution of alpine ecosystems.

8.5 Concluding remarks for future research

Recent phenological studies suggest that flowering phenologies can serve as adaptive strategies, especially when considered in the context of the entire reproductive process. However, phenologies have not been subject to either artificial selection or analysis of phenotypic selection under natural conditions. Evaluations of the direction and intensity of individual selection are needed for an inclusive understanding of flowering phenology as a reproductive strategy, especially given the plasticity of phenologies in response to the variable environments experienced by every plant (e.g., Stinchcombe *et al.* 2004a).

Clarification of biological interactions is crucial to evaluate the evolutionary significance of flowering phenologies. The influence of flowering phenologies in interactions among sympatric plant species must be studied with particular care, because the extent of flowering overlap and synchrony may not reflect present biological relations. Flowering behaviour is just one of various axes for character displacement among co-flowering plant species. Clarification of pollinator behaviour is extremely important to assess the effectiveness of flowering variation during pollination, and this aspect of interaction between co-flowering species has received the most attention. In contrast, interactions through floral herbivory and seed predation, such as apparent competition, have scarcely been studied, even though pre-dispersal seed predation strongly affects the reproductive success of many plants.

Studies of phenology from a strategic perspective have usually considered plant fitness as equivalent to seed production (parental component), whereas Lacey *et al.* (2003) demonstrated that seed quality is also an important fitness component (offspring component). Specifically, Lacey *et al.* found that early flowers of *Plantago lanceolata* produce more seeds, whereas late

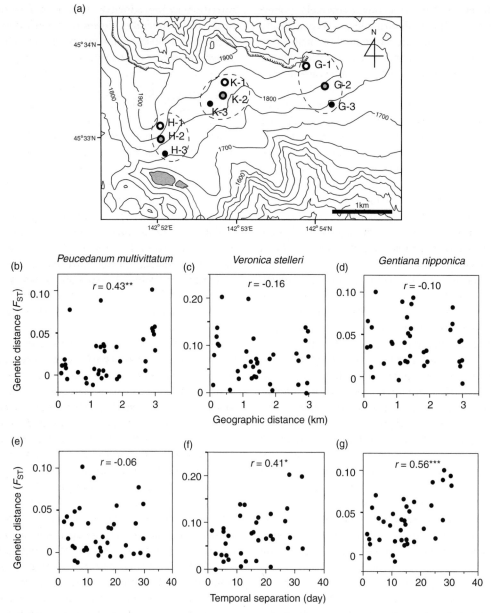

Figure 8.7 Effects of geographic distance and separation in flowering time on genetic differentiation of three alpine snowbed herbs, based on allozymes. (a) Location of the study sites, H, K, and G, in which (1) early-, (2) middle-, and (3) late-snowmelt plots were arranged along snowmelt gradients. Relations between genetic distance (pairwise F_{ST}) and geographic (b, c, and d) or phenological distance (e, f, and g) are shown for all pairs of nine populations in each species. The statistical significance of correlation coefficients was determined by partial Mantel tests (*$P < 0.05$, **$P < 0.01$, ***$P < 0.001$). Refer to Hirao and Kudo (2004) for details.

flowers produce fewer, but heavier, seeds with high seedling survival, reflecting contrasting environmental conditions during the reproductive season of the same plants. Whenever environmental conditions, such as temperature, water availability and day length, change predictably through the reproductive season, reproductive timing changes environmental conditions for maternal plants and it can influence seed quality. Such possible cross-generational fitness effects on the adaptive evolution of reproductive phenologies warrant more consideration.

Studies of phenology have traditionally considered extant variation within and among natural populations, so that much opportunity exists for experimental analysis. Manipulation of species composition would be particularly instructive in studies of interspecific influences on reproductive phenologies. In particular, the effects of invasive alien species on the reproductive success of native species may provide valuable insights into selection on phenologies (e.g., Brown *et al.* 2002). In addition to serving as a community manipulation, invasive species are interesting in their own right, because their phenologies may evolve as they occupy new environments. Common garden experiments across a wide geographic range similar to those performed on native and invasive genotypes of *Hypericum perforatum*, which found an altitudinal cline in plant size and fecundity that did not reflect traits in the original location of the invasive genotypes (Maron *et al.* 2004), would be particularly instructive in this regard. Finally, recent advances in molecular genetics offer powerful tools for studying phenology evolution. For example, the specific loci that determine flowering schedule and are responsible for a latitudinal cline in flowering time have been identified for *Arabidopsis thaliana* (Stinchcombe *et al.* 2004a, b). Such genetic material raises exciting prospects for further exploration of selection on phenological traits and the associated genetic constraints that shape phenological evolution (see Chapter 14).

The flowering phenologies of many wild plants may respond weakly to interspecific interactions, either because of abiotic restrictions on the period amenable for reproduction or because they interact in diverse biological networks, involving generalist pollinators, herbivores, and frugivores, which diffuses the intensity and direction of selection. In such cases, the main evolutionary impact of flowering phenology may involve its influence on the selection of other reproductive traits. In contrast, flowering phenologies may function as reproductive strategies most strongly for species involved in specialized relations with their pollinators or seed dispersers. These hypotheses await testing.

Our research on snowbed plants has demonstrated several ways in which flowering phenologies govern reproductive success and the selection of other traits. First, differences in the timing of flowering along the snowmelt gradient, which expose individuals to different pollination regimes, affect seed quality and paternal diversity, in addition to seed production. Second, spatio-temporal variation in interspecific competition for pollination can cause very local mating-system evolution. Third, flowering synchrony and asynchrony among populations induced by snowmelt timing create directional gene flow, thereby influencing spatial genetic structure. These results for individual species suggest that local flowering sequences determined by snowmelt gradients may govern the dynamic structure of alpine plant communities and ecosystems by affecting pollination interactions.

Nevertheless, ecological relations among multiple species within communities and their inclusive effects on reproductive outcomes and gene flow remain to be clarified. Gene flow obviously depends on flowering patterns in conspecific populations, but it may also be affected by interspecific interactions for pollination, including facilitation (e.g., Johnson *et al.* 2003a, b). Furthermore, the differential contributions of pollen dispersal and seed dispersal to gene flow need to be clarified to understand the formation and maintenance of genetic structure within and between populations. Although the biological interactions among all co-flowering species have not been studied for any community, this review demonstrates that they probably determine the ecological significance of flowering dynamics within communities and along environmental gradients, and the adaptive evolution of flowering phenologies.

Acknowledgements

I am grateful to Lawrence Harder, Spencer Barrett, James Thomson, and an anonymous reviewer for their critical comments and revision of the manuscript, to Tetsuya Kasagi and Akira Hirao for their collaboration, and to Yoko Nishikawa for her support during the writing of this chapter.

References

Aizen MA (2001). Flower sex ratio, pollinator abundance, and the seasonal pollinator dynamics of a protandrous plant. *Ecology*, **82**, 127–44.

Aizen MA (2003). Influences of animal pollination and seed dispersal on winter flowering in a temperate mistletoe. *Ecology*, **84**, 2613–27.

Ashman T-L (2004). Flower longevity. In LD Nooden, ed. *Cell death in plants*, pp. 349–62. Elsevier, London.

Augspurger CK (1981). Reproductive synchrony of a tropical shrub: experimental studies on effects of pollinators and seed predators on *Hybanthus prunifolius* (Violaceae). *Ecology*, **62**, 775–88.

Baker AM, Burd M, and Climie KM (2005). Flowering phenology and sexual allocation in single-mutation lineages of *Arabidopsis thaliana*. *Evolution*, **59**, 970–8.

Bell JM, Karron JD, and Mitchell RJ (2005). Interspecific competition for pollination lowers seed production and outcrossing in *Mimulus ringens*. *Ecology*, **86**, 762–71.

Bertin RI and Newman CM (1993). Dichogamy in angiosperms. *Botanical Review*, **59**, 112–52.

Brody AK (1997). Effects of pollinators, herbivores, and seed predators on flowering phenology. *Ecology*, **78**, 1624–31.

Brown BJ, Mitchell RJ, and Graham SA (2002). Competition for pollination between an invasive species (purple loosestrife) and a native congener. *Ecology*, **83**, 2328–36.

Cariveau D, Irwin RE, Brody AK, Garcia-Mayeya S, and von der Ohe A (2004). Direct and indirect effects of pollinators and seed predators to selection on plant and floral traits. *Oikos*, **104**, 15–26.

Charlesworth D, Schemske DW, and Sork VL (1987). The evolution of plant reproductive characters: sexual versus natural selection. In SC Stearns, ed. *The evolution of sex and its consequences*, pp. 317–36. Birkhauser Verlag, Boston.

Crone EE and Lesica P (2004). Causes of synchronous flowering in *Astragalus scaphoides*, an iteroparous perennial plant. *Ecology*, **85**, 1944–54.

de Jong TJ and Klinkhamer PGL (1991). Early flowering in *Cynoglossum officinale* L. constraint or adaptation? *Functional Ecology*, **5**, 750–6.

Diggle PK (1999). Heteroblasty and the evolution of flowering phenologies. *International Journal of Plant Sciences*, **160**, S123–34.

Ehlers BK and Thompson JD (2004). Temporal variation in sex allocation in hermaphrodites of gynodioecious *Thymus vulgaris* L. *Journal of Ecology*, **92**, 15–23.

Fox GA (2003). Assortative mating and plant phenology: evolutionary and practical consequences. *Evolutionary Ecology Research*, **5**, 1–18.

Fox GA and Kelly CK (1993). Plant phenology: selection and neutrality. *Trends in Ecology and Evolution*, **8**, 34–5.

Gómez JM (1993). Phenotypic selection on flowering synchrony in a high mountain plant, *Hormathophylla spinosa* (Cruciferae). *Journal of Ecology*, **81**, 605–13.

Harder LD and Johnson SD (2005). Adaptive plasticity of floral display size in animal-pollinated plants. *Proceedings of the Royal Society of London, Series B*, **272**, 2651–7.

Harder LD, Williams NM, Jordan CY, and Nelson WA (2001). The effects of floral design and display on pollinator economics and pollen dispersal. In L Chittka and JD Thomson, eds. *Cognitive ecology of pollination*, pp. 297–317. Cambridge University Press, Cambridge.

Herrera CM (1995). Floral biology, microclimate, and pollination by ectothermic bees in an early-blooming herb. *Ecology*, **76**, 218–28.

Hirao AS and Kudo G (2004). Landscape genetics of alpine-snowbed plants: comparisons along geographic and snowmelt gradients. *Heredity*, **93**, 290–8.

Hirao AS, Kameyama Y, Ohara M, Isagi Y, and Kudo G (2006). Seasonal changes in pollinator activity influence pollen dispersal and seed production of the alpine shrub *Rhododendron aureum* (Ericaceae). *Molecular Ecology*, **15**, 1165–73.

Huntly N (1991). Herbivores and the dynamics of communities and ecosystems. *Annual Review of Ecology and Systematics*, **22**, 477–504.

Johnson SD (1994). Evidence for Batesian mimicry in a butterfly-pollinated orchid. *Biological Journal of the Linnean Society*, **53**, 91–104.

Johnson SD and Steiner KE (2000). Generalization versus specialization in plant pollination systems. *Trends in Ecology and Evolution*, **15**, 140–3.

Johnson SD, Alexandersson R, and Linder HP (2003a). Experimental and phylogenetic evidence for floral mimicry in a guild of fly-pollinated plants. *Biological Journal of the Linnean Society*, **80**, 289–304.

Johnson SD, Peter CI, Nilsson LA, and Ågren J (2003b). Pollination success in a deceptive orchid is enhanced by co-occurring nectar plants: evidence for the magnet species effect. *Ecology*, **84**, 2919–27.

Johnston MO (1991). Natural selection on floral traits in two species of *Lobelia* with different pollinators. *Evolution*, **45**, 1468–79.

Jones KN (2001). Pollinator-mediated assortative mating: causes and consequences. In L Chittka and JD Thomson, eds. *Cognitive ecology of pollination*, pp. 259–73. Cambridge University Press, Cambridge.

Keller F and Körner C (2003). The role of photoperiodism in alpine plant development. *Arctic, Antarctic, and Alpine Research*, **35**, 361–8.

Kelly MG and Levin DA (2000). Directional selection on initial flowering date in *Phlox drummondii* (Polemoniaceae). *American Journal of Botany*, **87**, 382–91.

Kimura K, Yumoto T, and Kikuzawa K (2001). Fruiting phenology of fleshy-fruited plants and seasonal dynamics of frugivorous birds in four vegetation zones on Mt. Kinabalu, Borneo. *Journal of Tropical Ecology*, **17**, 833–58.

Kimura M, Seiwa K, Suyama Y, and Ueno N (2003). Flowering system of heterodichogamous *Juglans ailanthifolia*. *Plant Species Biology*, **18**, 75–84.

Kirkpatrick M and Nuismer SL (2004). Sexual selection can constrain sympatric speciation. *Proceedings of the Royal Society of London, Series B*, **271**, 687–93.

Kochmer JP and Handel SN (1986). Constraints and competition in the evolution of flowering phenology. *Ecological Monographs*, **56**, 303–25.

Kudo G (1991). Effects of snow-free period on the phenology of alpine plants inhabiting snow patches. *Arctic and Alpine Research*, **23**, 436–43.

Kudo G (1993). Relationship between flowering time and fruit set of the entomophilous alpine shrub, *Rhododendron aureum* (Ericaceae), inhabiting snow patches. *American Journal of Botany*, **80**, 1300–4.

Kudo G and Hirao AS (2006). Habitat-specific responses in the flowering phenology and seed set of alpine plants to climate variation: implications for global-change impacts. *Population Ecology*, **48**, 49–58.

Kudo G and Kasagi T (2005). Microscale variations in the mating system and heterospecific incompatibility mediated by pollination competition in alpine snow-bed plants. *Plant Species Biology*, **20**, 93–103.

Kudo G and Suzuki S (2002). Relationships between flowering phenology and fruit-set of dwarf shrubs in alpine fellfields in northern Japan: a comparison with a subarctic heathland in northern Sweden. *Arctic, Antarctic, and Alpine Research*, **34**, 185–90.

Lacey EP, Roach DA, Herr D, Kincaid S, and Perrot R (2003). Multigenerational effects of flowering and fruiting phenology in *Plantago lanceolata*. *Ecology*, **84**, 2462–75.

Levin DA (1987). Local differentiation and the breeding structure of plant populations. In LD Gottlieb and SK Jain, eds. *Plant evolutionary biology*, pp. 305–29. Chapman and Hall, London.

Linhart YB and Grant MC (1996). Evolutionary significance of local genetic differentiation in plants. *Annual Review of Ecology and Systematics*, **27**, 237–77.

Lloyd DG and Webb CJ (1986). The avoidance of interference between the presentation of pollen and stigmas in angiosperms. I. Dichogamy. *New Zealand Journal of Botany*, **24**, 135–62.

Lobo JA, Quesada M, Stoner KE, Fuchs EJ, Herrerías-Diego Y, Rojas J, and Saborío G (2003). Factors affecting phenological patterns of Bombacaceous trees in seasonal forests in Costa Rica and Mexico. *American Journal of Botany*, **90**, 1054–63.

Mahoro S (2002). Individual flowering schedule, fruit set, and flower and seed predation in *Vaccinium hirtum* Thunb. (Ericaceae). *Canadian Journal of Botany*, **80**, 82–92.

Maron JL, Vila M, Bommarco R, Elmendorf S, and Beardsley P (2004). Rapid evolution of an invasive plant. *Ecological Monographs*, **74**, 261–80.

Marquis RJ (1988). Phenological variation in the neotropical understory shrub *Piper arieianum*: causes and consequences. *Ecology*, **69**, 1552–65.

McNeilly T and Antonovics J (1968). Evolution in closely adjacent plant populations. IV. Barriers to gene flow. *Heredity*, **23**, 205–18.

Meagher TR and Delph LF (2001). Individual flower demography, floral phenology and floral display size in *Silene latifolia*. *Evolutionary Ecology Research*, **3**, 845–60.

Melampy MN (1987). Flowering phenology, pollen flow and fruit production in the andean shrub *Befaria resinosa*. *Oecologia*, **73**, 293–300.

Morales MA, Dodge GJ, and Inouye DW (2005). A phenological mid-domain effect in flowering diversity. *Oecologia*, **142**, 83–9.

Motten AF (1986). Pollination ecology of the spring wildflower community of a temperate deciduous forest. *Ecological Monographs*, **56**, 21–42.

Murray KG, Feinsinger P, Busby WH, Linhart YB, Beach JH, and Kinsman S (1987). Evaluation of character displacement among plants in two tropical pollination guilds. *Ecology*, **68**, 1283–93.

Ollerton J and Diaz A (1999). Evidence for stabilizing selection acting on flowering time in *Arum maculatum* (Araceae): the influence of phylogeny on adaptation. *Oecologia*, **119**, 340–8.

Ollerton J and Lack AJ (1992). Flowering phenology: an example of relaxation of natural selection? *Trends in Ecology and Evolution*, **7**, 274–6.

Ollerton J and Lack AJ (1998). Relationship between flowering phenology, plant size and reproductive success in *Lotus corniculatus* (Fabaceae). *Plant Ecology*, **139**, 35–47.

O'Neil P (1997). Natural selection on genetically correlated phenological characters in *Lythrum salicaria* L. (Lythraceae). *Evolution*, **51**, 267–74.

Pavón NP and Briones O (2001). Phenological patterns of nine perennial plants in an intertropical semi-arid Mexican scrub. *Journal of Arid Environments*, **49**, 265–77.

Petanidou T, Ellis WN, Margaris NS, and Vokou D (1995). Constraints on flowering phenology in a phryganic (East Mediterranean shrub) community. *American Journal of Botany*, **82**, 607–20.

Pilson D (2000). Herbivory and natural selection on flowering phenology in wild sunflower, *Helianthus annuus*. *Oecologia*, **122**, 72–82.

Primack RB (1987). Relationships among flowers, fruits, and seeds. *Annual Review of Ecology and Systematics*, **18**, 409–30.

Ramsey M (1995). Causes and consequences of seasonal variation in pollen limitation of seed production in *Blandfordia grandiflora* (Liliaceae). *Oikos*, **73**, 49–58.

Rathcke B (1983). Competition and facilitation among plants for pollination. In L Real, ed. *Pollination ecology*, pp. 305–29. Academic Press, Orlando, FL.

Rathcke B (1988). Interactions for pollination among coflowering shrubs. *Ecology*, **69**, 446–57.

Rathcke B and Lacey EP (1985). Phenological patterns of terrestrial plants. *Annual Review of Ecology and Systematics*, **16**, 179–214.

Real L (1983). Microbehavior and macrostructure in pollinator-plant interactions. In L Real, ed. *Pollination ecology*, pp. 287–304. Academic Press, Orlando, FL.

Sakai S, Momose K, Yumoto T, et al. (1999). Plant reproductive phenology over four years including an episode of general flowering in a lowland Dipterocarp forest, Sarawak, Malaysia. *American Journal of Botany*, **86**, 1414–36.

Schemske DW (1984). Population structure and local selection in *Impatiens pallida* (Balsaminaceae), a selfing annual. *Evolution*, **38**, 817–32.

Schlessman MA, Underwood NC, and Graceffa LM (1996). Floral phenology of sex-changing dwarf ginseng (*Panax trifolium* L., Araliaceae). *American Midland Naturalist*, **135**, 144–52.

Stanton ML, Galen ML, and Shore JS (1997). Population structure along a steep environment gradient: consequences of flowering time and habitat variation in the snow buttercup *Rannunculus adoneus*. *Evolution*, **51**, 79–94.

Stinchcombe JR, Dorn LA, and Schmitt J (2004a). Flowering time plasticity in *Arabidopsis thaliana*: a reanalysis of Westerman and Lawrence (1970). *Journal of Evolutionary Biology*, **17**, 197–207.

Stinchcombe JR, Weinig C, Ungerer M, et al. (2004b). A latitudinal cline in flowering time in *Arabidopsis thaliana* modulated by the flowering time gene *FRIGIDA*. *Proceedings of the National Academy of Sciences of the United States of America*, **101**, 4712–7.

Thompson JD (2001). How do visitation patterns vary among pollinators in relation to floral display and floral design in a generalist pollination system? *Oecologia*, **126**, 386–94.

Thompson JN and Willson MF (1979). Evolution of temperate fruit/bird interactions: phenological strategies. *Evolution*, **33**, 973–82.

van Doorn WG (1997). Effects of pollination on floral attraction and longevity. *Journal of Experimental Botany*, **48**, 1615–22.

Weis AE and Kossler TM (2004). Genetic variation in flowering times induces phenological assortative mating: quantitative genetic methods applied to *Brassica rapa*. *American Journal of Botany*, **91**, 825–36.

Wieder RK, Bennett CA, and Lang GE (1984). Flowering phenology at Big Run Bog, West Virginia. *American Journal of Botany*, **71**, 203–9.

Wolfe LM and Burns JL (2001). A rare continual flowering strategy and its influence on offspring quality in a gynodioecious plant. *American Journal of Botany*, **88**, 1419–23.

Wright JW and Meagher TR (2003). Pollination and seed predation drive flowering phenology in *Silene latifolia* (Caryophyllaceae). *Ecology*, **84**, 2062–73.

Yamagishi H, Taber AD, and Ohara M (2005). Effect of snowmelt timing on the genetic structure of an *Erythronium grandiflorum* population in an alpine environment. *Ecological Research*, **20**, 199–204.

Zimmerman M (1980). Reproduction in *Polemonium*: pre-dispersal seed predation. *Ecology*, **61**, 502–6.

CHAPTER 9

Flower performance in human-altered habitats

Marcelo A. Aizen[1] and Diego P. Vázquez[2,3]

[1] Laboratorio Ecotono, CRUB, Universidad Nacional del Comahue, Río Negro, Argentina
[2] Instituto Argentino de Investigaciones de las Zonas Áridas, Centro Regional de Investigaciones Científicas y Tecnológicas, Mendoza, Argentina
[3] National Center for Ecological Analysis and Synthesis, University of California, Santa Barbara, CA, USA

Outline

The functioning and performance of flowers and their associated pollinators are susceptible to human-driven habitat alteration. Although habitat alteration is increasingly perceived as an important threat to the integrity of the pollination process with practical and economic consequences, the relative importance of the mechanisms mediating the response of plant reproduction to habitat disturbance is not understood clearly. Here we provide a conceptual framework to help identify critical variables and guide the design of more process- and mechanism-oriented studies of the effects of anthropogenic habitat disturbances on flower performance. With a series of qualitative matrices, we summarize the effects of different disturbance types on different plant and pollinator attributes and evaluate how these attributes affect different aspects of pollination and plant reproduction. Although different disturbances can have distinctive immediate effects on plants and pollinators, they mediate their responses by affecting a series of common environmental, plant, and pollinator attributes. Our characterization of disturbance effects and their consequences could be translated easily into a path-analysis or other structural-model-building approach, which can help stimulate a more mechanistic focus for future research. Last, we identify some plant and animal attributes whose roles in different aspects of pollination have been little studied or not addressed directly in the context of habitat alteration. We also discuss the role of plant sexual system and pollination specialization in modulating the reproductive response of plants to habitat alteration, and structural features of plant–pollinator networks that may buffer pollination function against extinction of individual species.

9.1 Introduction

The habitats where flowering plants grow and reproduce are increasingly altered to different degrees by human activities. Habitat fragmentation, fire, clearcut and selective logging, invasion by alien species (plants, pathogens, and herbivores), and different types of chemical alteration (herbicide and pesticide use, pollution) are among the most common disturbances associated with humans that can disrupt plant–pollinator interactions (Aizen and Feinsinger 1994a; Kearns *et al.* 1998). Although flowering plants evolve and diversify in an ever-changing world (including, for instance, natural habitat fragmentation, fire, and species exchange), the rate, scale, and intensity of anthropogenic disturbances probably exceed those previously experienced by plants (Kearns et al. 1998).

The functioning and performance of flowers in these human-altered, strongly modified landscapes probably differs from that in less modified landscapes. Although considerable empirical evidence has accumulated during the past two decades documenting the effects of different anthropogenic disturbances on pollinator communities and plant reproduction (e.g., Aizen and Feinsinger 1994a; Renner 1998; Steffan-Dewenter and Tscharntke 1999; Cunningham 2000; Vázquez and Simberloff 2004), a unified framework that allows integrated consideration of the mechanisms involved and their relative importance remains elusive. On the other hand, despite analyses of the effects of plant individual, population, or community attributes on pollination and reproductive success, these effects have usually been assessed without consideration of the disturbance context that is likely to modify them (Ghazoul 2005a).

The primary goal of this chapter is to provide a comprehensive framework to aid in identifying the variables and mechanisms that determine the effects of different anthropogenic perturbations on pollination and plant reproduction, as well as organizing and designing studies of disturbance effects on flower function in human-altered landscapes. Fundamental to this framework is the recognition that disturbance can modify either plant or pollinator attributes, which in turn may affect different aspects of plant pollination and reproductive success (Fig. 9.1). Although many studies have addressed the effects of anthropogenic disturbance (particularly habitat fragmentation) on one or more components of plant reproduction, they mostly provide little guidance in understanding the processes and mechanisms behind the cause–effect relationships depicted in Fig. 9.1 (but see Vázquez and Simberloff 2004; Larsen et al. 2005). This shortcoming results in part because key variables that can be measured easily and provide critical information (e.g., pollen receipt) are usually overlooked, but more generally because of the lack of a unified conceptual framework that may help in identifying key variables.

To delve deeper into the processes involved, we develop a series of qualitative matrices based on each of the links depicted in Fig. 9.1, which summarize the effects of different disturbance types on plant and pollinator attributes and their consequences for pollination and plant reproduction. Through this systematic exercise we demonstrate that although different disturbance types can have distinctive immediate effects on plants and pollinators, they affect a common set of plant and pollinator attributes through a relatively few environmental parameters. We also emphasize that the magnitude and direction of changes on plants, pollinators, and their interaction differ not only among but also within types of human-induced alterations, depending on disturbance frequency and intensity. We also illustrate how our matrix representation can be translated into a path-analysis or other model-building approach to guide future research focused on mechanisms, rather than on patterns (see Section 9.8). Last, our examination uncovers some plant and animal

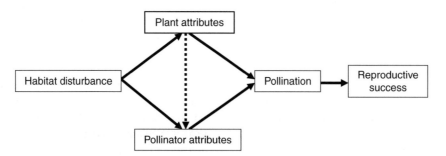

Figure 9.1 Conceptual diagram showing the links between habitat disturbance, plant and pollinator attributes, and their effects on pollination and plant attributes. Some effects of habitat perturbation on pollinator attributes can be mediated by the effects of habitat perturbation on plant attributes (dotted line).

attributes whose roles in different aspects of pollination have been little studied in the context of habitat alteration. Other attributes, such as sexual system and pollination specialization, are not contemplated in our matrices, because they are not expected to be modified by disturbance, at least in the short term. However, because of their critical importance in the pollination process we discuss their roles as modulators of the response of plant reproduction to disturbance.

Traditionally, studies of floral biology and its response to human-caused perturbations have focused on individual plant species and their associated pollinators, or individual pollinator species and their plant hosts. An emerging perspective adopts a broader focus, centring on entire, or large subsets of, local plant–pollinator interaction networks (see Jordano 1987; Waser et al. 1996; Bascompte et al. 2003; Vázquez and Simberloff 2003; Vázquez and Aizen 2004). Although this broader view overlooks many of the details revealed by studies of particular pollination interactions, it exposes a deeper understanding of the structure of plant–pollinator interactions (also see Chapter 6) and the effects of disturbance on pollination function. In addition, the study of anthropogenic effects on pollination webs may identify better measures of the integrity of the pollination function than those provided by single-species examples (Aizen and Feinsinger 2003). Within our general conceptual framework, we review recent developments on the effects of disturbance on plant–pollinator webs and highlight new directions in this area of active and promising research.

Like all studies on the effects of disturbance on plant pollination and reproduction, our chapter is based on the unstated assumption that seed production governs the population dynamics of plant populations. When populations are seed-limited, anthropogenic perturbation that impairs pollination function and increases pollen limitation will decrease population growth (Ashman et al. 2004; Knight et al. 2005). Although populations of short-lived species (Silvertown 1993) and some invasive species (Parker 1997) seem to be seed-limited, seed production does not seem to influence the demography of trees and other long-lived species strongly. However, complete disruption of plant–pollinator interactions may eventually cause the demographic demise of any plant population that depends on sexual reproduction for long-term survival. In addition, pollination-mediated effects of anthropogenic disturbance on progeny quality can erode a plant's evolutionary plasticity (Aizen and Feinsinger 2003).

9.2 General effects of disturbance and their reproductive consequences

To characterize the consequences of different disturbance types on pollination and plant reproduction we constructed five matrices, which relate the direct and indirect qualitative effects of disturbance on plant populations as depicted in Fig. 9.1. The first two matrices identify direct effects of disturbance types on plant and pollinator attributes, although we also acknowledge the indirect effects of disturbance on pollinator attributes mediated by changes in plant attributes. The next two matrices relate the effects of the modified plant and pollinator attributes on different aspects of stigmatic pollen deposition, which also mirror aspects of pollen export and thus of male function (e.g., Harder and Barrett 1995; Chapter 4). Finally, the fifth matrix relates pollination success to reproductive output. Each matrix cell contains a symbol indicating the currently understood effect of a row factor or attribute on a column factor or attribute. These symbols include: ↑, a positive relation; ↓, a negative relation; ↑↓, both positive and negative effects are possible; 0, no expected net change; Δ, a change that lacks directionality (e.g., flowering phenology) or involves multiple dimensions (e.g., species composition); and ?, a causal relation for which the expected direction of change is uncertain, despite a presumed effect (e.g., invasion by alien plants could change the quantity and quality of flower rewards through, for instance, changes in the resource status of focal native plants). Rather than reviewing the evidence on the effects of disturbance on flower function exhaustively, we present a series of predictions based on the most likely change(s) that we expect.

Our analysis considers the most common types of disturbances that have relatively immediate impacts on pollination and plant reproduction: (1) habitat fragmentation, (2) fire, (3) selective plant harvesting (e.g., selective logging), (4) introduction of herbivores (e.g., cattle, defoliating insects), (5) introduction of plants (e.g., crops, invasive plants), and (6) chemical disturbance (e.g., contamination with heavy metals, use of pesticides and herbicides). We did not include global-scale disturbances, such as climate change, because their pollination impacts are probably subtle and long term. We recognize that many of the disturbances listed above can be interrelated (e.g., fire may cause habitat fragmentation, or habitat fragmentation could foster invasion of aliens), but they have distinctive impacts on landscapes and individual plant features and so can be studied in isolation.

9.3 Effects of human-caused perturbations on plant attributes

Anthropogenic disturbances modify diverse attributes of individual plants, their populations, and communities. In our analysis and discussion, we consider the following features of individual plants: (1) display size (i.e., number of open flowers), (2) flower morphology (including flower size and shape), (3) flower rewards (mainly pollen and nectar), (4) flower physiology (including aspects of floral metabolism that could influence flower lifespan, pollen viability and performance, etc.), and (5) flowering phenology (including onset of flowering, duration, and intensity). The population attributes that we consider include (6) abundance, (7) density, and (8) relative density (i.e., the ratio of number of flowering individuals of a focal species to the number of individuals of other species that flower simultaneously). Finally, we consider the following community attributes: (9) plant species diversity (including both species richness and relative abundance) and (10) plant species composition (Table 9.1).

Except for the introduction of exotic plant species and pesticide application, the perturbations that we consider primarily change local environmental conditions by removing plant biomass, immediately increasing light availability, diurnal temperature, and evapotranspiration and changing nutrient pools. However, the graininess and spatial heterogeneity of the changes in environmental conditions depend on disturbance type. For instance, fragmentation of forest habitats increases light availability by increasing the edge:interior ratio (Fahrig 2003), whereas fires increase light availability by reducing the cover of fire-susceptible species (Waltz and Covington 2004).

Increased light availability and a sudden release of nutrients caused by massive disturbance could in turn trigger immediate physiological changes in remaining and newly recruited individuals (Table 9.1). For instance, higher insolation can increase flower production and display (e.g., Cunningham 1997), which could occur at the expense of, or be accompanied by, increased flower size (e.g., Sato and Yahara 1999), altered flowering phenology (Rathcke and Lacey 1985), enhanced nectar secretion (Rathcke 1992), and

Table 9.1 Predicted effects of different disturbance types on the attributes of individual plants (floral display, flower morphology, rewards, flower physiology, and flowering phenology), plant populations (size, absolute density, and relative density) and plant communities (species diversity and composition).

Disturbance type	Individual					Population			Community	
	Floral display	Flower morphology	Rewards	Flower physiology	Flower Phenology	Size	Absolute density	Relative density	Diversity	Composition
Fragmentation	↑	?	↑	?	Δ	↓	↑↓	↑↓	↑↓	Δ
Fire	↑	?	↑	↑	Δ	↑↓	↑↓	↑↓	↑↓	Δ
Selective harvesting	0/↑	?	0/↑	?	0/Δ	↓	↓	↓	↑↓	Δ
Herbivores	↑↓	↓	↓	↓	Δ	↓	↓	↑↓	↑↓	Δ
Exotic plants	↓	?	↓	↓	Δ	↓	↓	↓	↑↓	Δ
Chemical agents	↓	↓	↓	↓	?	↓	↓	↑↓	↑↓	Δ

changes in pollen production (e.g., Etterson and Galloway 2002). Changes in environmental conditions caused by disturbance can also affect paternal reproductive success through changes in pollen quality. For instance, pollen grains produced by water-stressed plants have a lower capacity for siring seeds than grains from plants in benign conditions (Young and Stanton 1990).

Generalist herbivores could also increase light availability; however, because they may affect flowering directly by eating flowers, or indirectly by eating vegetative tissues and reducing photosynthetic capacity, their net effect on floral display depends on the relative effects of tissue removal and increased light availability on plant performance (Chapter 7). Sometimes, herbivores may increase floral display by stimulating meristem production (Paige and Whitham 1987). Herbivory may also affect flowering phenology (Brody 1997) and reduce flower size (Strauss *et al.* 1996; Mothershead and Marquis 2000), nectar secretion (Krupnick *et al.* 1999), and pollen quantity and quality (Quesada *et al.* 1995; Strauss *et al.* 1996; Aizen and Raffaele 1998).

The effects of plant harvesting depend on the abundance of the target species. For instance, if harvesting focuses on a relatively rare species whose removal has limited impact on habitat structure, then attributes of individual plants in the remaining vegetation will be little affected. However, when harvesting causes substantial habitat destruction or targets an abundant species (e.g., logging of a dominant tree species), the resulting widespread habitat modification could induce all the physiological changes associated with increased light availability (e.g., Ghazoul and McLeish 2001; Table 9.1).

Other disturbances are more likely to have negative effects on individual plant attributes, but through different mechanisms (see also Chapter 7). Invasion of exotic plants may have an overall negative effect on plant performance by increasing competition for either light or resources, whereas chemical agents such as herbicides may have similar effects through their direct effects on plant metabolism.

Table 9.1 also identifies some influences of disturbance on individual-level attributes which are difficult to predict, primarily because they have been little studied. For instance, recent studies demonstrate that flower symmetry and shape can influence pollination (Neal *et al.* 1998), but almost nothing is known about how different types of disturbances modify these floral traits. Developmental stability, expressed as fluctuating asymmetry (Palmer 1996), may indicate the degree of environmental stress experienced by an organism. Although much research has addressed this topic for a variety of organisms and organs (including flowers; see Møller 2000), the evolutionary significance of fluctuating asymmetry remains controversial (Palmer 2000).

The clearest and most consistent predictions involve the population consequences of disturbance. For instance, disturbance immediately decreases plant abundance, because of either reductions in the size and number of habitat patches (fragmentation) or increased individual mortality (other perturbation types), although in frequently disturbed areas the abundance of light- or fire-tolerant species might increase in the long term. Population density also probably decreases through increased mortality caused by most perturbation types, except fragmentation and fire (Table 9.1). For instance, in the case of fragmentation the net effect on population density depends on complex indirect effects relating the magnitude and scale of habitat fragmentation to different life-history traits of focal species (e.g., light tolerance). The effect on relative population density (i.e., density of the focal plant species relative to that of simultaneously flowering plant species that share pollinators) depends on both the change in absolute density of the focal species and the response of other species to perturbation (Ghazoul 2005a). Thus, if fragmentation, fire, herbivores, introduced plants, and chemical agents favour perturbation-resistant species, the relative density of the focal species should decrease, whereas if the focal species is itself perturbation-resistant, its relative density could actually increase (Table 9.1). Selective harvesting should decrease the relative density of the focal species as long as it does not affect the density of other species substantially.

Community effects of disturbance depend on the individual responses of the focal and other plant species. The community response depends strongly on the intensity, frequency, and spatial scale of perturbations (Sousa 1984; Chesson and Huntly 1997), and so it is difficult to predict. However, a decrease in species diversity and a relative increase in dominance by one or a few disturbance-resistant species are expected under the intense, frequent, and large-scale disturbances characteristic of many anthropogenic habitat alterations.

9.4 Effects of human-caused perturbations on pollinator attributes

Pollination of most plant species depends, to different degrees, on flower visitation by animals, so we consider three aspects of pollinator visits that disturbance can modify independently of changes in plant attributes: (1) total visit frequency, (2) pollinator diversity, and (3) the composition of the pollinator fauna (Table 9.2). Although this list is not exhaustive, these plant and pollinator attributes are discussed most commonly in the pollination ecology literature and are probably most susceptible to anthropogenic perturbations.

Habitat disturbance can strongly affect pollinator assemblages. Most pollinators are short-lived insects with a fine-grained perception of their environment and are thus quite susceptible to local changes in resource supply and habitat structure (Didham *et al.* 1996; Aizen and Feinsinger 2003). Changes in the abundance of individual pollinator species and the composition of pollinator assemblages can result directly from altered environmental and structural characteristics of the habitats (e.g., Ghazoul and McLeish 2001; Burgess *et al.* 2006)

Table 9.2 Predicted effects of different disturbance types on the overall abundance, diversity, and composition of pollinator communities.

Disturbance type	Abundance	Diversity	Composition
Fragmentation	↑↓	↑↓	Δ
Fire	↑↓	↑↓	Δ
Selective harvesting	↑↓	↑↓	0/Δ
Herbivores	↑↓	↑↓	Δ
Exotic plants	↑↓	↑↓	Δ
Chemical agents	↓	↓	Δ

or be mediated by changes in plants which provide their food (indicated by a dotted line in Fig. 9.1). For instance, invasion of alien plants, which provides accessible and abundant floral resources, can facilitate the invasion of alien flower visitors independent of habitat disturbance (Morales and Aizen 2006).

In general, all disturbance types can decrease total pollinator abundance through increased mortality and habitat destruction, including a reduction in nesting sites. However, pollinator abundance may also increase in the short term, because some perturbations can increase individual plant floral display and promote the encroachment of mass-flowering, light-demanding species (Table 9.2; Westphal *et al.* 2003). All else being equal, a change in pollinator abundance should cause a change of the same direction and magnitude in pollinator visitation frequency. However, the net effect of disturbance on visitation frequency to flowers of the focal species depends on a complex interplay among direct effects of disturbance on pollinator abundance, changes in plant population size and absolute and relative density, the degree of pollination specialization, and other plant traits that determine the interaction with pollinators (Ghazoul 2005a). Thus, excluding cases such as intense, large-scale anthropogenic perturbations (e.g., urbanization) or the use of chemical agents for biological control (e.g., insecticides), disturbances may not affect either total pollinator abundance or visitation frequency (see also Ghazoul 2005b).

Net effects on the diversity and composition of pollinator assemblages depend on three factors: the structural complexity of habitats, which can be greatly simplified by human-driven disturbances; nectar and pollen availability (Westphal *et al.* 2003); and the diversity and composition of flowering species (Morales and Aizen 2006). For instance, mass flowering of an invading plant species with a generalist pollination system could sustain a pollinator community as rich as, or richer than, a plant assemblage composed of several, relatively specialized native plants. However, changes in habitat structure, light and temperature, resource availability, and plant community composition should be, and usually are,

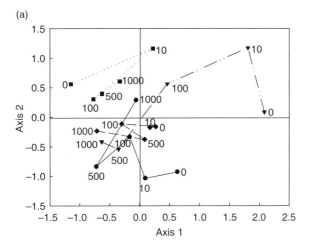
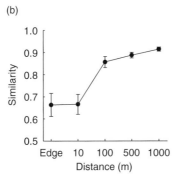

Figure 9.2 Effects of distance from native vegetation on pollinator species richness at 0, 10, 100, 500 and 1000m from premontane subtropical forest in four grapefruit (*Citrus paradisi*) plantations in northwest Argentina, based on a 1 Bray–Curtis distance coefficient. (a) Non-metric multidimensional scaling (NMDS) ordination of the pollinator assemblages. Axes 1 and 2 explain 43.7 and 25.3% of total variance in pollinator assemblage composition, respectively. The NMDS is based on a matrix of 50 species per 20 site × distance classes. Line segments link flower-visiting faunas at increasing distances from the forest edge within the same plantations. The four plantations are represented by different symbols. (b) Mean (±SE) similarity (averaged over all plantation pairs) versus distance to the edge. Reproduced with permission from Chacoff and Aizen (2006).

accompanied by strong changes in the composition of pollinator faunas (e.g., Potts *et al.* 2001). In many instances, anthropogenic disturbance increases the dominance of one or a few pollinator species (Aizen and Feinsinger 1994b; Morales and Aizen 2006). The most striking example is the increasing domination of disturbed pollinator communities by Africanized *Apis mellifera* throughout most of the Neotropics (Goulson 2003). This bee has become the dominant visitor to the flowers of many native plants that previously sustained rich assemblages of many pollinator species (e.g., Aizen and Feinsinger 1994b).

In addition to impoverishing pollinator faunas, intense and frequent disturbances can homogenize faunas over space, thus decreasing β-diversity. For instance, bee assemblages in grapefruit plantations in northwest Argentina become increasingly similar with distance from the forest edge (Fig. 9.2). Similarly, disturbed forest areas of northwest Patagonia tend to sustain convergent pollinator assemblages, independent of disturbance type (Morales and Aizen 2006).

9.5 Relation of pollination to modified plant attributes

Modification of individual, population, and community attributes of plants mediate indirect pollination responses to anthropogenic disturbance

Table 9.3 Predicted effects of the individual, population, and community-level attributes of plants listed in Table 9.1 on pollination variables.

Attribute level	Attribute	Quantity	Quality	Purity
Individual	Floral display	↑	↑↓	↑
	Floral morphology	↑	↑↓	↑
	Rewards	↑	↑↓	↑
	Reproductive physiology	↑	↑	?
	Phenology	Δ	?	Δ
Population	Size	↑↓	↑	0
	Absolute density	↑↓	↑	0
	Relative density	↑	↑	↑
Community	Diversity	?	?	?
	Composition	?	?	?

(Table 9.3). We consider three specific pollination outcomes: (1) pollen receipt; (2) the quality of received conspecific pollen, including genetic aspects, such as the proportion of self-pollen and the diversity of pollen donors, and/or physiological status affecting pollen viability; and (3) the purity of pollen deposition (i.e., number of conspecific versus heterospecific pollen grains). All of these outcomes can strongly influence reproductive success, including seed quantity and seed quality (e.g., seed size, germination rates, seedling vigour). Although these pollination components explicitly involve female function (i.e., seed set), they parallel male performance (i.e., seed siring), because poor cross-pollen receipt results from limited pollen export, self-pollination may limit pollen export and siring success on other plants (pollen discounting), and heterospecific pollination also causes lost siring opportunities (Chapter 4).

Changes in floral display and floral traits affect pollination success through various mechanisms. Indeed, much work during recent decades has considered the consequences of individual phenotypic variation in various floral attributes on pollinator attraction and pollination quantity and quality, mostly in an evolutionary context (e.g., Nilsson 1988; Herrera 1993; Neal *et al*. 1998; Chapters 2, 6, 14, and 15). Some of these studies show that variability in floral traits, including flowering phenology, often has contrasting effects on different pollination components. For instance, increased floral display or nectar production may enhance pollen removal and deposition by increasing pollinator attraction to attractive and rewarding plants. However, these traits may also increase self-pollination and pollen discounting by increasing the number of flowers visited per plant by individual pollinators (geitonogamy) and, in the case of nectar, the time that each pollinator spends visiting individual flowers (autogamy) (Harder and Barrett 1995; Eckert 2000). Thus, whereas these individual-level attributes may initially enhance both pollination quantity and quality by increasing pollinator attraction, they may eventually decrease pollination quality through the transfer of self-pollen while still increasing pollen deposition (Table 9.3).

Effects of perturbation on plant population attributes may also influence pollination quantity. Pollen receipt increases in a decelerating manner with population size or density, because of increasing pollinator attraction, until it decreases because of competition between neighbouring conspecifics and heterospecifics for a limited pollinator pool (Rathcke 1983; Kunin 1997; Brown *et al*. 2002; Table 9.3). Thus, perturbation effects on pollination quantity mediated by population size and density will depend on the pre- and post-perturbation levels of these attributes. However, this curvilinear relation indicates that two populations with contrasting sizes or densities could experience similar low pollination for contrasting reasons: limited attraction in small or low-density populations and intraspecific competition for pollinator service in large, high-density populations.

Population size and density may also affect pollination quality. Larger or denser populations may experience improved pollination quality (Table 9.3), if an increase in the number of conspecific individuals enhances either the genetic diversity represented in the pollen loads or the intensity of competition among male gametophytes (Mulcahy *et al*. 1996). Pollination quality commonly declines in fragmented populations of different tree species due to increased inbreeding (e.g., Aizen and Feinsinger 1994a; Cascante *et al*. 2002). These changes result principally from reduced population size or density, but more direct tests that account for the confounding effect of changing pollinator assemblages would be valuable.

Perturbations are likely to affect pollination quality and purity through changes in relative population density, particularly when generalist pollinators also visit other co-flowering plant species (Rathcke 1983; Kunin 1997; Brown et al. 2002). In this case, pollinators visit flowers of other species more frequently as the relative density of the focal species decreases, thus depositing proportionally more heterospecific pollen grains (Table 9.3). In extreme cases, the deposition of a large amount of heterospecific pollen could usurp space on the stigma, interfering with either the germination or tube growth of conspecific pollen. In addition, foreign pollen can have allelopathic effects on germination and tube growth of conspecific pollen (e.g., Murphy and Aarssen 1995). Although the deposition of heterospecific pollen is highly variable in nature, usually representing a low fraction of all pollen deposited (McLernon et al. 1996), it might become important when disturbance involves, or is accompanied by, the invasion of an alien flowering plant (Brown et al. 2002).

Effects of plant diversity and composition are important to the extent that they affect the relative density of co-flowering plant species that share pollinators with the focal plant species, although these effects are difficult to predict (Table 9.3). The hypothesis that increased floral diversity sustains richer pollinator assemblages which provide more efficient and predictable pollination services in terms of both quality and quantity (Aizen and Feinsinger 2003) awaits formal testing. However, co-flowering species can facilitate each other's pollination at low population densities (Rathcke 1983; Moeller 2004; Chapter 6), and the mere presence of flowering plants of a few rewarding species can facilitate the pollination of rewardless species (Johnson et al. 2003).

9.6 Relation of pollination to modified pollinator attributes

Pollinator attributes usually have direct and strong implications for pollination (Table 9.4). The effectiveness of a pollinator species, or its quantitative

Table 9.4 Predicted effects of the attributes of pollinator communities listed in Table 9.2 on pollination variables.

Attribute	Quantity	Quality	Purity
Abundance	↑	↑	0
Diversity	?	↑↓	↑↓
Composition	↑↓	↑↓	↑↓

contribution to the pollination of a given plant species, is the product of its visit frequency and the amount of pollen deposited per visit. Although both factors are important, visit frequency predicts total pollinator efficiency most closely, because its variation overwhelms that of per-visit effectiveness (Vázquez et al. 2005). This result indicates that quantitative aspects of pollination (both pollen removal and pollen deposition) depend strongly on pollinator abundance, provided that abundance and visit frequency vary positively. Pollinator abundance could also influence the quality of pollination, because large stigmatic pollen loads increase genetic diversity and opportunities for selection among germinating pollen grains (Mulcahy et al. 1996).

Pollinator diversity and composition can also influence quantitative and qualitative aspects of pollination. For instance, coffee fruit production is enhanced by increases in bee species richness, particularly that of solitary bees, independent of pollinator abundance (Klein et al. 2003). For a given visitation frequency, the effects of increased pollinator diversity or changes in its composition on pollination could range from negative to positive, depending on both plant characteristics and pollinator traits (Table 9.4). For a highly outcrossing plant species pollinated efficiently by a large-bodied, mobile bee, increased pollinator diversity or changes in assemblage composition could degrade pollination quantity and quality. However, the opposite trend could be quite common. For instance, *Apis mellifera* usually forage preferentially on highly localized nectar and pollen sources, despite being able to fly several kilometres from their nests. Thus, the replacement of diverse pollinator assemblages by the Africanized honeybee throughout the Neotropics (Goulson 2003) could decrease cross-pollination and increase self-pollination without net changes in total pollination (Aizen and Feinsinger 1994a). Another benefit of a diverse pollinator guild is the more predictable pollination service that it provides through time (Pettersson 1991). This is an issue that deserves more attention in the context of ecosystem services (e.g., Kremen *et al*. 2002; Larsen *et al*. 2005).

9.7 Relation of plant reproduction to modified pollination

All attributes of pollination loads, including their quantity, quality, and purity, should affect seed quantity and quality. Although Table 9.5 seems trivial (positive effects are predicted for all cause–effect relationships), it stresses the role of post-pollination processes that might be altered by habitat perturbation, but which are typically overlooked in the context of the effects of anthropogenic disturbance on plant reproduction (Aizen and Feinsinger 2003). Through these effects, increases in pollination quantity and quality can enhance both seed quantity and quality (Ramsey and Vaughton 2000).

The relation of seed production to pollen receipt is straightforward: increased pollen receipt usually enhances fruit and seed set until a threshold is reached at which resource, rather than pollen, availability limits fecundity. Two recent reviews demonstrate that pollen-limited reproduction is more the rule than the exception in nature (Ashman *et al*. 2004; Knight *et al*. 2005), in contrast to theoretical expectations (e.g., Chapter 4). The authors of these reviews proposed that pollen limitation is becoming increasingly common, because most plants currently live in human-disturbed environments where pollinators have become scarce or their abundances vary extensively. This conjecture is supported by Aguilar's (2005) meta-analysis of the reproductive response by 85 plant species to fragmentation, which found an overall negative effect on both pollination and seed output, despite heterogeneous responses by individual species (Fig. 9.3a). Most interesting, species showing strong pollination decreases in habitat fragments were also likely to exhibit sharp declines in seed output, suggesting that the reproductive decline of plants in disturbed environments can be explained largely by pollen limitation (Fig. 9.4).

Although pollen limitation is usually interpreted in terms of reduced visit frequency due to a scarcity of efficient pollinators, it can also arise from poor pollen quality, including pollination with either self-pollen, or cross-pollen loads with low genetic diversity that pre-emptively fertilize ovules that fail to mature into seeds (Ramsey and Vaughton 2000). Regrettably, the common protocol to evaluate pollen limitation, involving supplemental pollination with cross-pollen, does not allow discrimination between limitation from pollen quantity and quality (also see Chapter 4). These alternatives could be distinguished, and the magnitude of each of them measured, by knowing (1) the response curve of seed number to pollen receipt under natural conditions, which allows the estimation of the quantitative component of pollen limitation, and (2) the number of seeds produced by virgin flowers receiving unlimited, pure cross-pollen, which allows the estimation of the qualitative component of pollen limitation (M. A. Aizen and L. D. Harder in press).

In addition to determining fecundity, the amount of pollen deposited on stigmas and its genetic diversity, composition, and physiological status affect seed quality, including seed size and germination potential, and seedling and plant performance. Increased stigmatic pollen loads may enhance competition for access to ovules among pollen tubes growing in the style. Because of overlap in gene expression between the paternal sporophyte and the male gametophyte, fast-growing pollen tubes may sire vigorous seedlings (Mulcahy *et al*. 1996). Although conditions for gametophytic competition may be erratic (Herrera 2002), the potential effects of this phenomenon on both ecological and evolutionary time scales may

Table 9.5 Predicted effects of pollination variables on quantitative and qualitative aspects of plant reproductive success.

Pollination attribute	Quantity	Quality
Quantity	↑	↑
Quality	↑	↑
Purity	↑	↑

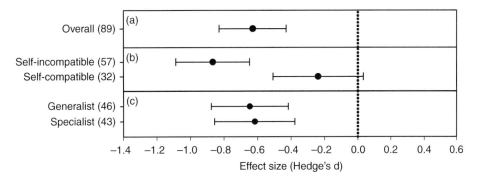

Figure 9.3 Weighted mean (±95% confidence interval) effect size (Hedge's *d* coefficient) from a meta-analysis of plant reproductive success (either fruit set, seed set, or total seed output) in habitat fragments and more continuous expanses of the same habitat type. Panel (a) shows the overall response, panel (b) depicts the comparative response between self-incompatible and self-compatible plant species, and panel (c) illustrates the comparative response between plant species with generalized or specialized pollination systems. Numbers in parentheses indicate the number of plant species included in each subgroup. The dotted line indicates effect size = 0. Redrawn from Aguilar (2005).

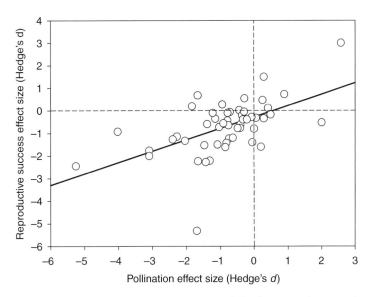

Figure 9.4 Mean effect sizes (Hedge's *d*) for the relation of reproductive success (either fruit set, seed set, or total seed output) to pollination (either number of pollen grains of pollen tubes) of plants in habitat fragments and more continuous expanses of the same habitat type ($r = 0.55$, $n = 50$ species, $P < 0.001$). The horizontal and vertical dotted lines indicate effect sizes = 0. Redrawn from Aguilar (2005).

be great (Niesenbaum and Casper 1994). In addition to pollen-tube competition, the extent of inbreeding, including selfing, can impact fecundity and seed quality directly (Charlesworth and Charlesworth 1987). As a general rule, inbreeding depression in outcrossing species is commonly expressed early during seed development, whereas in mostly selfing species those effects are expressed much later in mature plants, if at all (Husband and Schemske 1996).

9.8 Translation into a path-analysis framework: an example

It is useful at this point to consider an example of how our general matrix approach can be translated into a statistical modelling framework that can be used to evaluate specific causal hypotheses relating human-caused perturbations with floral biology. We base this discussion on work conducted by Vázquez and Simberloff (2004) evaluating the effects of introduced ungulates (cattle) on the pollination and reproduction of an understory herb, *Alstroemeria aurea*.

To consider the effects of particular perturbation types on specific systems, the general matrices (Tables 9.1–9.5) must be adapted to the problem at hand. This process necessarily involves selecting a subset of candidate variables thought to cause the hypothetical effects. Ideally, such variable selection should be based on existing biological knowledge of the system (Shipley 2000; Mitchell 2001). In our example, we first identify the type of perturbation in Table 9.1, namely "herbivory." Vázquez and Simberloff hypothesized that the effects of cattle on pollination and reproduction were mediated entirely by population-level plant attributes, particularly the absolute and relative density of the focal plant species. Using these variables, we characterize a "path diagram," as shown in Fig. 9.5a. In this diagram, variables connected with one-headed arrows are hypothesized to be linked causally; for example, the "browsing index" (a surrogate of the general perturbation caused by cattle) is hypothesized to affect the absolute and relative population densities of the focal plant species (Fig. 9.5a). According to Table 9.1, herbivory is expected to reduce absolute density, so we add an arrow with a dashed line (in path analysis, dashed lines represent negative effects, whereas solid lines represent positive effects). In contrast, the effect of herbivory on relative density could be either positive or negative (Table 9.1), so we tag that arrow with a question mark.

Vázquez and Simberloff's study did not include pollinator responses to perturbation explicitly in the causal model; however, it did include responses of pollinator visitation frequency to absolute plant density, so we add that link in Fig. 9.5a. Vázquez and Simberloff hypothesized that pollinator visitation frequency could be influenced by

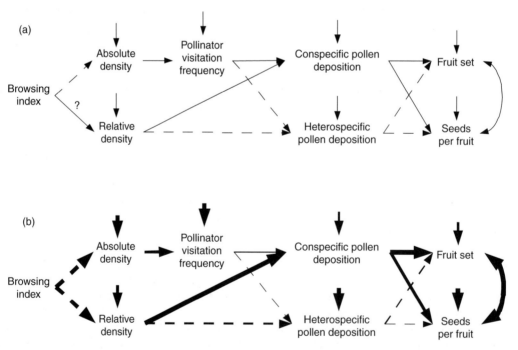

Figure 9.5 Path diagrams structured on the general framework outlined by Tables 9.1–9.5 depicting (a) hypothesized and (b) observed effects of cattle-caused perturbations on pollination and reproduction of the herbaceous understory plant *Alstroemeria aurea*. One-headed arrows linking variables represent unidirectional causal effects, the two-headed arrow represents bidirectional (i.e., correlational) effects between the two reproductive variables, and vertical one-headed arrows represent unexplained variation in the endogenous (dependent) variables. Line dashing indicates the direction of effects (solid, positive; dashed, negative); line thickness in (b) represents effect magnitude. Modified from Vázquez and Simberloff (2004).

cattle only through an effect on absolute population density, but not through an effect on relative density (see their Fig. 1), so we construct our path diagram to reflect this hypothesized mechanism. Vázquez and Simberloff predicted that this effect is positive, so we add a solid arrow.

We now move on to Table 9.3, which relates plant population attributes with pollination variables. Vázquez and Simberloff included two pollination variables, the numbers of conspecific and heterospecific pollen grains, representing the components of pollination quantity and purity. Notice that the effect of absolute density on pollen deposition is mediated by visitation frequency, which comes from Table 9.4. Here, greater absolute and relative density should promote increased conspecific pollination (solid line), but decreased heterospecific pollination (dashed line).

Moving on to the last matrix, we now relate reproductive outcomes to pollination variables. Vázquez and Simberloff included two such variables, fruit set (proportion of flowers producing fruit) and seeds per fruit. Both variables are quantitative, but seeds per fruit can also be used as a rough estimate of the "quality" of reproduction (see Vázquez and Simberloff 2004). Finally, we must account for both unexplained variability, which in path analysis is represented by vertical arrows pointing to all endogenous ("dependent") variables, and the likely correlation between the two reproductive outcomes (represented by a two-headed arrow). Again, line characteristics reflect the predicted direction of effects, so that conspecific pollen increases reproduction, whereas heterospecific pollen reduces it. This model can be assessed and compared with alternative models following the methods outlined in Shipley (2000) and Mitchell (2001).

The resulting path diagram represents a specific hypothesis depicting the impact of a particular perturbation type on plant reproduction through its effects on pollination. Figure 9.5b presents an evaluation of this causal hypothesis, adapted from Vázquez and Simberloff (2004). The general direction of effects matches predictions, but some effects (represented by arrow thickness) are weaker than expected. For example, pollinator visitation frequency affects pollen deposition weakly, indicating that cattle affect pollen deposition primarily through their effects on the relative density of *A. aurea*, rather than on absolute density. Similarly, whereas conspecific pollen deposition affects both reproductive outcomes strongly, the effect of heterospecific pollen deposition is rather weak. Thus, these data suggest that cattle affect the pollination and reproduction of *A. aurea* by decreasing its density relative to other species in the community, which in turn reduces conspecific pollen deposition and decreases reproductive success.

9.9 Modulators of plant reproductive response

The sensitivity of plant reproduction to the negative effects of habitat disturbance may depend on several plant traits. However, traits linked most directly to a plant's pollination and reproductive systems are expected to be most influential (Bond 1994; Aizen *et al*. 2002). In particular, two traits have been identified as primary modulators of the pollination and reproductive responses of plants to anthropogenic disturbance: sexual system and pollination specialization.

Plant sexual systems range from those that enforce outbreeding to those that ensure sexual reproduction via autonomous, within-flower selfing and autogamous seed set (Lloyd 1992). The most common outbreeders include species with hermaphroditic flowers and a genetically based self-incompatibility system and those with distinctive male and female individuals (i.e., dioecy). On the other hand, many self-compatible hermaphroditic species can set seed via selfing (Goodwillie *et al*. 2005). In animal-pollinated species, this inbreeding–outbreeding gradient establishes, beyond its genetic consequences, the overall dependence on the pollination mutualism for plant reproduction (Bond 1994; Chapter 10). Whereas reproduction of obligate outbreeders requires other mates, that of inbreeders can occur mostly independently of other plant individuals and pollinators. Plants also differ in their pollination specialization, from extreme specialists to extreme generalists. Pollination specialists are pollinated by one or a few ecologically similar animal species, whereas generalists are pollinated by several to many species, usually of diverse taxonomic affinities (Renner 1998). The yucca/yucca moth and fig/fig wasp mutualisms are classic cases of extreme specialization. However, flowers of most species are pollinated by a few to more than 100 animal species (Waser et al. 1996). This gradient in pollination specialization may relate to the likelihood of mutualism failure: pollination specialists should be more vulnerable than generalists, because the loss of one pollinator species could cause complete plant reproductive failure (Bond 1994).

Figure 9.6 portrays a graphical model showing the predicted differential responses of pollination and plant reproductive success with increasing disturbance (assuming that the range of disturbance frequency or intensity affects these variables negatively) in relation to sexual system and pollination specialization. All else being equal, the pollination and reproductive success of a self-incompatible species (i.e., an obligate outbreeder) is more likely to decline with increasing habitat disturbance than that of a phylogenetically related self-compatible species (i.e., a facultative inbreeder) (Fig. 9.6a). Similarly, but based on the likelihood of the disruption of the plant–pollinator link, pollination and reproduction of a specialist plant species should be more sensitive to the effects of habitat disturbance than that of a generalist plant (Fig. 9.6b).

Two recent reviews explored the effects of habitat fragmentation on pollination and reproductive success on the differential response of species to anthropogenic disturbance. A survey of 45 species (Aizen et al. 2002) found no evidence that either plant sexual system or degree of specialization influences the probability of negative responses to habitat fragmentation in terms of either pollination or reproductive success. Also, this probability was independent of whether species occupied tropical or temperate areas, or their growth form. In contrast, Aguilar's (2005) more detailed and complete meta-analysis supported one of our predictions (Fig. 9.6a). Whereas reproduction by self-incompatible species declined significantly in fragmented habitats, the pollination and reproduction of the self-compatible species were, on average, not particularly impaired (Fig. 9.3b). However, this meta-analysis agrees with Aizen et al.'s (2002) conclusion that the degree of pollination specialization does not affect a species' differential reproductive response to habitat fragmentation (Fig. 9.3c). This apparent contradiction between expectations and reality may reflect the structure of plant–pollinator interaction webs (Vázquez and Simberloff 2002; Ashworth et al. 2004), a subject that we develop in the following section.

9.10 Anthropogenic disturbance and the structure of pollination interaction networks

As discussed in preceding sections, human-caused perturbations can alter the structure and

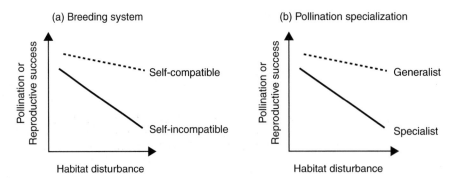

Figure 9.6 Expected pollination and reproductive responses by plants to increasing habitat disturbance as influenced by (a) plant sexual system and (b) pollination specialization. Although sexual systems and pollination specialization vary continuously, they are each divided into two contrasting categories for simplicity.

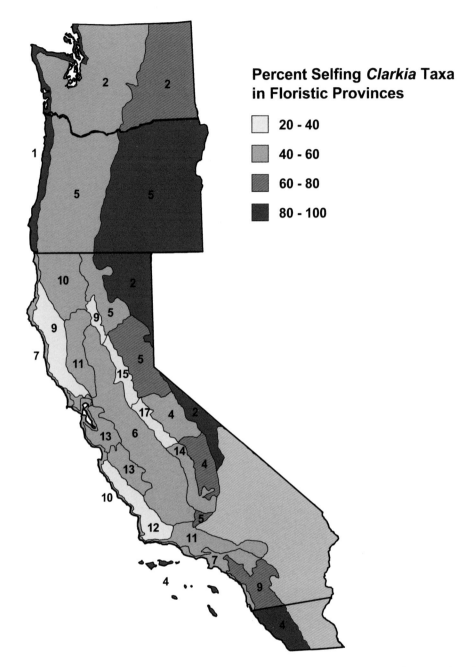

Plate 1 Biogeography of floral traits and mating-system frequency in *Clarkia* (see Chapter 6). The percentage of selfing taxa is shown for floristic provinces of the western United States of America and northern Mexico. Virtually all *Clarkia* taxa occur within this geographical region. The number shown in each floristic province indicates the total number of species (both outcrossing and selfing) in that province.

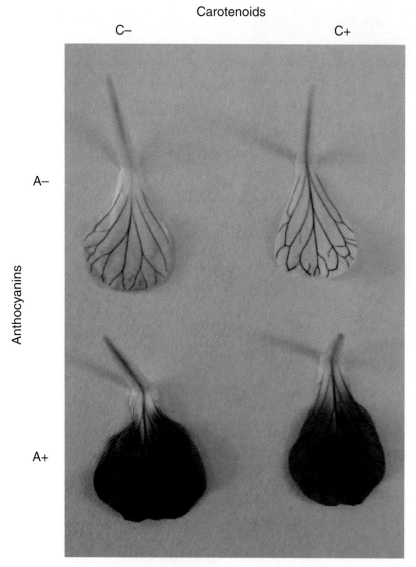

Plate 2 The four flower colour morphs of *Raphanus sativus* (Brassicaceae) are determined by two loci, each with two alleles (see Chapter 7). One locus determines anthocyanin expression, whereas the other controls carotenoid expression. The presence of at least one dominant allele at the anthocyanin locus results in petals expressing anthocyanins (purple and bronze forms). At least one dominant allele at the carotenoid locus produces white flowers, as long as anthocyanin alleles are both recessive. Yellow flowers are the product of the double recessive genotype and express carotenoids only. Mutualistic pollinators prefer carotenoid-producing morphs, but so do antagonistic herbivores. Yellow-flowered individuals are selected against by non-pollinator agents during the non-flowering phase of the plant's life history. Photograph by Sharon Y. Strauss.

Plate 3 Flowers of the *Narcissus* species featured in Chapter 13: (a) *Narcissus longispathus*; (b) Variation in herkogamy in *N. longispathus*; (c) *N. assoanus*, a species with stigma-height dimorphism; and (d) *N. triandrus*, a tristylous species. Photographs (a) and (b) by Carlos M. Herrera; (c) and (d) by Spencer C. H. Barrett.

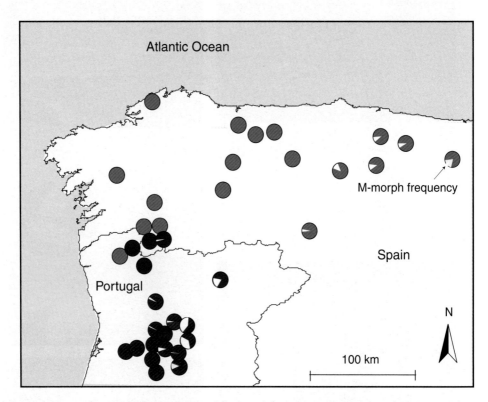

Plate 4 Geographical structure of *cp*-DNA haplotype variation and floral morph frequencies in 37 populations of *Narcissus triandrus* var. *triandrus* from northern Spain and Portugal (see Chapter 13). Each circle represents a population, with the colours distinguishing the four haplotypes found in this variety, and the white segment indicating the frequency of the M-morph in each population. Populations with no white segment are dimorphic; the remaining populations are trimorphic. Further details of this survey, including results from throughout the geographical range of *N. triandrus* and details on methods, will be published elsewhere (K. A. Hodgins and S. C. H. Barrett unpublished data).

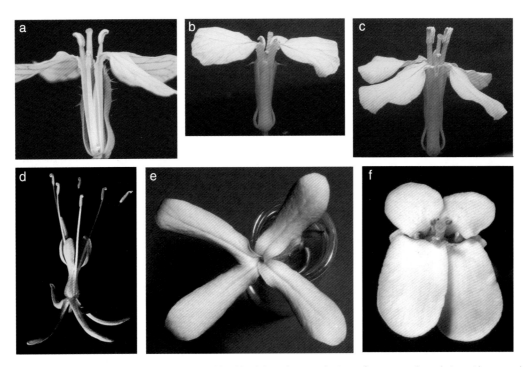

Plate 5 Flowers of selected species in the Brassicaceae: (a) wild radish *Raphanus raphanistrum* from a natural population, with one petal and sepal removed to show the filament and corolla tube lengths; (b) wild radish flower from a line subject to artificial selection for reduced anther exsertion (Section 14.5.1); (c) wild radish flower from a line selected for increased anther exsertion; (d) *Stanleya pinnata*, with naturally highly exserted anthers; (e) *Matthiola*, with highly inserted anthers; and (f) *Iberis*, with intermediate anther exsertion, typical of most species in the Brassicaceae (see Chapter 14). Photographs by Jeffrey K. Conner.

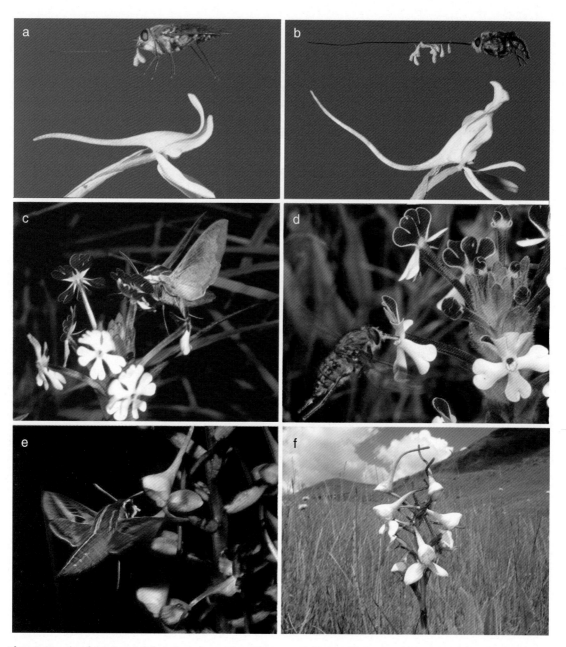

Plate 6 Examples of sister taxa that have diverged across the pollinator mosaic illustrated in Fig. 16.1. (a) Southern mountain forms in the *Disa draconis* complex (Orchidaceae) have short floral spurs (~30-40 mm) and are pollinated solely by the tabanid fly *Philoliche rostrata* (Johnson and Steiner 1997). (b) Lowland sandplain forms in the *Disa draconis* complex have longer floral spurs (~50-60 mm) and are pollinated solely by the large sandplain fly *Moegistorynchus longirostris*. (c) *Zaluzianskya natalensis* (Scrophulariaceae) occurs at mid-elevations and has night opening flowers pollinated by hawk moths (Johnson et al. 2002b). (d) Its sister taxon *Zaluzianskya microsiphon* occurs mostly at higher elevations in the Drakensberg Mountains and has day-opening flowers pollinated by the long-proboscid fly *Prosoeca gangbaueri*. The two species form occasional hybrids in contact zones and show evidence of introgressive gene flow (Archibald et al. 2004). (e) *Disa cooperi* (Orchidaceae) has evening-scented flowers pollinated by hawk moths (Johnson et al. 2005). (f) Its allopatric sister taxon *Disa scullyi* has unscented flowers pollinated by the long-proboscid fly *Prosoeca ganglbaueri* in the Drakensberg Mountains. Photographs by Steven D. Johnson.

Plate 7 Illustrations of characters hypothesized to affect flowering plant diversity (see Chapter 17). (a) An inflorescence of zygomorphic *Lupinus* flowers and actinomorphic *Eschscholzia californica*. (b) A single female flower of the dioecious, and wind-pollinated, *Simmondsia chinensis*. (c) A cluster of male flowers of *S. chinensis*. (d) A hawk moth *Hyles lineata* visiting the spurred flower of an *Aquilegia formosa* x *A. pubescens* hybrid. Note this spurred flower is also actinomorphic. (e) The spurred, and zygomorphic, flower of *Linaria triornithophora*. Photographs by Scott A. Hodges.

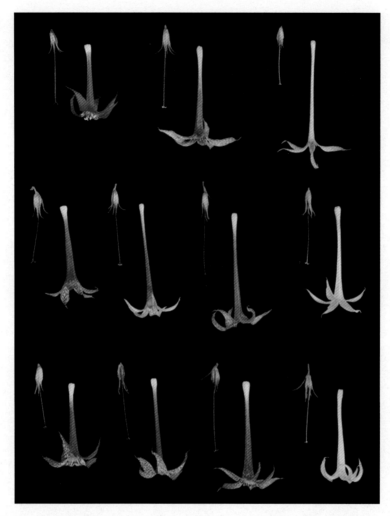

Plate 8 Flowers of *Ipomopsis aggregata* (top left). *I. tenuituba* (top right), F_1 (top middle), and F_2 hybrids (center and bottom rows). The flowers have been dissected to show style length and stigma position. Photographs by George Aldridge.

functioning of pairwise interactions between a plant and a pollinator species through a variety of mechanisms. However, these effects occur within a community context, rather than in isolation, thus having the potential to impact many pairwise interactions simultaneously (also see Chapter 6). Therefore, the magnitude and direction of the effects of perturbations will be determined partly by community structure.

Interactions among a set of species that coexist in a local environment are frequently represented as networks or "webs" in which species or populations are represented as nodes and interspecific interactions are represented as links. Studies in community ecology have long sought to identify regularities in network topology and the underlying mechanisms (Cohen and Newman 1985; Jordano 1987; Williams and Martinez 2000; Dunne et al. 2002; Jordano et al. 2003). These structural patterns may have dynamic implications for the populations that compose the network and thus may influence the responses of interacting species to perturbations (Dunne et al. 2002; Melián and Bascompte 2002; Memmott et al. 2004).

Recent studies have identified some apparently pervasive structural features of plant–pollinator interaction webs. First, whereas the classical view assumes that specialist plants interact differentially with specialist pollinators, and generalist plants with generalist plants (i.e., symmetric interactions), specialist plants actually interact with generalist pollinators more frequently than expected by chance, whereas generalist plants interact with a mix of generalist and specialist pollinators (i.e., asymmetric interactions; Bascompte et al. 2003; Vázquez and Aizen 2004, 2006). Thus, even if an intense perturbation caused the differential loss of disturbance-sensitive, specialist pollinators, most plants would be buffered against this loss, because of their tendency to interact with some generalized pollinators independently of their degree of pollination specialization (Vázquez and Simberloff 2002; Ashworth et al. 2004: Fig. 9.7). Second, generalists form a network "core" consisting of a densely connected subset of diffusely interacting species (Bascompte et al. 2003). This structural trait may buffer networks against extinctions, because high connectivity and abundance may increase the

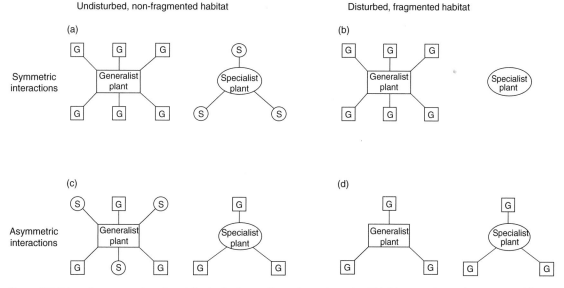

Figure 9.7 Schematic representation of specialization in plant–pollinator interaction webs. With (a) symmetric specialization, many different generalist pollinators (G) pollinate generalist plants, whereas a few taxa of specialist pollinators (S) pollinate specialist plants, so (b) habitat fragmentation has a stronger negative effect on specialist plants than on generalist plants. Under (c) asymmetric specialization, many specialist and generalist animal taxa pollinate generalist plants, whereas one or a few taxa of generalist pollinators serve specialist plants, resulting in (d) similar reproductive susceptibility to habitat fragmentation for specialist and generalist plants. From Ashworth et al. (2004).

probability of persistence of generalists, and many species depend on generalists. Third, a small subset of pollinator species may be sufficient for a plant to achieve reproductive success, and this importance is strongly, positively related to pollinator visitation frequency (Morris 2003; Vázquez et al. 2005) and a pollinator's degree of generalization (Vázquez and Aizen 2006). Therefore, this skewed distribution of pollinator effects on plants may further buffer networks against perturbations.

A few studies have explicitly examined the interaction between the structure of plant–pollinator networks and human-caused perturbations. These studies addressed two major classes of questions: whether perturbations affect structural features of plant–pollinator interaction networks, and how some of the above structural features of plant–pollinator networks may determine species' responses to perturbations.

Among the first class of questions (i.e., whether perturbations affect network structure), Vázquez and Simberloff (2003) found systematic changes in network structure resulting from cattle grazing in the understory of native forests. This effect resulted mainly from the modification of a few very frequent interactions, which are probably functionally important (see Morris 2003; Vázquez et al. 2005a). Unfortunately, this study did not identify which structural aspects of networks were affected by perturbations. This issue is important, because some structural feature of a network (e.g., nestedness or degree of asymmetry) could remain unchanged despite strong changes in the absolute and relative participation of species in the network.

A few studies provide tentative answers to whether network structure influences species' responses to perturbations. Memmott et al. (2004) simulated pollinator extinctions in two of the largest plant–pollinator networks available in the literature to date. Assuming that plants depend entirely on pollinators to reproduce (which is arguable; Bond 1994), they simulated secondary extinctions of plants as a result of extinctions of their pollinators. Memmott et al. found that when pollinators went extinct in decreasing order of generalization (i.e., from the most generalized to the most specialized), secondary extinctions of plants occurred earlier and faster than when pollinator extinctions occurred randomly or in increasing order of generalization. This result was explained by the highly nested structure of the networks analysed. Thus, the interaction of a few extremely generalized pollinators with most plant species in the community prevents their extinction, which in turn prevents the secondary extinction of the many plants that depend on these pollinators for reproduction (see also Renner 1998; Vázquez and Simberloff 2002; Ashworth et al. 2004). Furthermore, highly connected species tend to be more abundant than species with few connections (see Vázquez and Aizen 2006), and rare populations tend to suffer greater extinction risks than abundant populations (Lawton and May 1995), so that highly connected species may be particularly resistant to extinction, further enhancing the robustness of plant–pollinator networks.

Memmott et al.'s (2004) study was unrealistic in that they assumed complete dependence of plants on pollinators and did not include information about pollinators' effectiveness, so they implicitly assumed that all pollinators are equally effective. Morris (2003) attempted to overcome this limitation by explicitly incorporating data on pollinator effectiveness. Based on published data on pollinator effectiveness for 24 plant species, he simulated the loss of total pollinator service (i.e., the added contribution of pollinator species to plant reproductive success) as pollinator species went extinct. Morris' results indicated that a large proportion of pollinator species could be lost before substantial reproductive service to plants was lost. This result was explained by the highly uneven distribution of interaction frequency and of per-interaction effectiveness among pollinator species: frequent pollinators tend to contribute most to plant reproduction, regardless of their per-interaction effectiveness (see Section 9.6 above; Vázquez et al. 2005a). The findings of Memmott et al. (2004) and Morris (2003) suggest that plant–pollinator networks are highly resistant to perturbations because (1) plant–pollinator networks tend to be organized in a nested, asymmetrically specialized fashion, (2) the most frequent pollinators tend to contribute most to plant reproduction, and (3) the most frequent pollinators are probably highly connected and rather resistant to extinction due to their abundance.

9.11 Prospects

Pollination is an essential service in both natural and agricultural ecosystems, so understanding pollination mechanisms and their susceptibility to different habitat disturbances is of paramount importance in different applied fields (e.g., Kremen *et al.* 2002, 2004). Despite considerable progress during the past two decades, much remains to be done. Many studies assume unidirectional effects of anthropogenic disturbance on plant and pollinator attributes and flower function (reviewed in Aizen *et al.* 2002; Ghazoul 2005a). However, many matrix cells (ca. 25%) in Tables 9.1–9.4 were filled with ↑↓, implying that those effects could be positive, negative, or represented by a modal function such as a quadratic. Furthermore, the effects of anthropogenic disturbance are complex, because they modify many environmental variables simultaneously. In general, these effects do not differ qualitatively from those triggered by natural disturbance, although they may differ in frequency and intensity (Sousa 1984; Chesson and Huntly 1997).

The question marks in Tables 9.1–9.4 (~15% of all matrix cells) also identify particular unresolved issues whose predictions are elusive and await research. For instance, despite many studies considering how variation in floral morphology affects pollination and reproductive success (particularly in the context of phenotypic selection), few have examined how different types of disturbances modify flower traits. Relevant questions in this area include how and to what extent a given environmental disturbance (e.g., fire) modifies floral morphology (including flower size, shape, and symmetry) and its role in pollen dispersal and reproductive performance. Also, although some studies have focused on how certain disturbances, such as herbivory and fire, affect the reproductive physiology of flowers (e.g., post-pollination pollen performance), the relations of these aspects of flower performance to other kinds of disturbances remain unknown. For example, can resource competition with invasive plants affect the physiological status of flowers of native plants, and can this in turn affect pollination?

Another key group of largely unexplored questions concern whether changes induced by habitat disturbance on pollinator community characteristics, particularly diversity and composition, reduce pollination significantly. For instance, will the loss of specialist pollinators due to anthropogenic disturbance decrease pollination quantity and/or quality beyond any effect on overall visit frequency? Also, the replacement of complex pollinator assemblages by Africanized honeybees throughout the Neotropics may have consequences for pollination and patterns of gene flow, which remain largely unexplored. On the other hand, the agricultural service provided by managed or unmanaged populations of native pollinators as an alternative to *Apis mellifera* is a topic of much conservation and economic value, which is still in its infancy (Kremen *et al.* 2002).

Despite evidence for many of the links portrayed in our matrices, understanding the relative importance of the different mechanisms affecting the pollination process in altered landscapes is limited. We advocate the use of path analysis and associated structural equation modelling (Shipley 2000), which allows assessment and comparison of alternative, progressively complex causal models.

Many relevant questions about the structure of plant–pollinator networks and how it is affected by anthropogenic disturbance await answers. For instance, which structural aspects of these webs (e.g., connectance, distribution of specialization, asymmetry, nestedness) change most commonly under disturbance and which are apt to remain invariant? In addition, what consequences have those changes for individual species persistence? Most important, a more direct link between these structural modifications and their consequences for pollination function must be identified.

Geographic information systems also offer increasing opportunities to extrapolate site-specific pollination models to the regional scale (Kremen *et al.* 2004). We expect that this interface between landscape and pollination ecology will be crossed more frequently in the near future, increasing our ability to predict how different aspects of pollination will change in altered habitats with specific landscape configurations.

To conclude, we hope the framework outlined in this chapter will help in developing a more mechanistic approach to the study of anthropogenic perturbations on pollination and plant reproduction. More than providing conclusive answers, our goal has been to show how this approach can be used to identify open questions and future avenues of research, and to aid in organizing and designing studies of disturbance effects on flower function in human-altered landscapes.

Acknowledgements

We thank L. D. Harder, S. C. H. Barrett, and two anonymous reviewers for useful suggestions and encouragement while writing this chapter and R. Aguilar for allowing us the use of unpublished material. Research leading to this work was supported by a postdoctoral fellowship to DPV at the National Center for Ecological Analysis and Synthesis (funded by the University of California, Santa Barbara, and the National Science Foundation, USA). MAA and DPA are researchers with the National Research Council of Argentina (CONICET).

References

Aguilar R (2005). Efectos de la fragmentación de hábitat sobre el éxito reproductivo de especies nativas del bosque chaqueño serrano de Córdoba. Doctoral Dissertation. Universidad Nacional de Córdoba, Córdoba, Argentina.

Aizen MA and Feinsinger P (1994a). Forest fragmentation, pollination, and plant reproduction in a Chaco dry forest, Argentina. *Ecology*, **75**, 330–51.

Aizen MA and Feinsinger P (1994b). Habitat fragmentation, native insect pollinators, and feral honeybees in Argentine "Chaco Serrano." *Ecological Applications*, **4**, 378–92.

Aizen MA and Feinsinger P (2003). Bees not to be? Responses of insect pollinator faunas and flower pollination to habitat fragmentation. In GA Bradshaw and PA Marquet, eds. *How landscapes change: human disturbance and ecosystem fragmentation in the Americas*, pp. 111–29. Springer-Verlag, Berlin.

Aizen MA and Harder LD. In press. Expanding the limits of the pollen-limitation concept: effects of pollen quantity and quality. *Ecology*.

Aizen MA and Raffaele E (1998). Flowering shoot defoliation affects pollen grain size and postpollination pollen performance in *Alstroemeria aurea*. *Ecology*, **79**, 2133–42.

Aizen, MA, Ashworth L, and Galetto L (2002). Reproductive success in fragmented habitats: do compatibility systems and pollination specialization matter? *Journal of Vegetation Science*, **13**, 885–92.

Ashman T, Knight TM, Steets JA, *et al.* (2004). Pollen limitation of plant reproduction: ecological and evolutionary causes and consequences. *Ecology*, **85**, 2408–21.

Ashworth L, Aguilar R, Galetto L, and Aizen MA (2004). Why do pollination generalist and specialist plant species show similar reproductive susceptibility to habitat fragmentation? *Journal of Ecology*, **92**, 717–19.

Bascompte J, Jordano P, Melián CJ, and Olesen JM (2003). The nested assembly of plant–animal mutualistic networks. *Proceedings of the National Academy of Sciences of the United States of America*, **100**, 9383–7.

Bond WJ (1994). Do mutualisms matter? Assessing the impact of pollinator and disperser disruption on plant extinction. *Philosophical Transactions of the Royal Society of London, Series B*, **344**, 83–90.

Brody AK (1997). Effects of pollinators, herbivores, and seed predators on flowering phenology. *Ecology*, **78**, 1624–31.

Brown BJ, Mitchell RJ, and Graham SA (2002). Competition for pollination between and invasive species (Purple loosestrife) and a native congener. *Ecology*, **83**, 2328–36.

Burgess VJ, Kelly D, Robertson AW, and Ladley JJ (2006). Positive effects of forest edges on plant reproduction: literature review and a case study of bee visitation to flowers of *Peraxilla tetrapetala* (Loranthaceae). *New Zealand Journal of Ecology*.

Cascante A, Quesada M, Lobo JJ, and Fuchs EA (2002). Effects of dry tropical forest fragmentation on the reproductive success and genetic structure of the tree *Samanea saman*. *Conservation Biology*, **16**, 137–47.

Chacoff NP and Aizen MA (2006). Edge effects on flower-visiting insects in grapefruit plantations bordering premontane subtropical forest. *Journal of Applied Ecology* **43**, 18–27.

Charlesworth D and Chalesworth B (1987). Inbreeding depression and its evolutionary consequences. *Annual Review of Ecology and Systematics*, **18**, 237–68.

Chesson P and Huntly N (1997). The roles of harsh and fluctuating conditions in the dynamics of ecological communities. *American Naturalist*, **150**, 519–53.

Cohen JE and Newman CM (1985). A stochastic theory of community food webs. I. Models and aggregated data.

Proceedings of the Royal Society of London, Series B, **224**, 421–48.

Cunningham SA (1997). The effect of light environment, leaf area, and stored carbohydrates on inflorescence production by a rain forest understory palm. *Oecologia*, **111**, 36–44.

Cunningham SA (2000). Depressed pollination in habitat fragments causes low fruit set. *Proceedings of the Royal Society of London, Series B*, **267**, 1149–52.

Didham RK, Ghazoul J, Stork NE, and Davis AJ (1996). Insects in fragmented forests: a functional approach. *Trends in Ecology and Evolution*, **11**, 255–60.

Dunne JA, Williams RJ, and Martinez ND (2002). Food-web structure and network theory: The role of connectance and size. *Proceedings of the National Academy of Sciences of the United States of America*, **99**, 12917–22.

Eckert CG (2000). Contributions of autogamy and geitonogamy to self-fertilization in a mass-flowering, clonal plant. *Ecology*, **81**, 532–42.

Etterson JR and Galloway LF (2002). The influence of light on paternal plants in *Campanula americana* (Campanulaceae): pollen characteristics and offspring traits. *American Journal of Botany*, **89**,1899–906.

Fahrig L (2003). Effects of habitat fragmentation on biodiversity. *Annual Review of Ecology, Evolution and Systematics*, **34**, 487–515.

Ghazoul J (2005a). Pollen and seed dispersal among dispersed plants. *Biological Reviews*, **80**, 413–43.

Ghazoul J (2005b). Buzziness as usual? Questioning the global pollination crisis. *Trends in Ecology and Evolution*, **20**, 367–73.

Ghazoul J and McLeish M (2001). Reproductive ecology of tropical forest trees in logged and fragmented habitats in Thailand and Costa Rica. *Plant Ecology*, **153**, 335–45.

Goodwillie C, Kalisz S, and Eckert CG 2005 The evolutionary enigma of mixed mating systems in plants: occurrence, theoretical explanations, and empirical evidence. *Annual Review of Ecology, Evolution and Systematics*, **36**, 47–79.

Goulson D (2003). Effects of introduced bees on native ecosystems. *Annual Review of Ecology, Evolution and Systematics*, **34**, 1–26.

Harder LD and Barrett SCH (1995). Mating cost of large floral displays in hermaphrodite plants. *Nature*, **373**, 512–15.

Herrera CM (1993). Selection on floral morphology and environmental determinants of fecundity in a hawk moth-pollinated violet. *Ecological Monographs*, **63**, 251–75.

Herrera CM (2002). Censusing natural microgametophyte populations: variable spatial mosaics and extreme fine graininess in winter-flowering *Helleborus foetidus*. *American Journal of Botany*, **89**, 1570–8.

Husband BC and Schemske DW (1996). Evolution of the magnitude and timing of inbreeding depression in plants. *Evolution*, **50**, 54–70.

Johnson SD, Peter CI, Nilsson LA, and Ågren J (2003) Pollination success in a deceptive orchid is enhanced by co-occurring rewarding magnet plants. *Ecology*, **84**, 2919–27.

Jordano P (1987). Patterns of mutualistic interactions in pollination and seed dispersal: connectance, dependence asymmetries, and coevolution. *American Naturalist*, **129**, 657–77.

Jordano P, Bascompte J, and Olesen JM (2003). Invariant properties in coevolutionary networks of plant-animal interactions. *Ecology Letters*, **6**, 69–81.

Kearns CA, Inouye DW, and Waser NM (1998). Endangered mutualisms: the conservation of plant-pollinator interactions. *Annual Review of Ecology and Systematics*, **29**, 83–112.

Klein AM, Steffan-Dewenter I, and Tscharntke T (2003). Fruit set of highland coffee increases with the diversity of pollinating bees. *Proceedings of the Royal Society of London, Series B*, **270**, 955–61.

Knight TM, Steets JA, Vamosi JC, et al. (2005). Pollen limitation of plant reproduction: pattern and process. *Annual Review of Ecology, Evolution and Systematics*, **36**, 467–97.

Kremen C, Williams NM, and Thorp RW (2002). Crop pollination from native bees at risk from agricultural intensification. *Proceedings of the National Academy of Sciences of the United States of America*, **99**, 16812–6.

Kremen C, Williams NM, Bugg RL, Fay JP, and Thorp RW (2004). The area requirements of an ecosystem service: crop pollination by native bee communities in California. *Ecology Letters*, **7**, 1109–19.

Krupnick G, Weis A, and Campbell D (1999). The consequences of floral herbivory for pollinator service to *Isomeris arborea*. *Ecology*, **80**, 125–34.

Kunin WE (1997). Population size and density effects in pollination: pollinator foraging and plant reproductive success in experimental arrays of *Brassica kaber*. *Journal of Ecology*, **85**, 225–34.

Larsen T H, Williams NM, and Kremen C (2005). Extinction order and altered community structure rapidly disrupt ecosystem functioning. *Ecology Letters*, **8**, 538–47.

Lawton JH and May RM, eds. (1995). *Extinction rates*. Oxford University Press, Oxford.

Lloyd DG (1992). Self- and cross-fertilization in plants. II. The selection of self-fertilization. *International Journal of Plant Sciences*, **153**, 370–80.

McLernon SM, Murphy SD and Aarssen LW(1996). Heterospecific pollen transfer between sympatric species in a mid successional old-field community. *American Journal of Botany*, **83**, 1168–74.

Melián CJ and Bascompte J (2002). Complex networks: two ways to be robust? *Ecology Letters*, **5**, 705–8.

Memmott J, Waser NM, and Price MV (2004). Tolerance of pollination networks to species extinctions. *Proceedings of the Royal Society of London, Series B*, **271**, 2605–11.

Mitchell RJ (2001). Path analysis: pollination. In SM Scheiner and J Gurevitch, eds. *Design and analysis of ecological experiments*, pp. 217–234. Oxford University Press, Oxford.

Moeller DA (2004). Facilitative interactions among plants via shared pollinators. *Ecology*, **85**, 3289–301.

Møller AP (2000) Developmental stability and pollination. *Oecologia*, **123**, 149–57.

Morales CL and Aizen MA (2006). Invasive mutualisms and the structure of plant–pollinator interactions in the temperate forest of NW Patagonia, Argentina. *Journal of Ecology*, **94**, 171–80.

Morris WF (2003). Which mutualists are most essential? Buffering of plant reproduction against the extinction of pollinators. In P Kareiva and SA Levin, eds. *The importance of species: perspectives on expendability and triage*, pp. 260–80. Princeton University Press, Princeton.

Mothershead K and Marquis RJ (2000). Fitness impacts of herbivory through indirect effects on plant-pollinator interactions in *Oenothera macrocarpa*. *Ecology*, **81**, 30–40.

Mulcahy DL, Sari-Gorla M, and Mulcahy GB (1996). Pollen selection: past, present and future. *Plant Sexual Reproduction*, **9**, 353–6.

Murphy SD and Aarssen LD (1995). Reduced seed set in *Elytrigia repens* caused by allelopathic pollen from *Phleum pratense*. *Canadian Journal of Botany*, **73**, 1417–22.

Neal PR, Dafni A, and Giurfa M (1998). Floral symmetry and its role in plant-pollinator systems: terminology, distribution, and hypothesis. *Annual Review of Ecology and Systematics*, **29**, 345–73.

Niesembaum RA and Casper BB (1994). Pollen tube numbers and selective fruit maturation in *Lindera benzoin*. *American Naturalist*, **144**, 184–91.

Nilsson LA (1988). The evolution of flowers with deep corolla tubes. *Nature*, **334**, 147–9.

Paige KN and Whitham TG (1987). Overcompensation in response to mammalian herbivory: the advantage of being eaten. *American Naturalist*, **129**, 407–16.

Palmer AR (1996). Waltzing with asymmetry. *BioScience*, **46**, 518–32.

Palmer AR (2000). Quasireplication and the contract of error: lessons from sex ratios, heritabilities and fluctuating asymmetry. *Annual Review of Ecology and Systematics*, **31**, 441–80.

Parker IM (1997). Pollinator limitation of *Cytisus scoparius* (Scotch broom), an invasive exotic shrub. *Ecology*, **78**, 1457–70.

Pettersson MW (1991). Pollination by a guild of fluctuating moth populations: option for unspecialization in *Silene vulgaris*. *Journal of Ecology*, **79**, 591–604.

Potts SG, Dafni A, and Ne'eman G (2001). Pollination of a core flowering shrub species in Mediterranean phrygana: variation in pollinator diversity, abundance and effectiveness in response to fire. *Oikos*, **92**, 71–80.

Quesada M, Bollman K, and Stephenson AG (1995). Leaf damage decreases pollen production and hinders pollen performance in *Cucurbita texana*. *Ecology*, **76**, 437–43.

Ramsey M and Vaughton G (2000). Pollen quality limits seed set in *Burchardia umbellata* (Colchicaceae). *American Journal of Botany*, **87**, 845–52.

Rathcke B (1983). Competition and facilitation among plants for pollination. In L Real, ed. *Pollination biology*, pp. 305–29. Academic Press, London.

Rathcke B (1992). Nectar distributions, pollinator behavior, and plant reproductive success. In MD Hunter, T Ohgushi, and PW Price, eds. *Effects of resource distribution on animal-plant interactions*, pp. 113–38. Academic Press, New York.

Rathcke BJ and Lacey EP (1985). Phenological patterns of terrestrial plants. *Annual Review of Ecology and Systematics*, **16**, 179–214.

Renner SS (1998). Effects of habitat fragmentation of plant pollinator interactions in the tropics. In DM Newbery, HHT Prins, and ND Brown, eds. *Dynamics of tropical communities*, pp. 339–60. Blackwell Science, London.

Ricketts TH (2004). Tropical forest fragments enhance pollinator activity in nearby coffee crops. *Conservation Biology*, **18**, 1262–71.

Sato H and Yahara T (1999). Trade-offs between flower number and investment to a flower in selfing and outcrossing varieties of *Impatiens hypophylla* (Balsaminaceae). *American Journal of Botany*, **86**, 1699–707.

Shipley B (2000). *Cause and correlation in biology*. Cambridge University Press, Cambridge.

Silvertown J, Franco M, Pisanty I, and Mendoza A (1993). Comparative plant demography—relative importance of life-cycle components to the finite rate of increase in woody and herbaceous perennials. *Journal of Ecology*, **81**, 465–76.

Sousa WP (1984). The role of disturbance in natural communities. *Annual Review of Ecology and Systematics*, **15**, 353–91.

Steffan-Dewenter I and Tscharntke T (1999). Effects of habitat isolation on pollinator communities and seed set. *Oecologia*, **121**, 432–40.

Strauss SY, Conner JK, and Rush SL (1996). Foliar herbivory affects floral characters and plant attractiveness to pollinators: implications for male and female plant fitness. *American Naturalist*, **147**, 1098–107.

Vázquez DP and Aizen MA (2004). Asymmetric specialization: a pervasive feature of plant-pollinator interactions. *Ecology*, **85**, 1251–7.

Vázquez DP and Aizen MA (2006). Community-wide patterns of specialization in plant–pollinator interactions revealed by null-models. In NM Waser and J Ollerton, eds. *Specialization and generalization in plant–pollinator interactions* pp. 200–19. University of Chicago Press, Chicago.

Vázquez DP and Simberloff D (2002). Ecological specialization and susceptibility to disturbance: conjectures and refutations. *American Naturalist*, **159**, 606–23.

Vázquez DP and Simberloff D (2003). Changes in interaction biodiversity induced by introduced ungulate. *Ecology Letters*, **6**, 1077–83.

Vázquez DP and Simberloff D (2004). Indirect effects of an introduced ungulate on pollination and reproduction. *Ecological Monographs*, **74**, 281–308.

Vázquez DP, Morris WF, and Jordano P (2005). Interaction frequency as a surrogate for the total effect of animal mutualists on plants. *Ecology Letters*, **8**, 1088–94.

Waltz, AEM and Covington WW (2004). Ecological restoration treatments increase butterfly richness and abundance: mechanisms of response. *Restoration Ecology*, **12**, 85–96.

Waser N, Chittka L, Price M, Williams N, and Ollerton J (1996). Generalization in pollination systems, and why it matters. *Ecology*, **77**, 1043–60.

Westphal C, Steffan-Dewenter I, and Tscharntke T (2003). Mass flowering crops enhance pollinator densities at landscape scale. *Ecology Letters*, **6**, 961–5.

Williams RJ and Martinez ND (2000). Simple rules yield complex food webs. *Nature*, **404**, 180–3.

Young HJ and Stanton ML (1990). Influence of environmental quality on pollen competitive ability in wild radish. *Science*, **29**, 1631–3.

PART 3

Mating strategies and sexual systems

The diversification in form and function of flowers among angiosperm species is associated with an extensive variety of mating strategies and sexual systems. For flowering plants, mating involves both the pollination and post-pollination processes that result in ovule fertilization. Consequently, a plant's *mating system* refers to its participation in fertilization as a maternal and/or paternal parent, including the incidence of self- versus cross-fertilization, the diversity of outcrossed mates, and their characteristics (e.g., assortative versus disassortative mating). Plants influence mating through floral traits that affect the movement of pollen within and among their own flowers and those of other plants and through physiological mechanisms that govern the fate of pollen after it reaches stigmas. In particular, a population's *sexual system*, or qualitative differences among flowers within and between plants in the production of pollen and ovules and compatibility/incompatibility status, strongly influences who mates with whom.

In contrast to most animal groups, flowering plants exhibit considerable sexual diversity, often among closely related species. Understanding the selective influences on sexual systems and their mating consequences, including the ecological context in which they operate, represents a major challenge for reproductive botany. Part 3 of *Ecology and Evolution of Flowers* focuses on three particular topics in this area, which have been the subject of intensive work since Charles Darwin's seminal explorations of plant mating strategies and sexual systems. The four chapters illustrate active research on mating and sexual systems; however, they represent a fraction of the diversity of mating and sexual systems in flowering plants. References at the end of this section should be consulted for an expanded introduction to this diversity.

Most flowering plants are outcrossing hermaphrodites and possess diverse floral adaptations that limit the harmful effects of inbreeding. Prompted by Darwin's work, the study of outcrossing mechanisms dominated early research in floral biology, and novel means of promoting cross-pollination continue to be discovered in angiosperms. However, despite the advantages of outcrossing, the shift to predominant self-fertilization is the most frequent evolutionary transition among plant mating strategies. Although this transition has been well documented in many herbaceous groups and it has stimulated the largest body of theoretical work in plant population biology, understanding of how and why selfing evolves remains incomplete. In Chapter 10, Christopher Eckert, Karen Samis, and Sara Dart examine evidence for one of the most widely invoked explanations for the evolution of selfing, namely, that selfing assures reproduction when pollen vectors deliver insufficient pollen to fertilize all ovules. Eckert *et al*. illustrate how this reproductive-assurance hypothesis can be tested by examining intraspecific variation and using experimental field manipulations. Darwin proposed that reproductive assurance was the main reason for the evolution of selfing, and considerable biogeographical and ecological evidence is consistent with this hypothesis. Nevertheless, Chapter 10 exposes some of the complexities involved with testing the reproductive-assurance hypothesis rigorously and also highlights how little is known about the role of reproductive assurance in the evolution of asexuality in plants.

Another recurring evolutionary transition in plant reproduction is the evolution of separate sexes (gender dimorphism) from combined sexes (gender monomorphism). Theoretical models of this transition generally identify three central factors: the relative fitness of progeny arising from self- versus cross-fertilization, the optimal allocation of resources to female and male function, and the inheritance of sex determination. Although genetic aspects of this transition are understood reasonably well, much less is known about the ecological mechanisms favouring the spread and maintenance of unisexual individuals in co-sexual populations. In Chapter 11, Tia-Lynn Ashman considers the abiotic and biotic context in which dioecy evolves from cosexuality via the gynodioecy pathway. She examines Darwin's original idea that harsh environments may promote the evolution of gender dimorphism and provides evidence that stress conditions can promote sexual-system evolution by influencing sex-differential plasticity and size-dependent allocation.

Migration among populations counteracts the tendency of individual populations to diverge from each other as a result of genetic drift and adaptation to the local environment. Although the effects of gene flow for the genetic structure of populations have been appreciated since Sewall Wright's work in the 1930s and 1940s, the implications for phenotypic evolution have received little attention until recently. In Chapter 12, John Pannell explores the consequences of migration for the evolution of mating and sexual systems in plants from two perspectives: a single migration that founds a new isolated population and more frequent exchange among local populations within a larger metapopulation. Pannell clarifies that migration and successful establishment often involves individuals with specific characteristics, so that the resulting gene flow does not draw a random sample of variation from the source population, which predisposes contrasting evolution from what would occur in a large, mixed population. Through a series of empirical and theoretical examples, Pannell illustrates many unexpected consequences of reproductive evolution for populations subject to migration.

Since Darwin's early experiments on *Lythrum* and *Primula*, studies of heterostyly have contributed greatly to our understanding of genetics and morphological adaptation between flowers and their animal pollinators. These polymorphisms are maintained in populations by negative frequency-dependent selection and their functional significance in promoting cross-pollination is well understood. One of the advantages of heterostyly is the relatively direct linkage between floral morphology, intermorph mating, and morph ratios, providing a visible signature of aggregate mating patterns in preceding generations. More recently, heterostyly and several related stylar polymorphisms have been investigated in a wider range of angiosperms, exposing considerable variation in the expression of heterostyly and in the types of mating that can occur in populations. In Chapter 13, Spencer Barrett and Kathryn Hodgins contrast the symmetrical mating and equal morph ratios of typical heterostylous populations with the asymmetrical mating patterns and biased morph ratios that characterize *Narcissus* species with stylar polymorphisms. Their work demonstrates how *Narcissus* provides a rare opportunity to expose the population-level consequences of small, but functionally significant, variation in sex-organ deployment within and between flowers.

Selected key references

Barrett SCH, ed. (1992). *Evolution and function of heterostyly*. Spring-Verlag, Berlin.

Barrett SCH (2002). The evolution of plant sexual diversity. *Nature Reviews Genetics*, **3**, 274–84.

Darwin CR (1877). *The different forms of flowers on plants of the same species*. John Murray, London, UK.

de Nettancourt D (2001). *Incompatibility and incongruity in wild and cultivated plants*. 2nd edition. Springer, New York.

Geber MA, Dawson TE, and Delph LF, eds. (1999). *Gender and sexual dimorphism in flowering plants*. Springer-Verlag, New York.

Goodwillie CS, Kaliz S, and Eckert CG (2005). The evolutionary enigma of mixed mating systems in plants: occurrence, theoretical explanations and empirical evidence. *Annual Review of Ecology, Evolution and Systematics*, **36**, 47–79.

Richards AJ (1996). *Plant breeding systems*. Chapman and Hall, London, UK.

CHAPTER 10

Reproductive assurance and the evolution of uniparental reproduction in flowering plants

Christopher G. Eckert, Karen E. Samis, and Sara Dart

Department of Biology, Queen's University, Kingston, Ontario, Canada

Outline

The assurance of reproduction when outcrossing is unpredictable is a venerable and widely invoked explanation for uniparental reproduction via self-fertilization or asexuality. This hypothesis is supported by evidence that seed production by outcrossing plants is frequently pollen limited. However, its intuitive simplicity belies its considerable complexity. Theory cautions that selfing may increase geometric mean fitness in unpredictable pollination environments, but may also compromise current and future outcrossing through physiological and demographic trade-offs. Moreover, reproductive assurance (RA) may affect population and metapopulation dynamics, introducing complex feedbacks plus alternative equilibria and multiple levels of selection. Experimental manipulations of pollination and mating have tested whether RA explains widespread mixed-mating systems, although definitive evidence of this is still lacking. Species exhibiting geographic variation in floral traits have been exploited to determine the role of RA in transitions between outcrossing and selfing, a very common evolutionary trend in plants. Transplant experiments are particularly useful for detecting divergent selection on floral traits underlying mating-system variation. Recent work by our research group suggests that small, selfing flowers benefit from RA by avoiding parasitism rather than pollen limitation, emphasizing that self-pollination and associated floral traits can ameliorate an array of ecological pressures faced by outcrossing plants. Although evolution of asexuality and selfing can be studied using a common theoretical framework, whether asexuality evolves because it provides RA remains virtually unstudied. Wide variation in sexuality within diverse species provides underexploited opportunities for experimentation, which could ultimately provide a better understanding of the general role of RA in plant reproductive evolution.

10.1 Introduction

Uniparental reproduction has evolved from outcrossing via self-fertilization or asexuality a great many times in plants (Stebbins 1974; Carman 1997). Changes in reproductive mode strongly influence important population-genetic processes as well as the trajectory of life-history evolution (Lloyd 1980a; Takebayashi and Morrell 2001). Thus, the evolution of uniparental reproduction has been a subject of sustained interest since the dawn of evolutionary biology (Darwin 1876).

Uniparental reproduction seems highly advantageous. Lloyd (1980b) showed that, in terms of transmission between generations, alleles for either selfing or asexuality enjoy an equivalent advantage over an allele for outcrossing in a population of hermaphrodites. The two modes of uniparental reproduction are genetically and developmentally distinct in that selfing involves recombination and syngamy whereas asexuality does not. Yet, both double the probability of an allele being transmitted from mother to offspring; hence they avoid the so-called cost of meiosis, or gene sharing (Lloyd 1988).

Selfing and asexuality also assure successful reproduction when pollen vectors and/or mating partners are scarce (Darwin 1876). Comparative analyses indicate that insufficient pollination often limits the seed production of individuals in outcrossing populations (Burd 1994; Ashman *et al.* 2004) and that the ability to self-fertilize reduces pollen limitation (Larson and Barrett 2000). The same phenomenon is evident in the geographic distributions of outcrossing versus selfing or asexual taxa (Lloyd 1980a; Chapter 6). Baker (1955) and others since have argued that self-compatible genotypes disperse long distances more successfully, presumably because the ability to self-fertilize facilitates population establishment and population persistence during subsequent periods of low density (Cox 1989; Rambuda and Johnson 2004). Likewise, selfing species or selfing populations of otherwise outcrossing species occur more often than outcrossers in geographically and/or ecologically marginal habitats, where outcross pollination may be uncertain (Jain 1976; Lloyd 1980a; Elle 2004; Chapter 6). The same contrast is evident between asexual taxa and their sexual relatives (Bierzychudek 1987a; Richards 2003). Reproductive assurance (RA) has also commonly been invoked to explain the widespread occurrence of mating strategies involving a mix of selfing and outcrossing (Jain 1976; Holsinger 1996; Goodwillie *et al.* 2005). Similarly, some plants engage in both asexual and sexual reproduction, perhaps in an adaptive mixture (Grimanelli *et al.* 2001; Richards 2003; Bicknell and Koltunow 2004).

Despite the intuitive appeal of RA as an adaptive explanation for the evolution of uniparental reproduction, and its long-standing consideration in the literature, this hypothesis has received serious theoretical and empirical testing only recently. This chapter adopts a critical perspective on RA and its role in the evolution of selfing and asexuality. We begin with a brief synopsis of theory that considers how RA influences the selection of self-fertilization. We then review empirical work testing the importance of RA in two contexts: (1) the evolution of mixed mating systems and (2) evolutionary transitions between outcrossing and selfing. We present results from recent work by our research group to highlight some useful empirical approaches and the challenges involved in application and interpretation. In doing so, we discuss other forms of reproductive assurance that do not involve pollen limitation directly. We finish by contrasting asexuality with self-fertilization in terms of both theoretical framework and empirical approaches to highlight the substantial gaps in current knowledge of the ecology and evolution of asexuality as a mechanism of RA in plants.

10.2 Reproductive assurance and self-fertilization: theoretical context

Despite the obvious advantages of uniparental reproduction, most flowering plants predominantly outcross and relatively few reproduce exclusively via self-fertilization (Goodwillie *et al.* 2005). Inbreeding depression, the reduced vigour of inbred compared with outbred individuals, is one of the few selective factors strong enough to oppose the automatic selection of selfing (Charlesworth and Charlesworth 1987). Early mathematical models pitting the two-fold gene transmission advantage of selfing against inbreeding depression (e.g., Nagylaki 1976) indicated that selfing should be selected only if the fitness of selfed progeny (ω_s) compared with that of outcrossed progeny (ω_x) exceeds one-half ($\omega_s/\omega_x > 0.5$). These models assume that selfing affects neither seed production nor pollen export. If floral modifications that cause self-pollination also reduce the transfer of pollen to stigmas of conspecifics (pollen discounting), then the male fitness advantage of siring one's own seeds is offset by reduced siring of seeds produced by other individuals (Holsinger 1996; Harder and Wilson 1998).

The balance between the advantages of selfing and inbreeding depression is dynamic, because the increased homozygosity caused by selfing allows selective reduction in the frequencies of deleterious recessive mutations, thereby purging genetic load (Lande and Schemske 1985; Charlesworth and Charlesworth 1987; although see Porcher and

Table 10.1 A summary of theoretical studies exploring the importance of reproductive assurance (RA) for the evolution of self-fertilization

Main theme/reference	Modes of self-pollination	Model type	δ	Major insights
Lloyd 1979	gfbd	P	C	Introduced modes of selfing. Autonomous selfing (especially d) selected because it provides RA
Lloyd 1992	gfbcd	P	C	Benefit of RA relative to D_f and D_m depends on the mode of self-pollination. D_f plus low ω_s/ω_x erodes benefit of RA
Schoen et al. 1996	gfbcd	P	C	Selection on floral variation depends on the net fitness effect of covariation with multiple modes of self-pollination. RA may cause selection of floral traits that simultaneously cause autonomous and facilitated self-pollination
Inbreeding depression coevolves with selfing				
Johnston 1998	Many*	P	V†	RA decreases threshold ω_s/ω_x required to prevent evolution of full selfing, but does not lead to stable MM
Porcher and Lande 2005b	c	G	V	RA decreases threshold ω_s/ω_x but only leads to stable MM if it is associated with D_m. RA cannot explain low stable selfing
Temporal variation in pollinator visitation				
Schoen and Brown 1991	d	P	C	Drastic fluctuation in pollinator visitation selects for "induced" autogamy and yields stable MM
Morgan and Wilson 2005	bcd	P	C	Variation in pollination does not affect selection of c or d, but strongly decreases threshold ω_s/ω_x above which p is selected because RA increases geometric mean fitness. MM can be stable
Ramifications of seed discounting				
Sakai and Ishii 1999	d	P	C	Seed size/number trade-off reduces selective value of selfing (resource-based D_f). Pollinator uncertainty yields MM, but outcrossing and seed set < 1 is ESS when benefits of fewer, bigger seeds > more, smaller seeds
Morgan et al. 1997	cd	P	C	Benefit of RA readily eroded via reduced residual reproductive value, resulting in a between-season D_f in perennial populations
Population dynamics included				
Lloyd 1980a	bd	P	C	Autonomous self-pollination provides RA at low density, which then increases density. ω_s/ω_x declines with increasing density. This negative feedback leads to stable MM
Cheptou 2004	c	P	C	Potential for complex, counterintuitive feedback between selfing, RA, ω_s/ω_x and population growth, but RA does not lead to stable MM
Morgan et al. 2005	fd	P	C	RA allows populations to exist below outcrossing extinction threshold. MM is common and b is often selected, even when $\omega_s/\omega_x < 0.5$
Metapopulation dynamics included				
Pannell and Barrett 2001	fb‡	G‡	C	Higher b did not enhance colonization ability relative to context-dependent f (where f = 1/N). High colony turnover favoured outcrossing when $\omega_s/\omega_x < 0.5$ because of higher seed production and hence dispersal

For each study we list the mode(s) of self-pollination considered (g = geitonogamy, f = facilitated autogamy, b = prior autogamy; c = simultaneous autogamy, d = delayed autogamy); whether a phenotypic (P) or genetic (G) model was used; whether inbreeding depression ($\delta = 1 - [\omega_s/\omega_x]$) was held constant (C) or coevolved with self-fertilization via purging (V); and the major insight gained with respect to how RA influences the evolution of the mating system, especially whether it allows the evolution of stable mixed mating (MM). D_f and D_m are seed and pollen discounting, respectively.
* Particular mode(s) of self-pollination was not specified, but sets of curves relating the proportion of ovules fertilized to self-pollination were used to model the effect of selfing on seed production. Different curves can be anticipated to represent different selfing modes or combinations of modes.
† This model simulates purging by allowing inbreeding depression to decline with increased selfing, but does not depict the purging process *per se*.
‡ A single locus controlled the mating system, with a dominant allele for prior selfing. All three genotypes were self-compatible and thus experienced facilitated selfing as a result of random mating (f = 1/N, where N is population size).

Lande 2005a). Positive feedback results as selection for selfing becomes progressively stronger in a population that is already partially selfing. This process yields only two endpoints of mating-system evolution: predominant outcrossing with $\omega_s/\omega_x < 0.5$, and predominant selfing with $\omega_s/\omega_x > 0.5$. Increasing evidence that empirical estimates of selfing do not exhibit the predicted bimodal distribution has motivated much theoretical work aimed at explaining the widespread occurrence of mixed-mating systems (reviewed in Goodwillie et al. 2005). Many of these models explore ecological selective factors associated with pollination (also see Chapter 4), especially reproductive assurance.

RA was first considered mathematically by Lloyd (1979) in a model that also introduced the selective importance of how and when selfing occurs in relation to outcrossing. Self-pollination occurs via the transfer of pollen between anthers and stigmas within flowers (autogamy) or between flowers on the same plant (geitonogamy). Autogamy, in turn, occurs autonomously in the absence of pollinators, or is facilitated by pollinators. Autonomous autogamy can occur before, during, or after opportunities for outcrossing (prior, simultaneous, and delayed autogamy, respectively). All forms of autonomous autogamy provide RA, thus they are selected more readily than facilitated autogamy and geitonogamy, both of which require pollinator visitation and occur at the same time as outcrossing. Lloyd (1992) further examined the selection of various modes of selfing via their effects on pollen discounting and seed discounting, the latter of which occurs when ovules that are self-fertilized could have been outcrossed. The fitness costs of seed discounting become significant when $\omega_s/\omega_x < 0.5$ and thus selfed seed are less valuable genetically than outcrossed seed. Some modes of autonomous autogamy may provide RA with little pollen or seed discounting and are, therefore, readily selected. In particular, delayed selfing is usually advantageous because it involves only pollen and ovules that, by definition, can no longer participate in outcrossing (although see Chapter 4). Hence floral mechanisms causing delayed selfing may be favoured even when $\omega_s/\omega_x \ll 0.5$. In contrast, prior and simultaneous autogamy provide RA, but likely cause seed discounting.

Subsequent theory explored several different aspects of RA related to four general issues: (1) selective purging of inbreeding depression; (2) temporal variability in pollination environments; (3) physiological and demographic costs associated with seed discounting; and (4) metapopulation dynamics. Table 10.1 summarizes the relevant theory: Goodwillie et al. (2005) provided a more detailed discussion. Taken together, recent theory reveals that the role of RA in mating-system evolution might be much more complicated than previously thought. For instance, spatio-temporal variation in pollination environments enhances the fitness benefits of autonomous selfing, because RA increases geometric mean fitness (Schoen and Brown 1991; Morgan and Wilson 2005; Chapter 2). However, the potential cost of seed discounting extends well beyond the simple usurpation of ovules by self-fertilization. Because developing selfed seeds consume maternal resources, seed discounting can involve costly trade-offs between offspring quantity and quality, as well as reduced survival and future reproduction (Lloyd 1979; Morgan et al. 1997; Sakai and Ishii 1999; Chapter 4). RA can also affect the growth, density and persistence of populations and metapopulations, which introduces significant feedbacks, alternative equilibria, and multiple levels of selection that have just begun to be explored theoretically (Lloyd 1980a; Pannell and Barrett 1998, 2001; Cheptou 2004; Morgan et al. 2005; Chapters 2 and 12).

10.3 Reproductive assurance and self-fertilization: empirical approaches

Empirical investigation of the role that RA plays in mating-system evolution has increased during the past decade, but still lags well behind theoretical advances, and has only scratched the surface of the potential complexities revealed by theory. This slow progress is surprising, because the core of the RA hypothesis is that selfing alleviates pollen limitation, and the occurrence of pollen limitation in flowering plants has received much attention, including considerable discussion of its detection, and how and why it might occur (Burd 1994;

Larson and Barrett 2000; Ashman *et al.* 2004). We now consider how the importance of RA in maintaining mixed mating can be tested, and then assess how RA can be isolated as a factor causing the transition from outcrossing to selfing in species with extensive mating-system variation.

10.3.1 Reproductive assurance and the evolution of mixed-mating systems

What evidence would support the hypothesis that RA has influenced the evolution of a stable mixed-mating system? Assuming evolutionary equilibrium, the fitness benefits of selfing (gene transmission, RA) should be matched by fitness costs (inbreeding depression, pollen and seed discounting), so that no other phenotype with a slightly different mating strategy can achieve higher fitness (Lloyd 1979; Eckert and Herlihy 2004). This hypothesis can be tested most directly by manipulating the mating system and quantifying fitness changes.

A simple test of RA involves reducing the capacity for autogamy by removing anthers before they dehisce, and comparing the seed production of these emasculated flowers (F_E) with that of intact flowers (F_I) to test whether $F_E < F_I$. Although Cruden and Lyon (1989) and Schoen and Lloyd (1992) recommended this approach more than a decade ago, results have appeared slowly. We reviewed the literature and found experimental results for 29 taxa, all animal-pollinated, and most (55%) published during the past five years. This is a small number compared with the hundreds of species that have been subject to pollen supplementation experiments to quantify pollen limitation (Ashman *et al.* 2004). Figure 10.1 plots the proportional reduction in seed production caused by emasculation ($[F_I - F_E]/F_I$), a measure of RA, against the capacity for autonomous autogamy, or autofertility (AF), estimated by comparing seed production of flowers excluded from pollinators (F_C) with that of naturally or hand-pollinated flowers (F_C/F_I, following Lloyd and Schoen 1992). We expect RA ≤ AF because the contribution of autonomous autogamy to the seed set of naturally pollinated flowers should not exceed their capacity to set seed via autonomous autogamy. This expectation is violated by only four species, perhaps because the floral traits that cause modest autonomous self-pollination (e.g., low herkogamy or dichogamy) also allow facilitated autogamy (Lloyd 1979; Schoen *et al.* 1996). In addition, emasculation may, as a confounding effect, reduce seed set by damaging flowers, shortening their life span, or making them less attractive to pollinators, thereby causing RA to be overestimated (Schoen and Lloyd 1992; Eckert and Herlihy 2004). These unintended effects of emasculation are rarely quantified (Electronic appendix 10.1, http://www.eeb.utoronto.ca/EEF/).

Figure 10.1 illustrates wide variation in both AF and RA, but no covariation between them. The capacity for autonomous autogamy is not matched to its importance for seed production in natural populations. Whereas some species (e.g., *Silene noctiflora*, Sn) have high AF and high RA, as expected (Davis and Delph 2005), 11 of 20 species with AF > 0.4 do not seem to benefit from AF, as autogamy makes little or no contribution to seed production (RA ~ 0). This result may be explained by recent theory emphasizing that RA can be favoured even if pollen limitation occurs rarely (Morgan and Wilson 2005). Therefore, empirical tests of RA must assess temporal variation in pollination and mating (Herrera *et al.* 2001). In contrast, 62% of species in Fig. 10.1 were studied during only one year in a single population.

Whether spatio-temporal variation in the pollination environment favours RA has been addressed experimentally in only two species, and neither case provides compelling evidence that the particular mixture of selfing and outcrossing exhibited by these species is evolutionarily stable. *Collinsia verna* (Plantaginaceae) is a winter annual that flowers during early spring when cross-pollination might vary dramatically both within and between years. Conspicuous white and blue flowers suggest predominant outcrossing, yet flowers become capable of autonomous self-pollination just before senescence, as the style elongates, bringing the stigma close to the dehisced anthers (Kalisz *et al.* 1999). Using floral emasculation, supplemental hand-pollination, and the isolation of flowers from pollinators, Kalisz and Vogler

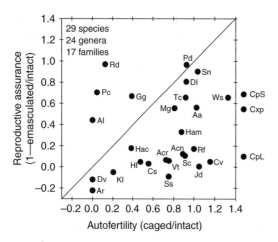

Figure 10.1 Variation in fertility of autonomous selfing (reproductive assurance) and the degree of autofertility via selfing among 29 species of flowering plants. Intact and emasculated flowers experienced natural pollinator visitation. Emasculated flowers had anthers removed before dehiscence to eliminate autogamy. Caged flowers were excluded from pollinators and hence could set seed only via autonomous selfing. The diagonal shows the maximum expected reproductive assurance given the level of autofertility. Points to the right of the plot are species for which autofertility estimates are not available. Reproductive assurance does not correlate with autofertility among species (Pearson $r = 0.17$, $P > 0.3$). Data points are identified by species code (in parentheses) as follows: Agalinus auriculata (Aa), Anthericum liliago (Al), Anthericum ramosum (Ar), Aquilegia caerulea (Acr), Aquilegia canadensis (Acn), Calyophus serrulatus (Cs), Clarkia xantiana parviflora (Cxp), Collinsia parviflora large-flowered populations (CpL), Collinsia parviflora small-flowered populations (CpS), Collinsia verna (Cv), Decodon verticillatus (Dv), Drosophyllum lusitanicum (Dl), Gentianella germanica (Gg), Hepatica acutiloba (Hac), Hepatica americana (Ham), Hibiscus laevis (Hl), Jeffersonia diphylla (Jd), Kalmia latifolia (Kl), Mimulus guttatus (Mg), Pedicularis dunniana (Pd), Pulsatilla cernua (Pc), Rhododendron ferrugineum (Rf), Roridula dentata (Rd), Sanguinaria canadensis (Sc), Scilla sibirica (Ss), Silene noctiflora (Sn), Tacca chantrieri (Tc), Verbascum thapsus (Vt), Werauhia sintenisii (Ws). Raw data are available in electronic appendix 10.1 (http://www.eeb.utoronto.ca/EEF/).

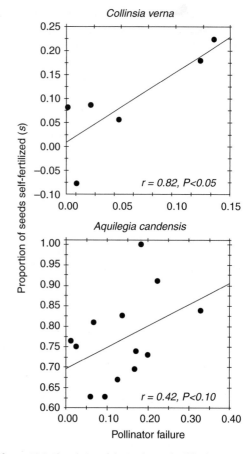

Figure 10.2 The relation of the incidence of self-fertilization to pollinator failure in two species with autonomous autogamy. Data for Collinsia verna are from Kalisz et al. (2004); those for Aquilegia canadensis are recalculated from Herlihy and Eckert (2002). Pollinator failure equals 1 − (fruit or seed production of emasculated flowers/fruit or seed production of intact flowers). Intact flowers were hand-outcrossed in the Collinsia study and naturally pollinated in the Aquilegia study. Hand-outcrossing does not increase seed production in these populations of Aquilegia.

(2003) showed that cross-pollination varied significantly among populations in Pennsylvania, USA, between years and during flowering seasons. Delayed selfing provided substantial RA: an 8% average increase in fruit set. Kalisz et al. (2004) then showed that, as expected, the proportion of seeds self-fertilized, estimated using genetic markers, differed between populations and years and correlated positively with the degree of pollinator failure (Fig. 10.2). These results are consistent with the RA hypothesis. However, the maintenance of mixed mating requires a cost of selfing, such as inbreeding depression, to balance the combined benefits of gene transmission and RA, presumably in some frequency-dependent manner (Goodwillie et al. 2005; Table 10.1). Yet, inbreeding depression in C. verna is far too weak ($\omega_s/\omega_x \sim 0.9$; Kalisz et al. 2004) to account for the maintenance of high outcrossing (mean outcrossing = 0.91). The selective factors preventing the spread of alleles that increase selfing remain unknown.

The opposite problem emerged from work with *Aquilegia canadensis* (Ranunculaceae), a short-lived, spring-flowering perennial that occurs in small, patchy populations on barren rocky outcrops, and also flowers during spring. Emasculations conducted during one reproductive season in a single population in Ontario, Canada, failed to show that high AF provided RA (Eckert and Schaefer 1998). However, subsequent work indicated much spatio-temporal variation in the proportion of self-fertilized seeds among populations and between years within populations (Fig. 10.3). A more extensive emasculation experiment conducted across 10 Ontario populations indicated that autogamy usually provided substantial RA: a 14% average increase in seed production. Like *C. verna*, self-fertilization correlated positively with the degree of pollinator failure, though not quite significantly (Fig. 10.2). However, further dissection of the mating system using emasculation and transplant experiments combined with marker-gene analysis revealed that some apparent selfing was actually crossing between closely related plants (Griffin and Eckert 2003; Herlihy and Eckert 2004). Direct estimates of autogamous selfing, the only component that provides RA, did not correlate with pollinator failure (Fig. 10.4), suggesting that RA cannot account for high variance in selfing among *A. canadensis* populations.

Unlike the delayed selfing in *C. verna* caused by reduced herkogamy towards the end of floral life, autonomous selfing in *A. canadensis* occurs simply because stigmas are close to dehiscing anthers throughout floral life, including the period of cross-pollination (Eckert and Schaefer 1998; Griffin et al. 2000). By using genetic markers to estimate the absolute number of selfed and outcrossed seeds produced by intact versus emasculated flowers, Herlihy and Eckert (2002) showed that RA caused severe seed discounting in *A. canadensis*. This displacement of outcrossed seeds by selfed seeds is costly, because of high inbreeding depression (mean $\omega_s/\omega_x = 0.07$), so that the benefits of gene transmission and RA appear to be more than offset (Herlihy and Eckert 2002). The obvious conclusion, that high selfing is strongly disadvantageous in this species, is especially perplexing given that populations of *A. canadensis* contain substantial genetic variation in herkogamy, which effectively reduces selfing (Eckert and

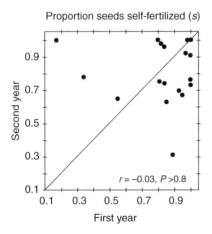

Figure 10.3 Extensive year-to-year variation in self-fertilization within 19 natural populations of *Aquilegia canadensis* from Ontario, Canada (C.G. Eckert, B. Ozimec, and C.R. Herlihy unpublished data). The proportion of seed produced through self-fertilization (s) was estimated from the segregation of two allozyme polymorphisms among at least 30 progeny arrays per population per year. Despite considerable variation among populations in each year, s did not correlate between years among populations ($r = -0.03$, $P > 0.8$), indicating substantial temporal variation in the mating system.

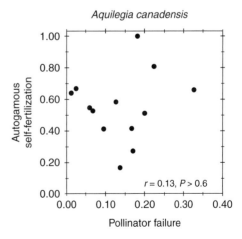

Figure 10.4 No relation of the proportion of seeds self-fertilized autogamously to pollinator failure among 13 populations of *Aquilegia canadensis*. Data are recalculated from Herlihy and Eckert (2002). Autogamy was estimated genetically from the difference in apparent self-fertilization between intact and emasculated flowers, following Schoen and Lloyd (1992).

Herlihy 2004). The selective factors opposing the spread of alleles that increase herkogamy, and thereby reduce selfing, remain unidentified.

Herlihy and Eckert (2002) measured seed discounting in terms of seed numbers for individual flowers, yet theory emphasizes that the use of resources to make low-quality selfed seed may have indirect and longer-term costs (Table 10.1). Seed discounting probably operates between flowers in *A. canadensis*, as experimental manipulations of investment in seed revealed dynamic allocation of resources among flowers within inflorescences (Eckert and Herlihy 2004). Resources spent on selfed seeds in early flowers are unavailable to outcrossed seeds in later flowers. Reproductive expenditures on selfed seed during one year could also compromise survival, reproductive output, and hence opportunities for outcrossing during subsequent years (Morgan *et al*. 1997). Although these costs are widely appreciated in the pollen limitation literature (Ashman *et al*. 2004), the ramifications of seed discounting throughout the life cycle for the evolution of selfing remain uninvestigated empirically.

These two detailed studies of the adaptive significance of RA in natural populations confirm that, as expected from work on pollen limitation, autogamous selfing can regularly boost seed production in habitats where pollinators and/or mates are sometimes scarce. Work on both *Collinsia* and *Aquilegia* also suggests that the value of RA is increased by variation of outcross pollination in time and space, as suggested by theory (Table 10.1). However, putting RA in a broader cost–benefit analysis, including the cost of meiosis, inbreeding depression, and seed discounting, does not yield evidence that the mating system is evolutionarily stable for either species. Neither case study includes estimates of pollen discounting owing to the technical challenges of estimating male fitness in natural plant populations. Although theory suggests that pollen discounting is a key factor maintaining mixed mating when RA is advantageous (Porcher and Lande 2005b), a trade-off between selfing and male outcross success is unlikely to resolve either conundrum (Eckert and Herlihy 2004). Perhaps these results reveal that current theory does not incorporate essential influences on mating-system evolution. How can partial selfing be maintained when inbreeding depression is very strong (*Aquilegia*) or very weak (*Collinsia*)? The existence of only two detailed case studies emphasizes that empirical work on RA lags well behind theoretical progress, and underscores the need for experimental analyses of RA in a broader range of plants.

10.3.2 Exploiting intraspecific variation to test the reproductive assurance hypothesis

Wide variation in selfing versus outcrossing among closely related plant taxa and among populations within species has been used profitably for comparative analysis of mating-system evolution, as exemplified by Lloyd's (1965) landmark work on *Leavenworthia*. Variable taxa in *Amsinckia*, *Arenaria*, *Clarkia*, *Eichhornia*, *Gilia*, *Lycopersicon*, *Mimulus*, *Phlox*, *Primula*, and several other genera have provided insight into key issues, including the genetic basis and purging of inbreeding depression (e.g., Holtsford and Ellstrand 1990; Johnston and Schoen 1996; Busch 2005), the developmental and genetic bases of floral changes causing selfing (e.g., Holtsford and Ellstrand 1992; Fenster and Barrett 1994; Fishman *et al*. 2002), and the occurrence of pollen discounting (e.g., Ritland 1991; Kohn and Barrett 1994; Fishman 2000).

These variable taxa are useful because the mating system probably differentiated recently, so mating variation is not confounded by unrelated variation in genetics, ecology, and life history. The selective factors originally involved are also probably still maintaining variation in the mating system. Recently diverged populations also offer practical advantages, such as more direct comparison of the fitness of alternative mating phenotypes, replicated comparative analysis of covariation between ecology and the mating system, and the wide phenotypic variation that can be created with controlled crosses to quantify the strength and form of natural selection on floral traits.

Below, we present recent results from our study of the influence of RA in the evolution of striking geographical variation in floral morphology and self-compatibility in a coastal dune plant. Our results are preliminary, thus the discussion that follows is sometimes speculative. Nevertheless, working through the intriguing patterns we found illustrates the approaches to, and problems with, exploiting intraspecific mating-system variation to test the RA hypothesis, and serves as a vehicle for considering a broader definition of RA.

Variation in floral morphology and self-incompatibility in a coastal dune plant
Camissonia cheiranthifolia (Onagraceae) is a short-lived perennial endemic to Pacific coastal dunes from northern Baja California, Mexico, to southern Oregon, USA. Taxonomic studies by Raven (1969) and our geographical surveys have revealed the full range of mating-system variation along the species' essentially one-dimensional range (Fig. 10.5). In southern Californian populations of *C. cheiranthifolia*, plants produce large, herkogamous flowers that secrete nectar and are strongly self-incompatible. Further north, plants are large-flowered, but highly self-compatible. North of Point Conception, California, towards the northern range limit, plants are usually small-flowered and self-compatible. However, two populations (CGN and CSP) just north of Point Conception vary extensively in flower size, probably as a result of inadvertent transplanting of large-flowered genotypes from more southerly populations during dune restoration 20–30 years ago, followed by the introgression of alleles that increase flower size into otherwise small-flowered populations. Plants are also small-flowered in populations at the southern range limit in Baja California and on the Channel Islands off southern California. One northern population contains high

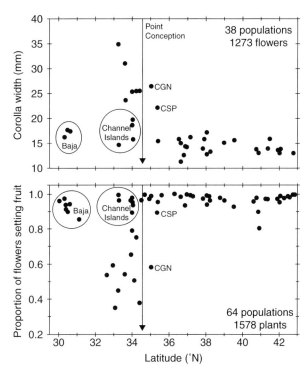

Figure 10.5 Extensive variation in flower size and fruit set across the geographic range of the Pacific-coast dune endemic *Camissonia cheiranthifolia*. Points are population means for samples of ~35 flowers for corolla width and ~25 plants for fruit set (K.E. Samis, E. Austen, and C.G. Eckert unpublished data). Circled points indicate small-flowered populations at the southern edge of the range in Baja California, Mexico, and on the Channel Islands (San Nicholas, Santa Cruz, and Santa Rosa Islands). Populations CGN and CSP had high frequencies of large-flowered plants in a geographic region dominated by small-flowered populations. The transition zone between large- and small-flowered populations occurs at Point Conception. Comparison of the floral morphology exhibited by 15 of these populations under both field conditions and in a common glasshouse environment revealed a very strong genetic basis to this geographic variation (correlation of population means across environments: $r = 0.97$ for corolla width and $r = 0.93$ for herkogamy, both $P < 0.0001$).

frequencies of plants that produce only fully selfing, cleistogamous (closed) flowers.

Our surveys throughout the range of *C. cheiranthifolia* (Fig. 10.5) confirm striking variation in corolla width, which covaries strongly with the spatial separation of anthers and stigmas within flowers (herkogamy), a trait expected to regulate self-pollination (Eckert and Herlihy 2004). We are developing genetic markers to estimate self-fertilization in these populations, but there is little doubt that this will reveal extensive mating-system variation. Raven (1969) hypothesized that ancestral populations were large-flowered, from which small-flowered populations evolved as the species spread northward; a scenario we are testing with population-genetic analysis. For now, we assume that large-flowered populations in central California represent the ancestral morphological, genetic and perhaps ecological conditions from which the small-flowered phenotype evolved.

Conflicting evidence for the reproductive-assurance hypothesis

Evolutionary analysis of intraspecific mating-system variation involves two steps: (1) identify possible selective factors by comparing genetic and ecological characteristics of divergent populations; and (2) determine which factor(s) maintain outcrossing in one part of the range but tip the balance of costs and benefits to allow the evolution of selfing elsewhere. The RA hypothesis is usually framed in terms of pollen limitation caused by low pollinator visitation. The obvious first test compares pollinator abundance and species composition in selfing versus outcrossing populations, although this has rarely been done (Inoue *et al.* 1996; Fausto *et al.* 2001; Herrera *et al.* 2001).

Geographic surveys of insects visiting *C. cheiranthifolia* identified small, oligolectic bees as the major pollinators, which seem to be more abundant (or at least more active) south than north of Point Conception. This observation led Linsley *et al.* (1973) to hypothesize that small flowers and selfing were selected to provide RA north of Point Conception, where populations experience dense morning fog, which inhibits pollinators. Consistent with this hypothesis, plants in small-flowered populations set almost twice as many fruit per flower as those in large-flowered populations, and the variance in fruit set among populations is seven-fold higher for large- than small-flowered populations (Figs 10.5 and 10.6). This striking difference between floral phenotypes does not arise indirectly from general latitudinal variation in fruit set (Fig. 10.5). Moreover, small-flowered plants exhibit higher and less variable fruit set than large-flowered plants when both phenotypes inhabit the same population (Fig. 10.6). These results are consistent with evidence from other species that outcrossers are prone to pollen limitation, a necessary precondition for the evolution of selfing via RA (e.g., Fausto *et al.* 2001; Goodwillie 2001).

To confirm that pollinator scarcity causes low fruit set in large-flowered populations, we compared modes of self-pollination and the extent of pollen limitation between large- and small-flowered populations. Tests of variation in pollen limitation in taxa exhibiting wide variation in floral morphology, autofertility, or self-fertilization are few, and have yielded mixed results (Piper *et al.* 1986; Goodwillie 2001; Herrera *et al.* 2001). In *C. cheiranthifolia*, the anthers of small flowers dehisce in the bud, causing substantial prior autogamy, whereas anthers of large flowers shed pollen only after anthesis. To test whether this mechanism coupled with geographic differences in flower size resulted in geographic variation in self-compatibility and autofertility we used standard pollination treatments on plants in a pollinator-free glasshouse (Fig. 10.7). All plants from one southern population were strongly self-incompatible, as suggested by Raven (1969). As expected, plants from 10 small-flowered populations were all strongly autofertile, setting 78% as many seeds autonomously as after hand self-pollination. Unexpectedly, plants from six self-compatible, large-flowered populations were also highly autofertile, setting 85% as many seeds autonomously as after hand self-pollination. Autonomous autogamy occurs in both phenotypes when flowers close and press anthers onto the stigma. However, large flowers open for at least another day, whereas small flowers often stay closed (S. Dart and C.G. Eckert unpublished data). Thus small flowers seem to engage in prior and delayed autogamy, whereas large flowers engage primarily in simultaneous,

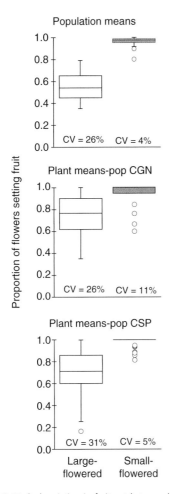

Figure 10.6 Marked variation in fruit set between large- and small-flowered *Camissonia cheiranthifolia* (K.E. Samis, S. Dart, and C.G. Eckert unpublished data). The top panel shows the distributions of population means for 10 large-flowered and 42 small-flowered populations across the mainland geographic range ($n > 25$ plants per population). The bottom two panels show the distributions of plant means within populations CGN and CSP, which exhibit wide floral variation ($n > 30$ plants per flower type per population). The bottom line in each box plot is the 25th percentile, the midline is the median, and the top is the 75th percentile. The whiskers extend 1.5 interquartile distances from the median or to the extreme points, whichever is closest. Points beyond 1.5 interquartile distances from the median are shown as circles. Small-flowered phenotypes set significantly more fruit than large-flowered phenotypes, both among populations and among plants within populations CGN and CSP (*t*-tests: all $P < 0.0001$). The variance in fruit set, as summarized by the coefficient of variation (CV), was higher for large- than small-flowered phenotypes (Levene's tests: all $P < 0.001$).

autonomous autogamy and perhaps pollinator-facilitated autogamy. Although both combinations of selfing modes provide some RA, they probably differ in their contributions to the mating system and the extent to which they cause seed and pollen discounting. We are testing this prediction with emasculation experiments. Supplemental hand-pollination experiments (following Ashman *et al.* 2004) did not increase fruit set in either three large-flowered or three small-flowered populations (S. Dart and C.G. Eckert unpublished data), further suggesting high autofertility of both small- and large-flowered plants. Together, these results cast serious doubt on the hypothesis that small flowers and self-pollination were selected to alleviate insufficient cross-pollination in central Californian populations of *C. cheiranthifolia*.

Other forms of reproductive assurance provided by self-fertilization and/or small flowers

A major goal of recent mating-system theory is to identify and evaluate selective factors associated with pollination (Barrett and Harder 1996; Chapter 4). However, self-pollination and associated floral traits can ameliorate some other ecological challenges faced by outcrossing plants. For instance, selfing may evolve as an indirect consequence of selection for rapid development and reproduction where suitable ecological conditions occur briefly, which is consistent with the association between selfing and stressful or ephemeral habitats (reviewed in Lloyd 1980a; Elle 2004; Chapter 8). Several studies have explored this hypothesis by comparing development rates between large-flowered outcrossing and small-flowered selfing populations (Runions and Geber 2000; Elle 2004; Mazer *et al.* 2004). In *C. cheiranthifolia*, reproductive plants tend to be smaller and less woody in small- than in large-flowered populations. However, small-flowered plants flower later, not earlier, when grown from seed along with large-flowered plants in a common glasshouse (K. E. Samis, S. Dart, and C. G. Eckert unpublished data). Moreover, the timing of environmental conditions suitable for growth and reproduction does not change abruptly in association with the transition from large- to small-flowered populations. In fact,

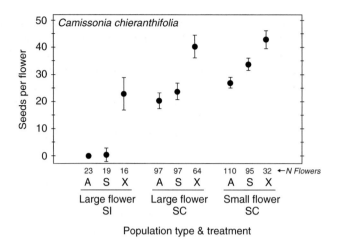

Figure 10.7 Variation in mean (±SE) self-compatibility and autofertility among populations of *Camissonia cheiranthifolia* revealed by a crossing experiment conducted under glasshouse conditions for 553 flowers on 180 plants from 17 populations (K.E. Samis, C.E. Inglis, and C.G. Eckert unpublished data). Flowers were unpollinated to test for autonomous self-fertilization (treatment A), self-pollinated (treatment S), or cross-pollinated with pollen from an unrelated plant within the same population (treatment X). One population, from San Diego County, is strongly self-incompatible (SI). All plants from the other 16 populations were self-compatible (SC). Two-way, fixed-effects ANOVA conducted on data from the SC populations revealed significant variation in seed production among pollination treatments ($P < 0.0001$), but not between population types ($0.10 > P > 0.05$), with no significant interaction ($P > 0.7$).

climatic variation in coastal dune habitat is strongly dampened by the maritime influence. An analysis of 30-year normals for a variety of climatic parameters indicates that only precipitation varies discontinuously within the range of *C. cheiranthifolia*, and that the transition from drier climates in the south (~1000 mm rain/year) to wetter climates in the north (~2000 mm rain/year) occurs hundreds of kilometres north of the flower-size transition. Analysis of variation in demographic parameters further supports this conclusion (K.E. Samis and C.G. Eckert unpublished data).

Self-fertilization may also be a strategy to escape competition for pollinators and/or hybridization with closely related, sympatric species (Levin 1972), thereby providing RA for both offspring quantity and quality. For example, experiments using artificial populations of *Arenaria uniflora* indicated that selfing guards against reduced fertility caused by interspecific pollination (Fishman and Wyatt 1999). However, in *Clarkia xantiana*, selfing may provide RA in habitats where cross-pollination via specialist bees is lower due to the absence of co-flowering congeners (Moeller and Geber 2005; Chapter 6). In contrast to other species in section *Holostigma* of *Camissonia*, most of which exhibit a history of hybridization, *C. cheiranthifolia* appears well differentiated, and is not known to interact with congeners via pollinators, especially in dune habitat north of Point Conception (Raven 1969). The pollination environment of *C. cheiranthifolia* may be influenced by the many unrelated co-flowering species in the wonderful Pacific-coast dune habitat, but the species composition of dune communities does not change noticeably in the flower-size transition zone.

Although flowers are sites of mutualism between plants and animals, they are also involved in other biotic interactions (Chapter 7) that sometimes vary geographically (Herrera et al. 2002). Many floral traits that attract pollinators also attract foliar herbivores, floral herbivores, and pre-dispersal seed predators, with drastic fitness consequences (Adler and Bronstein 2004). How simultaneous interactions between flowers, their benefactors, and antagonists generate correlational selection favouring genetic and biochemical associations between traits that attract pollinators and traits that deter parasites has attracted considerable recent interest (Herrera et al. 2002; Chapter 7). These interactions may cause direct or indirect selection on the mating system, if the changes to floral morphology and development associated with selfing also reduce parasitism. This possibility has not been explored seriously, even though it can be viewed as simply another form of RA or, as Lloyd (1980a) put it, an "autogamy of defense."

Striking geographical variation in floral traits and fruit set in *C. cheiranthifolia* is matched only by parallel geographic variation in floral parasitism (Fig. 10.8). Plants in large-flowered populations suffered four times more bud damage, and two times more damage to flowers and fruits, than those in small-flowered populations. Mean damage drops from 32% in the most northerly large-flowered population to 10% in the most southerly small-flowered population, even though these populations are <0.3° latitude apart. Moreover, large-flowered plants experience three times more floral parasitism than small-flowered plants in the two populations north of Point Conception where they co-occur. Much of this damage seems to be caused by a single, as yet unidentified, beetle species, the larvae of which burrow into developing buds and consume pollen and other floral organs, as well as seeds if the damaged flower becomes a fruit. We conclude tentatively that floral parasitism, not pollen limitation, primarily causes low fruit set of plants in large-flowered populations.

Mating-system evolution in response to floral parasitism?
Floral parasitism could affect mating-system evolution in at least three ways. First, small flowers may be selected because they experience lower parasitism. If herbivory differences between large- and small-flowered plants of *C. cheiranthifolia* create female and/or male fitness differentials of similar magnitude, then a small-flowered mutant might have enjoyed a large enough advantage in ancestral large-flowered populations to compensate for inbreeding depression. Although this process is conceptually similar to the evolution of selfing via RA in unpredictable pollination environments, the targets of selection and sequence of evolutionary changes differ.

The most likely initial steps in the evolution of selfing via RA involve subtle changes in the positioning of anthers and stigmas, causing delayed selfing with minimal seed and pollen discounting (Lloyd 1992). Smaller flowers, shorter floral lifespan, and reduced nectar reward should evolve secondarily in response to the reduced fitness gains of pollinator attraction in selfing populations (Lloyd 1980a). In contrast, insect parasitism may select for flowers that are less conspicuous or rewarding to ovipositing female parasites or their progeny. For instance, small flowers of *C. cheiranthifolia* secrete no nectar and are open less than half as long as large flowers. A reduction in overall flower size probably reduces herkogamy (Elle 2004). Although mutants with inconspicuous flowers would likely suffer strong seed and pollen

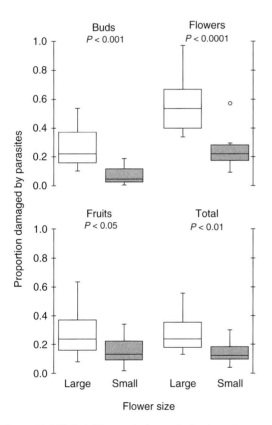

Figure 10.8 Marked differences in damage by floral parasites to buds, flowers, and fruits between 12 small-flowered and 12 large-flowered populations of *Camissonia cheiranthifolia*. Box plots show the distributions of population means (see Fig. 10.6 for details). Differences between large- and small-flowered populations were assessed using *t*-tests (*P* values). These results are based on the fates of 13 820 buds, flowers, or fruits on 711 plants in 24 populations across the species' geographic range (S. Dart and C.G. Eckert unpublished data).

discounting, a floral variant with complete discounting can be selected whenever ω_s/ω_x exceeds the probability of successful pollen export (Lloyd 1988: see Fishman and Wyatt 1999). Floral parasitism may reduce pollen export dramatically in *C. cheiranthifolia*. We are estimating ω_s/ω_x and the fitness consequences of parasitism in both large- and small-flowered populations to complete this cost–benefit analysis.

A fundamentally different hypothesis is that high parasitism maintains outcrossing in large-flowered populations. In the short term, parasite damage could increase the expression of inbreeding depression, which is generally stronger in more stressful environments (Roff 1997). However, the few studies of the effect of parasites on inbreeding depression in plants have produced mixed results. Some parasites attack outcrossed offspring with high vigour preferentially over low-vigour selfed offspring, thereby facilitating the selection of selfing (Hull-Sanders and Eubanks 2005). Outcrossing may also be maintained over the longer term through coevolution with parasites (Levin 1975), which may explain why outcrossing plant species tend to be infected with a higher diversity of fungal parasites than selfing species (Busch *et al.* 2004). Although parasites can, in theory, reduce selection for selfing, the opposite result can occur depending on the genetic basis of infection and the breeding system of the parasite (Agrawal and Lively 2001). Detailed knowledge of host and parasite(s) is required to test this hypothesis conclusively.

10.4 Asexual reproduction: a neglected mechanism of reproductive assurance

Lloyd (1979, 1980b, 1988) and others have noted repeatedly the evolutionary communalities between self-fertilization and the various forms of asexual reproduction. Like selfing, asexuality has evolved repeatedly in plants. Asexual seed production (apomixis) has been described in >400 flowering plant taxa representing >40 families (Carman 1997), and vegetative or clonal asexual reproduction occurs in up to 80% of plant species (Klimes *et al.* 1997). Yet, compared with the large body of work on the evolution of selfing versus outcrossing, the evolutionary balance between sex and asexuality in plants has received little attention, even though the predominant occurrence of sex in most complex organisms remains one of the most enduring puzzles in evolutionary biology (Bierzychudek 1987b). In fact, self-fertilization and asexuality tend to be studied by different scientists asking different questions. ISI Web of Science classifies most papers (63%) published on self-fertilization during the past 25 years as Ecology and Evolutionary Biology, whereas few (18%) on apomixis appear in those subject categories, and most (98%) were classified as either Plant Science or categories related to agriculture, plant breeding, or biotechnology. Most research on apomixis investigates its genetic and developmental bases, and is aimed largely at producing apomictic crop plants that breed true and restrict the "escape" of engineered genes (Richards 2003; Bicknell and Koltunow 2004).

Similarities in the fitness costs and benefits of apomixis and self-fertilization allow both reproductive systems to be considered in a unified theoretical framework developed largely to study the evolution of selfing (Table 10.1). Both apomixis and selfing are selected automatically because an allele for either form of uniparental reproduction experiences an equivalent transmission advantage by avoiding the cost of meiosis (Nagylaki 1976; Lloyd 1980b, 1988; Marshall and Brown 1981). However, the progeny produced by apomixis are genetic replicas of their parents and, therefore, do not express the inbreeding depression associated with the increased homozygosity caused by selfing. This feature removes a major fitness cost from the evolutionary equation governing the selection of apomixis. The main presumed short-term fitness disadvantage to asexuality lies in the lack of genetic variability among progeny, which may reduce the fitness of an asexual parent in temporally and spatially variable environments (Lloyd 1980b). However, this fitness disadvantage has often eluded detection in empirical studies on plants (Bierzychudek 1987b), whereas a wide variety of plant species exhibit strong inbreeding depression under a wide range of conditions (Charlesworth and Charlesworth 1987; Goodwillie *et al.* 2005). Furthermore, any fitness cost associated with asexuality is unlikely to decline as asexuality evolves, unlike the purging of inbreeding depression during generations of selfing, hence a short-term balance between the fitness costs and benefits of apomixis may lead to an evolutionarily stable mixed strategy combining sex and asexuality.

Like selfing, the ecological costs and benefits of apomixis are likely to depend on how and when it occurs (Lloyd 1979, 1992). Both apomictic seed production and clonal vegetative reproduction involve a bewildering variety of mechanisms, some of which are poorly understood (Klimes et al. 1997; Spielman et al. 2003). For example, apomictic progeny may be derived from diploid maternal cells in the ovule, or from egg nuclei in the embryo sac that either do not undergo meiosis or return to a diploid state after meiosis. Sometimes, seed development requires conventional fusion of a pollen sperm cell with the polar nuclei to form endosperm tissue (pseudogamy), whereas in other taxa the endosperm develops spontaneously. Apomixis that involves both spontaneous embryo and endosperm development provides RA. However, pseudogamous apomixis, which is thought to be much more widespread taxonomically (Richards 2003), requires pollination and is, therefore, unlikely to provide much RA (Lloyd 1980b). Apomixis is generally thought to involve deregulation in the timing of developmental processes involved in sex (Richards 2003; Bicknell and Koltunow 2004), hence partially apomictic genotypes may commit ovules to apomixis before opportunities for sexual outcrossing. As a result, RA provided by apomixis will often be associated with seed discounting, much like prior selfing. Of course, the fitness consequences of seed discounting depend on whether apomixis reduces progeny fitness (see above). If not, apomictic seed discounting is of little consequence. Note in this regard that self-fertilization is more prevalent in annual plants than perennials, possibly because an annual life history affords less opportunity to suffer the costs of seed discounting (Morgan et al. 1997; Sakai and Ishii 1999). In contrast, apomixis and vegetative reproduction frequently evolve in association with perenniality (Bicknell and Koltunow 2004). Does this life-history difference reflect lower short-term costs of seed discounting for apomixis than for self-fertilization?

The effect of apomixis on outcrossed male fitness will also vary depending on mechanism. Pollen discounting may occur if apomixis requires pollination (pseudogamy) or if the genetic mechanism that causes apomixis also reduces the production or functionality of male gametes. However, some apomictic species produce fully functional pollen and the floral structures that encourage pollen export (Grimanelli et al. 2001; Bicknell and Koltunow 2004). In some groups, apomixis has evolved via the invasion of apomictic females into dioecious populations; thus the concept of pollen discounting applies only indirectly in that an allele causing apomixis spreads solely through female transmission (Lloyd 1980b). If female transmission is less variable than transmission through pollen, an allele for apomixis may actually realize higher geometric mean fitness than an alternative allele for sexuality, although this possibility has not been explored theoretically.

Asexual reproduction via clonal vegetative propagation can also be viewed in terms of the same fitness costs and benefits. Cloning increases gene transmission at the expense of progeny variability, and will always provide RA because vegetative propagation occurs independently of pollination. Trade-offs with sexual reproduction (discounting) are mediated through competition for resources or, in some cases, meristems, which can be committed to either vegetative propagules or sexual structures (Thompson and Eckert 2004). The developmental details of vegetative propagation in terms of when it occurs in relation to sex are likely to influence the strength of discounting. Vegetative propagules and sexual progeny are also likely to differ in ecological attributes that affect their performance. In some cases, tiny propagules (e.g., bulbils) may be very similar to seeds in their dispersal distances and the conditions required for successful recruitment. In contrast, larger vegetative propagules may enjoy high recruitment owing to a substantial and sometimes ongoing resource subsidy from the parent plant, although this reduces their capacity for dispersal. These ecological differences between progeny types probably influence selection of clonal reproduction (Bengtsson and Ceplitis 2000).

Despite this common framework for thinking about the selection of uniparental reproduction via either selfing or asexuality, progress in understanding the evolution of either clonal or apomictic reproduction has been limited. One big challenge involves the underlying genetic mechanism. Self-pollination can be achieved simply by an increase in the proximity of dehiscing anthers and receptive stigmas. This change may involve straightforward alterations to flowers with a relatively simple genetic basis, and many models for the selection of selfing versus outcrossing adopt an evolutionary stable strategy (ESS) approach based on straightforward polygenic variation in the mating system. In contrast, pollination-independent, autonomous apomixis requires at least three steps: (1) generation of an unreduced cell capable of developing into an embryo without meiosis, (2) development of the embryo without fertilization, and (3) spontaneous development of endosperm tissue. Hence, the genetic basis of apomixis in many taxa remains elusive, despite intensive investigation (Richards 2003). The most successful genetic studies indicate that changes at multiple loci are required for apomixis (Spielman *et al.* 2003; Bicknell and Koltunow 2004). The genetic and developmental complexity of apomixis may be a major constraint on its evolution (Lloyd 1980b; Marshall and Brown 1981). In addition, autonomous apomixis is usually associated with polyploidy (Carman 1997), which introduces a confounding factor in its evolutionary analysis. Because an increase in ploidy may enhance tolerance of extreme environments, polyploidy may, on its own, favour apomixis in ecologically or geographically marginal habitats compared with sexuals: a pattern also predicted by the RA hypothesis (Bierzychudek 1987a).

10.4.1 Challenges and opportunities for investigating the evolution of asexuality

Studies of selfing have used simple floral manipulations (e.g., emasculations) and controlled crosses effectively to generate phenotypic variation in key floral traits with which to measure the fitness consequences of variation in the mating system. In contrast, the level of apomixis cannot be manipulated easily via simple physical alterations of flowers, and the complex inheritance and genetic systems of most apomicts may severely hamper the generation of reproductive variation via controlled crosses. However, several taxa exhibit wide variation in sexuality among closely related species and sometimes within species. In some cases, the morphology, life history, and ploidy of sexual versus asexual genotypes are sufficiently similar to enable "natural experiments" investigating the selective costs and benefits of asexuality.

Antennaria (Asteraceae) is one such group. This genus of dioecious perennial herbs is distributed throughout temperate and arctic northern hemisphere and includes several taxa with both sexual and asexual forms. Apomixis is fully autonomous and, therefore, provides RA (Bierzychudek 1990). In the two taxa studied in most detail, *A. parviflora* and *A. parlinii*, sexual and apomictic individuals also differ in reproductive ecology, seed fertility, and habitat distribution. Asexual females produce much more seed than sexual females (Bierzychudek 1990; O'Connell and Eckert 2001), in part because sexual females suffer pollen limitation when males are locally scarce (Bierzychudek 1990; O'Connell and Eckert 1999). In *A. parviflora*, apomicts are distributed more widely and occur at higher altitudes than sexuals (Bierzychudek 1990). Apomicts of *A. parlinii* are also distributed more widely and tend to occur in more ephemeral and disturbed habitats than sexual individuals (O'Connell and Eckert 2001). These distributional differences are consistent with an advantage of apomixis where adequate pollination is infrequent or variable, especially because reproductive mode is not confounded with ploidy differences in either species. In the only experimental test of the RA hypothesis to date, Bierzychudek (1990) transplanted populations of sexual and apomictic individuals to high altitude where only apomicts occur normally, and found that, as predicted, the seed fitness of sexual individuals was low and strongly dependent on population size, whereas apomicts realized much higher seed fitness, independent of population size. These and other variable taxa provide excellent, but underexploited, opportunities for fine-grained comparative analysis of whether RA facilitates the evolution of asexuality.

10.5 More experiments needed

In nature, diverse selective forces are likely to be simultaneously influencing the mating pattern of a population. Detailed observations of the actual selective forces involved will be required to determine the importance of the various factors contributing to multiple correlations between mating patterns and ecological parameters. (Lloyd 1980a, pp. 85–6)

The theoretical advances summarized in Table 10.1 strongly suggest that the intuitive appeal of reproductive assurance as a selective factor in the evolution of uniparental reproduction belies the challenges involved in testing it rigorously. Plenty of scope remains for new models to explore the genetic, physiological, demographic, and ecological feedbacks involved the evolution of the mating systems in environments with limited opportunities for outcross pollination. However, the need for empirical tests of the assumptions and predictions of theory with biologically realistic experimental manipulations is even greater. Further empirical progress will first require estimates of the contributions of the various modes of selfing to the mating system and male and female fitness (Schoen and Lloyd 1992; Kalisz and Vogler 2003; Herlihy and Eckert 2004). In particular, the increased seed production afforded by autonomous selfing must be weighed against seed discounting, recognizing that the discount may apply within and between flowers, as well as between reproductive episodes (Morgan *et al.* 1997; Eckert and Herlihy 2004). Experimental work on *Collinsia* and *Aquilegia* also emphasizes that accurate estimates of inbreeding depression are essential for evaluating both the costs of seed discounting and the stability of mixed mating in nature (Goodwillie *et al.* 2005). The next step will be to expand the temporal and spatial context of empirical studies. Do opportunities for cross-pollination vary in time and space to the extent that RA could be selected as a bet-hedging strategy (Morgan and Wilson 2005)? Does the RA provided by selfing affect the dynamics of natural populations and metapopulations in a way that can feed back on the evolution of the mating system (Pannell and Barrett 2001; Cheptou 2004; Morgan *et al.* 2005)?

Intraspecific variation in floral biology and mating also offers opportunities for continued empirical progress. For example, the next step in exploring the evolutionary origin and maintenance of mating-system variation in *C. cheiranthifolia* will involve experimental identification of the factor(s) that maintain outcrossing south of Point Conception, but promote the spread of selfing to the north. Reciprocally transplanting large- and small-flowered phenotypes into large- and small-flowered populations would test the two key predictions generated from our preliminary work: small-flowered *C. cheiranthifolia* benefit from avoiding parasitism, and more frequent parasitism north of Point Conception selects indirectly for self-fertilization.

The reciprocal transplant experiment is a venerable and powerful tool for studying local adaptation (Kawecki and Ebert 2004). Moreover, phenotypically variable experimental populations, created by crossing between divergent phenotypes, can be used in a reciprocal transplant experiment to quantify selection on key traits and to contrast the mode of selection between environments (Moeller and Geber 2005; Chapter 6). These experimental approaches have just begun to be used to examine selective factors, such as RA, that may underlie what is expected to be local adaptation in plant reproductive systems (Bierzychudek 1990; Fishman and Wyatt 1999; Fishman 2000; Elle and Carney 2003; Moeller and Geber 2005). Given that multiple interactions among ecological factors (e.g., pollinators and parasites) probably influence selection, individual factors must be isolated by controlling other possible selective factors. For example, floral parasites may be controlled with judicious insecticide application, whereas the influence of pollinators can be manipulated by excluding them or rendering them superfluous by manual cross-pollination. We anticipate that new theory and the ongoing development of innovative experimental methods will continue to provide opportunities to delve more rigorously into the tremendous diversity of mating systems in plants, which Jain (1976) referred to as an "embarrassment of riches."

Acknowledgements

Our work on the evolution of uniparental reproduction has involved collaborations with many students and research assistants. We acknowledge, in particular, Chris Herlihy, Barbara Ozimec, and Amy Schaefer for collecting much of the *Aquilegia* data reanalysed above, and Emily Austen, Colleen Inglis, Sarah Yakimowski, Sarah Chan, and Yona Gellert for participating in studies on *Camissonia*. We thank Lawrence Harder and Spencer Barrett for their relentless hard work as editors and for helpful comments on the manuscript, Charlie Fenster and David Moeller for sharing unpublished work, and the Natural Sciences and Engineering Research Council of Canada for funding this work through Discovery Grants to CGE and a scholarship to KES.

References

Adler LS and Bronstein JL (2004). Attracting antagonists: does floral nectar increase leaf herbivory? *Ecology*, **85**, 1519–26.

Agrawal AF and Lively CM (2001). Parasites and the evolution of self-fertilization. *Evolution*, **55**, 869–79.

Ashman TL, Knight TM, Steets JA, et al. (2004). Pollen limitation of plant reproduction: Ecological and evolutionary causes and consequences. *Ecology*, **85**, 2408–21.

Baker HG (1955). Self-compatibility and establishment after "long-distance" dispersal. *Evolution*, **9**, 347–48.

Barrett SCH and Harder LD (1996). Ecology and evolution of plant mating. *Trends in Ecology and Evolution*, **11**, 73–79.

Bengtsson BO and Ceplitis A (2000). The balance between sexual and asexual reproduction in plants living in variable environment. *Journal of Evolutionary Biology*, **13**, 415–22.

Bicknell RA and Koltunow AM (2004). Understanding apomixis: recent advances and remaining conundrums. *Plant Cell*, **16**, S228–45.

Bierzychudek P (1987a). Patterns in plant parthenogenesis. In SC Stearns, ed. *The evolution of sex and its consequences*, pp. 197–217. Birkhäuser Verlag, Basel, Switzerland.

Bierzychudek P (1987b). Resolving the paradox of sexual reproduction: A review of experimental tests. In SC Stearns, ed. *The evolution of sex and its consequences*, pp. 163–74. Birkhäuser Verlag, Basel, Switzerland.

Bierzychudek P (1990). The demographic consequences of sexuality and apomixis in *Antennaria*. In S Kawano, ed. *Biological approaches and evolutionary trends in plants*, pp. 293–307. Academic Press, Tokyo, Japan.

Burd M (1994). Bateman's principle and plant reproduction: the role of pollen limitation in fruit and seed set. *Botanical Review*, **60**, 83–139.

Busch JW (2005). Inbreeding depression in self-incompatible and self-compatible populations of *Leavenworthia alabamica*. *Heredity*, **94**, 159–65.

Busch JW, Neiman M, and Koslow JM (2004). Evidence for maintenance of sex by pathogens in plants. *Evolution*, **58**, 2584–90.

Carman JG (1997). Asynchronous expression of duplicate genes in angiosperms may cause apomixis, bispory, tetraspory, and polyembryony. *Biological Journal of the Linnean Society*, **61**, 51–94.

Charlesworth D and Charlesworth B (1987). Inbreeding depression and its evolutionary consequences. *Annual Review of Ecology and Systematics*, **18**, 237–68.

Cheptou PO (2004). Allee effect and self-fertilization in hermaphrodites: reproductive assurance in demographically stable populations. *Evolution*, **58**, 2613–21.

Cox PA (1989). Baker's law: Plant breeding systems and island colonization. In JH Bock and YB Linhart, eds. *The evolutionary ecology of plants*, pp. 209–24. Westview Press, Boulder, CO, USA.

Cruden RW and Lyon DL (1989). Facultative xenogamy: Examination of a mixed mating system. In JH Bock and YB Linhart, eds. *The evolutionary ecology of plants*, pp. 171–208. Westview Press, Boulder, CO, USA.

Darwin CR (1876). *The effects of cross and self-fertilization in the vegetable kingdom*. John Murray, London, UK.

Davis SL and Delph LF (2005). Prior selfing and gynomonoecy in *Silene noctiflora* L. (Caryophyllaceae): opportunities for enhanced outcrossing and reproductive assurance. *International Journal of Plant Sciences*, **166**, 475–80.

Eckert CG and Herlihy CR (2004). Using a cost-benefit approach to understanding the evolution of self-fertilization in plants: the perplexing case of *Aquilegia canadensis* (Ranunculaceae). *Plant Species Biology*, **19**, 159–73.

Eckert CG and Schaefer A (1998). Does self-pollination provide reproductive assurance in wild columbine, *Aquilegia canadensis* (Ranunculaceae)? *American Journal of Botany*, **85**, 919–24.

Elle E (2004). Floral adaptations and biotic and abiotic selection pressures. In QCB Cronk, J Whitton, RH Lee, and IEP Taylor, eds. *Plant adaptation: molecular genetic and ecology*, pp. 111–18. NRC Press, Ottawa, Ontario, Canada.

Elle E and Carney R (2003). Reproductive assurance varies with flower size in *Collinsia parviflora* (Scrophulariaceae). *American Journal of Botany*, **90**, 888–96.

Fausto JA Jr, Eckhart VM, and Geber MA (2001). Reproductive assurance and the evolutionary ecology of self-pollination in *Clarkia xantiana* (Onagraceae). *American Journal of Botany*, **88**, 1794–800.

Fenster CB and Barrett SCH (1994). Inheritance of mating-system modifier genes in *Eichhornia paniculata* (Pontederiaceae). *Heredity*, **72**, 433–45.

Fishman L (2000). Pollen discounting and the evolution of selfing in *Arenaria uniflora* (Caryophyllaceae). *Evolution*, **54**, 1558–65.

Fishman L and Wyatt R (1999). Pollinator-mediated competition, reproductive character displacement, and the evolution of selfing in *Arenaria uniflora* (Caryophyllaceae). *Evolution*, **53**, 1723–33.

Fishman L, Kelly AJ, and Willis JH (2002). Minor quantitative trait loci underlie floral traits associated with mating system divergence in *Mimulus*. *Evolution*, **56**, 2138–55.

Goodwillie C (2001). Pollen limitation and the evolution of self-compatibility in *Linanthus* (Polemoniaceae). *International Journal of Plant Sciences*, **162**, 1283–92.

Goodwillie C, Kalisz S, and Eckert CG 2005 The evolutionary enigma of mixed mating systems in plants: occurrence, theoretical explanations, and empirical evidence. *Annual Review of Ecology, Evolution and Systematics*, **36**, 47–79.

Griffin CAM and Eckert CG (2003). Experimental analysis of biparental inbreeding in a self-fertilizing plant. *Evolution*, **57**, 1513–19.

Griffin SR, Mavraganis K, and Eckert CG (2000). Experimental analysis of protogyny in *Aquilegia canadensis* (Ranunculaceae). *American Journal of Botany*, **87**, 1246–56.

Grimanelli D, Leblanc O, Perotti E, and Grossniklaus U (2001). Developmental genetics of gametophytic apomixis. *Trends in Genetics*, **17**, 597–604.

Harder LD and Wilson WG (1998). A clarification of pollen discounting and its joint effects with inbreeding depression on mating-system evolution. *American Naturalist*, **152**, 684–95.

Herlihy CR and Eckert CG (2002). Genetic cost of reproductive assurance in a self-fertilizing plant. *Nature*, **416**, 320–23.

Herlihy CR and Eckert CG (2004). Experimental dissection of inbreeding and its adaptive significance in a flowering plant, *Aquilegia canadensis* (Ranunculaceae). *Evolution*, **58**, 2693–703.

Herrera CM, Sanchez LAM, Medrano M, *et al.* (2001). Geographical variation in autonomous self-pollination levels unrelated to pollinator service in *Helleborus foetidus* (Ranunculaceae). *American Journal of Botany*, **88**, 1025–32.

Herrera CM, Medrano M, Rey PJ, *et al.* (2002). Interaction of pollinators and herbivores on plant fitness suggests a pathway for correlated evolution of mutualism- and antagonism-related traits. *Proceedings of the National Academy of Sciences of the United States of America*, **99**, 16823–28.

Holsinger KE (1996). Pollination biology and the evolution of mating systems in flowering plants. *Evolutionary Biology*, **29**, 107–49.

Holtsford TP and Ellstrand NC (1990). Variation in inbreeding depression among families and populations of *Clarkia tembloriensis* (Onagraceae). *Heredity*, **76**, 83–91.

Holtsford TP and Ellstrand NC (1992). Genetic and environmental variation in floral traits affecting outcrossing rate in *Clarkia tembloriensis* (Onagraceae). *Evolution*, **46**, 216–25.

Hull-Sanders HM and Eubanks MD (2005). Plant defense theory provides insight into interactions involving inbred plants and insect herbivores. *Ecology*, **86**, 897–904.

Inoue K, Maki M, and Masuda M (1996). Evolution of *Campanula* flowers in relation to insect pollination on islands. In DG Lloyd and SCH Barrett, eds. *Floral biology. Studies on floral evolution in animal–pollinated plants*, pp. 377–400. Chapman and Hall, New York, USA.

Jain SK (1976). The evolution of inbreeding in plants. *Annual Review of Ecology and Systematics*, **7**, 69–95.

Johnston MO (1998). Evolution of intermediate selfing rates in plants: pollination ecology versus deleterious mutations. *Genetica*, **102/103**, 267–78.

Johnston MO and Schoen DJ (1996). Correlated evolution of self-fertilization and inbreeding depression: An experimental study of nine populations of *Amsinckia* (Boraginaceae). *Evolution*, **50**, 1478–91.

Kalisz S and Vogler DW (2003). Benefits of autonomous selfing under unpredictable pollinator environments. *Ecology*, **84**, 2928–42.

Kalisz S, Vogler DW, Fails B, *et al.* (1999). The mechanism of delayed selfing in *Collinsia verna* (Scrophulariaceae). *American Journal of Botany*, **86**, 1239–47.

Kalisz S, Vogler DW, and Hanle KM (2004). Context-dependent autonomous self-fertilization yields reproductive assurance and mixed mating. *Nature*, **430**, 884–87.

Kawecki TJ and Ebert D (2004). Conceptual issues in local adaptation. *Ecology Letters*, **7**, 1225–41.

Klimes L, Klimesová J, Hendriks R, and van Groenendael J (1997). Clonal plant architecture: a comparative analysis of form and function. In H de Kroon and J van Groenendael, eds. *The ecology and evolution of clonal plants*, pp. 1–29. Backhuys, Leiden, The Netherlands.

Kohn JR and Barrett SCH (1994). Pollen discounting and the spread of a selfing variant in tristylous *Eichhornia paniculata*: Evidence from experimental populations. *Evolution*, **48**, 1576–94.

Lande R and Schemske DW (1985). The evolution of self-fertilization and inbreeding depression in plants. I. Genetic models. *Evolution*, **39**, 24–40.

Larson BMH and Barrett SCH (2000). A comparative analysis of pollen limitation in flowering plants. *Biological Journal of the Linnean Society*, **69**, 503–20.

Levin DA (1972). Competition for pollinator service: A stimulus for the evolution of autogamy. *Evolution*, **26**, 668–69.

Levin DA (1975). Pest pressure and recombination systems in plants. *American Naturalist*, **109**, 437–51.

Linsley EG, MacSwain JW, Raven PH, and Thorp RW (1973). Comparative behavior of bees and Onagraceae. V. *Camissonia* and *Oenothera* bees of cismontane California and Baja California. *University of California Publications in Entomology*, **71**, 1–68.

Lloyd DG (1965). Evolution of self-compatibility and racial differentiation in *Leavenworthia* (Cruciferae). *Contributions of the Gray Herbarium, Harvard University*, **195**, 3–133.

Lloyd DG (1979). Some reproductive factors affecting the selection of self-fertilization in plants. *American Naturalist*, **113**, 67–79.

Lloyd DG (1980a). Demographic factors and mating patterns in angiosperms. In OT Solbrig, ed. *Demography and evolution in plant populations*, pp. 67–88. Blackwell, Oxford, UK.

Lloyd DG (1980b). Benefits and handicaps of sexual reproduction. *Evolutionary Biology*, **13**, 69–111.

Lloyd DG (1988). Benefits and costs of biparental and uniparental reproduction in plants. In RE Michod and BR Levin, eds. *The evolution of sex. An examination of current ideas*, pp. 233–52. Sinauer Associates, Sunderland, MA, USA.

Lloyd DG (1992). Self- and cross-fertilization in plants. II. The selection of self-fertilization. *International Journal of Plant Sciences*, **153**, 370–80.

Lloyd DG and Schoen DJ (1992). Self- and cross-fertilization in plants. I. Functional dimensions. *International Journal of Plant Sciences*, **153**, 358–69.

Marshall DR and Brown AHD (1981). The evolution of apomixis. *Heredity*, **47**, 1–15.

Mazer SJ, Paz H, and Bell MD (2004). Life history, floral development and mating system in *Clarkia xantiana* (Onagraceae): do floral and whole-plant rates of development evolve independently. *American Journal of Botany*, **91**, 2041–50.

Moeller DA and Geber MA (2005). Ecological context of the evolution of self-pollination in *Clarkia xantiana*: Population size, plant communities, and reproductive assurance. *Evolution*, **59**, 786–99.

Morgan MT and Wilson WG (2005). Self-fertilization and the escape from pollen limitation in variable pollination environments. *Evolution*, **59**, 1143–8.

Morgan MT, Schoen DJ, and Bataillon TM (1997). The evolution of self-fertilization in perennials. *American Naturalist*, **150**, 618–38.

Morgan MT, Wilson WG, and Knight TM (2005). Plant population dynamics, pollinator foraging, and the selection of self-fertilization. *American Naturalist*, **166**, 169–83.

Nagylaki T (1976). A model for the evolution of self-fertilization and vegetative reproduction. *Journal of Theoretical Biology*, **58**, 55–8.

O'Connell LM and Eckert CG (1999). Differentiation in sexuality among populations of *Antennaria parlinii* (Asteraceae). *International Journal of Plant Sciences*, **160**, 567–75.

O'Connell LM and Eckert CG (2001). Differentiation in reproductive strategy between sexual and asexual populations of *Antennaria parlinii* (Asteraceae). *Evolutionary Ecology Research*, **3**, 311–30.

Pannell JR and Barrett SCH (1998). Baker's law revisited: reproductive assurance in a metapopulation. *Evolution*, **52**, 657–68.

Pannell JR and Barrett SCH (2001). Effects of population size and metapopulation dynamics on a mating-system polymorphism. *Theoretical and Applied Biology*, **59**, 145–55.

Piper JG, Charlesworth B, and Charlesworth D (1986). Breeding system evolution in *Primula vulgaris* and the role of reproductive assurance. *Heredity*, **56**, 207–17.

Porcher E and Lande R (2005a). Reproductive compensation in the evolution of plant mating systems. *New Phytologist*, **166**, 673–84.

Porcher E and Lande R (2005b). The evolution of self-fertilization and inbreeding depression under pollen discounting and pollen limitation. *Journal of Evolutionary Biology*, **18**, 497–508.

Rambuda TD and Johnson SD (2004). Breeding systems of invasive alien plants in South Africa: does Baker's rule apply? *Diversity and Distributions*, **10**, 409–16.

Raven PH (1969). A revision of the genus *Camissonia*. *Contributions of the US National Herbarium*, **37**, 161–396.

Richards AJ (2003). Apomixis in flowering plants: an overview. *Philosophical Transactions of the Royal Society of London, Series B*, **358**, 1085–93.

Ritland K (1991). A genetic approach to measuring pollen discounting in natural plant populations. *American Naturalist*, **138**, 1049–57.

Roff DA (1997). *Evolutionary quantitative genetics*. Chapman & Hall, New York, USA.

Runions CJ and Geber MA (2000). Evolution of the self-pollinating flower in *Clarkia xantiana* (Onagraceae). I. Size and development of floral organs. *American Journal of Botany*, **87**, 1439–51.

Sakai S and Ishii HS (1999). Why be completely outcrossing? Evolutionarily stable outcrossing strategies in an environment where outcross-pollen availability is unpredictable. *Evolutionary Ecology Research*, **1**, 211–22.

Schoen DJ and Brown AHD (1991). Whole- and within-flower self-pollination in *Glycine clandestina* and *G. argyrea* and the evolution of autogamy. *Evolution*, **45**, 1651–65.

Schoen DJ and Lloyd DG (1992). Self- and cross-fertilization in plants. III. Methods for studying modes and functional aspects of self-fertilization. *International Journal of Plant Sciences*, **153**, 381–93.

Schoen DJ, Morgan MT, and Bataillon T (1996). How does self-pollination evolve? Inferences from floral ecology and molecular genetic variation. *Philosophical Transactions of the Royal Society of London, Series B*, **351**, 1281–90.

Spielman M, Vinkenoog R, and Scott RJ (2003). Genetic mechanisms of apomixis. *Philosophical Transactions of the Royal Society of London, Series B*, **358**, 1095–103.

Stebbins GL (1974). *Flowering plants: evolution above the species level*. Belknap Press, Cambridge, USA.

Takebayashi N and Morrell P (2001). Is self-fertilization a dead end? Revisiting an old hypothesis with genetic theories and a macroevolutionary approach. *American Journal of Botany*, **88**, 1143–50.

Thompson FL and Eckert CG (2004). Trade-offs between sexual and clonal reproduction in an aquatic plant: experimental manipulations versus phenotypic correlations. *Journal of Evolutionary Biology*, **17**, 581–92.

CHAPTER 11

The evolution of separate sexes: a focus on the ecological context

Tia-Lynn Ashman

Department of Biological Sciences, University of Pittsburgh, PA, USA

Outline

Sexual-system evolution has a long rich history of theoretical study, which has guided empirical explorations and the recent expansion of ideas concerning the role of ecological context in gender separation. In this chapter, I first provide a brief overview of theoretical understanding of a prominent sexual-system transition, namely, the gynodioecy pathway for the evolution of dioecy from hermaphroditism. I then review empirical support for several predictions based on theory and use these as a springboard to both highlight the strengths of existing theory and show how its limitations have helped reveal the importance of ecological context for the evolution of dioecy. I specifically review work that suggests that harsh environments mediate sexual-system evolution via sex-differential plasticity and size-dependent allocation, as well as consider less-studied mechanisms by which resource limitation can affect sexual-system evolution, including plant–pollinator interactions, mating system, and inbreeding depression. I then address the emerging role of enemies, and of multi-species interactions, in modifying fundamental parameters of gender models and thus selection for separate sexes. In doing so, I illustrate similarities and dissimilarities between underlying mechanisms and identify areas of interaction that demand empirical investigation. I conclude with a discussion of two unresolved issues: the stability of subdioecy and the dual role of pollinators in the evolution of separate sexes.

11.1 Introduction

Evolution has produced a fantastic array of sexual systems in plants. One end of this continuum involves gender monomorphism (e.g., hermaphroditism and monoecy), whereby the genetic contributions plants make to the next generation vary continuously within populations, but on average individuals contribute equally as males and females. The other end involves gender dimorphism, with individuals grouped into two distinct morphs that contribute to the next generation primarily as males or females, including dioecy (males and females), gynodioecy (females and hermaphrodites) or androdioecy (males and hermaphrodites) (Lloyd 1980).

Dioecy is relatively uncommon among angiosperms (6% of species; Renner and Ricklefs 1995), but it has evolved repeatedly from hermaphroditism (at least 100 transitions; Charlesworth 2002). Although several pathways for the transition to dioecy have been proposed, they largely fall into two types: those that involve the invasion of a male- or female-sterile mutant (androdioecy or gynodioecy, respectively), and those that involve disruptive selection on existing variation (distyly, heterodichogamy, monoecy). Transitions between these general pathways are also possible (i.e., monoecy or distyly to dioecy might also involve invasion by a male-sterile mutant: Sarkissian *et al.* 2001; Rosas *et al.* 2005). Here, I focus on the gynodioecy pathway, because of the potential similarities in the evolutionary dynamics, and because current evidence suggests that it is an important and particularly common pathway (Weiblen *et al.* 2000).

The influence of ecological context in the evolution of dioecy, regardless of pathway, is a central theme which has emerged during the past decade. Research in this area is at a turning-point, as the traditional descriptions of the ecological correlates of dioecy (Darwin 1877; Thomson and Brunet 1990; Weiblen *et al.* 2000; Vamosi *et al.* 2003) are supplemented by the recognition and elucidation of the myriad mechanisms through which ecological context can affect the evolution of dioecy.

Thus, in this chapter I briefly review theoretical understanding of the gynodioecy pathway, and then discuss its instrumental role in both guiding studies of evolutionary dynamics and revealing the importance of the ecological context for sexual-system evolution. I then focus on two features of ecological context (harsh environments and enemies), evaluating current understanding of the mechanisms by which these features affect sexual-system evolution, which I illustrate with my research on *Fragaria virginiana*. Finally, I discuss two unresolved issues in the evolution of separate sexes—stability of subdioecy and the dual role of pollinators—and draw attention to additional subjects in need of work.

11.2 The gynodioecy pathway to dioecy

11.2.1 Hermaphroditism to gynodioecy

During the first step in the gynodioecy pathway, females (male steriles) invade and are maintained in a population of hermaphrodites. To invade, females must compensate for their loss of male function. Most models of the evolution of gynodioecy include compensation through increased female seed fertility and/or inbreeding avoidance (e.g., Lloyd 1975; Charlesworth and Charlesworth 1978). Females may reallocate resources not spent on pollen to seed production. The exact seed fertility advantage required of females depends on the inheritance of sex type (reviewed in Charlesworth 1999). For example, when nuclear genes determine male sterility and hermaphrodites do not self-fertilize, females need a two-fold advantage to invade, and a greater advantage to increase in frequency (Lewis 1941; Lloyd 1975). In contrast, when cytoplasmic mutation causes male sterility, females need only a slight advantage (Lewis 1941), Finally, if both cytoplasmic and nuclear genes are responsible for sexual identity, the necessary magnitude of compensation for the spread of females is the same as with pure cytoplasmic male sterility; however, the evolutionary dynamics depend on positive pleiotropic effects of cytoplasmic genes and negative effects of nuclear restorer genes, which generally creates dynamical systems in which sex ratio is determined alternatively by cytoplasmic and nuclear genes (reviewed by Bailey *et al.* 2003).

Despite increased ovule production by females, their seed fertility may be limited by insufficient pollen receipt, because they require outcross pollination for seed production. As the frequency of females increases (or pollinator visits decline), pollen limitation becomes more likely and, if it occurs, will limit female spread (Lewis 1941; Lloyd 1974). Maurice and Fleming (1995) revealed that pollen limitation restricts the conditions under which females (or males) are maintained, and favours hermaphroditism, even under strong inbreeding depression.

Inbreeding avoidance is another major mechanism of female compensation (reviewed in Charlesworth 1999). For example, under nuclear inheritance females can invade if hermaphrodites produce a large fraction of selfed offspring (s) and inbreeding depression (δ) is intense (i.e., $s\delta > 0.5$), even in the absence of enhanced ovule production (Charlesworth and Charlesworth 1978). This model assumes that selfing is constant and does not affect pollen export, as could occur through autonomous autogamy. Early work by Lloyd (1975) demonstrated that the "mode" of selfing influences the equilibrium frequency of females. He showed that for a given selfing frequency and inbreeding depression, autonomous selfing before outcrossing ("prior selfing") allows lower equilibrium female frequencies than does simultaneous self- and cross-pollination ("competing selfing"). In contrast, females cannot be maintained via an outbreeding advantage if autonomous self-pollination occurs after cross-pollination ("delayed selfing": Lloyd 1975). In addition, when the selfing rate is not fixed, hermaphroditism is favoured when selfing and pollen export increase with male allocation (e.g., with floral display size), as could be the case with geitonogamous selfing (de Jong et al. 1999). Under these conditions, females are favoured only with severe inbreeding depression. Whatever the mechanism of female maintenance, the presence of females fundamentally changes how hermaphrodites gain fitness and sets the stage for the second step in the evolution of dioecy.

11.2.2 Gynodioecy to dioecy

In contrast to the equitable contributions through male and female functions made by hermaphrodites, on average, in purely hermaphroditic populations, the presence of females causes hermaphrodites of gynodioecious populations to contribute genes through pollen more often than through seeds (Lloyd 1974). Thus, during the second step of the gynodioecy pathway (gynodioecy to dioecy), hermaphrodites experience frequency-dependent selection for enhanced male function, which is strongest when female frequencies and fertilities are high relative to those of hermaphrodites (Charlesworth 1989; e.g., McCauley and Brock 1998). If male allocation is heritable and male investment varies negatively with female investment (reviewed by Ashman 2003), then gynodioecious populations should evolve toward dioecy (Charlesworth 1999). These selective requirements for the evolution of dioecy are generally the same under nuclear and nucleo-cytoplasmic gynodioecy, although dioecy may sometimes evolve more easily with cytoplasmic-nuclear than nuclear inheritance (Maurice et al. 1993; Maurice et al. 1994; Shultz 1994). Whether dioecy evolves in these systems depends strongly on the genetic conditions.

Recent models by de Jong et al. (1999) extend the work of Lloyd (1975) and Charlesworth and Charlesworth (1978) by showing both that the mode of selfing is important for the evolution of females (see Section 11.2.1), and that the propensity for dioecy to evolve depends additionally on whether selfing reduces pollen export (i.e., pollen discounting). A positive relation between pollen discounting and floral display retards the evolution of dioecy, regardless of the magnitude of inbreeding depression.

11.2.3 Predicted relations of female frequency and key model parameters

The preceding models (based primarily on nuclear gynodioecy) reveal five key predictions about the associations of female frequency, which can be tested empirically (Table 11.1)—although note that the sex-ratio dynamics may differ with cytoplasmic-nuclear gynodioecy. The prediction supported most strongly involves the positive association between female frequency and the seed fertility of females relative to hermaphrodites (Table 11.1, prediction 1). Eleven of the 14 among-population analyses of this relation demonstrated tight control of population sex ratio by female seed fertility advantage. Second, three of six studies that measured mating system directly supported the predicted positive relation between female frequency and hermaphrodite selfing rate (Table 11.1, prediction 2). An additional study of species of two genera (Delph 1990a), inferred that the relation might exist, based on knowledge of the pollinating fauna, although the mating system was not measured. Third, female frequency should increase with the severity of inbreeding depression (Table 11.1, prediction 3); however, only Thompson et al. (2000) have measured inbreeding depression in multiple populations and they found no association with female frequency, possibly owing to other mitigating factors. Fourth, five of nine studies have found that pollen limitation of seed production by females increases with increased female frequencies (Table 11.1, prediction 4). Four of these studies examined local patches within natural populations or small experimental populations, suggesting either heterogeneous pollen limitation within populations, depending on local female frequencies, or that pollen limitation depends on additional factors, such as an interaction between female frequency and pollinator availability (A.L. Case and T.-L. Ashman manuscript submitted). Last, the prediction that male allocation by hermaphrodites varies positively with female frequencies has been studied in six species (Table 11.1, prediction 5). The predicted relation exists among three species of *Thymus*, but not within individual *Thymus* species, or among populations in the three other species studied (*Fragaria virginiana*, *Geranium*

Table 11.1 Empirical support for five predicted relations among features of gynodioecious populations.

Species	Prediction					Citations
	(1)	(2)	(3)	(4)	(5)	
Astilbe biternata	No			No		Olson 2001
Bidens spp.		Yes				Sun and Ganders 1988
Chionohebe spp.		Yes[a]				Delph 1990a
Daphne laureola		No		No		Medrano et al. 2005
Fragaria virginiana	Yes			Yes[b]	No[c]	Ashman 1999; Ashman and Diefenderfer 2001
Geranium richardsonii	Yes			Yes (No)[d]		Williams et al. 2000
Geranium sylvaticum				No	No[c]	Asikainen and Mutikainen 2005; Asikainen and Mutikainen 2004
Gingidia spp.	Yes[f]					Webb 1981
Glechoma hederecea	No			Yes[e]		Widen 1992; Widen and Widen 1990
Hebe spp.		Yes[a]				Delph 1990a
Hebe strictissima	Yes					Delph 1990b
Lignocapa spp.	Yes[f]					Webb 1981
Ochradenus baccatus	Yes					Wolfe and Shmida 1997
Pachycereus pringlei		No				Molina-Freaner et al. 2003
Plantago coronopus		Yes				Wolff et al. 1988
Salvia pratensis		Yes				Van Treuren et al. 1993
Scandia rosaefolia	Yes[f]					Webb 1981
Sidalcea hendersonii	Yes[g]					Marshall and Ganders 2001
Sidalcea malviflora				Yes[e]		Graff 1999
Silene acaulis	Yes					Delph and Carroll 2001
Silene vulgaris				Yes[f]		McCauley and Brock 1998
Thymus mastichina	Yes (No)[h]			Yes (No)[h]		Manicacci et al. 1998
Thymus vulgaris	Yes (No)[h]	No	No	Yes (No)[h]		Manicacci et al. 1998/Thompson and Tavagre 2000
Thymus zygis	Yes (No)[h]			Yes (No)[h]		Manicacci et al. 1998
Trifolium hirtum	No			Yes		Molina-Freaner and Jain 1992
Wurmbea biglandulosa	Yes					Vaughton and Ramsey 2002, 2004
Wurmbea dioica	Yes				No[c]	Barrett 1992

Predictions: (1) female (F) frequency increases with decreased hermaphrodite (H) seed fertility, increased female seed fertility, or increased F:H fertility ratio; (2) female frequency increases with increased hermaphrodite selfing rate; (3) female frequency increases with increased severity of inbreeding depression of the progeny of hermaphrodites; (4) pollination of females decreases (pollen limitation increases) with increased female frequency; (5) allocation to male traits by hermaphrodites increases with increased female frequency.
Complete citation details are available at http://www.eeb.utoronto.ca/EEF/.
[a] Altitude is a surrogate for pollinator type, which is a surrogate for outcrossing rate.
[b] Experimental populations.
[c] Pollen per flower.
[d] For pollen receipt, but pollen limitation unlikely.
[e] Local patches in natural population.
[f] Across species or subspecies.
[g] Seed that survived predation.
[h] Among, but not within, species.

sylvaticum, Wurmbea dioica). These results may reflect ecological or genetic constraints on the evolution of enhanced maleness of hermaphrodites (see Ashman 2002, 2003; Ashman *et al.* 2004), at least in the characters studied (e.g., pollen production per flower). Note that the patterns observed in many of these studies reflect the evolutionary feedback between female presence and sexual-system parameters (i.e., selfing rates, inbreeding depression, seed production) and thus cannot be used to infer the conditions responsible for the original appearance of females.

11.2.4 Comparisons of observed and predicted female frequencies

That models of the gynodioecy pathway depict essential features of the evolution of dioecy is also evident in comparisons of the predicted and observed frequencies of females in nature. For instance, Weller and Sakai (2005) used average values of inbreeding depression (δ), selfing rate (s), and relative seed fertility (k) among populations of *Schiedea salicaria* in Charlesworth and Charlesworth's (1978) formula for the equilibrial frequency of females (Z),

$$Z = (k + 2s\delta - 1)/2(k + 2s\delta).$$

They predicted an average female frequency of 8.1%, which compared well to the 12–13% observed in the field. Wolfe and Shmida (1997) predicted female frequency (p) in *Ochradenus baccatus* based on relative seed fertilities of the sex morphs (C) using Lloyd's (1976) formula,

$$p = (1 - 2C)/(2[1 - C]).$$

Their prediction of 43% females was very close to the 41% observed in the focal population (Wolfe and Shmida 1997). Similar tests have been conducted successfully in a few other species, and the high congruence overall indicates that these models capture well the essential features of the systems (Fig. 11.1). However, situations in which observed and predicted frequencies differ may be even more informative than those in which they do not.

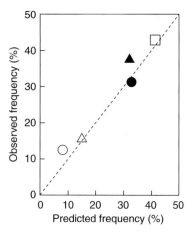

Figure 11.1 The relation between observed and predicted female frequencies (%) for several gynodioecious taxa. Predicted frequencies are based on female–hermaphrodite relative seed fertilities, or seed fertilities, selfing rates, and inbreeding depression. *Limnanthes douglasii* (open triangle), mean of two populations (Kesseli and Jain 1984); *Cucurbita foetidissima* (closed triangle; Kohn and Biardi 1995); *Schiedea salicaria* (open circle), mean of two populations (Weller and Sakai 2005); *Fragaria virginiana* (closed circle), mean of five populations (Ashman 1999); *Ochradenus baccatus* (open square; Wolfe and Shmida 1997).

If populations are in equilibrium and the mechanism of sex determination is identified correctly, deviations from model predictions indicate that the chosen model overlooks important features of dioecy in these specific cases. Two studies provide particularly vivid examples. First, Marshall and Ganders (2001) found a much higher female frequency (average 44%) than had been predicted (0%) for *Sidalcea hendersonii*. However, after accounting for hermaphrodite-biased seed predation by weevils (raising the female: hermaphrodite seed fertility ratio from 1.0 to 1.6) the predicted female frequency (27%) was closer to the observed frequency. Second, in *Fragaria virginiana* population-specific estimates of expected sex ratio and data on fruit production by females and hermaphrodites in the absence of herbivores indicated lower female frequencies in the field than predicted (Ashman 1999). Moreover, the deviation of predicted and observed female frequencies increased significantly with the incidence of damage to flowers of hermaphrodites by weevils in the field ($r_s = 0.93$; $P < 0.04$; $N = 5$; Fig. 11.2). In particular, the sex ratio in the

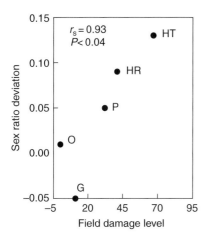

Figure 11.2 Positive association between incidence (percentage of plants) of damage to hermaphrodite *Fragaria virginiana* in five populations and the difference between the observed sex ratio and that predicted based on relative female and hermaphrodite seed fertilities in the absence of herbivores.

population without weevil herbivory matched the prediction exactly, whereas the greatest disparity occurred in the population with the most herbivory (HT). Moreover, these deviations were unrelated to soil resources (moisture: $P > 0.25$; nitrogen: $P > 0.8$) that influence sex ratio (Ashman 1999). These results suggest that morph-specific weevil herbivory maintains the gynodioecious sexual system in what would otherwise be near-dioecious populations of *F. virginiana*.

These are but two examples of the role of theory in heightening awareness of the importance of ecological context for gender evolution. In the following sections, I further evaluate evidence pointing to the centrality of ecological context for sexual-system evolution.

11.3 Importance of ecological context

Modelling sexual-system dynamics is only one aspect of understanding when and how gender dimorphism evolves. Related research has long sought patterns among sexual system, plant phenotype, and ecological context (e.g., Renner and Ricklefs 1995; Vamosi *et al.* 2003). Recognition that ecological factors mediate and influence all important aspects of sexual-system evolution described above (reviewed in Table 11.2) has stimulated workers to go beyond documenting patterns and to consider underlying mechanisms. I now highlight three ecological features that influence gender evolution: direct effects of harsh, dry environments, indirect effects of harsh environments mediated by pollinators, and by enemies. In the following sections, I describe the contribution of ecology to patterns of sexual-system variation and explore similarities in underlying mechanisms. I also illustrate why this framework will have to be expanded to multi-species interactions (see also Chapter 7) for the context of sexual-system evolution to be understood fully.

11.3.1 Harsh abiotic environment

Darwin (1877) first noted the association between gender dimorphism and stressful or harsh habitats, stating "a very dry station apparently favours the presence of the female form" (p. 301). This association has been documented by a few comparative and phylogenetic studies across taxa (e.g., Hart 1985; Weller *et al.* 1995) and by numerous studies of variation among populations (or subspecies) in gynodioecious (or subdioecious) species (Table 11.2, prediction 1). Studies of 14 species have all found higher female frequencies in environments with low resource availability (soil water, nutrients). Three non-exclusive mechanisms could account for this recurring pattern: (1) resource limitation of hermaphrodite seed fertility, (2) resource limitation intensifies inbreeding depression, and (3) resource-based changes in pollination or plant morphology increase selfing by hermaphrodites. All three mechanisms could facilitate female invasion and spread, and the subsequent evolution of males. I now consider the direct environmental effects on seed production and inbreeding depression, leaving the indirect effects on pollination to the next section.

Resource limitation of seed production by hermaphrodites

Hermaphrodites allocate resources to both pollen and seeds, whereas females invest only in seeds. Consequently, in resource-poor environments some hermaphrodites may be unable to maintain both sex

Table 11.2 Empirical support for six predicted relations between features of gynodioecious populations and ecological factors.

Species	Prediction						Citations
	(1)	(2)	(3)	(4)	(5)	(6)	
Alsinidendron species					Yes[a]		Weller et al. 1998
Daphne laureola	Yes[b]						Alonso and Herrera 2001
Ecballium elaterium	Yes[c]						Costich 1995
Eritrichum aretioides	Yes[b]						Puterbaugh et al. 1997
Fragaria virginiana	Yes	Yes	Yes	Yes			T.-L. Ashman 1999, unpublished data
Geranium sylvaticum	Yes[b]						Asikainen and Mutikainen 2004
Hebe strictissima	Yes		Yes				Delph 1990b, c
Lignocapa carnosula	Yes						Webb 1979
Minuartia obtusiloba	Yes[b]						Schrader 1986
Nemophila menziesii	Yes[d]	Yes				Yes	Ashman 2002; Barr 2004
Ochradenus baccatus	Yes[b]	Yes[e]					Wolfe and Shmida 1997
Pachycereus pringlei					No[f]		Molina-Freaner et al. 2003
Sagittaria latifolia			Yes[e]				Sarkissian et al. 2001
Schiedea species					Yes[a]		Weller et al. 1998
Sidalcea hendersonii						Yes[d]	Marshall and Ganders 2001
Silene acaulis	Yes (No)	Yes				No	Delph and Carroll 2001; Hermanutz and Innes 1994
Thymus vulgaris	Yes						Darwin 1877
Wurmbea biglandulosa	Yes	Yes					Vaughton and Ramsey 2002, 2004
Wurmbea dioica	Yes	Yes[e]			Yes		Case 2000; Case and Barrett 2004a

Predictions: (1) female (F) frequency increases with reduced habitat nutrients or water; (2) seed fertility of hermaphrodites (H) decreases (F:H ratio increases) with reduced habitat nutrients or water; (3) seed fertility of hermaphrodites decreases with reduced plant size (vigour); (4) plant size declines with reduced habitat nutrients or water; (5) pollination/pollinators change with habitat (or female frequency); (6) female frequency increases with increased herbivore damage to hermaphrodites, or reduced tolerance to herbivory by hermaphrodites.
Complete citation details are available at http://www.eeb.utoronto.ca/EEF/.

[a] Wind pollination increased with increased female frequency.
[b] Elevation or latitude is a surrogate for drought, harshness, or rainfall amount.
[c] Among species or subspecies.
[d] Not significant.
[e] Perfect or female flowers, or in monoecious populations.
[f] Female frequency decreased with reduced (poorer) pollinator service.

functions, instead allocating resources preferentially to male function, which is less costly than, and precedes, female function (reviewed in Case and Ashman 2005). Thus, resource stress can cause a plastic reduction of hermaphrodite seed production (e.g., Delph 1990c; Ashman 1999; Chapter 3), either directly or via changes in vegetative plant size. The latter result may occur if male function entails opportunity costs for vegetative growth (Eckhart and Chapin 1997), or because male-biased sex allocation is favoured at small plant size owing to a deceleration of male, but not female, fitness accrual with plant size (Klinkhamer et al. 1997). In either case, females may not be as susceptible to resource tradeoffs as hermaphrodites, either because they invest only in seed production, or because of pleiotropic effects of male-sterility genes (Lloyd 1975; Ross and Weir 1976) that help maintain female seed fitness under low-resource conditions. The preceding arguments lead to five predictions: (1) the seed fertility of hermaphrodites declines with resources (Fig. 11.3a); (2) pollen production stays constant or decreases less sharply with lower resources than seed production (Fig. 11.3a); (3) the level of resources (or size) that allow hermaphrodites to start producing seeds equalizes marginal fitness through female and male function (Fig. 11.3b); (4) seed production of hermaphrodites responds more plastically to nutrient/water availability than that of females (Fig. 11.3c); and (5) hermaphrodite seed production is more size dependent than that of females (Fig. 11.3c). I now review data relevant to testing these predictions.

About half of the among-population comparisons described above (Table 11.2, prediction 1) also show that hermaphrodites produce fewer seeds in dry or nutrient-poor habitats (Table 11.2, prediction 2). Whether this pattern reflects resource- (or size-) based plasticity of seed production, or genetic differentiation of hermaphrodites as a result of divergent selection along the resource gradient is unknown. Specifically, because female frequency increases as resources decline (Table 11.2, prediction 1) and high frequencies of females select for increased maleness of hermaphrodites (Charlesworth 1989), hermaphrodites in low-resource populations may be genetically more male (i.e., low or no fruit set) than those in high-resource sites. Moreover, if plasticity is costly and environment dependent, then selection may eliminate highly plastic types in low-resource conditions. For example, if maintaining flexible fruit production reduces fitness through pollen (e.g., "genetic costs" of plasticity caused by negative genetic correlations between plasticity and fitness as a consequence of pleiotropy; van Kleunen and Fischer 2005), then hermaphrodites with limited plasticity in fruit set would be favoured when hermaphrodites derive fitness primarily via male function (e.g., Conner et al. 2003; T.-L. Ashman unpublished data). Such a scenario could create a negative correlation between plasticity in fruit set and local resource status. Whether an observed gradient in hermaphrodite seed fertility reflects plasticity, genetic differentiation, or both can be determined definitively by common garden studies or studies that manipulate plants from populations at contrasting locations on the gradient. In addition, demonstration of any costs of plasticity and whether they occur in all environments or locally would shed light on whether plastic types are favoured universally or only in certain resource regimes (e.g., Sultan and Spencer 2002).

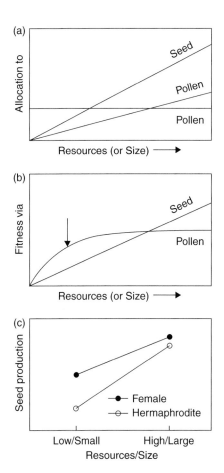

Figure 11.3 Schematic representations of five predictions for resource-mediated effects on allocation to seeds or pollen, and fitness of hermaphrodites in gynodioecious populations (Section 11.3.1). (a) Seed production by hermaphrodites, but not pollen production, declines with declining resources. Two possible relations between pollen and resources are shown (invariant or declining but at a lower rate than seed). (b) The relations of fitness via seed and pollen to resources or plant size. The arrow denotes the resource availability or size at which hermaphrodites should initiate allocation to seeds. (c) Seed production by hermaphrodites responds plastically to size or resource availability, but to a greater extent than that by females.

I conducted such an experiment with *Fragaria virginiana* plants from five populations that differed in sex ratio (see Ashman 1999 for details). I scored fruit set on two to four clonal replicates of 29 to 35 hermaphroditic genotypes per population which were grown either in the glasshouse or in a field garden. Plasticity in fruit set differed significantly among populations (population × location interaction from mixed-model ANOVA: $F_{4,148} = 7.67$; $P < 0.001$) and plasticity correlated negatively with female frequency ($r = -0.80$; $P = 0.05$, one-tailed test), with hermaphrodites from resource-poor populations being least plastic (Fig. 11.4). Furthermore, the degree of plasticity in fruit set exhibited a negative genetic correlation with pollen production per flower across all genotypes studied ($r = -0.17$; $P < 0.05$, one-tailed test; $N = 124$), indicating a potential pollen fitness cost to plasticity. If the lower pollen production of plastic types causes a resource-dependent cost to siring success, then selection would favour fixed genotypes under low resources, but more plastic genotypes under high resources.

A few studies present evidence to test the prediction that male allocation should be invariant, or increase to a lesser degree than female allocation, with resources (or plant size) (Fig. 11.3a). First, Klinkhamer et al. (1999) found that pollen per flower was unrelated to plant size, whereas seeds per flower increased with increasing plant size in *Echium vulgare*. Likewise, Sarkissian et al. (2001) found female, but not male, flower production increased with increasing plant size in monoecious *Sagittaria latifolia*. Ashman and colleagues also found that seed production by hermaphrodite *F. virginiana* responded more to high resources or larger plant size than did pollen production (Ashman et al. 2001; A.L. Case and T.-L. Ashman unpublished data). However, evidence relating the resource level (or size) at which hermaphrodites initiate seed production to the point where their marginal fitness returns per unit investment in female and male function are equal (Fig. 11.3b) is currently lacking for any species.

In addition to the above predictions, fruit set of females should be less plastic than that of hermaphrodites (Fig. 11.3c). Females may be more able to maintain fruit set under low resources as a consequence of not producing pollen or via plasticity in

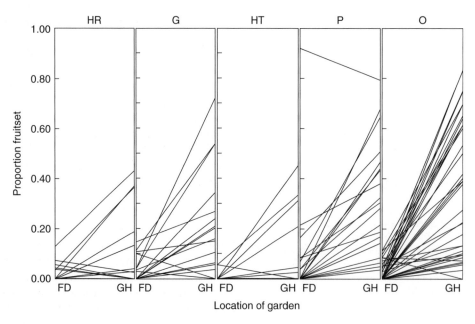

Figure 11.4 Variation in plasticity in fruit set in five populations of *Fragaria virginiana* that differ in site quality and the frequency of females. Populations (HR, G, HT, P, O) are ordered by increasing site resource availability (low to high) from left to right. Lines connect the fruit set by ramets from the same genet grown in a field garden (FD) or a glasshouse (GH).

other traits, i.e., photosynthetic rates. Delph (2003) recently reviewed observational evidence for sex-differential plasticity in several gynodioecious species. However, a few studies have also explored sex-differential plasticity by resource manipulation. Findings are somewhat mixed, possibly owing to genetic variability in the subject material (see above) or difficulties of maintaining appropriately divergent growth conditions. In an elegant *in situ* manipulation of *Nemophila menziesii*, Barr (2004) found that supplemental water increased seed production by hermaphrodites but not by females, with no sex-differential changes in plant size in the drier of two sites. In contrast, Dorken and Barrett (2004) found more variable flower production by females from dioecious populations of *Sagittaria latifolia* than by males in response to high nutrients. In other studies, the direction of sex-differential plasticity depended on the trait measured. For example, Eckhart and Chapin (1997) manipulated nutrient availability in *Phacelia linearis* and found significantly less plasticity in females than in hermaphrodites for flower production, but not for seed production.

Last, several studies have shown that seed production (or fruit set) by hermaphrodites, but not females, depends on plant size (Table 11.2, prediction 3). In addition to studies reviewed by Delph (2003), greater size dependence of female allocation (seeds or flowers) in hermaphrodites than females has been observed in *Echium vulgare* (Klinkhamer *et al.* 1994) and in *Wurmbea dioica* (Case and Barrett 2004a). In addition, I have explored how sex-differential effects of plant size may contribute to population sex ratio in *F. virginiana* (Ashman 1999). Because fruit set of hermaphrodites, but not of females, increases significantly with plant size, and plant size increases with soil resources ($r = 0.85$; $N = 5$, $P = 0.07$, one-tailed test), the observed female frequency (33%) was more similar to the frequency predicted from size-dependent seed fertility (33%) than to predictions based on seed fertility once the effects of plant size were removed statistically (41%).

A difficulty of any study of phenotypic associations between plant size and gender variation arises from the confounding of genetic and environmental sources of variation in sex allocation. Manipulation of the growth environment to modify the sizes of different ramets with the same genotype can reveal conclusively the degree to which plants adjust sex allocation according to size (e.g., Ashman *et al.* 2001). Additionally, such studies can determine whether populations differ in size-dependent sex allocation as described above for resource-based plasticity. It will also be valuable to understand the shape of the relation between size and sex allocation (Chapter 3), so studies that produce a range of sizes will be the most informative.

Severity of inbreeding depression
Harsh habitats could facilitate the evolution of dioecy by aggravating inbreeding depression, facilitating both the first and second steps in the gynodioecy pathway to a dioecious sexual system (Ganders 1978; Weller *et al.* 1990; Ashman 1999). Although inbreeding depression is often claimed to be exacerbated by harsh conditions, supporting evidence is rare and equivocal (Armbruster and Reed 2005). Several studies have interpreted greater inbreeding depression in the field than in glasshouses as a response to environmental harshness (e.g., Ramsey and Vaughton 1998), but these environments differ in many biotic and abiotic factors. Far fewer studies of inbreeding depression have characterized the nature of the environment or manipulated specific stressors. In these controlled studies, enemies or reduced resources increased inbreeding depression (Hauser and Loeschcke 1996; Carr and Eubanks 2002). However, this effect often depends on the fitness trait measured or the population studied, and some studies detect no effect or counter-intuitive effects (Delph and Lloyd 1996; Norman *et al.* 1996; Mustajärvi *et al.* 2005).

Future work
The preceding discussion identifies several poorly understood aspects of the role of habitat quality-mediated sex expression in the evolution of dioecy. Although plasticity of fruit set by hermaphrodites may facilitate the spread of females, it may also preclude the evolution of males. Therefore, experimental work exploring sex and population

differentials in plasticity, and the costs of plasticity and how male gain curves change with resource availability (e.g., Chapter 3), would be particularly valuable (see also Section 11.4.1). In addition, both controlled manipulative studies that assess whether low resources aggravate inbreeding depression and studies demonstrating that inbreeding depression varies with habitat harshness would be valuable. Thus, the importance of direct influences of harsh environments on the evolution of dioecy remains unclear (Renner and Ricklefs 1995).

11.3.2 Harsh environments alter pollination, pollen movement, and mating system

The female outcrossing rate of hermaphrodites is a key parameter in sexual-system evolution, and studies indicate a range of mating systems, primarily involving a mixture of selfing and outcrossing (reviewed in Collin 2003). However, the factors causing mating-system variation or the genesis of mixed mating are poorly known (Goodwillie et al. 2005; Chapters 4 and 6). Several authors have suggested that harsh environments either support different pollinator faunas (Ganders 1978; Delph 1990a; Weller et al. 1990) or alter plant phenotypes in ways that change pollen movement or the mode of selfing, thereby modifying the mating system (Case and Barrett 2004b; Vaughton and Ramsey 2004; Chapter 7). I now review the limited existing data concerning these effects and then illustrate how resource-mediated changes in plant phenotype can alter the sufficiency of pollination, the mode of self-pollination, and the potential for pollen discounting, and thus ultimately the sexual system.

Changes in pollinators in harsh habitats
Several correlative studies have indicated that increased habitat harshness reduces pollinator service, or changes the composition and/or abundance of the pollinator fauna (Table 11.2, prediction 5), which could increase selfing. For instance, species of *Alisinidendron* and *Schiedea* that occupy wet habitats are pollinated by animals, whereas wind pollination is more common in dry habitats (Weller et al. 1998). Delph (1990a) suggested that altitudinal changes in the pollinator faunas of *Hebe* species from mainly bees to flies and beetles might

contribute to increased gender dimorphism at high elevations. Case and Barrett (2004b) also suggested that differences in pollinators and flower morphology may have facilitated gender evolution in *Wurmbea dioica*. The only direct evidence that resource availability changes pollinator abundance comes from a *Fragaria virginiana* experiment by A.L. Case and T.-L. Ashman (submitted). Plants grown under low-resource conditions displayed significantly fewer and smaller flowers which received significantly fewer pollinator visits than

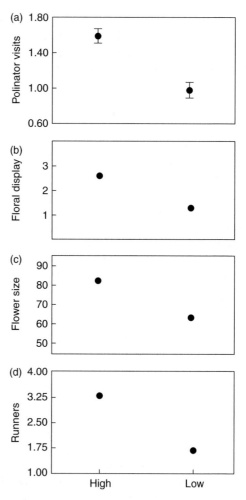

Figure 11.5 Effects of resource availability on mean (±SE) (a) pollinator visits (visits per flower per hour), (b) floral display (open flowers per plant), (c) flower size (petal area in mm^2), and (d) number of runners per plant for *Fragaria virginiana* (A.L. Case and T.-L. Ashman, unpublished data).

plants grown under high resources (Fig. 11.5a–c). Such reduced visitation could increase autogamous self-pollination in this self-compatible species. Unfortunately, none of the studies described above measured the direct effect of pollinator fauna/ abundance on the mating system of hermaphrodites.

Studies that have measured female outcrossing rates and detected variation among populations have not linked this variation to specific features of the pollination environment. For instance, Sun and Ganders (1988) found no relation between floral features and outcrossing rate for several gynodioecious species of *Bidens* and instead concluded that mating-system variation must result from unidentified environmental differences among habitats. Clearly, direct evidence linking selfing rate variation to pollinator abundance or type is desperately needed to test the validity of this mechanism.

Consequences of plant phenotype in harsh habitats for self-pollination
Changes in plant phenotypes with resource availability could alter the selfing rate and/or the mode of selfing, even without changing pollinator fauna/abundance. Self-pollination might increase in harsh habitats if poor growth conditions reduce flower size, bringing stigmas and anthers in closer proximity in individual flowers and increasing autogamy (Elle and Carney 2003). Case (2000) suggested that such increased selfing might have facilitated the evolution of dioecy in *Wurmbea dioica* in dry sites. However, other features also vary with resources and can affect the mode of selfing. For instance, plants grown in low resources display fewer flowers simultaneously and produce fewer ramets (Costich 1995; Elle and Hare 2002; Fig. 11.2b and d), both of which should reduce geitonogamy and increase autogamy in hermaphrodite species (e.g., Harder and Barrett 1995).

The role of resources in mediating modes of selfing is relevant to the evolution of sexual systems, because they both determine the selfing rate and may dictate the degree of pollen discounting (discussed in Harder and Barrett 1995; Harder *et al.* 2000), which negatively affects the evolution of dioecy (de Jong *et al.* 1999). Specifically, if autogamy involves less pollen discounting and is more likely under low resources because of small flowers, floral displays, and/or clones, then dioecy should evolve more commonly in poor environments. In contrast, hermaphroditism should be favoured in rich environments if selfing primarily involves geitonogamy, and pollen discounting is high because of large floral displays and clones. Furthermore, elevated selfing via autogamy in poor environments might combine with high inbreeding depression to select for dioecy in harsh habitats, whereas low inbreeding depression combined with geitonogamy and high pollen discounting might select against dioecy in moderate or luxuriant habitats. Currently no data are available to address these alternative scenarios, but the first experimental test is under way (T.-L. Ashman unpublished data).

At this time, evidence that harsh habitats affect mating is largely inferential. Although pollinator faunas vary with habitat and limited resources alter floral phenotypes in ways that could promote autogamous selfing, no evidence links the selfing rate or the mode of selfing by hermaphrodites directly to habitat quality.

11.3.3 Enemies—analogous to harsh environments?

In contrast to the emphasis on resource availability, the influence of plant enemies in sexual-system evolution has received little attention (Chapter 7). Recently, Ashman (2002) highlighted the diverse roles that enemies (e.g., herbivores or pathogens) could play in sexual-system variation. In addition to explaining inconsistencies between observations and predictions based on traditional approaches (see Section 11.2.4), consideration of the effects of enemies reveals similarities with the effects of resources on sexual-system evolution (Table 11.3) and motivates assessment of the consequences of dissimilarities. Here, I consider a few examples: a more complete analysis of enemy effects can be found in Ashman (2002).

By reducing the resources available to plants, herbivory may induce sex-differential tolerance to damage (Table 11.3). The direction of this sex difference may vary with reproductive costs at the time of damage (Ashman *et al.* 2004). Specifically, high investment in flowers by hermaphrodites may reduce their tolerance of damage to

Table 11.3 Summary of the key parameters, potential effects, expected consequences, and potential mechanisms by which low resources ($-R$) or the presence of enemies ($+E$) can effect the evolution of the first (A) and second (B) steps in the evolution of dioecy.

Parameter	Effect	Expected consequence	Mechanism
(A) Hermaphroditism $\rightarrow$ Gynodioecy			
H seed fertility	Decrease	Increase F advantage; increase FF; favour GD	$-R$: limit H seed production (plasticity in sex allocation) $+E$: preferential damage to H seed $+E$: reduces H attractiveness, leads to increased pollen limitation of H seed $+E$: H less tolerant of damage than F
F seed fertility	Decrease	Decrease F advantage; decrease FF; favour H	$+E$: reduces pollen availability/or export; increases pollen limitation of females $-R$: reduces pollinator visitation to population; F are more pollen limited
H selfing rate	Increase autogamy (competing)	Increase F advantage; increase FF; favour GD	$+E$ and $-R$: leads to altered floral phenotype $+E$ and $-R$: damaged plants or harsh environments support different/fewer pollinator fauna.
	Increase autogamy (prior)	Decrease F advantage; decrease FF; favour H	$+E$ and $-R$: causes flowers not to open or anthers to dehisce prior to opening
	Increase autogamy (delayed)	Decrease F advantage; decrease FF; favour H	$+E$ and $-R$: reduces pollinator visitation
	Decrease geitonogamy	Decrease F advantage; decrease FF; favour H	$-R$: support different/fewer pollinator fauna. $+E$ and $-R$: reduce number of simultaneously open flowers/plant and ramets/genotype
Inbreeding depression	Increase	Increase F advantage; increase FF; favour GD	$+E$ and $-R$: increase severity of inbreeding depression
(B) Gynodioecy $\rightarrow$ Dioecy			
Pollen fitness gain curve	Decelerating	Decrease FF; select against D	$+E$: damage increases with increased pollen production $+E$: damage random; tolerance decreases with increased pollen production $-R$: high resources lead to greater flower number, greater selfing through geitonogamy and pollen discounting
H sex allocation	To femaleness	Decrease FF; select against D	$+E$: damage to flowers leads to reduced male allocation
	To maleness	Increase FF; select for D	$+E$: damage leads to reduced female allocation $-R$: low resources lead to smaller vegetative size and reduced female allocation
F seed fertility	Decrease	Decrease FF; select against D	$+E$: damage to H reduces pollen availability and increases pollen limitation of F
H seed fertility	Decrease	Increase M frequency; select for D	$+E$: increased pollen export reduces seed production of H

D: dioecy; F: female; FF: female frequency; GD: gynodioecy; H: hermaphrodite; M; male.

photosynthate "sources" (foliage or vasculature) during flowering compared with that of females. In contrast, females may tolerate less damage to sources during fruiting. The consequences of herbivory for sexual-system evolution may also depend on the type of damage. For example, in *F. virginiana,* seed set by hermaphrodites was less tolerant to damage by spittlebugs, but more tolerant to damage by flower-bud clipping weevils than that of females (Ashman 2004; Cole and Ashman 2005).

The impact of enemies is even more complex, because the susceptibility to damage is often sex differential, especially when enemies attack flowers and fruits (reviewed in Ashman 2002). The sexes of *F. virginiana* suffer damage by spittlebugs equally, but hermaphrodites are more susceptible to weevils. Thus, resistance and tolerance probably combine to determine the net effect of herbivory on sexual-system parameters. As a result, consequences of plant–enemy interactions for sexual-system evolution may be more diverse than those with abiotic resources. Nevertheless, female frequency may vary with the severity of damage to hermaphrodites or their tolerance (Table 11.2, prediction 6; Fig. 11.2).

Like abiotic resources, enemies can affect sexual-system dynamics by altering plant interactions with pollinators, changing the mating system of hermaphrodites, and increasing the severity of inbreeding depression (Ashman 2002). These effects have not been considered for gender-dimorphic species, but studies of hermaphrodite species show that damage can reduce/alter the composition of the pollinator fauna (e.g., Steets *et al.* in press), reduce the number of open flowers per plant and decrease geitonogamy (Elle and Hare 2002; Steets *et al.* in press), and intensify the expression of inbreeding depression (e.g., Carr and Eubanks 2002).

Because exploration of the role of enemies in the evolution of sexual systems is just beginning, research on the effect of enemies on any of the key parameters outlined in Table 11.3, as well as information on how damage varies with sexual phenotype and how it varies among populations, will greatly enhance knowledge of the environmental context of sexual-system evolution.

11.3.4 Multi-species interactions and little-studied aspects of ecological context

The effect of any ecological factor (e.g., herbivores) on plant reproduction probably depends on other biotic (pollinators, pathogens) and abiotic (resource availability) factors (also see Chapter 7), so the roles of resources, mutualists, and enemies in the evolution of sexual systems are linked. Thus, studies that consider sex-differential multi-species interactions will ultimately be crucial for a full understanding of the role of ecological context in sexual-system evolution.

Two recent studies clarify this point. Cole and Ashman (2005) found that spittlebug infestation of *F. virginiana* increased damage by fungal pathogens in females, but not in hermaphrodites. They concluded that the secondary effect of fungal damage may be more detrimental to female fertility than the relatively transient direct effect of spittlebugs. In contrast, spittlebug infestation reduced damage by weevils that clip flower buds in hermaphrodites, an interaction which may increase plant fitness, because weevils may reduce male fitness more than spittlebugs. Similarly, Collin *et al.* (2002) found that even though hermaphrodites of *Dianthus sylvestris* suffered higher predispersal seed predation than females, seed predators had a net positive effect on hermaphrodite fitness, because larval herbivory prevented infection by a more damaging smut fungus, which sterilized anthers.

These studies highlight the importance of considering multi-species interactions in the quest to understand the ecological context for sexual-system evolution. However, the net effect of these interactions on the key parameters or their effects on female frequency remain unclear. Finally, several other types of interactions deserve more attention, such as systems in which pollinators transmit pathogens (Lopez-Villavicencio *et al.* 2005), which may mediate sex-differential pollen limitation, and the role of below-ground mutualists (e.g., mycorrhizae; Pendleton 2000), which may mediate sex-differential responses to low resources or the expression of inbreeding depression (see Section 11.3.1).

11.4 Spotlight on unresolved issues

The preceding review and discussion illustrate both that theory provides a powerful framework for studying sexual-system evolution and that recent empirical emphasis on mechanistic relations reveals the centrality of ecological context for these dynamics. They also shed light on unresolved issues. Here, I focus on two issues which will require a synthetic approach for resolution.

11.4.1 Subdioecy: reflective of constraints or adaptation?

The evolution of dioecy from gynodioecy requires the evolution of males from hermaphrodites. However, the prediction that male allocation of hermaphrodites increases with female frequency has received very little support (Table 11.1, prediction 5). In addition, populations of many species with variable sexual expression seem "stalled" in a subdioecious state containing females, males, and hermaphrodites. Such cases appear to satisfy the conditions for the evolution of males, namely, that hermaphrodites have a genetically determined reduction in female function (e.g., Delph and Lloyd 1991; Ashman 2003) resulting from a genetic trade-off between male and female function (reviewed in Ashman 2003). These conflicting data raise the question "Do constraints limit the evolution of full dioecy, or is subdioecy an adaptive state?"

If selection in populations with high female frequency favours males, then genetic or ecological constraints could limit the evolution of increased male allocation in hermaphrodites. For example, limited genetic variation for pollen production, positive genetic correlations between pollen and ovule production per flower within hermaphrodites, and/or between ovule production across the sex morphs may slow the evolution of increased pollen production and the elimination of female function in hermaphrodites of *F. virginiana* (Ashman 2003; A.L. Case and T.-L. Ashman manuscript submitted). Alternatively, selection against male allocation may impede the evolution of males. For instance, weevils that preferentially damage the flowers of hermaphroditic plants with high pollen and flower production select against the most male plants of *F. virginiana* (Ashman 2002; Ashman *et al.* 2004).

Likewise, the tendency for hermaphrodites not to be replaced entirely by males in many populations with high female frequency may indicate that retaining female function is adaptive, or that plasticity in fruit set may stabilize subdioecy, as suggested by Delph and Wolfe (2005). For instance, Charlesworth (1999) showed that in partially selfing hermaphrodites, some residual female allocation in hermaphrodites persists at equilibrium. Alternatively, as noted above, plasticity may preclude the evolution of males if the costs (male or otherwise) of plasticity are negligible or environment specific. Resource-based plasticity may also change male fertility gain curves in a way that maintains subdioecy (see Chapter 3). For example, de Laguerie (1993) modelled sexual-system evolution with resource-dependent sigmoidal male fitness gain curves and showed that intermediate resource levels maintain gynodioecy, whereas either extreme, high or low, favoured hermaphroditism or dioecy, respectively. However, see Chapter 3 for an alternative view of the importance of sigmoidal gain curves.

Last, the role of metapopulation dynamics (Pannell 1997; Chapter 12) in stabilizing subdioecy should be explored, because metapopulation structure can favour plasticity over local adaptation (Sultan and Spencer 2002). In particular, the benefits of producing fruit by founding hermaphrodites may outweigh the costs of plasticity.

Additional research into each of these possibilities is required to resolve the seemingly anomalous occurrence of subdioecy. Specifically, integrative studies that measure both phenotypic selection directly, while manipulating potential agents (resources or enemies), and the response to selection are needed to determine whether genetic constraints or opposing selection limit the evolution of males. In addition, more information on population differentiation in plasticity and male costs of plasticity (e.g., Section 11.3.1), and how male gain curves change with resource availability is sorely needed. New theory that incorporates plasticity, its costs, habitat heterogeneity, and metapopulation dynamics will also aid in evaluating these ideas.

11.4.2 Pollinators: facilitators or inhibitors of sexual-system evolution?

Pollinators have been ascribed two important, but conflicting, roles in gender evolution. On one hand, reductions or shifts in faunal composition of pollinators may lead to insufficient or inferior pollination, respectively (see Harder and Barrett 1996), increasing autonomous or competing selfing of hermaphrodites and in turn facilitating the spread of females. On the other hand, if newly established females attract fewer pollinators, receive insufficient pollination, and thus suffer greater pollen limitation of their seed fertility, then they may not spread. No studies address the first scenario directly. Several studies have tested, but found limited evidence supporting, the second prediction (Shykoff et al. 2003), suggesting that pollinators are important in the initial invasion by females, but that pollen limitation does not impede female spread. This outcome could arise for two reasons. First, pollinators may disperse outcross pollen more efficiently to females than among hermaphrodites, perhaps because vestigial anthers of females interfere less with styles, or because male-sterility genes have pleiotropic effects that enhance pollen capture, such as longer styles, more papillae or greater floral longevity (Ashman and Stanton 1991). Alternatively, these traits or attractive traits may evolve quickly enough to eliminate pollen limitation of female reproduction (Ashman and Diefenderfer 2001). Studies that relate pollinator characteristics (e.g., fauna abundances, or behaviours) to pollen receipt and selfing rate among populations or taxa are needed to clarify the role of pollinators in sexual-system evolution. Identification of how pollinators interact with plant phenotype (especially that modified by resource availability) to mediate selfing rate will also be crucial, in light of the importance of the mode of selfing during both the first and second steps in the gynodioecy pathway to dioecy.

Acknowledgements

I thank numerous members of the Ashman laboratory for help in the field or glasshouse during the past decade, and S. C. H. Barrett, L. D. Harder and two anonymous reviewers for comments on an earlier draft. This research was supported by the National Science Foundation (DEB 9903802, 0108099, and 0449488).

References

Armbruster P and Reed DH (2005). Inbreeding depression in benign and stressful environments. *Heredity*, **95**, 235–42.

Ashman T-L (1999). Determinants of sex allocation in a gynodioecious wild strawberry: implications for the evolution of dioecy and sexual dimorphism. *Journal of Evolutionary Biology*, **12**, 648–61.

Ashman T-L (2002). The role of herbivores in the evolution of separate sexes from hermaphroditism. *Ecology*, **83**, 1175–84.

Ashman T-L (2003). Constraints on the evolution of dioecy and sexual dimorphism: Field estimates of quantitative genetic parameters for reproductive traits in three populations of gynodioecious *Fragaria virginiana*. *Evolution*, **57**, 2012–25.

Ashman T-L and Diefenderfer C (2001). Sex ratio represents a unique context for selection on attractive traits: consequences for the evolution of sexual dimorphism. *American Naturalist*, **157**, 334–47.

Ashman T-L and Stanton ML (1991). Seasonal variation in pollination dynamics of the sexually dimorphic species, *Sidalcea oregana* ssp. *spicata* (Malvaceae). *Ecology*, **72**, 993–1003.

Ashman T-L, Pacyna J, Diefenderfer C, and Leftwich T (2001). Size-dependent sex allocation in a gynodioecious wild strawberry: the effects of sex morph and inflorescence architecture. *International Journal of Plant Sciences*, **162**, 327–34.

Ashman T-L, Cole D, and Bradburn M (2004). Sex-differential resistance and tolerance to florivory in a gynodioecious wild strawberry: Implications for floral and sexual system evolution. *Ecology*, **85**, 2550–9.

Bailey MF, Delph LF, and Lively CM (2003). Modeling gynodioecy: novel scenarios for maintaining polymorphism. *American Naturalist*, **161**, 762–76.

Barr CM (2004). Soil moisture and sex ratio in a plant with nuclear-cytoplasmic sex inheritance. *Proceedings of the Royal Society of London, Series B*, **271**, 1935–9.

Carr DE and Eubanks MD (2002). Inbreeding alters resistance to insect herbivory and host plant quality in *Mimulus guttatus* (Scrophulariaceae). *Evolution*, **56**, 81–109.

Case AL and Ashman T-L (2005). Gender dimorphism: implications for reproductive cost and gender-specific physiology. In E Reekie and F Bazzazz, eds. *The allocation of resources to reproduction in plants.* pp 126–14. Elsiever Science Press, New York, NY.

Case AL and Barrett SCH (2004a). Environmental stress and the evolution of dioecy: *Wurmbea dioica* (Colchicaceae) in Western Australia. *Evolutionary Ecology*, **18**, 145–64.

Case AL and Barrett SCH (2004b). Floral biology of gender monomorphism and dimorphism in *Wurmbea dioica* (Colchicaceae) in Western Australia. *International Journal of Plant Sciences*, **165**, 289–301.

Charlesworth B and Charlesworth D (1978). A model for the evolution of dioecy and gynodioecy. *American Naturalist*, **112**, 975–97.

Charlesworth D (1989). Allocation to male and female function in hermaphrodites, in sexually polymorphic populations. *Journal of Theoretical Biology*, **139**, 327–42.

Charlesworth D (1999). Theories on the evolution of dioecy. In MA Geber, TE Dawson, and LF Delph, eds. *Gender and sexual dimorphism in flowering plants*, pp. 33–60. Springer-Verlag, Berlin.

Charlesworth D (2002). Plant sex determination and sex chromosomes. *Heredity*, **88**, 94–101.

Collin CL (2003). Outcrossing rates in the gynomonoecious-gynodioecious species *Dianthus sylvestris* (Caryophyllaceae). *American Journal of Botany*, **90**, 579–85.

Collin CL, Pennings PS, Rueffler C, Widmer A, and Shykoff JA. (2002). Natural enemies and sex: how seed predators and pathogens contribute to sex-differential reproductive success in a gynodioecious plant. *Oecologia*, **131**, 94–102.

Conner JK, Rice AM, Stewart C, and Morgan MT (2003). Patterns and mechanisms of selection on a family diagnostic trait: evidence from experimental manipulation and lifetime fitness selection gradients. *Evolution*, **57**, 480–6.

Costich DE (1995). Gender specialization across a climatic gradient: Experimental comparison of monoecious and dioecious *Ecballium*. *Ecology*, **76**, 1036–50.

Couvet D, Atlan A, Belhassen E, Gliddon C, Gouyon PH, and Kjellberg F (1990). Co-evolution between two symbionts: the case of cytoplasmic male-sterility in higher plants. In D Futuyma and J Antonovics, eds. *Oxford surveys in evolutionary biology*, pp 225–47. Oxford University Press.

Darwin CR (1877). *The different forms of flowers on plants of the same species.* J Murray, London.

de Jong TJ, Klinkhamer PGL, and Rademaker MCJ (1999). How geitonogamous selfing affects sex allocation in hermaphrodite plants. *Journal of Evolutionary Biology*, **12**, 166–76.

de Laguerie P, Olivieri I, and Gouyon P-H (1993). Environmental effects on fitness-sets shape and evolutionarily stable strategies. *Journal of Theoretical Biology*, **163**, 113–25.

Delph LF (1990a). The evolution of gender dimorphism in New Zealand *Hebe* (Scrophulariaceae) species. *Evolutionary Trends in Plants*, **4**, 85–98.

Delph LF (1990b) Sex differential resource allocation patterns in the subdioecious shrub *Hebe subalpina*. *Ecology*, **71**, 1342–51.

Delph LF (1990c) Sex-ratio variation in the gynodioecious shrub *Hebe strictissima* (Scrophulariaceae). *Evolution*, **44**, 134–42.

Delph LF (2003). Sexual dimorphism in gender plasticity and its consequences for breeding system evolution. *Evolution and Development*, **5**, 34–9.

Delph LF and Carroll SB (2001). Factors affecting relative seed fitness and female frequency in a gynodioecious species, *Silene acaulis*. *Evolutionary Ecology Research*, **3**, 487–505.

Delph LF and Lloyd DG (1991). Environmental and genetic control of gender in the dimorphic shrub *Hebe subalpina*. *Evolution*, **45**, 1957–64.

Delph LF and Lloyd DG (1996). Inbreeding depression in the gynodioecious shrub *Hebe subalpina* (Scrophulariaceae). *New Zealand Journal of Botany*, **34**, 241–7.

Delph LF and Wolf DE (2005). Evolutionary consequences of gender plasticity in genetically dimorphic breeding systems. *New Phytologist*, **166**, 119–28.

Dorken ME and Barrett SCH (2004). Phenotypic plasticity of vegetative and reproductive traits in monoecious and dioecious populations of *Sagittaria latifolia* (Alismataceae): a clonal aquatic plant. *Journal of Ecology*, **92**, 32–44.

Dorken ME, Friedman J, and Barrett SCH (2002). The evolution and maintenance of monoecy and dioecy in *Sagittaria latifolia* (Alismataceae). *Evolution*, **56**, 31–41.

Eckhart VM and Chapin FS (1997). Nutrient sensitivity of the cost of male function in gynodioecious *Phacelia linearis* (Hydrophyllaceae). *American Journal of Botany*, **84**, 1092–98.

Elle E and Carney R (2003). Reproductive assurance varies with flower size in *Collinsia parviflora* (Scrophulariaceae). *American Journal of Botany*, **90**, 888–96.

Elle E and Hare JD (2002). Environmentally induced variation in floral traits affects the mating system in *Datura wrightii*. *Functional Ecology*, **16**, 79–88.

Ganders FR (1978). The genetics and evolution of gynodioecy in *Nemophila menziesii* (Hydrophyllaceae). *Canadian Journal of Botany*, **56**, 1400–8.

Goodwillie C, Kalisz S, and Eckert CG 2005 The evolutionary enigma of mixed mating systems in plants: occurrence, theoretical explanations, and empirical evidence. *Annual Review of Ecology, Evolution and Systematics*, **36**, 47–79.

Harder LD and Barrett SCH (1995). Mating cost of large floral displays in hermaphrodite plants. *Nature*, **373**, 512–5.

Harder LD and Barrett SCH (1996). Pollen dispersal and mating patterns in animal-pollinated plants. In DG Lloyd and SCH Barrett, eds. *Floral biology: Studies on floral evolution in animal-pollinated plants.* pp. 140–91. Chapman and Hall, New York.

Harder LD, Barrett SCH, and Cole WW (2000). The mating consequences of sexual segregation within inflorescences of flowering plants. *Proceedings of the Royal Society of London, Series B*, **267**, 315–20.

Hart JA (1985). Evolution of dioecism in *Lepechinia* Willd. Sect. Parviflorae (Lamiaceae). *Systematic Botany*, **10**, 147–54.

Hauser TP and Loeschcke V (1996). Drought stress and inbreeding depression in *Lychnis flos-cuculi* (Caryophyllaceae). *Evolution*, **50**, 1119–26.

Kesseli R and Jain SK (1984). An ecological genetic study of gyndioecy in *Limnanthes douglasii* (Limnanthaceae). *American Journal of Botany* **71**, 775–86.

Klinkhamer PGL, de Jong TJ, and Nell HW (1994). Limiting factors for seed production and phenotypic gender in the gynodioecious species *Echium vulgare* (Boraginaceae). *Oikos*, **71**, 469–78.

Klinkhamer PGL, de Jong TJ, and Metz H (1997). Sex and size in cosexual plants. *Trends in Ecology and Evolution*, **12**, 260–5.

Kohn JR and Biardi JE (1995). Outcrossing rates and inferred levels of inbreeding depression in gynodioecious *Cucurbita foetidissima* (Cucurbitaceae). *Heredity*, **75**, 77–83.

Lewis D (1941). Male sterility in natural populations of hermaphrodite plants. *New Phytologist*, **40**, 56–63.

Lloyd DG (1974). Theoretical sex ratios of dioecious and gynodioecious angiosperms. *Heredity*, **31**, 11–34.

Lloyd DG (1975). The maintenance of gynodioecy and androdioecy in angiosperms. *Genetica*, **45**, 1–15.

Lloyd DG (1976). The transmission of genes via pollen and ovules in gynodioecious angiosperms. *Theoretical Population Biology*, **9**, 299–316.

Lloyd DG (1980). Sexual strategies in plants. III. A quantitative method for describing the gender of plants. *New Zealand Journal of Botany* **18**, 103–8.

Lopez-Villavicencio M, Branca A, Giraud T, and Shykoff JA (2005). Sex-specific effect of *Microbotryum violaceum* (Uredinales) spores on healthy plants of the gynodioecious *Gypsophila repens* (Caryophyllaceae). *American Journal of Botany*, **92**, 896–900.

Marshall M and Ganders FR (2001). Sex-biased seed predation and the maintenance of females in a gynodioecious plant. *American Journal of Botany*, **88**, 1437–43.

Maurice S, Charlesworth D, Desfeux C, Couvet D, and Gouyon P-H (1993). The evolution of gender in hermaphrodites of gynodioecious populations with nucleo-cytoplasmic male-sterility. *Proceedings of the Royal Society of London, Series B*, **251**, 253–61.

Maurice S, Belhassen E, Couvet D, and Gouyon P-H (1994). Evolution of dioecy: can nuclear-cytoplasmic interactions select for maleness? *Heredity*, **73**, 346–54.

McCauley DE and Brock MT (1998). Frequency-dependent fitness in *Silene vulgaris*, a gynodioecious plant. *Evolution*, **52**, 30–6.

Mustajärvi K, Shikamäki P, and Åkerberg A (2005). Inbreeding depression in perennial *Lychnis viscaria* (Caryophyllaceae): effects of population mating history and nutrient availability. *American Journal of Botany* **92**, 1853–61.

Norman JK, Sakai AK, Weller SG, and Dawson TE (1996). Inbreeding depression in morphological and physiological traits of *Schiedea lydgatei* (Caryophyllaceae) in two environments. *Evolution*, **49**, 297–306.

Pannell JR (1997). The maintenance of gynodioecy and androdioecy in a metapopulation. *Evolution*, **51**, 10–20.

Pendleton RL (2000). Pre-inoculation by arbuscular mycorrhizal fungus enhances male reproductive output of *Cucurbita foetidissima*. *International Journal of Plant Sciences*, **161**, 683–9.

Ramsey M and Vaughton G (1998). Effect of environment on the magnitude of inbreeding depression in seed germination in a partially self-fertile perennial herb (*Blandfordia grandiflora*, Liliaceae). *International Journal of Plant Sciences*, **159**, 98–104.

Renner SS and Ricklefs RE (1995). Dioecy and its correlates in the flowering plants. *American Journal of Botany*, **82**, 596–606.

Rosas LF, Perez-Alquicira J, and Dominguez CA (2005). Environmentally induced variation in fecundity compensation in the morph-biased male-sterile distylous shrub *Erythroxylum havanense* (Erythroxylaceae). *American Journal of Botany*, **92**, 116–22.

Ross MD and Weir BS (1976). Maintenance of males and females in hermaphroditic populations and the evolution of dioecy. *Evolution*, **30**, 425–41.

Sarkissian TS, Barrett SCH, and Harder LD (2001). Gender variation in *Sagittaria latifolia* (Alismataceae): is size all that matters? *Ecology*, **82**, 360–73.

Shultz ST (1994). Nucleo-cytoplasmic male sterility and alternative routes to dioecy. *Evolution*, **48**, 1933–45.

Shykoff JA, Kolokotronis S-O, Collin CL, and Lopez-Villavicencio M (2003). Effects of male sterility on reproductive traits in gynodioecious plants: a meta-analysis. *Oecologia*, **135**, 1–9.

Sultan SE and Spencer HG (2002). Metapopulation structure favors plasticity over local adaptation. *The American Naturalist*, **160**, 271–83.

Steets JA, Hamrick JL, and Ashman T-L (in press). The consequences of vegetative herbivory for the maintenance of intermediate outcrossing in an annual plant. *Ecology*.

Sun M and Ganders FR (1988). Mixed mating systems in Hawaiian *Bidens* (Asteraceae). *Evolution*, **42**, 516–27.

Thomson JD and Brunet J (1990). Hypothesis for the evolution of dioecy in seed plants. *Trends in Ecology and Evolution*, **5**, 11–6.

Thompson JD and Tarayre M (2000). Exploring the genetic basis and proximate cause of female fertility advantage in gynodioecious *Thymus vulgaris*. *Evolution*, **54**, 1510–20.

Vamosi JC, Otto SP, and Barrett SCH (2003). Phylogenetic analysis of the ecological correlates of dioecy in angiosperms. *Journal of Evolutionary Biology*, **16**, 1006–18.

Van Kleunen M and Fischer M (2005). Constraints on the evolution of adaptive phenotypic plasticity in plants. *New Phytologist* **166**, 49–60.

Van Noordwijk AJ and de Jong G (1986). Acquisition and allocation of resources: their influence on variation in life-history tactics. *American Naturalist*, **128**, 137–42.

Vaughton G and Ramsey M (2004). Dry environments promote the establishment of females in monomorphic populations of *Wurmbea biglanulosa* (Colchicaceae). *Evolutionary Ecology*, **18**, 323–41.

Weiblen GD, Oyama RK, and Donoghue MJ (2000). Phylogenetic analysis of dioecy in monocotyledons. *American Naturalist*, **155**, 46–57.

Weller SG and Sakai AK (2005). Selfing and resource allocation in *Schiedea salicaria* (Caryophyllaceae), a gynodioecious species. *Journal of Evolutionary Biology*, **18**, 301–8.

Weller SG, Sakai AK, Wagner WL, and Herbst DR (1990). Evolution of dioecy in *Schiedea* (Caryophyllaceae: Alsinoideae) in the Hawaiian Islands (USA): Biogeographical and ecological factors. *Systematic Botany*, **15**, 266–76.

Weller SG, Wagner WL, and Sakai AK (1995). A phylogenetic analysis of *Schiedea* and *Alsinidendron* (Caryophyllaceae: Alsinoideae): implications for the evolution of breeding systems. *Systematic Botany*, **20**, 315–37.

Weller SG, Sakai AK, Rankin AE, Golonka A, Kutcher B, and Ashby KE (1998). Dioecy and the evolution of pollination systems in *Schiedea* and *Alsinidendron* (Caryophyllaceae: Alsinideae) in the Hawaiian islands. *American Journal of Botany*, **85**, 1377–88.

Wolfe LM and Shmida A (1997). The ecology of sex expression in a gynodioecious Israeli desert shrub (*Orchradenus baccatus*). *Ecology*, **78**, 101–10.

Wolff K, Friso B, and van Damme JMM (1988). Outcrossing rates and male sterility in natural populations of *Plantago coronopus*. *Theoretical and Applied Genetics*, **76**, 190–6.

CHAPTER 12

Effects of colonization and metapopulation dynamics on the evolution of plant sexual systems

John R. Pannell

Department of Plant Sciences, University of Oxford, UK

Outline

The colonization of unoccupied habitat has unique effects on the evolution of plant sexual systems. Viewed broadly, colonization encompasses both single-event, long-distance dispersal and the recurrent colonizations by which species persist in a metapopulation. In both cases, colonization acts as a selective sieve, favouring some genotypes or phenotypes over others. In addition, colonization alters the balance between selection and drift, and disrupts the biotic interactions among individuals. However, the two scenarios differ in the opportunity for newly colonized populations to evolve in the absence of gene flow from the source population. I review current understanding of the evolution of sexual systems during and after long-distance dispersal, focusing on sex allocation and sexual-system polymorphisms maintained by frequency-dependent selection. Colonization shapes the phenotypic composition of populations and the genetic architecture underlying this variation in ways that differ from selection in large populations of constant size. I conclude with a metapopulation model that predicts the evolution of geitonogamy under circumstances that would prevent its evolution in the absence of colonization dynamics.

12.1 Introduction

At one time or another, all extant plant lineages are probably subject to evolutionary processes that accompany the colonization of new habitats. These processes can be grouped usefully into three categories. First, colonization favours individuals or species capable of successful establishment following long-distance dispersal (Baker 1955; Pannell and Barrett 1998); colonization therefore acts as an ecological sieve or selective agent. Second, colonization dramatically restricts a newly founded population's effective size, and thus increases the importance of drift over selection (Wright 1940; Whitlock and Barton 1997). Genetic bottlenecks also tend to convert dominance and epistatic variance into additive genetic variance, thereby creating new adaptive opportunities (Robertson 1952; Willis and Orr 1993; Naciri-Graven and Goudet 2003). Third, the evolution of a population following colonization may occur under very different selective regimes from that experienced in the source population (Darwin 1859; Stebbins 1950). For example, colonization frequently alters the context of selection on phenotypes in their new environment. This contrast between selective regimes is nowhere better illustrated than by the success of invasive species after their introduction to a new area, due, for example, to escape from specialist pests, predators, and competitors (Keane and Crawley 2002).

In this chapter, I consider these ideas in the context of the evolution of plant sexual systems. I begin by recognizing a distinction between, on the one hand, single-event colonization following long-distance dispersal (e.g., to oceanic islands) and, on the other hand, recurrent colonization in species with a metapopulation structure and dynamic. I then explore the effects of single-event and recurrent colonization in turn. First, I suggest that certain aspects of the distribution of sexual systems on islands might be understood by distinguishing between the effects of colonization as an ecological or biogeographical sieve in assembling particular traits on islands and the autochthonous evolution of sexual-system traits subsequent to establishment. I then consider the evolution of the sexual system in a metapopulation, focusing primarily on sexual-system polymorphisms maintained by frequency-dependent selection. Here, I discuss the effects of metapopulation dynamics both on sex-allocation strategies and on the dominance versus recessivity of traits affecting the sexual system. Finally, I adopt a metapopulation perspective to re-analyse selection on geitonogamy, a process likely to be uniformly disadvantageous in large demographically stable populations, but which can be selected during colonization.

12.2 Single-event versus recurrent colonization

Colonization broadly includes both single-event, long-distance dispersal (e.g., colonization of oceanic islands) and recurrent dispersal characteristic of ruderal lineages maintained in a matrix of periodically disturbed habitat. These scenarios are similar in important ways, but they differ in others. Whether colonization is a single or a recurrent event, it should always act as a sieve on traits that affect dispersal and establishment and it tends to increase the role played by genetic drift. In addition, colonization is very likely to alter the context of biotic and abiotic interactions, by transporting individuals to a new habitat, by changing the population density and the frequency of different life-history and sexual-system strategies encountered by individuals, or both. However, single-event and recurrent colonization differ in the extent to which long-distance dispersal isolates the colonized population from other populations of the species.

Colonization of an island as a single event isolates a lineage completely from its source and all other populations, whereas populations established by dispersal in a metapopulation may continue to be linked following colonization by migration. This difference has important implications. Both single-event and recurrent colonization will act as a sieve that favours propagules capable of establishment following long-distance dispersal (Baker 1955; Pannell and Barrett 1998). Similarly, both processes are also likely to displace the genetic architecture of colonized populations away from evolutionary equilibria established at their source. However, this potential displacement may be more temporary following recurrent than single-event colonization, because only in the former instance can subsequent migration of genotypes not present among the original colonizers allow selection to restore a population quickly to its evolutionary equilibrium. Restoration of new evolutionary equilibria might be much more protracted in island populations, with colonized populations evolving trait combinations not found in their source.

Colonization is obviously a fundamental process in the establishment of plants on islands. However, although metapopulation analysis has strongly influenced understanding of many animal systems (Hanski and Gaggiotti 2004), the concept remains controversial for plants, largely due to inherent difficulties in assessing habitat occupancy and extinctions (Freckleton and Watkinson 2003, and references cited therein). Nevertheless, it is clear that many ruderal plant species persist through a balance between local colonizations and extinctions. In addition, a large literature documents levels of genetic differentiation among populations that vary with life-history and sexual-system traits that influence dispersal (reviewed in Hamrick and Godt 1996; Charlesworth and Pannell 2001), and much of this variation is consistent with a history of colonization and limited migration. Several case studies are reviewed briefly in Section 12.4.

12.3 Effects of single-event colonization on the sexual system

12.3.1 Long-distance dispersal to oceanic islands

Sexual strategies are distributed unevenly among the world's floras. In particular, the floras of oceanic islands have more self-compatibility amongst hermaphrodites and a higher proportion of dioecious species than those on continents (reviewed in Barrett 1996). The low frequency of self-incompatibility on islands is consistent with Baker's law, namely that colonization should select for uniparental reproduction, including self-fertilization, because it offers reproductive assurance in the absence of mates (Baker 1955, 1967). In contrast, the high relative incidence of dioecy on islands would seem to contradict Baker's law.

Several explanations of this paradox have been suggested that invoke colonization as a kind of biogeographical or ecological sieve. First, the high frequency of dioecy on islands may to some extent represent its frequency in the particular (tropical) floras from which island floras have been assembled, and in which dioecy is also more common than it is globally (Baker and Cox 1984). Second, because dioecious species tend to have fleshy fruits with bird-dispersed seeds (Renner and Ricklefs 1995), they may be more likely to reach islands than non-dioecious species that are not dispersed by birds (Bawa 1980). Third, dioecious species, which tend to be pollinated by small generalist insects (Bawa and Opler 1975; Charlesworth 1993), may be pre-adapted to establish on islands, where generalist pollinators are also prevalent (reviewed in Barrett 1996). Fourth, because dioecious species tend to be long-lived perennials (Renner and Ricklefs 1995), or are dispersed in multi-seeded fruits, they may establish on islands either via an accumulation of successive dispersal events during the lifetime of single individuals or as several individuals together (Baker and Cox 1984). Finally, constraints imposed by Baker's law may be relaxed because dioecy is often "leaky," with males or females able to self-fertilize progeny by producing a few flowers of the opposite gender (Lloyd 1980; Baker and Cox 1984).

Another explanation for the high frequency of dioecy on islands accepts that colonization should favour self-compatible hermaphrodites, but invokes the autochthonous evolution of dioecy from hermaphroditism subsequent to establishment (Barrett 1996). First, following dispersal to an island, reliance on new generalist pollinators selects for floral structures associated with dioecy, such as small white or green flowers, or for wind pollination (Charlesworth 1993). Second, dioecy might evolve as an outcrossing mechanism following an earlier loss of self-incompatibility and/or following the evolution of reduced control over the position and movement of pollinators within and between flowers, which can increase selfing (Thomson and Barrett 1981). Thus, dioecy may be prevalent on islands either as a result of a biogeographical or ecological sieve during the assembly of an island flora via long-distance dispersal or through the evolution of dioecy *in situ* subsequent to colonization, or to both processes in concert (as has been found for the Hawaiian genus *Schiedea*: Sakai *et al.* 1995). The evolution of outcrossing versus selfing under the influence of colonization is explored in more detail in Section 12.6.

12.3.2 Long-distance dispersal and the evolution of dioecy in *Cotula*

Lloyd's (1975) analysis and interpretation of reversions to monoecy from dioecy in several *Cotula* species in New Zealand (syn., *Leptinella*, Asteraceae) provides an interesting illustration of how colonization as a single event can shape the evolution of plant sexual systems. Sex expression in *C. dioica*, *C. dispersa*, and *C. rotundata* varies considerably among populations in both the relative number of male and female florets per capitula and several secondary sexual characters, including floret size (Lloyd 1975). Typically, dioecy is more common in these species. However, Lloyd described several monomorphic populations, which he interpreted as descendants from allopatric dioecious populations of the same species, and his model for this evolution nicely illustrates how colonization can alter a population's gender distribution and thus the direction of selection on sex allocation (Fig. 12.1).

Three life-history and sex-expression traits observed in *Cotula* are fundamental to Lloyd's (1975) scheme. First, individuals reproduce clonally by rhizomatous growth, allowing single colonists to establish unisexual "populations," which may persist vegetatively. Second, females are heterogametic, so that populations established by a single female receive the unexpressed male-determining allele.

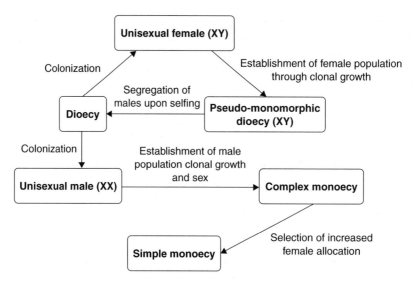

Figure 12.1 Lloyd's (1975) model for the evolution of monoecy from dioecy in New Zealand species of *Cotula* (syn: *Leptinella*). See text for a detailed explanation.

Third, as in many other dioecious species, both males and (less frequently) females produce the occasional flower of the opposite gender. Lloyd (1975) proposed that populations established by long-distance dispersal of a single female remained monomorphic only until such sex inconstancy allowed self-fertilization, yielding male and female progeny and initiating a normal dioecious population. He thus called these all-female populations "pseudo-monomorphic dioecious." In contrast to populations established by single females, selfing in populations colonized by males would yield only male progeny. Lloyd referred to these populations as "complex monoecy," because the male-biased sex allocation of constituent individuals does not reflect their cosexual functional gender (i.e., they are not at a sex-allocation equilibrium). Only subsequent evolution towards increased production of female florets can return these populations to an equilibrium; indeed, this appears to have occurred in several *Cotula* populations (Lloyd 1975).

12.4 Evolution of sexual systems in a metapopulation

Lloyd's (1975) scheme invokes long-distance dispersal that completely isolates the colonized population from all other populations of the species. In this sense, it resembles long-distance colonization of oceanic islands, and it differs from colonization in a metapopulation, in which populations may be linked by migration. The evolution of sexual systems and other life-history traits in a metapopulation is particularly interesting, because it brings into focus the relative importance of three processes: local evolution, which takes a population to a new equilibrium; migration, which restores an old one; and the stochastic ecological sieve imposed by colonization itself, which introduces a novel level of selection (Barrett and Pannell 1999; Ronce and Olivieri 2004; Pannell et al. 2005; Pannell and Dorken 2006).

These interactions are well illustrated in their effects on the selection of uniparental reproduction, which confers reproductive assurance during colonization (Lloyd 1979; Pannell 1997a; Pannell and Barrett 1998; see also Chapter 10). The evolution of self-fertilization in colonizing lineages is akin to Baker's law, described above (Stebbins 1957; Baker 1967; Cox 1989). Pannell and Barrett (1998) modelled the outcome of selection on selfers versus outcrossers in a metapopulation and concluded that the intensity of selection for an ability to self-fertilize increases dramatically with increasing rates of population turnover and decreasing average numbers of colonists, particularly when little available habitat is occupied. This conclusion is consistent with the observation, for example, that self-fertilization is often most common in species polymorphic for the mating system at the margins of their geographical range (reviewed in Pannell and Barrett 1998).

Selection in a metapopulation should also alter the sex allocation of individuals and populations. During colonization and population establishment, mating must occur either through self-fertilization or between close relatives. Selection under these circumstances favours female-biased sex allocation (Hamilton 1967; Frank 1998). Interestingly, although colonizing female animals commonly bias the sex ratio of their progeny towards daughters and against sons (Hamilton 1967), no similar evidence has been found in dioecious plants. This difference probably has three causes: colonizing dioecious plants are rare; female plants have limited information on the likelihood of consanguineous mating among their progeny; and enhanced metapopulation dynamics is more likely to select for self-fertile hermaphroditism. By contrast, colonizing hermaphrodites that self-fertilize their progeny are more likely to respond to selection for female-biased sex allocation (Pannell 2001). Indeed, populations of the European plant *Mercurialis annua* in regions dominated by monoecy have strongly female-biased sex allocation, probably in response to selection during bouts of colonization (Pannell 1997b).

Theoretical analysis indicates that the selection of sexual systems varies with population turnover in metapopulations (Pannell 1997a, 2001). With frequent turnover, hermaphroditism replaces dioecy quickly, whereas with very low rates of extinction and recolonization, or when gene flow among populations outweighs the effects of colonization, dioecy may be stable (J. R. Pannell unpublished data). Finally, at intermediate rates of population turnover, females or males can be maintained indefinitely with hermaphrodites through a balance between selection for separate sexes in local populations (e.g., due to negative frequency dependence and the advantages of sexual specialization), and selection for self-fertile hermaphroditism at the metapopulation level (through selection for reproductive assurance during colonization). In this case, the frequency of males or females maintained in the metapopulation depends on both the rate of population turnover and the rate of gene flow among population (Fig. 12.2), in addition to factors acting locally, such as relative pollen or seed productivities, selfing rates, and inbreeding depression (Charlesworth and Charlesworth 1978).

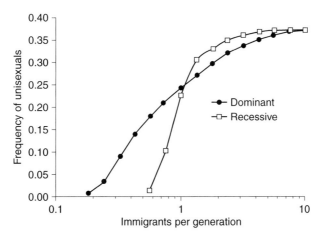

Figure 12.2 The effect of mode of expression (dominant versus recessive) of an allele for maleness or femaleness on the relation of the proportion of males or females (unisexuals) maintained in a metapopulation with hermaphrodites to the average number of individuals immigrating into local populations per generation. The extinction rate per deme is 0.05 and number of colonists follows a Poisson distribution, with the mean plotted on the *x*-axis (modified after Pannell 1997a). Note that unisexual individuals can be maintained over a larger range of immigration rates when they are determined by a dominant allele, but that the equilibrium frequency of unisexuals is higher under recessive sex determination at relatively high immigration rates.

In *Mercurialis annua*, variation in habitat occupancy and population size in areas with different sexual systems and patterns of genetic diversity support the metapopulation hypothesis. Populations in areas with monoecious populations, where males and females are entirely absent, have patterns of occupancy and abundance consistent with high rates of population turnover, whereas unisexuals are present where turnover rates are likely to be low (Eppley and Pannell in press). Monoecious populations also possess significantly less neutral genetic diversity and are substantially more differentiated from each other than dimorphic populations (Obbard *et al.* 2006; Fig. 12.3), as predicted by population genetic theory for populations subject to high turnover (reviewed in Pannell and Charlesworth 2000). In *M. annua*, and several other animal and plant species (reviewed in Pannell 2002), androdioecy appears to have evolved from dioecy following selection of hermaphroditism for reproductive assurance. Pannell's (2001) model for the breakdown of dioecy in a metapopulation suggests that androdioecy, rather than gynodioecy, may be the intermediate state, because females producing a small amount of pollen would colonize better than males producing a few seeds (Fig. 12.4).

Population structure and turnover are probably also responsible for the high variance in sex ratios observed in several gynodioecious species for which sex allocation is determined by interactions between maternally inherited, cytoplasmic genes and biparentally inherited, nuclear genes (Lewis 1941; Charlesworth and Ganders 1979; Jacobs and Wade 2003). First, colonization creates variance in the proportion of hermaphrodites and females among populations (McCauley and Taylor 1997; Olson *et al.* 2005), so that females in female-biased demes suffer reduced seed set (Fig. 12.5). This effect, which has been observed in gynodioecious *Sidalcea malviflora* (Graff 1999) and *Silene vulgaris* (McCauley *et al.* 2000), may slow the spread of cytoplasmic male-sterility elements (McCauley and Taylor 1997). Second, variation in the proportion of females amongst populations following colonization results from a mismatch between cytoplasmically inherited male-sterility genes and nuclear genes that restore male fertility (Frank 1989). This scenario is probably quite general and has been invoked to explain sex-ratio variation in *Thymus vulgaris* (Belhassen *et al.* 1989) and *Plantago lanceolata* (van Damme 1986; Frank and Barr 2001). Couvet *et al.* (1998) have shown that such sex-ratio variation can give rise to selection at the

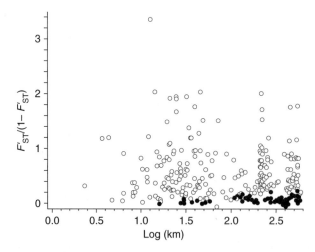

Figure 12.3 Relation of genetic differentiation at isozyme loci ($F'_{ST}/[1 - F'_{ST}]$) between pairs of populations of *Mercurialis annua* in the Iberian Peninsula to their geographic separation (ln-transformed). Differentiation between androdioecious populations (solid circles) was significantly lower than that between monoecious populations (open symbols), irrespective of the distance separating them. The data support the hypothesis that monoecious populations were colonized more recently than androdioecious populations, between which ongoing migration has eroded any genetic differentiation caused during colonization. From Obbard *et al.* (2006).

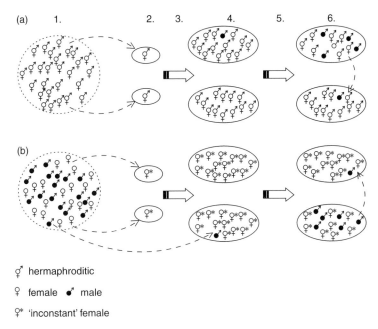

♂ hermaphroditic
♀ female   ♂ male
♀* 'inconstant' female

Figure 12.4 Scenario for the evolution of androdioecy in a metapopulation from (a) a hermaphroditic (meta)population, and (b) a dioecious (meta)population. (1) The ancestral state, which may be either a large population or a metapopulation consisting of hermaphrodites (a) or males and females (b). (2) Colonization, corresponding to a high selfing rate: self-fertile, female-biased hermaphrodites are selected in (a); whereas self-fertile "inconstant" females that produce occasional hermaphroditic flowers are selected in (b). (3) Population growth, which coincides with decreasing selfing rates. (4) Male invasion by mutation (a), or immigration or mutation (b). (5) Local spread of males. (6) Males migrate throughout the metapopulation. From Pannell (2001)

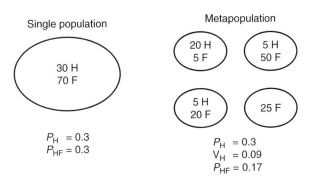

Figure 12.5 Schematic representation of a single panmictic population and a subdivided metapopulation of hermaphroditic (H) and female (F) individuals. The overall frequency of hermaphrodites (P_H) is 0.3 in both cases, but the subjective frequency of hermaphrodites from the perspective of females (P_{HF}) is lower in the structured metapopulation. Colonization bottlenecks strongly influence the variance in the frequency of hermaphrodites among demes (V_H), which set the context for subsequent mating and selection. After McCauley and Taylor (1997).

metapopulation level, such that female-biased patches have a colonization advantage if females produce more seeds than hermaphrodites. Pannell and Barrett (2001) pointed out a similar effect of variance among populations in seed production.

12.5 Evolution of dominant versus recessive traits in a metapopulation

Although much can be learned about the evolution of sexual-system polymorphisms, a full understanding may often require knowledge of the underlying genetic architecture. Such genetic knowledge is particularly important for explaining the maintenance of sexual polymorphisms in subdivided populations or metapopulations, where phenotypic variation within local demes is subject to fluctuating interactions between frequency-dependent selection and genetic drift (Pannell et al. 2005). In this case, local morph frequencies can vary unpredictably among recently colonized populations, whereas at the metapopulation level they may deviate predictably from panmictic expectations in ways that depend on the dominance or recessivity of gene expression (Pannell et al. 2005). Two processes have been invoked to explain such deviations.

First, genetic drift in small local populations or demes favours fixation of the morph determined by a recessive allele at the expense of the morph determined by a dominant allele (Barrett et al. 1989; see also Pannell and Barrett 2001). For example, consider an outcrossing gynodioecious population with nuclear sex determination, and suppose that the equilibrium frequency of females is x. If the female-determining allele is dominant, its equilibrium frequency is $x/2$, whereas if it is recessive, its equilibrium frequency is $1-(x/2)$, which exceeds $x/2$ for all $x < 1$. Being less common, the dominant allele is more susceptible to loss by genetic drift. Thus, more small populations in a polymorphic metapopulation should be fixed for the recessive allele than for the dominant allele. This local fixation bias will contribute to a global deficit of the dominant allele and its associated morph (Barrett et al. 1989; Pannell et al. 2005).

Second, the biased loss of dominant over recessive alleles might be countered by the repeated action of "Haldane's sieve" in a metapopulation (Pannell et al. 2005). Haldane (1924) noted that a new advantageous allele is more likely to be fixed in a population if it is not completely recessive, because recessive alleles are expressed (and thus selected) only in homozygotes, which are rare initially. In contrast, because dominant alleles are expressed in heterozygotes, they respond to selection immediately, and should quickly become common. Haldane's sieve is usually considered in the context of new mutations and has been invoked to account for the relative importance of dominant alleles in adaptive evolution (e.g., Turner 1977; Charlesworth 1998). From the perspective of the evolution of sexual systems, Haldane's sieve explains why the short-styled morph in distylous species is typically determined by a dominant, rather than a recessive allele, if distyly evolves by the spread of a short-styled variant, as seems likely (e.g., Lloyd and Webb 1992; Chapter 13). However, the greater prospect of dominant alleles spreading when rare also applies in the context of gene flow among populations. Thus, in a subdivided population or metapopulation, Haldane's sieve should operate repeatedly on immigrants to local demes, such that dominant alleles spread towards their local equilibria more quickly than recessive alleles (Pannell et al. 2005).

Interactions between negative frequency-dependent selection, genetic drift and gene expression have been invoked to explain morph frequencies predicted by computer simulations of three reproductive polymorphisms. First, stochastic simulations of single tristylous populations predicted the biased loss of the short-styled morph in small populations, due to loss of the dominant allele (and fixation of the recessive allele) at the *S-locus* (Barrett et al. 1989). These results are broadly consistent with observations of the distribution of tristylous morph frequencies in the aquatic annual plant *Eichhornia paniculata* (Barrett et al. 1989; Pannell et al. 2005). Second, in simulations of androdioecious and gynodioecious metapopulations, Pannell (1997a) found that the frequencies of males and females at equilibrium depended on whether the respective sterility mutations were recessive or dominant (Fig. 12.2). Finally, simulations of a subdivided population in the absence of extinction and recolonization indicated that the relative frequencies of sporophytic self-incompatibility alleles depended on their positions in the dominance hierarchy (Schierup et al. 2000). Only this last study accounted for the effects of both local genetic drift and the repeated action of Haldane's sieve on migrants; it showed that the interactions between these two processes in a subdivided population are complex, and suggested that more detailed analysis of other floral polymorphisms is likely to be revealing (Pannell et al. 2005).

12.6 Modes of selfing and the evolution of geitonogamy

Models for the evolution of self-fertilization indicate that selfing variants should be selected over outcrossers as long as selfed progeny are at least half as fit as outcrossed progeny (Lande and Schemske 1985; Charlesworth and Charlesworth 1987; Lloyd 1988). This outcome assumes that seed set is not pollen-limited and requires that self-pollination does not reduce outcrossed siring by the selfing variant compared with outcrossing parents (no "pollen discounting": reviewed in Harder and Wilson 1998). When pollination is limited by the availability of mates or pollinators, self-fertilization may evolve even in the presence of high inbreeding depression, as long as the selfed ovules would not otherwise have been outcrossed (no "seed discounting": Lloyd 1992; Herlihy and Eckert 2002).

The relative importance of both pollen and seed discounting depends largely on the mode of self-fertilization (Lloyd 1992; also see Chapters 2 and 4). Delayed selfing, which occurs without the need of pollinator service, should confer the greatest advantage of reproductive assurance (Chapter 10), particularly in monocarpic species in which future reproductive resources are not wasted through selfing. In contrast, geitonogamy cannot evolve in populations at demographic equilibrium, except as the indirect outcome of selection to increase outcrossing, because geitonogamy involves complete pollen and seed discounting. Geitonogamy also requires the action of a pollination vector and so offers no reproductive assurance in the absence of pollinators (Lloyd 1992).

Importantly, reproductive assurance can be selected both in the absence of pollinators and in the absence of mates, whether or not pollinators abound. The implications of mate limitation have been under-appreciated in discussions of the evolution of geitonogamy, perhaps because models have been framed in terms of populations with large mating neighbourhoods. For example, Lloyd's (1992) insightful analysis implicitly assumed an infinite number of mates. In contrast, in small populations mate availability can severely constrain reproduction, particularly during colonization by one or a few individuals. Under such conditions we might expect even geitonogamy to be advantageous, because pollen discounting cannot occur when there is no prospect for siring outcrossed progeny. However, if an individual's ability to self-fertilize depends on its floral phenology and/or aspects of its inflorescence architecture and display, then an ability to self in the absence of mates will also increase its selfing rate in the presence of mates. Examples of this scenario include the extent to which individuals of an enantiostylous species produce both left-handed and right-handed flowers in the same inflorescence, or the extent to which monoecious individuals produce staminate and pistillate flowers simultaneously. Intuitively, the advantage of geitonogamy in the absence of mates during colonization must be counteracted by the costs of seed and pollen discounting when mates are abundant.

Pannell and Barrett (1998) analysed the relative general benefits of selfing versus outcrossing in a metapopulation, but they did not consider mating between selfers and outcrossers, nor the implication of specific modes of selfing, so several interesting questions remain. First, under what conditions can a genetic variant invade and spread in a self-compatible population comprising individuals with floral traits that limit the possibility of geitonogamy? Second, having invaded a population, should such a variant then spread to fixation, completely replacing the outcrosser, or be maintained at an intermediate frequency in a polymorphic population? This second question impinges on the more general question concerning the maintenance of selfing–outcrossing polymorphisms, which models have shown to be difficult (Pannell and Barrett 2001). Third, when a geitonogamy–outcrossing polymorphism is maintained in a metapopulation, to what extent does the dominance or recessivity of the variant for increased geitonogamy affect the equilibrium frequencies? As reviewed above, we expect the balance between local genetic drift and the effect of Haldane's sieve to cause phenotypes governed by dominant alleles to settle at different frequencies from those governed by recessive ones. Below, I address these questions using computer simulations of a metapopulation that incorporate rates of extinction, the numbers of colonists and immigrants into extant populations, population growth rates, levels of inbreeding depression, and the dominance of a trait that increases self-fertilization.

12.6.1 Model of geitonogamy in a metapopulation

Consider a metapopulation of D populations, in which extant populations become extinct with probability E and vacant patches are colonized by c individuals drawn randomly from the metapopulation. Further assume that an average of I individuals migrate into extant populations each generation. Populations grow exponentially to a carrying capacity K, with the growth rate depending on the numbers of viable progeny produced per individual in the population. Suppose that each individual produces g ovules and p pollen grains and that mating in populations containing more than one individual involves either outcrossing or selfing. Assume that the metapopulation comprises two phenotypes: an obligate outcrosser, which possesses a morphology or phenology that prevents geitonogamy, and a facultative mixed mater, which experiences geitonogamy, resulting in self-fertilization of some of its ovules. Sufficient pollinators visit this phenotype that it always self-fertilizes a proportion S of its ovules in the absence of mates (with proportion $1-S$ left unfertilized). The same phenotype can outcross in the presence of mates; however, in this case morphology and/or phenology causes self-fertilization of a fraction $S\alpha$ of its ovules, so that only a fraction $1-S\alpha$ are outcrossed. Thus, in the presence of mates, all ovules of both phenotypes are fertilized, whereas in the absence of mates only the phenotype capable of geitonogamy produces offspring.

Self-fertilization also reduces the amount of pollen contributed by each plant to the outcrossing pollen pool, so that selfing plants export $p(1-Sb)$ pollen grains in the presence of mates. For geitonogamous selfing, we suppose that $b=1$. Finally, a fraction d of selfed progeny suffer from inbreeding depression and are non-viable. For simplicity, I assume that d is fixed, although this is clearly unrealistic for two reasons: the genetic load of populations evolves with the mating system (Lande and Schemske 1985; Uyenoyama and Waller 1991; Lande et al. 1994), and "outcrossing" may involve different frequencies of biparental inbreeding in populations of different sizes and with different colonization histories. The implications of this latter assumption for the maintenance of different sexual systems are not well understood, and further theoretical work is needed (e.g., see Ronfort and Couvet 1995; Roze and Rousset 2004). Nevertheless, despite these uncertainties, a model with fixed d illustrates important general principles.

Extinction occurs stochastically after mating and seed production, and empty sites are re-colonized immediately by seeds sampled randomly from the metapopulation. Density-dependent competition regulates deme sizes prior to mating during the next generation, with truncation of the population size to K if $N > K$. In polymorphic demes, the numbers of selfing and outcrossing individuals are reduced in proportion to their local frequencies. Because dominance can affect a trait's fate in a metapopulation (Section 12.5), we consider two scenarios: when a dominant allele confers an ability for geitonogamy, only homozygotes for the alternative allele retain mechanisms to prevent geitonogamy; whereas when the allele for geitonogamy is recessive, heterozygotes outcross obligately.

I explored the behaviour of the simulation model under a range of parameter values. Simulations began with the allele for increased selfing at a frequency of 0.01 in all populations, and ran until either one allele spread to fixation or a mixed-mating equilibrium had been reached. Mixed-mating equilibria were checked by determining whether the same end point was reached in simulations that began with the allele for obligate outcrossing at a frequency of 0.01. Allele frequencies were recorded as means over 1000 successive generations of the metapopulation at equilibrium and were checked by comparing with means calculated during the next 1000 generations. For all results reported, these two measures did not differ by more than 0.01.

12.6.2 Model results and discussion

As with classic mating-system models (Lande and Schemske 1985; Charlesworth and Charlesworth

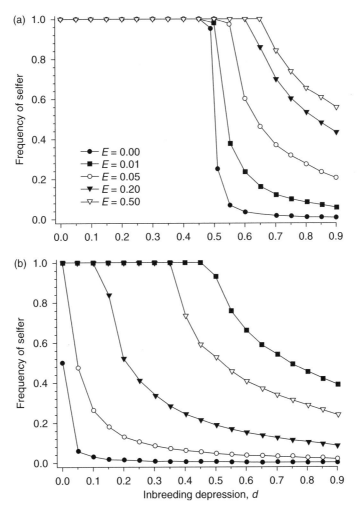

Figure 12.6 The relation of the equilibrium frequency of a selfing phenotype in a metapopulation to the severity of inbreeding depression, d, with (a) no ($b=0$) or (b) complete pollen discounting ($b=1$). Curves illustrate a range of extinction rates, E (indicated). All simulations modelled a dominant allele for selfing, with $S=0.5$, $\alpha=1.0$, $K=1000$, $g=10$, $l=1.0$, and $c=1$.

1987; Lloyd 1988), increased selfing was always advantageous for mild inbreeding depression ($d < 0.5$) in the absence of pollen discounting, regardless of the rate of population turnover (Fig. 12.6a). Also like classic models, alleles for increased selfing were lost with severe inbreeding depression ($d > 0.5$) and no population extinction ($E=0$). However, for $E > 0$, the allele for increased selfing persisted at an intermediate frequency, so that a mixed-mating polymorphism was maintained (Fig. 12.6a).

As Lloyd (1992) reported, in the absence of extinction ($E=0$) geitonogamous selfing was dis-advantageous with complete pollen discounting ($b=1.0$) if inbreeding depression $d > 0$ (Fig. 12.6b). Note that the frequency of selfers is not exactly 0 for $d > 0$ when $E=0$ in Fig. 12.6b. This effect of mating in populations of finite size arises because outcrossing individuals are assumed not to contribute to the pollen on their own stigmas. The pollen of mixed maters thus has access to the ovules of N individuals, whereas that of outcrossers has access to $N-1$ individuals. With increasing population size, this difference becomes negligible. In a metapopulation with population turnover ($E > 0$)

some geitonogamous selfing was selected, even for $d > 0$. Indeed, with sufficiently high E and low d, the allele for geitonogamous selfing spread to fixation. For intermediate E and d, a mixed-mating polymorphism was maintained in the metapopulation (Fig. 12.6b). Thus, metapopulation dynamics allowed the selection of geitonogamy when it would otherwise be selected against in a demographically stable population.

The fate of the selfing morph in a metapopulation depended strongly on the extent to which its selfing rate was elevated above that of the outcrosser. For low to intermediate extinction probabilities, the selfing morph was maintained at a higher frequency if its selfing rate was relatively low (Fig. 12.7a and b). This negative relation reflects a compromise between the benefits of reproductive assurance during colonization and the genetic and mating costs in established populations. In contrast, for high extinction rates the frequency of the selfing morph increased with S (i.e., the trend was reversed; Fig. 12.7a and b), because the advantage of reproductive assurance during colonization outweighed the costs of pollen discounting and inbreeding depression in short-lived, established populations.

As expected, the balance between the costs of geitonogamy and the benefits of reproductive assurance, and thus the threshold selfing rate, depended on the number of colonizing propagules; with $c > 1$, reproductive assurance was less advantageous (compare Fig. 12.7a with 12.7b, and Fig. 12.7c with 12.7d: note that with $c = 2$, for

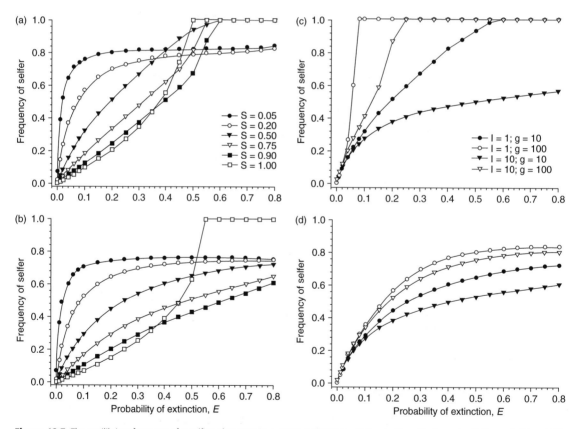

Figure 12.7 The equilibrium frequency of a selfing phenotype in a metapopulation in relation to the extinction rate, E. Curves in (a) and (b) illustrate a range of selfing rates expressed by the selfing phenotype (indicated). Curves in (c) and (d) correspond to metapopulations in which the average number of immigrants per population is $I = 1$ (circles) or $I = 10$ (triangles), and plants produce $g = 10$ (closed symbols) or $g = 100$ (open symbols) ovules. The number of colonists, c, is 1 in panels (a) and (c) and 2 in panels (b) and (d). All simulations modelled a dominant allele for selfing, with $\alpha = 1.0$, $d = 0.5$, and $K = 1000$. In panels (a) and (b), $g = 10$ and $I = 1.0$. In panels (c) and (d), $S = 0.5$.

example, two individuals capable of geitonogamy may successfully establish a new population, but two outcrossers will fail). Similarly, high ovule production, g, which determines the rate of population growth, increased the advantage of selfers during colonization. This advantage apparently outweighed the potential disadvantage of rapid growth to a population size at which selfing ability is no longer a benefit (Fig. 12.7c and d). In contrast, geitonogamous selfing became an increasing disadvantage with increases in I, because alleles that prevent geitonogamy, and which are beneficial in established populations, can invade rapidly as migrants (Fig. 12.7c and d).

Finally, the outcome of selection on geitonogamy in the simulated metapopulation depended on whether the allele for increased selfing was dominant or recessive (compare Fig. 12.8a and 12.8b). Elevated migration among extant demes reduced the equilibrium frequency of the selfer, but for each migration rate investigated, the selfing morph was typically maintained at a higher frequency, or was fixed at a lower extinction rate, when it was controlled by a dominant rather than a recessive allele (Fig. 12.8). However, note that this pattern was reversed for very low migration rates. For example, with $I = 0.01$ and $I = 0.05$ in Fig. 12.8, the selfer was maintained at the higher frequency when it was determined by a recessive allele. Apparently, a dominant allele for selfing ability can be maintained at a higher frequency than a recessive allele in a metapopulation, because both homozygotes and

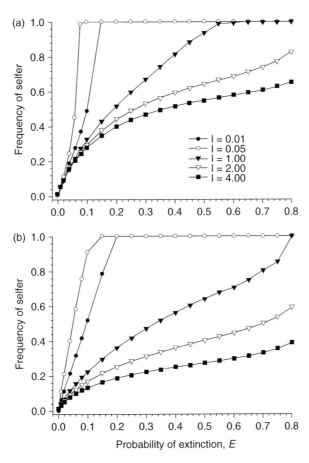

Figure 12.8 The equilibrium frequency of a selfing phenotype in a metapopulation in relation to the extinction rate, E, when selfing is controlled by (a) a dominant allele or (b) a recessive allele. Curves illustrate a range of immigration rates (indicated). In all simulations, $S = 0.5$, $\alpha = 1.0$, $d = 0.5$, $K = 1000$, and $c = 1$.

heterozygotes benefit from reproductive assurance under dominant expression. This outcome differs from the examples of Haldane's sieve described in Section 12.5, where negative frequency-dependent selection tends to maintain a polymorphism in established populations. In this case, the selfing phenotype is uniformly selected against when mates are available, because of the combined effects of pollen discounting and inbreeding depression. The results are thus complex and deserve more detailed analysis. However, they clearly indicate that the selection of phenotypes in a metapopulation depends substantially on the patterns of expression of the underlying genes.

12.7 Conclusions

In this chapter, I have interpreted the effects of colonization on plant sexual systems in three ways. First, colonization acts as a sieve that favours certain phenotypes over others in terms of their ability to disperse and establish new populations. Second, the genetic bottlenecks associated with colonization alter the balance between drift and selection. Third, colonization alters the mating context of individuals, thereby affecting the strength and direction of selection on sexual-system traits. The metapopulation framework has proven useful in modelling these effects. However, the structure and dynamics of putative plant metapopulations remain poorly understood, and this subject deserves further attention. In particular, how frequently do plant populations turn over in ways that meet the assumptions made by the simple models, and how might more complex population structures and dynamics affect the evolutionary outcomes? This question should apply especially to polymorphic traits, not least because the mechanisms for their maintenance (i.e., negative frequency-dependent selection) are well understood and provide quantitative predictions that are readily testable. There is a need for further work on the genetic architecture governing sexual-system traits and polymorphisms, particularly the among-population variance in the frequency of genes with different modes of expression. Can such patterns be explained adequately by the effects of selection and drift in local populations, as appears to be the case in *Eichhornia*

paniculata (Barrett *et al.* 1989), or does Haldane's sieve also need to be invoked (Pannell *et al.* 2005)?

Single-event and recurrent colonization affect sexual-system evolution in both similar and different ways. Essentially, whether their effects are comparable or different depends largely on the extent to which colonization isolates the colonizing lineage from the influence of subsequent migrants. The most interesting, but also the most complex, situations are likely to be those in which migrants play a role that is neither negligible nor overwhelming. Such cases should exhibit evidence for both the effect of the selective sieve imposed by colonization and the subsequent evolution of lineages in (partial) isolation from one another. The populations of dynamic metapopulations within a region are likely to show the former, but perhaps not the latter. In contrast, related species evolving in allopatry may show independent evolution, but signatures of the selective sieve will quickly become obscured. The most fertile ground for future empirical investigation might therefore be found in comparative studies between metapopulations (or regions) that differ in their rates of turnover and migration, rather than between populations, or species. Much can be learnt about sexual-system evolution from studies conducted at appropriately large geographic scales (reviewed in Barrett *et al.* 2001).

Acknowledgements

I thank the editors, Lawrence Harder and Spencer Barrett, for inviting me to contribute to this volume. The editors and two anonymous referees made valuable comments on an early version of the manuscript. I acknowledge the NERC, UK, for financial support.

References

Baker HG (1955). Self-compatibility and establishment after "long-distance" dispersal. *Evolution*, **9**, 347–8.
Baker HG (1967). Support for Baker's Law—as a rule. *Evolution*, **21**, 853–6.
Baker HG and Cox PA (1984). Further thoughts on dioecism and islands. *Annals of the Missouri Botanical Garden*, **71**, 244–53.

Barrett SCH (1996). The reproductive biology and genetics of island plants. *Philosophical Transactions of the Royal Society of London, Series B*, **351**, 725–33.

Barrett SCH and Pannell JR (1999). Metapopulation dynamics and mating-system evolution in plants. In P Hollingsworth, R Bateman, and R Gornall, eds. *Molecular systematics and plant evolution*, pp. 74–100. Chapman and Hall, London.

Barrett SCH, Morgan MT, and Husband BC (1989). The dissolution of a complex genetic polymorphism: the evolution of self-fertilization in tristylous *Eichhornia paniculata* (Pontederiaceae). *Evolution*, **43**, 1398–416.

Barrett SCH, Dorken ME, and Case AL (2001). A geographical context for the evolution of plant reproductive systems. In J Silvertown and J Antonovics, eds. *Integrating ecology and evolution in a spatial context*, pp. 341–64. Blackwell, Oxford.

Bawa KS (1980). Evolution of dioecy in flowering plants. *Annual Review of Ecology and Systematics*, **11**, 15–39.

Bawa KS and Opler P (1975). Dioecism in tropical trees. *Evolution*, **29**, 167–79.

Belhassen E, Traboud L, Couvet D, and Gouyon PH (1989). An example of nonequilibrium processes: gynodioecy of *Thymus vulgaris* L. in burned habitats. *Evolution*, **43**, 622–7.

Charlesworth B (1998). Adaptive evolution: the struggle for dominance. *Current Biology*, **8**, R502–4.

Charlesworth D (1993). Why are unisexual flowers associated with wind-pollination and unspecialised pollinators? *American Naturalist*, **141**, 481–90.

Charlesworth D and Charlesworth B (1978). A model for the evolution of dioecy and gynodioecy. *American Naturalist*, **112**, 975–97.

Charlesworth D and Charlesworth B (1987). Inbreeding depression and its evolutionary consequences. *Annual Review of Ecology and Systematics*, **18**, 273–88.

Charlesworth D and Ganders FR (1979). The population genetics of gynodioecy with cytoplasmic-genic male-sterility. *Heredity*, **43**, 213–8.

Charlesworth D and Pannell JR (2001). Mating systems and population genetic structure in the light of coalescent theory. In J Silvertown and J Antonovics, eds. *Integrating ecology and evolution in a spatial context*, pp. 73–95. Blackwell, Oxford.

Couvet D, Ronce O, and Gliddon C (1998). The maintenance of nucleocytoplasmic polymorphism in a metapopulation: The case of gynodioecy. *American Naturalist*, **152**, 59–70.

Cox PA (1989). Baker's Law: plant breeding systems and island colonization. In YB Linhart and JH Bock, eds. *The evolutionary ecology of plants*, pp. 209–24. Westview Press, Boulder, Colorado.

Darwin CR (1859). *The origin of species*. Murray, London.

Eppley SM and Pannell JR (in press). Sexual systems and measures of occupancy and abundance in an annual plant: testing the metapopulation model. *American Naturalist*.

Frank SA (1989). The evolutionary dynamics of cytoplasmic male sterility. *American Naturalist*, **133**, 345–76.

Frank SA (1998) *Foundations of social evolution*. Princeton University Press, Princeton, New Jersey.

Frank SA and Barr CM (2001). Spatial dynamics of cytoplasmic male sterility. In J Silvertown and J Antonovics, eds. *Integrating ecology and evolution in a spatial context*, pp. 219–43. Blackwell, Oxford.

Freckleton RP and Watkinson AR (2003). Are all plant populations metapopulations? *Journal of Ecology*, **91**, 321–4.

Graff A (1999). Population sex structure and reproductive fitness in gynodioecious *Sidalcea malviflora malviflora* (Malvaceae). *Evolution*, **53**, 1714–22.

Haldane JBS (1924). A mathematical theory of natural and artificial selection, part I. *Proceedings of the Cambridge Philosophical Society*, **23**, 19–41.

Hamilton WD (1967). Extraordinary sex ratios. *Science*, **156**, 477–88.

Hamrick JL and Godt MJW (1996). Effects of life history traits on genetic diversity in plant species. *Philosophical Transactions of the Royal Society of London, Series B*, **351**, 1291–8.

Hanski IA and Gaggiotti OE, eds (2004). *Ecology, genetics and evolution of metapopulations*. Elsevier, San Diego.

Harder LD and Wilson WG (1998). A clarification of pollen discounting and its joint effects with inbreeding depression on mating system evolution. *American Naturalist*, **152**, 684–95.

Herlihy CR and Eckert CG (2002). Genetic cost of reproductive assurance in a self-fertilizing plant. *Nature*, **416**, 320–3.

Jacobs MS and Wade MJ (2003). A synthetic review of the theory of gynodioecy. *American Naturalist*, **161**, 837–51.

Keane RM and Crawley MJ (2002). Exotic plant invasions and the enemy release hypothesis. *Trends in Ecology and Evolution*, **17**, 164–70.

Lande R and Schemske DW (1985). The evolution of self-fertilization and inbreeding depression in plants: I. Genetic models. *Evolution*, **39**, 24–40.

Lande R, Schemske DW, and Schultz ST (1994). High inbreeding depression, selective interference among loci, and the threshold selfing rate for purging recessive lethal mutations. *Evolution*, **48**, 965–78.

Lewis D (1941). Male sterility in natural populations of hermaphrodite plants. *New Phytologist*, **40**, 56–63.

Lloyd DG (1975). Breeding systems in *Cotula* IV. Reversion from dioecy to monoecy. *New Phytologist*, **74**, 125–45.

Lloyd DG (1979). Some reproductive factors affecting the selection of self-fertilization in plants. *American Naturalist*, **113**, 67–79.

Lloyd DG (1980). The distribution of gender in four angiosperm species illustrating two evolutionary pathways to dioecy. *Evolution*, **34**, 123–34.

Lloyd DG (1988). Benefits and costs of biparental and uniparental reproduction in plants. In BR Levin and RE Michod, eds. *The evolution of sex: an examination of current ideas*, pp. 233–52. Sinauer Associates, Sunderland, MA.

Lloyd DG (1992). Self- and cross-fertilization in plants. II. The selection of self-fertilization. *International Journal of Plant Science*, **153**, 370–80.

Lloyd DG and Webb CJ (1992). The evolution of heterostyly. In SCH Barrett, ed. *Evolution and function of heterostyly*, pp. 151–78. Springer Verlag, Berlin.

McCauley DE and Taylor DR (1997). Local population structure and sex ratio: evolution in gynodioecious plants. *American Naturalist*, **150**, 406–19.

McCauley DE, Olson MS, Emery SN, and Taylor DR (2000). Sex ratio variation in a gynodioecious plant: spatial scale and fitness consequences. *American Naturalist*, **155**, 814–9.

Naciri-Graven Y and Goudet J (2003). The additive genetic variance after bottlenecks is affected by the number of loci involved in epistatic interactions. *Evolution*, **57**, 706–16.

Obbard DJ, Harris SA, and Pannell JR (2006). Sexual systems and population genetic structure in an annual plant: testing the metapopulation model. *American Naturalist*, **167**, 354–66.

Olson MS, McCauley DE, and Taylor D (2005). Genetics and adaptation in structured populations: sex ratio evolution in *Silene vulgaris*. *Genetica*, **123**, 49–62.

Pannell JR (1997a). The maintenance of gynodioecy and androdioecy in a metapopulation. *Evolution*, **51**, 10–20.

Pannell JR (1997b). Variation in sex ratios and sex allocation in androdioecious *Mercurialis annua*. *Journal of Ecology*, **85**, 57–69.

Pannell JR (2001). A hypothesis for the evolution of androdioecy: the joint influence of reproductive assurance and local mate competition in a metapopulation. *Evolutionary Ecology*, **14**, 195–211.

Pannell JR (2002). The evolution and maintenance of androdioecy. *Annual Review of Ecology and Systematics*, **33**, 397–425.

Pannell JR and Barrett SCH (1998). Baker's Law revisited: reproductive assurance in a metapopulation. *Evolution*, **52**, 657–68.

Pannell JR and Barrett SCH (2001). Effects of population size and metapopulation dynamics on a mating system polymorphism. *Theoretical Population Biology*, **59**, 145–55.

Pannell JR and Charlesworth B (2000). Effects of metapopulation processes on measures of genetic diversity. *Philosophical Transactions of the Royal Society of London, Series B*, **355**, 1851–64.

Pannell JR and Dorken ME (2006). Colonisation as a common denominator in plant metapopulations and range expansions: effects on genetic diversity and sexual systems. *Landscape Ecology*.

Pannell JR, Dorken ME, and Eppley SM (2005). Haldane's Sieve in a metapopulation: sifting through plant reproductive polymorphisms. *Trends in Ecology and Evolution*, **20**, 374–9.

Renner SS and Ricklefs RE (1995). Dioecy and its correlates in the flowering plants. *American Journal of Botany*, **82**, 596–606.

Robertson A (1952). The effect of inbreeding on the variation due to recessive genes. *Genetics*, **37**, 189–207.

Ronce O and Olivieri I (2004). Life-history evolution in metapopulations. In: Hanski I and Gaggiotti OE, eds. *Metapopulation biology*, pp. 227–57. Academic Press, San Diego.

Ronfort J and Couvet D (1995). A stochastic model of selection on selfing rates in structured populations. *Genetical Research*, **65**, 209–22.

Roze D and Rousset FO (2004). Joint effects of self-fertilization and population structure on mutation load, inbreeding depression and heterosis. *Genetics*, **167**, 1001–15.

Sakai AK, Wagner WL, Ferguson DM, and Herbst DR (1995). Biogeographical and ecological correlates of dioecy in the Hawaiian flora. *Ecology*, **76**, 2530–43.

Schierup MH, Vekemans X, and Charlesworth D (2000). The effect of subdivision on variation at multi-allelic loci under balancing selection. *Genetical Research*, **76**, 51–62.

Stebbins GL (1950) *Variation and evolution in plants*, Columbia University Press, New York.

Stebbins GL (1957). Self-fertilization and population variability in the higher plants. *American Naturalist*, **91**, 337–54.

Thomson JD and Barrett SCH (1981). Selection for outcrossing, sexual selection, and the evolution of dioecy in plants. *American Naturalist*, **118**, 443–9.

Turner JRG (1977). Butterfly mimicry: the genetical evolution of an adaptation. *Evolutionary Biology*, **11**, 163–206.

Uyenoyama MK and Waller DM (1991). Coevolution of self-fertilization and inbreeding depression. 1. Mutation selection balance at one locus and two loci. *Theoretical Population Biology*, **40**, 14–46.

van Damme JMM (1986). Gynodioecy in *Plantago lanceolata* L. V. Frequencies and spatial distribution of nuclear and cytoplasmic genes. *Heredity*, **56**, 355–64.

Whitlock MC and Barton NH (1997). The effective size of a subdivided population. *Genetics*, **146**, 427–41.

Willis JH and Orr HA (1993). Increased heritable variation following population bottlenecks: the role of dominance. *Evolution*, **47**, 949–57.

Wright S (1940). Breeding structure of populations in relation to speciation. *American Naturalist*, **74**, 232–48.

CHAPTER 13

Floral design and the evolution of asymmetrical mating systems

Spencer C. H. Barrett and Kathryn A. Hodgins

Department of Botany, University of Toronto, Ontario, Canada

Outline

Floral traits that promote cross-pollination have a fundamental influence on the origin and maintenance of sexual polymorphisms in plants. This influence results from mating within and between sexual morphs, which determines the character of negative frequency-dependent selection and governs the evolution of morph ratios. The daffodil genus, *Narcissus*, exhibits exceptional diversity in floral design, pollination biology and sexual systems, and includes species with stylar monomorphism, stigma-height dimorphism, distyly and tristyly. *Narcissus* is unusual among heterostylous groups because species with stylar polymorphism possess a self-incompatibility system that permits intra-morph mating. This association results in imperfect sex-organ reciprocity, asymmetrical mating and biased morph ratios in populations. Here, we provide a synthesis of recent investigations on floral and sexual-system diversity in *Narcissus*, and consider the evolutionary causes and consequences of asymmetrical mating. We contrast three focal species that are the subjects of experimental studies on the relation between floral design and mating. *Narcissus longispathus*, *Narcissus assoanus*, and *Narcissus triandrus* exhibit contrasting pollination syndromes and possess stylar monomorphism, dimorphism and trimorphism, respectively. In these species, variation in sex-organ position significantly influences female selfing rates and fertility, patterns of asymmetrical mating, opportunities for self-interference, and the promotion of evolutionary transitions between sexual systems. *Narcissus* provides a rare opportunity in flowering plants to expose the population-level consequences of small, but functionally significant, variations in floral design.

13.1 Introduction

Non-random mating characterizes the reproductive biology of most flowering plants. Immobility, local pollen dispersal, and differences among individuals in flowering phenology and display size all foster structured variation in mating success within populations. The sexual strategies of plants also promote non-random mating. In species with sexual polymorphisms, populations are sub-divided into distinct mating groups with restricted mating options. For example, in populations of dioecious and gynodioecious species (Chapters 3, 11, and 12) a significant proportion of plants cannot mate with one another because they share the same sexual phenotype. Because of the co-existence of morphologically distinct mating groups within sexually polymorphic populations they can provide valuable experimental systems for investigating the influences of morphological, ecological and genetic factors on non-random mating.

Mating patterns in typical heterostylous populations are non-random because of restriction on physiological compatibility among plants. Populations with this sexual polymorphism include two (distyly) or three (tristyly) floral morphs (Darwin 1877; Ganders 1979; Barrett 1992), that differ in their possible mating partners. The morphological and physiological traits that distinguish morphs in heterostylous populations promote inter-morph (disassortative) mating, resulting in the negative frequency-dependent selection that maintains sexual polymorphism. In most heterostylous species, the reciprocal positioning of stigma and anther heights (reciprocal herkogamy) is associated with a heteromorphic incompatibility system that prevents self- and intra-morph mating (Barrett and Cruzan 1994). Heteromorphic incompatibility enforces both non-random mating and *symmetrical mating*, because on average each morph mates either exclusively with the alternate morph (distyly), or equally with the remaining two morphs (tristyly) in a population (Fig. 13.1a and b, respectively). Given symmetrical mating, frequency-dependent selection results in equilibrium morph frequencies of 1:1 and 1:1:1 in distylous and tristylous populations, respectively (Fisher 1941; Charlesworth and Charlesworth 1979; Heuch 1979), a condition known as isoplethy (Finney 1952).

Deviations from isoplethy occur commonly in non-equilibrium heterostylous populations following founder events and in species in which clonal propagation dominates (reviewed by Barrett 1993). However, biased morph frequencies (anisoplethy) can occur also in equilibrium populations with *asymmetrical mating*. This unequal equilibrium results from morph-specific differences

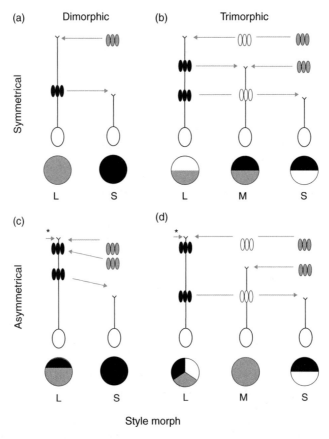

Figure 13.1 Symmetrical and asymmetrical mating patterns in plant populations with stylar polymorphisms. Typical (a) distyly and (b) tristyly; in both cases heteromorphic incompatibility allows only mating between stigmas and anthers of equivalent height, resulting in symmetrical disassortative mating, as indicated by the arrows. Panels (c) stigma-height dimorphism and (d) tristyly in *Narcissus*. In this genus, reciprocity between sexual organs is incomplete and species possess a self-incompatibility system that allows intra-morph (assortative) mating. As a result, mating patterns are asymmetrical. Circles illustrate the share of matings obtained by each morph (Black = L-morph, white = M-morph and grey = S-morph). L, M, S refer to the L-, M-, and S-morphs, respectively. The arrows below the asterisks in (c) and (d) represent cross-pollen transfer from the long-level anthers of the L-morph to stigmas of this morph, rather than self-pollination.

in selfing, intra-morph (assortative) mating, or inter-morph (disassortative) mating. Some of the best examples of asymmetrical mating occur in the daffodil genus, *Narcissus*, which is the focus of this chapter. Figure 13.1c and d illustrate asymmetrical mating in dimorphic and trimorphic populations of *Narcissus*, respectively. In both cases, the share of outcrossed matings by each morph differs, in contrast to typical distylous and tristylous species. This mating asymmetry is associated with the imperfect reciprocity of anthers and stigmas that characterize the morphs in *Narcissus*. Here we explore how asymmetrical mating evolves as a general feature of sexually polymorphic populations and consider the consequences of this mating system for the evolution of morph ratios.

In *Narcissus*, intra- and inter-morph mating occurs because polymorphic species possess a late-acting, ovarian self-incompatibility system that limits selfing but permits all classes of outcrossed mating (Bateman 1952; Dulberger 1964; Fernandes 1965; Sage *et al.* 1999). The absence of heteromorphic incompatibility in *Narcissus* allows differences in floral morphology among morphs to influence mating patterns more strongly than in other heterostylous plants, for which physiology dictates that most mating occurs between morphs.

Accordingly, plants need not mate symmetrically in *Narcissus* populations, depending on floral morphology and the local mating environment, resulting in corresponding anisoplethic variation in equilibrium morph frequencies.

Early studies of *Narcissus* identified "anomalous" features in comparison with "typical" heterostylous groups (Fernandes 1935, 1965; Bateman 1968; Yeo 1975). These anomalies include intra-morph compatibility, weakly developed sex-organ reciprocity, and anisoplethic morph ratios. These features provoked dispute as to whether heterostyly genuinely occurred in *Narcissus* (reviewed in Lloyd *et al.* 1990; Barrett *et al.* 1996). However, during the past decade this controversy has largely subsided and it is now recognized that *Narcissus* includes species with several distinct sexual systems, including stylar monomorphism, stigma-height dimorphism (Dulberger 1964), distyly (Arroyo and Barrett 2000) and tristyly (Barrett *et al.* 1997). These conditions are associated with striking differences among species in floral morphology and pollination biology (Graham and Barrett 2004; Barrett and Harder 2005), implicating floral design and its effect on pollen dispersal in promoting transitions among sexual systems, including the evolution of asymmetrical mating.

Table 13.1 Reproductive attributes of three *Narcissus* species (*N. longispathus*, *N. assoanus*, and *N. triandrus*) with contrasting sexual systems.

Attribute	*Narcissus longispathus*[a]	*Narcissus assoanus*[b]	*Narcissus triandrus*[c]
Stylar condition	Monomorphism	Dimorphism	Trimorphism and dimorphism
Geographic distribution	Restricted, Southeast Spain	Widespread, Southern France and Spain	Widespread, Spain and Portugal
Habitat	Stream margins, poorly drained meadows	Meadows and stony pastures on limestone	Oak and pine forests, acidic soils
Floral design	Short tube, large corona	Long tube, short corona	Long tube, large corona
Main pollinators	*Andrena bicolor*	*Gonepteryx cleopatra* *Macroglossum stellatarum* *Anthophora* spp.	*Anthophora* spp. *Bombus* spp.
Compatibility	Self-compatible	Self-incompatible	Self-incompatible
Mating system	mean $t_m = 0.63$ range $= 0.54$–0.77	mean $t_m = 0.99$ range $= 0.94$–1.00	mean $t_m = 0.75$ range $= 0.61$–0.87
Average morph ratios	N/A	$L = 0.62$, $S = 0.38$	Trimorphic $L = 0.58$, $M = 0.19$, $S = 0.23$ Dimorphic $L = 0.71$, $S = 0.29$

[a] Herrera (1995); Barrett *et al.* (2004a)
[b] Baker *et al.* (2000a and b)
[c] Barrett *et al.* (1997, 2004b); Hodgins and Barrett (2006a)

Here, we synthesize recent investigations on the ecology and evolution of floral and sexual-system diversity in *Narcissus* to illustrate how and why asymmetrical mating systems evolved in this genus. We first summarize comparative and phylogenetic evidence on the evolution of stylar conditions in *Narcissus*, and provide a conceptual framework for understanding floral evolution in the genus. We then review current research on three *Narcissus* species that have been the subject of detailed investigations on the role that floral morphology plays in pollen dispersal and mating patterns. The three species (Table 13.1, Plate 3) possess stylar monomorphism (*Narcissus longispathus*), dimorphism (*Narcissus assoanus*) and trimorphism (*Narcissus triandrus*) and floral designs associated with different degrees of mating complexity. Shifts from stylar monomorphism to polymorphism in *Narcissus* result in the evolution of mating asymmetries and we consider how floral traits influence this transition.

13.2 The evolution and functional basis of floral and sexual-system diversity

Floral morphology and sexual-system diversity are closely associated in *Narcissus*, implying a functional basis to their correlated evolution. *Narcissus* flowers are insect pollinated and comprise three basic components: the floral tube, free tepals, and the corona. Floral diversity within the genus evolved largely through changes in the relative sizes of these floral structures. Barrett and Harder (2005) distinguished several distinct floral designs in *Narcissus*, two of which are common. In the "daffodil" design (Plate 3a) the floral tube is short and the corona is large and cylindrical or trumpet-like, allowing bees to enter the flower completely while foraging for nectar and/or pollen. This morphology characterizes section *Pseudonarcissus* (trumpet daffodils) and all species possess stylar monomorphism. In contrast, the "paperwhite" design (Plate 3c) is found in diverse sections (e.g., *Apodanthi, Jonquillae, Narcissus, Tazettae*) and has a short flaring corona and a relatively long, narrow floral tube. Paperwhite flowers are pollinated by long-tongued insects, primarily Lepidoptera (*Sphingidae, Pieridae, Nymphalidae*), which feed on nectar and transport pollen on their faces or proboscides. Flowers with this morphology are presented horizontally and are usually fragrant. Species with the paperwhite design possess either stylar monomorphism or stigma-height dimorphism. A third floral form, the "triandrus" design (Plate 3d), is restricted to the two heterostylous species, *Narcissus albimarginatus* (distyly) and *N. triandrus* (tristyly) and combines features of the daffodil (large bell-shaped corona) and paperwhite (long narrow floral tube) designs. Flowers of these two species are pendulous and, at least in *N. triandrus* for which pollinator observations have been made, long-tongued bees enter the corona and probe deeply for nectar. This summary indicates that stylar polymorphisms in *Narcissus* have evolved only in lineages with narrow floral tubes that are pollinated by Lepidoptera and/or long-tongued bees, as Lloyd and Webb (1992a, b) predicted based on pollen-transfer models of the evolution of heterostyly.

Recent phylogenetic investigations provide insight into the evolution of floral and sexual-system diversity in *Narcissus* (Pérez et al. 2003; Graham and Barrett 2004). Character reconstructions indicate the likely pathways to sexual diversification and the associated morphological changes. Stylar polymorphisms have evolved independently from monomorphism at least five times. Because most monomorphic *Narcissus* species possess approach herkogamy, stigma-height dimorphism probably evolved repeatedly from this condition by the invasion of short-styled variants into long-styled populations. Concentrated-change tests indicate that long, narrow floral tubes preceded the evolution of stigma-height dimorphism (Graham and Barrett 2004). Distyly and tristyly originated independently in sections *Apodanthi* and *Ganymedes*, respectively, and represent an evolutionary convergence unique to the Amaryllidaceae (Pérez et al. 2003; Graham and Barrett 2004). Distyly evolved from stigma-height dimorphism, a sequence predicted by Lloyd and Webb's (1992a, b) theoretical models. In contrast, the sequence for the evolution of tristyly is unclear as the closest known relatives possess stylar monomorphism. Simultaneous invasion of an ancestral monomorphic population by two morphs

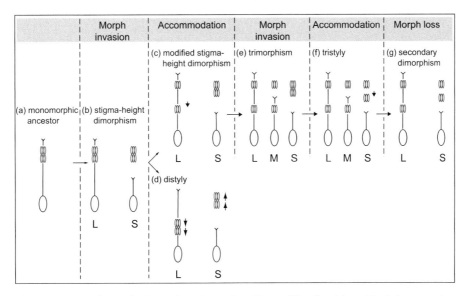

Figure 13.2 The diversity and evolution of stylar conditions in *Narcissus*. The transitions from left to right depict repeated stages involving the invasion of a novel variant into a population followed by accommodation of floral morphology resulting in increased mating proficiency. Arrows beside anthers indicate changes in position associated with morphological accommodation. L, M, S refer to the L-, M-, and S-morphs, respectively. All the stigma–anther arrangements (a–g) occur within and among *Narcissus* species. Reprinted with permission from Barrett and Harder (2005).

to create tristyly seems unlikely, and indeed theoretical models feature a dimorphic intermediate stage (Charlesworth 1979), so the most parsimonious explanation is that the dimorphic intermediate condition did not persist. Nevertheless, the independent origin of long, narrow tubes combined with deep coronas in *Apodanthi* and *Ganymedes* probably provided the morphological innovation that permitted segregated anther positions and the evolution of heterostyly.

Based on this phylogenetic evidence and using functional arguments, Barrett and Harder (2005) proposed hypothetical pathways for sexual-system diversification in *Narcissus* (Fig. 13.2). Their scheme begins with a monomorphic ancestor with approach herkogamy (Fig. 13.2a) from which evolves stigma-height dimorphism (Fig. 13.2b), distyly (Fig. 13.2d) and finally tristyly (Fig. 13.2f) through repeated sequences of the invasion of novel variants into populations followed by morphological adjustment of floral traits of resident morph(s), which increases their mating proficiency. As discussed in Section 13.4, relatively simple genetic changes in style length and anther height create the novel floral variants in *Narcissus* that initiate invasion. Subsequent adaptive accommodation in floral morphology of the resident morph(s) changes morph frequencies until all morphs realize equal mating success. In contrast to other heterostylous plants, equivalent mating success in *Narcissus* need not involve isoplethic morph ratios.

13.3 Mating in monomorphic populations

Most *Narcissus* species possess stylar monomorphism, approach herkogamy and are self-sterile. Whether the failure of monomorphic species to set seed after self-pollination results from self-incompatibility and/or early-acting inbreeding depression has not been investigated. Nevertheless, there is no evidence of distinct mating groups that would structure mating within populations.

A few monomorphic *Narcissus* species are self-compatible, providing opportunities to examine the influence of floral traits on mating patterns. *Narcissus longispathus* is a bee-pollinated trumpet daffodil endemic to mountain ranges in southeastern Spain (Table 13.1). Plants commonly produce a single, large (corolla length ~50 mm), long-lived (~16.5 days) flower on each inflorescence (Plate 3a), and controlled self- and cross-pollinations result in equivalent seed set (Herrera 1995). Flowering occurs during early spring (late February–April) when cool, wet weather frequently limits visitation by the primary pollinator, *Andrena bicolor* (Andrenidae). Investigations of *N. longispathus* have addressed two main questions: (1) does the association between long-lived flowers and unpredictable pollinator service result in significant selfing, and (2) does the conspicuous variation in stigma–anther separation (herkogamy; Plate 3b) cause variation in maternal outcrossing rates. These questions have been addressed by measuring mating patterns using allozyme markers in open-pollinated seeds sampled from natural populations.

Narcissus longispathus exhibits mixed mating, producing significant numbers of both outcrossed and selfed seeds (Barrett *et al.* 2004a; $t_m = 0.54 - 0.77$; $N = 6$ populations). Marker-based estimates of inbreeding depression (see Ritland 1990) indicate that few selfed seeds survive to maturity, a pattern also reported in *N. triandrus* (Hodgins and Barrett 2006a), so that selfing provides little reproductive assurance. Estimates of the proportion of outcrossed seed and the inbreeding coefficient for samples from the same population obtained 13 years apart yielded very similar values, indicating remarkable stability in mixed mating and inbreeding (Medrano *et al.* 2005). The strong selection against selfed offspring in *N. longispathus* raises the intriguing question of why selection has not produced floral mechanisms that reduce this significant waste of zygotes.

Given that *N. longispathus* populations exhibit among the widest variation in stigma–anther separation (range 1–10 mm) reported for flowering plants (Plate 3b), herkogamy might reasonably be expected to limit self-pollination. Experimental studies on diverse taxa commonly find a monotonically increasing relation between outcrossing rate and degree of herkogamy (reviewed in Medrano *et al.* 2005). In contrast, in *N. longispathus* plants with intermediate stigma–anther separation have higher outcrossing rates than those with either small or large herkogamy (Medrano *et al.* 2004). Absence of the expected monotonic relation between mating patterns and herkogamy may result because the stigma is close to the corona rim in flowers with well-developed herkogamy and bees bask on the rim to thermoregulate before departing (Herrera 1995; fig. 2 in Medrano *et al.* 2005). If this interpretation is correct, it emphasizes an important lesson: details of pollinator behaviour should be considered when assessing the influence of floral morphology on mating patterns.

13.4 Mating in dimorphic populations

Stigma-height dimorphism is the most common stylar polymorphism in *Narcissus*, even though this condition is rare among angiosperms (Barrett *et al.* 2000). This polymorphism occurs in at least a dozen *Narcissus* species, distributed among three sections (*Apodanthi, Jonquillae, Tazettae*). Populations with stigma-height dimorphism contain two floral morphs that differ in style length (hereafter L- and S-morphs); however, in both morphs the two anther levels within a flower occupy similar positions at the top of the floral tube. Therefore, unlike distyly, reciprocity between the stigma and anther heights is weak. Stigmas of the L-morph correspond in height to the upper-level anthers of both morphs, whereas stigmas of the S-morph are positioned below the lower-level anthers (Fig. 13.2b). Consequently, the S-morph exhibits large stigma–anther separation, whereas herkogamy is either weakly developed or absent in the L-morph. Weak sex-organ reciprocity has important consequences for pollen transfer and is the structural cause of asymmetrical mating and anisoplethic morph ratios in most dimorphic *Narcissus* populations. Investigations of stigma-height dimorphism have focused on determining the mechanisms responsible for the evolution and maintenance of the polymorphism, and the striking variation in morph frequencies among populations.

Narcissus assoanus, a diminutive geophyte of open habitats in southern France and Spain has been a particularly useful experimental system for investigating the ecology and evolution of stylar polymorphism (Table 13.1, Plate 3c). This species is pollinated primarily by Cleopatra butterflies

(*Gonepteryx cleopatra*), day-flying hawk moths (*Macroglossum stellatarum*) and solitary bees (*Anthophora* spp.) and exhibits a wide range of morph frequencies. Populations vary greatly in size and unreliable pollinator service results in considerable spatial and temporal variation in pollen limitation (Baker *et al.* 2000a, b, c). *Narcissus assoanus* populations typically exhibit L-morph biased morph ratios, although isoplethic morph ratios predominate in very large populations on the limestone plateau north of Montpellier in SW France (Baker *et al.* 2000b; fig. 5.9 in Thompson 2005). The occurrence of 1:1 morph ratios in *N. assoanus* is significant, because it demonstrates that symmetrical mating can occur in the absence of reciprocal herkogamy.

13.4.1 Evolution and maintenance of stigma-height dimorphism

Theoretical and experimental evidence from *Narcissus* indicates that stigma-height dimorphism increases the proficiency of cross-pollination and limits self-interference (Lloyd and Webb 1992b; Barrett *et al.* 1996; Thompson *et al.* 2003; Cesaro *et al.* 2004; Cesaro and Thompson 2004). The establishment of stylar dimorphism requires the invasion of a monomorphic population with approach herkogamy by a short-styled variant. In *Narcissus*, short styles are governed by a dominant allele at the style-length locus (L-morph—*ss*; S-morph—*Ss* or *SS*; Dulberger 1967). Therefore an advantageous short-styled variant should spread easily in populations, because all individuals with the dominant allele express the novel phenotype (Haldane 1927). This pattern of inheritance also occurs in most distylous species, indicating the important role of dominance in the invasion dynamics and inheritance of stylar polymorphisms (Lloyd and Webb 1992a).

Theoretical models indicate that the maintenance of stigma-height dimorphism requires negative frequency-dependent mating resulting from greater inter-morph than intra-morph pollen transfer (Lloyd and Webb 1992b; Barrett *et al.* 1996). Experimental manipulation of morph frequencies in natural populations of *N. assoanus* has demonstrated frequency-dependent reproductive success (Thompson *et al.* 2003). For example, minority S-morph plants set significantly more seed than L-morph plants in L-biased patches, as predicted. Also, S-morph plants in monomorphic patches set significantly fewer seeds than other treatments, including monomorphic L-morph patches, indicating that the extreme herkogamy in the S-morph limits intra-morph pollination. Direct measurement of pollen transfer further demonstrated more frequent inter-morph cross-pollination than intra-morph cross-pollination, particularly from the L-morph to the S-morph (Cesaro and Thompson 2004). The rates of pollen transfer observed in this study satisfy the theoretical conditions necessary for the establishment of stigma-height dimorphism under pollen limitation (Lloyd and Webb 1992b).

The role of self-interference in the evolution of stigma-height dimorphism in *Narcissus* remains enigmatic. Experimental pollinations of *N. assoanus* demonstrate that prior self-pollination, or simultaneous mixtures of self- and cross-pollen reduce seed set considerably compared with exclusive cross-pollination (Cesaro *et al.* 2004). Lower seed set results from the abortion of developing ovules after self-pollination (ovule discounting: Barrett *et al.* 1996; Sage *et al.* 1999). However, these effects do not appear to influence the female fertility of the morphs under field conditions (Baker *et al.* 2000b; Cesaro *et al.* 2004). This limited impact is surprising, because variation in the degree of herkogamy between the morphs cause strong differences in the incidence of autonomous self-pollination in bagged flowers. Pollen loads on stigmas of the S-morph were negligible, whereas the L-morph experienced a high degree of autonomous self-pollination. Self-interference may be difficult to detect in the field because the morphs have different mechanisms that limit the incidence of self-pollination: protandry in the L-morph and herkogamy in the S-morph (Cesaro *et al.* 2004). Conditions favouring establishment of stigma-height dimorphism may occur when pollinators visit infrequently, so that protandry is ineffective in limiting self-pollination. The long floral longevities typical of *Narcissus* flowers would aggravate this problem, permitting considerable self-pollination in the L-morph and could contribute to the invasion of monomorphic populations by short-styled variants.

13.4.2 Asymmetrical mating and biased morph ratios

Interactions between pollinators and flower morphology are probably the main cause of the variation in morph ratios among *N. assoanus* populations. Variation in selfing rates caused by differences in herkogamy or partial self-incompatibility do not account for the predominance of the L-morph in populations (Baker *et al.* 2000b). Populations are highly outcrossing (Table 13.1), so biased morph ratios probably result from differences between morphs in patterns of inter- and intra-morph mating. Specifically, the S-morph should be less proficient at intra-morph mating than the L-morph because of its well-developed herkogamy. In contrast, stamen positions in the L-morph have presumably been optimized for efficient cross-pollination in ancestral populations (Fig 13.2a), and in monomorphic populations of the L-morph that occur commonly in several species with stigma-height dimorphism (reviewed in Arroyo *et al.* 2002). Theoretical models of asymmetrical pollen transfer and mating confirm that greater assortative mating in the L-morph results in equilibrium ratios in which this morph predominates (Barrett *et al.* 1996; Baker *et al.* 2000b). The field experiments discussed in Section 13.4.1 and later in this section, which indicate reduced fertility of the S-morph in monomorphic patches, also support this hypothesis.

A recent investigation of the spatial distribution of floral morphs within *N. assoanus* populations provides further support for asymmetrical pollen transfer and mating. Stehlik *et al.* (2006) mapped the location of all L- and S-morphs in eight small populations to determine whether plant density and the morph composition of local neighbourhoods influenced female fertility. Using neighbourhood models they predicted the quantitative relations between the spatial clustering of morphs and variation in seed set quite accurately. The fertility of the L-morph increased significantly with the total number of plants in local neighbourhoods, regardless of their morph identity (Fig 13.3a). In contrast, the fertility of the S-morph increased significantly with the number of individuals of the L-morph, but was insensitive to the number of individuals of the S-morph (Fig. 13.3b). These patterns are expected if the S-morph receives pollen primarily from the L-morph, whereas for the L-morph both the L- and S-morphs are functionally equivalent as paternal mating partners. Pollen transfer and mating in *N. assoanus* is therefore context dependent, with the morphs responding differently to the density and morph identity of plants in local neighbourhoods.

The L-morph of *N. assoanus* shows evidence of morphological accommodation to the presence of the S-morph in some populations. The lower-level stamens are lower in the floral tube of the L-morph than in the S-morph (Fig. 13.2c). This change probably promotes more pollen transfer to stigmas of the S-morph, benefiting the male fertility of the L-morph. In contrast to typical distyly (Fig. 13.1a), the two morphs in *N. assoanus* differ in their paternal roles as mating partners: the S-morph mates primarily with the L-morph, whereas the L-morph divides its paternal contribution between both morphs (Fig. 13.1c). This functional differentiation in the L-morph is probably caused by disruptive selection on the two stamen levels, resulting in increased male mating proficiency. Parallel selection is unlikely in the S-morph, because a reduction in height of its lower-level stamens to promote intra-morph mating would increase the likelihood of self-pollination, causing pollen and ovule discounting.

Adaptive accommodation similar to that in the L-morph of *N. assoanus* probably culminated in the evolution of distyly from stigma-height dimorphism in *N. albimarginatus* (Fig. 13.2d). This shift between dimorphic sexual systems may have been promoted by morphological modifications to flowers associated with a shift from Lepidoptera to bee pollination (Arroyo and Barrett 2000). As a consequence of increased reciprocity of sex organs mating patterns become more symmetrical, with morph ratios closer to isoplethy. In *N. albimarginatus* this evolutionary sequence has not proceeded to complete reciprocal herkogamy, and as a consequence morph ratios exhibit a small bias in favour of the L-morph (Pérez *et al.* 2003).

13.5 Mating in trimorphic populations

Tristyly is a rare sexual polymorphism known from only six angiosperm families (reviewed in

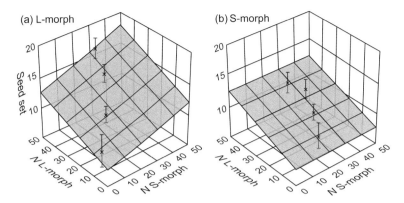

Figure 13.3 Differences in seed set associated with different mating neighbourhoods for the floral morphs of *Narcissus assoanus*, a species with stigma-height dimorphism. The two figures illustrate the fitted relations based on neighbourhood models between morph-specific seed set and the number and identity of neighbours averaged for eight populations in SW France. (a) L-morph; (b) S-morph. Data points indicate the mean (±SE) number of seeds and the mean number of neighbours (*N*) within each of four neighbourhood sizes. The sizes of the neighbourhood categories differ because they were chosen to illustrate positive relations between seed set and the number of neighbours. After Stehlik *et al.* (2006), which provides methods and computational details.

Barrett 1993; Thompson *et al.* 1996; Barrett *et al.* 2000). In *Narcissus*, tristyly occurs in a single, wide-ranging polymorphic species. *Narcissus triandrus* is native to the Iberian Peninsula and is distributed from Andalucía through central Spain and Portugal to northwestern Spain (Table 13.1, Plate 3d). Some workers subdivide this taxon into separate species (e.g., Pérez *et al.* 2003), but we follow Blanchard (1990) in recognizing two largely allopatric varieties: *N. triandrus* var. *cernuus* from central and southern parts of the range characterized by small stature and pale lemon flowers; and the taller *N. triandrus* var. *triandrus* from northern Portugal and northwestern Spain with white to cream flowers. As discussed in Section 13.2, phylogenetic analyses provide few clues concerning the evolutionary pathways leading to floral trimorphism, but we assume that an intermediate dimorphic condition was involved (Fig. 13.2e and f). The rarity of tristyly in *Narcissus*, and among angiosperms in general, implies that strong constraints must be overcome to enable its origin and maintenance (Charlesworth 1979; Barrett 1993). Our investigations have focused primarily on the ecological and evolutionary factors maintaining floral trimorphism and on morph-ratio evolution. We are also interested in why the expression of tristyly in *N. triandrus* differs from that of other tristylous species.

13.5.1 Evolution of morph ratios

Geographical surveys of morph frequencies can provide insights into how polymorphisms are maintained in the face of strong environmental gradients. Our surveys of *N. triandrus* have revealed patterns unlike those reported for other tristylous species (Barrett *et al.* 1997, 2004b), including: (1) a predominance of the L-morph in populations; (2) a negative relation between the frequencies of the L- and M-morphs; and (3) dimorphic populations missing the M-morph. Significantly, although populations of both *N. triandrus* varieties are usually L-morph biased, trimorphism is only a stable feature of var. *cernuus* populations. In *N. triandrus* var. *triandrus* the M-morph becomes relatively less common in the northwest of the Iberian Peninsula (fig. 3a in Barrett *et al.* 2004b), with several geographically separated transitions between trimorphic and dimorphic populations (Plate 4). Our work attempts to elucidate the mechanisms that account for these patterns.

Stylar dimorphism in *N. triandrus* is restricted to parts of the geographical range in which trimorphic populations exhibit low frequencies of the M-morph, whereas in southern and central parts of the range, the M-morph occurs at moderate frequencies. This pattern indicates that the ecological

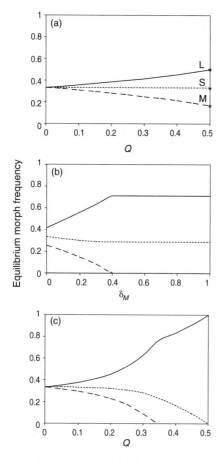

Figure 13.4 The influence of floral design, assortative mating and female fertility on equilibrium morph frequencies in tristylous populations, based on pollen-transfer models described in the text. (a) The consequence of elevating the upper-level stamens of the L-morph, where Q represents pollen transfer from the L-morph to other plants of the L-morph, and 0.5-Q represents pollen transfer from the L-morph to the M-morph ($Q = 0$ represents typical tristyly, $Q \approx 0.5$ represents *N. triandrus*). The asterisks indicate that the illustrated equilibrium is one of many possibilities. (b) The consequence of reduced female fertility in the M-morph when $Q = 0.30$, and δ_M represents the proportion of unfertilized ovules. (c) The consequence of elevating the upper-level stamens of the L-morph in a pollen-limited environment ($\alpha = 1$), where the female fertility of each morph is a function of the level of pollen receipt. L-morph = solid line, M-morph = dashed line, S-morph = dotted line. Panel (a) after Barrett et al. (2004b); panels (b) and (c) after Hodgins and Barrett (2006b).

conditions in the NW Atlantic region are often unfavourable for the persistence of tristyly. A survey in northwest Spain and northern Portugal of *cp*-DNA haplotype variation and morph frequencies in 37 populations of var. *triandrus* revealed four distinct haplotypes (Plate 4). The haplotypes exhibit significant geographical clustering corresponding to physiographic barriers, such as mountain ranges and river basins. The three common haplotypes each include both dimorphic and trimorphic populations. These regional patterns are unlikely to represent historical contingency, but probably arose because local ecological factors cause transitions between sexual systems. The patterns also suggest that dimorphic populations do not have a common origin, but evolved independently through gradual loss of the M-morph from populations. The alternate hypothesis that contemporary dimorphic populations of var. *triandrus* represent the ancestral dimorphic state is unlikely, because of their distinctive floral morphology, characterized by mid-level anthers in the S-morph (Figs 13.1d and 13.2g and see Section 13.5.2). It is hard to imagine why this anther level would be selected in the absence of mid-level stigmas in a population. However, if these populations do represent ancestral dimorphism, the variation in M-morph frequency among trimorphic populations within each of the three haplotypes implies independent invasions of the M-morph. We are currently attempting to distinguish between these alternate hypotheses using molecular markers.

In contrast to other tristylous species, the L- and M-morphs of *N. triandrus* possess equivalent anther positions (compare Fig. 13.1b and d). Absence of true "mid-level" stamens in the L-morph is the structural basis for the asymmetric mating patterns and L-biased morph ratios that occur in *N. triandrus*. The typical stigma–anther reciprocity found in tristylous species is unbalanced in *N. triandrus*, because different numbers of anther levels target the three stigma heights (L-stigma—3, M-stigma—1, S-stigma—2). This contrast results in unequal competition among floral morphs for outcrossed mating. With a polymorphic equilibrium maintained by negative frequency-dependent selection, all morphs realize equal mating success. Therefore, at the equilibrium, the intensity of competition should be balanced by the availability of mating opportunities. The observed frequencies of morphs in populations of *N. triandrus* reflect this balance between competition and mating opportunities (Barrett et al. 2004b). The L-morph is the most common morph and the M-morph is often the least common morph.

Pollen-transfer models developed by Charlesworth (1979) and Lloyd and Webb (1992b) are useful tools for exploring the evolution of morph ratios in heterostylous species. We have used this approach to investigate the influence of floral design and variation in mating patterns on equilibrium morph ratios in *Narcissus* (Barrett et al. 1996; 2004b). With the arrangement of sex organs that characterizes *N. triandrus*, the predicted outcome is a polymorphic equilibrium in which the L-morph predominates at the expense of the M-morph (Fig. 13.4a), a pattern entirely consistent with observed morph ratios. Here, we extend these models to explore the joint effects of female fertility and asymmetrical mating patterns on the evolution of morph frequencies. Further details of these models are presented in Hodgins Barrett (2006b).

In our pollen-transfer models the fitness of each morph when seed production is not pollen-limited is

$$w_i = \frac{1}{2}s_i + \frac{1}{2}\left[s_i^* \frac{q_{ii}f_i}{q_{ii}f_i + q_{ji}f_j + q_{ki}f_k} + s_j^* \frac{q_{ij}f_j}{q_{jj}f_j + q_{ij}f_i + q_{kj}f_k} + s_k^* \frac{q_{ik}f_k}{q_{kk}f_k + q_{ik}f_i + q_{jk}f_j}\right], \quad (1)$$

where s_i is the number of seeds produced by morph i, q_{ij} is the proportion of pollen exported from morph i to morph j, and f_i represents the frequency of the ith morph. We assume that individuals of all morphs export equivalent amounts of pollen and do not differ in their maternal selfing rates, which has been confirmed for *N. triandrus* (Hodgins and Barrett 2006a). The first term on the right side, $s_i/2$, represents a morph's contribution of genes as a maternal parent and the remaining terms represent a morph's paternal contribution realized through pollen competition in the pistils of the L-, M- and S-morphs, respectively. The matrix of pollen-transfer proficiencies is presented in Barrett et al. (2004b).

We model two possible causes of variation in female fertility, s_i. In the first case, fertility variation among morphs results from differences in ovule discounting caused by self-pollination, where $s_i = (1 - \delta_i)$, and δ_i represents the proportion of ovules of morph i that do not develop into seeds. Lower female fertility of the M-morph causes reduced equilibrium frequencies and loss of the M-morph from populations. More significantly, with assortative mating in the L-morph promoted by its long-level stamens, even slight reductions in female fertility of the M-morph result in reduced frequencies and loss of this morph from populations (Fig. 13.4b).

We now investigate the influence of pollen limitation as a cause of reduced female fertility. In this case, the fertility of a morph is a function of pollen receipt, $s_i = 1 - e^{-\alpha x_i}$, where $x_i = q_{ii}f_i + q_{ji}f_j + q_{ki}f_k$, which depends on the frequency of compatible mates, pollen-transfer probabilities between donor and recipient morphs, and the intensity of pollen limitation, represented by α. Higher values of α reduce pollen limitation. When mating patterns are symmetrical, as in typical tristylous populations, each morph realizes equal success through female function and morph frequencies are maintained at isoplethy. However, with the *N. triandrus* morphology the unbalanced stigma–anther reciprocity affects the female fertility of each morph differently. The L-morph has the highest fertility, because all three morphs have anthers targeting its stigma, whereas the M-morph has the lowest fertility, because only the S-morph has mid-level stamens. Thus, although asymmetrical mating causes anisoplethy, its interaction with female fertility differences resulting from pollen limitation can magnify this effect considerably (Fig. 13.4c). As assortative mating in the L-morph intensifies, its frequency increases with the frequencies of the M- and S-morphs declining at different rates. The M-morph is lost from a population first, with the L-morph going to fixation when at least half of its mating is assortative. L-morph monomorphy is also predicted from strong assortative mating in models concerning the evolution of morph ratios in populations with stigma-height dimorphism (Baker et al. 2000b).

Empirical data support the theoretical prediction that the M-morph can be lost from trimorphic populations of *N. triandrus* because of reduced female fertility. In nine populations with a wide range of morph frequencies, the M-morph was the most common in populations where it had the

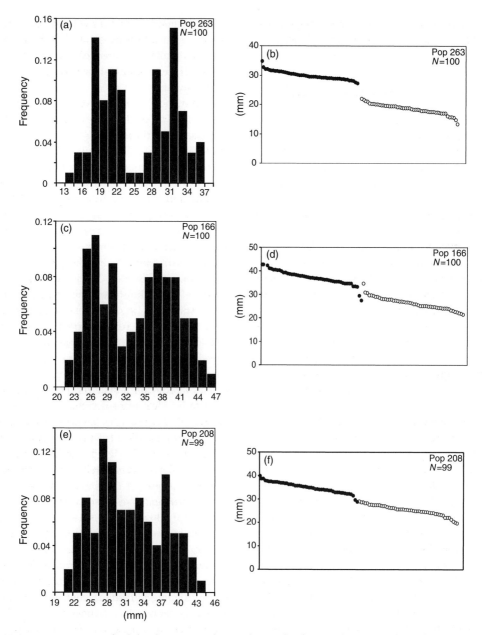

Figure 13.5 Contrasting patterns of style-length variation in the L- and M-morphs of *N. triandrus* var. *triandrus* in three trimorphic populations illustrating the range of variation among populations. The number of flowers sampled in each population is indicated. Measurements were made using digital calipers based on a single distal flower per plant. Rank distributions (b, d, f) are adjusted for flower size variation, with the L- and M-morphs represented by closed and open circles, respectively.

highest proportional seed set and least common in populations where its female fertility was significantly lower than the other two morphs (Hodgins and Barrett 2006b). This result contrasts with expectations for typical tristylous species. Negative frequency-dependent selection should involve a negative association between morph frequencies and fertility in non-equilibrium populations (Ågren and Ericson 1996). Therefore, these results showing the opposite relation suggest that variation in female fertility among the morphs plays a significant role in the evolution of dimorphism from trimorphism in *N. triandrus*.

13.5.2 Variation and evolution of sexual organs

Sex-organ position in *N. triandrus* varies extensively compared with most other tristylous species (fig. 12 in Eckert and Barrett 1994). This variation is manifested both within and among populations and suggests spatially variable selection. Bateman (1968 p. 645) proposed that the high variation in style length in *N. triandrus* represents "an early stage in speciation through divergent adaptation to different pollinators"; an improbable claim given the maintenance of stylar polymorphism throughout the species' range. We consider it more likely that geographical variation results, in part, from changes in the types of pollinators that service populations. For example, flower size increases along the southeast–northwest gradient of M-morph frequencies, with accompanying differences in allometric relations of sex-organ position (Barrett *et al.* 2004b). Although the proximate ecological mechanism responsible for the flower-size cline appears to be rainfall related changes in overall plant size, increased flower size is also associated with a switch from pollination by *Anthophora* in the drier south to *Bombus* in the wetter north. Unfortunately, our attempts to investigate the contribution of pollinator-mediated selection on this floral variation have been thwarted by the very low frequency of pollinator visits in *N. triandrus* populations.

A second component of floral variation in *N. triandrus* appears to be associated with the strength of selection maintaining sex-organ reciprocity, and the extent to which the adaptive accommodation of stigmas and anthers promotes mating efficiency. We next present two examples of sex-organ variation in *N. triandrus* to illustrate this hypothesis: the patterns of style-length variation in trimorphic populations, and variation in the position of stamens in trimorphic versus dimorphic populations. Both provide evidence that selection for precision and reciprocity in *N. triandrus* is considerably weaker than in typical tristylous species with heteromorphic incompatibility. This difference probably occurs because intra-morph pollen transfer in *N. triandrus* incurs no wastage of gametes, unlike typical heterostylous species.

Tristyly in *N. triandrus* is unique in part because the L- and M-morphs have similar anther heights (Fig. 13.1d), so that morph identity depends only on the position of the stigma relative to the two anther heights. Note this issue of identity does not pertain to the S-morph, which possesses a distinct combination of anther levels (Fig. 13.1d). A *N. triandrus* population could display continuous variation in style length and still be classified into discrete morphs. However, this pattern is not generally accepted for heterostylous species, for which bimodality or trimodality is the rule. Indeed, Bateman (1952, 1954) queried whether *N. triandrus* was genuinely tristylous because the small sample of plants that he examined displayed almost continuous variation in style length.

We measured style length in the L- and M-morphs from 11 trimorphic populations of var. *triandrus* to determine whether they exhibit the bimodality expected in a heterostylous species. The combined distributions of the two morphs within nine populations displayed significant bimodality, although the degree of bimodality varied considerably. However, style length varied continuously in two populations, with no hint of bimodality. Three representative populations from this sample are illustrated in Fig. 13.5. Two contrasting explanations could account for the continuous variation evident in population 208 (Fig. 13.5e and f). The observed pattern, neglecting environmental and developmental variation, may represent "ancestral" quantitative variation on which disruptive selection for bimodality could drive the evolution of stylar trimorphism with distinct stigma heights. Alternatively, disruptive selection by pollinators maintaining style-length bimodality could have been relaxed, resulting in the accumulation of style-length modifiers. Contemporary populations of *N. triandrus* appear to contain all of the standing variation necessary for either process. Because morphology and incompatibility are not associated in *N. triandrus*, the loss, and

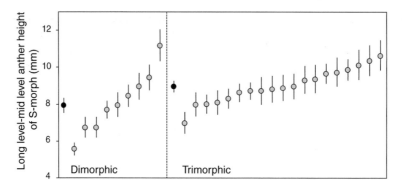

Figure 13.6 Variation in the separation between the anthers of long- and mid-level stamens of the S-morph in nine dimorphic and 19 trimorphic populations of *N. triandrus* var. *triandrus* (number of individuals sampled per population, range 14–35). We analyzed this variation with a general linear model that considered sexual system and population (sexual system) as fixed effects and flower length as a covariate. Both anther separation and flower length were ln-transformed. Black and grey circles represent back-transformed least-squares means (±SE) for each sexual system and population, respectively.

perhaps also the gain, of the M-morph in populations will be considerably easier than in tristylous species with heteromorphic incompatibility.

When the M-morph is lost in tristylous species, mid-level stamens of the L- and S-morphs are not deployed optimally for efficient inter-morph pollen transfer. In several groups, particularly *Lythrum* and *Oxalis* spp., such loss has caused evolutionary modification of the positions of mid-level anthers in the remaining morphs (reviewed in Weller 1992, but see Eckert and Mavraganis 1996). We measured variation in mid-level stamens in dimorphic and trimorphic populations of *N. triandrus* for evidence of evolutionary repositioning. Because only the S-morph possesses mid-level stamens, we restrict our analysis to this morph. We predicted that because stamens at the mid-level have lost their target stigmas, selection for pollen transfer to stigmas of the L-morph should elevate them closer to long-level stamens. However, if populations differ in their history of M-morph loss, or mid-level stamens experience relaxed selection in some dimorphic populations, mid-level stamen position should vary more among dimorphic populations than among trimorphic populations.

Our results support repositioning of mid-level stamens in some, but not all dimorphic populations of *N. triandrus*. On average, the anthers of mid-level stamens were significantly closer to those of long-level stamens in dimorphic compared to trimorphic populations, after controlling for differences in flower size ($F_{1,525} = 14.89$, $P < 0.001$; Fig. 13.6). Also, the separation between mid- and long-level stamens of the S-morph varied more among dimorphic than trimorphic populations (Levene's test on population means adjusted for flower size; $F_{1,26} = 4.63$, $P < 0.05$). This variation ranges from populations in which mid-level stamens are identical in position to those in trimorphic populations, to populations in which these stamens are closer to long-level stamens. Absence of heteromorphic incompatibility increases the functional significance of this variation.

13.6 Discussion

Sexual polymorphisms in *Narcissus* were first reported over a century ago (Wolley-Dod 1886; Henriques 1887) and later became a source of controversy (Fernandes 1935, 1965; Bateman 1952, 1968). However, not until David Lloyd and colleagues (Lloyd *et al.* 1990; Barrett *et al.* 1996) alerted floral biologists to the rich diversity of sexual systems in *Narcissus* did modern experimental investigations begin in earnest. Subsequent work during the past decade on the floral biology, pollination and mating systems of *Narcissus* species have been facilitated by collaboration among several research groups and has resulted in the publication of ∼20 journal articles (see References) and several doctoral dissertations.

The preceding overview highlights the merits of *Narcissus* for investigating the relations between floral design and mating patterns. For self-compatible species, such as *N. longispathus*, variation in sex-organ deployment affects the balance between outcrossing and selfing, which in sexually polymorphic species results in asymmetrical patterns of outcrossed mating and biased morph ratios. Stylar polymorphisms are often portrayed as textbook cases of morphological adaptation between flowers and animal pollinators. Depictions of heterostyly commonly emphasize the tight association between sex-organ reciprocity and inter-morph mating resulting in equal morph ratios. Our review of *Narcissus* mating demonstrates that alternatives to this elegant symmetry in form and function can occur, without compromising the fundamental adaptive role of the polymorphisms as floral mechanisms promoting animal-mediated pollen transfer among plants.

Most cases of asymmetrical mating in heterostylous populations are associated with morph-specific differences in selfing rate caused by the weakening or loss of heteromorphic incompatibility. In some cases, asymmetrical mating promotes the evolutionary breakdown of heterostyly and its replacement by alternative sexual systems (reviewed in Ganders 1979; Weller 1992; Barrett 1993). However, asymmetrical mating need not disrupt the functioning of heterostyly and alternative anisoplethic equilibria can arise when morphs differ in their ability to mate with each other, for example because floral morphs differ in pollen production (Barrett *et al.* 1983), or possess other incompatibility systems (Schou and Philipp 1984; Ornduff 1988).

Polymorphic *Narcissus* species exhibit imperfect sex-organ reciprocity, asymmetrical mating and biased morph ratios. These features are the evolutionary consequences of the atypical association between stylar polymorphisms and a self-incompatibility system that permits intra-morph mating. Evolution of stylar polymorphisms in self-incompatible groups with significant intra-morph compatibility (reviewed in Barrett and Cruzan 1994) should include imperfect sex-organ reciprocity, asymmetrical mating and biased morph ratios as general features. Further investigations of taxa which also display these associations, such as *Anchusa*, *Lithodora* and *Villarsia*, are warranted.

Why are floral morphology, mating patterns and morph ratios in *Narcissus* species so distinct from those of typical heterostylous species? The type of incompatibility in *Narcissus* limits possibilities for the evolution of balanced reciprocity of sexual organs and opportunities for symmetrical mating. For example, in *N. triandrus* selection reducing the height of long-level stamens in the L-morph to match mid-level stigmas is unlikely, because this change in position would reduce overall siring success. Because the L-morph is most common in populations and the M-morph is usually at a lower frequency, modification of the position of upper-level stamens in the L-morph would reduce the pool of target stigmas available for pollen export. Similar arguments apply to stamen modifications in the L-morph for species with stigma-height dimorphism, because populations usually exhibit L-biased morph ratios. It is important to emphasize that although asymmetric pollen transfer is possible in typical heterostylous species, and indeed pollen-flow studies demonstrate that it occurs commonly (Gander 1979; Paillier *et al.* 2002; Lau and Bosque 2003), heteromorphic incompatibility guarantees that the realized mating patterns in populations are symmetrical.

The diversity of sexual systems in *Narcissus* is closely associated with floral design and the pollination biology of species. This association suggests that transitions among stylar conditions are promoted by shifts in pollinator service that improve pollen transfer, although little information is as yet available to test this hypothesis. However, several *Narcissus* species possess significant geographical variation in floral morphology and morph ratios and some of this variation is associated with differences in pollinator fauna (e.g., Arroyo and Dafni 1995; Arroyo *et al.* 2002; Barrett *et al.* 2004b). This variation provides opportunities to investigate experimentally the detailed functional relations between floral design, pollen transfer and the evolution of sexual systems.

Acknowledgements

We thank Lawrence Harder, Sean Graham, Bill Cole, John Thompson, Carlos Herrera, Adeline Cesaro, Mònica Medrano, and Ivana Stehlik for advice and collaboration; Carlos Herrera for photographs of *N. longispathus*, and Bill Cole for field assistance and help with figures. This work was funded by NSERC Discovery Grants to SCHB, and graduate fellowships from NSERC to KAH.

References

Ågren J and Ericson L (1996). Population structure and morph-specific fitness differences in tristylous *Lythrum salicaria*. *Evolution*, **50**, 126–39.

Arroyo J and Barrett SCH (2000). Discovery of distyly in *Narcissus* (Amaryllidaceae). *American Journal of Botany*, **87**, 748–51.

Arroyo J and Dafni A (1995). Variations in habitat, season, flower traits and pollinators in dimorphic *Narcissus tazetta* L. (Amaryllidaceae) in Israel. *New Phytologist*, **129**, 135–45.

Arroyo J, Barrett SCH, Hidalgo R, and Cole WW (2002). Evolutionary maintenance of stigma-height dimorphism in *Narcissus papyraceus* (Amaryllidaceae). *American Journal of Botany*, **89**, 1242–9.

Baker AM, Thompson JD, and Barrett SCH (2000a). Evolution and maintenance of stigma-height dimorphism in *Narcissus*. I. Floral variation and style-morph ratios. *Heredity*, **84**, 502–13.

Baker AM, Thompson JD, and Barrett SCH (2000b). Evolution and maintenance of stigma-height dimorphism in *Narcissus*. II. Fitness comparisons between style morphs. *Heredity*, **84**, 514–24.

Baker AM, Barrett SCH, and Thompson JD (2000c). Variation of pollen limitation in the early flowering Mediterranean geophyte *Narcissus assoanus* (Amaryllidaceae). *Oecologia*, **124**, 529–35.

Barrett SCH, ed. (1992). *Evolution and function of heterostyly*. Springer-Verlag, Berlin.

Barrett SCH (1993). The evolutionary biology of tristyly. In D Futuyma and J Antonovics, eds. *Oxford surveys in evolutionary biology*, pp. 283–326. Oxford University Press, Oxford, UK.

Barrett SCH and Cruzan MB (1994). Incompatibility in heterostylous plants. In EG Williams, RB Knox, and AE Clark, eds. *Genetic control of incompatibility and reproductive development in flowering plants*, pp. 189–219. Kluwer Academic, The Netherlands.

Barrett SCH and Harder LD (2005). The evolution of polymorphic sexual systems in daffodils (*Narcissus*). *New Phytologist*, **165**, 45–53.

Barrett SCH, Price SD, and Shore JS (1983). Male fertility and anisoplethic population structure in tristylous *Pontederia cordata* (Pontederiaceae). *Evolution*, **37**, 745–59.

Barrett SCH, Lloyd DG, and Arroyo J (1996). Patterns of style length variation in *Narcissus*: implications for the evolution of heterostyly. In DG Lloyd and SCH Barrett, eds. *Floral biology: studies on floral evolution in animal pollinated plants*, pp. 339–76. Chapman and Hall, New York.

Barrett SCH, Cole WW, Arroyo J, Cruzan MB, and Lloyd DG (1997). Sexual polymorphisms in *Narcissus triandrus* (Amaryllidaceae): is this species tristylous? *Heredity*, **78**, 135–45.

Barrett SCH, Jesson LK, and Baker AM (2000). The evolution and function of stylar polymorphisms in flowering plants. *Annals of Botany*, **85** Supplement A, 253–65.

Barrett SCH, Cole WW, and Herrera CM (2004a). Mating patterns and genetic diversity in the wild Daffodil *Narcissus longispathus* (Amaryllidaceae). *Heredity*, **92**, 459–65.

Barrett SCH, Harder LD, and Cole WW (2004b). Correlated evolution of floral morphology and mating-type frequencies in a sexually polymorphic plant. *Evolution*, **58**, 964–75.

Bateman AJ (1952). Trimorphism and self-incompatibility in *Narcissus*. *Nature*, **170**, 496–7.

Bateman AJ (1954). The genetics of *Narcissus* I—Sterility. *Daffodil and tulip year book*, **19**, pp. 23–9. Royal Society of Horticultural Society, London.

Bateman AJ (1968). Role of heterostyly in *Narcissus* and *Mirabilis*. *Evolution*, **22**, 645–6.

Blanchard JW (1990). *Narcissus: a guide to wild daffodils*. Alpine Garden Society, Surrey, UK.

Cesaro AC and Thompson JD (2004). Darwin's cross-promotion hypothesis and the evolution of stylar polymorphism. *Ecology Letters*, **7**, 1209–15.

Cesaro AC, Barrett SCH, Maurice S, Vaissiere BE, and Thompson JD (2004). An experimental evaluation of self-interference in *Narcissus assoanus*: functional and evolutionary implications. *Journal of Evolutionary Biology*, **17**, 1367–76.

Charlesworth D (1979). The evolution and breakdown of tristyly. *Evolution*, **33**, 486–98.

Charlesworth D and Charlesworth B (1979). A model for the evolution of distyly. *American Naturalist*, **114**, 467–98.

Darwin CR (1877). *The different forms of flowers on plants of the same species*. John Murray, London, UK.

Dulberger R (1964). Flower dimorphism and self-incompatibility in *Narcissus tazetta* L. *Evolution*, **18**, 361–3.

Dulberger R (1967). Pollination systems in plants in Israel. Ph.D. Thesis, Hebrew University, Jerusalem.

Eckert CG and Barrrett SCH (1994). Tristyly, self-compatibility and floral variation in *Decodon verticillatus* (Lythraceae). *Biological Journal of the Linnean Society*, **53**, 1–30.

Eckert CG and Mavraganis K (1996). Evolutionary consequences of extensive morph loss in tristylous *Decodon verticillatus*: a shift from tristyly to distyly. *American Journal of Botany*, **83**, 1024–32.

Fernandes A (1935). Remarque sur l'hétérostylie de *Narcissus triandrus* et de *N. reflexus*. *Boletim Da Sociedade Broteriana Séries 2*, **10**, 5–15.

Fernandes A (1965). Contribution à la connaissance de la génétique de l'hétérostylie chez le genre *Narcissus* L. II. L'hétérostylie chez quelques populations de *N. triandrus* var cernuus et *N. triandrus* var. concolor. *Genética Ibérica*, **17**, 215–39.

Finney DJ (1952). The equilibrium of a self-incompatible species. *Genetica*, **26**, 33–64.

Fisher RA (1941). The theoretical consequences of polyploid inheritance for the mid style form in *Lythrum salicaria*. *Annals of Eugenics*, **11**, 31–8.

Ganders FR (1979). The biology of heterostyly. *New Zealand Journal of Botany*, **17**, 607–35.

Graham SW and Barrett SCH (2004). Phylogenetic reconstruction of the evolution of stylar polymorphisms in *Narcissus* (Amaryllidaceae). *American Journal of Botany*, **91**, 1007–21.

Haldane JBS (1927). A mathematical theory of natural and artificial selection. Part V. Selection and mutation. *Proceedings of the Cambridge Philosophical Society*, **23**, 838–44.

Henriques JA (1887). Observações sobre algumas especies de *Narcissus*, encontrados em Portugal. *Boletim Da Sociedade Broteriana*, **6**, 45–7.

Herrera CM (1995). Floral biology, microclimate, and pollination by ectothermic bees in an early-blooming herb. *Ecology*, **76**, 218–28.

Heuch I (1979). Equilibrium populations of heterostylous plants. *Theoretical Population Biology*, **15**, 43–57.

Hodgins KA and Barrett SCH (2006a). Mating patterns and demography in the tristylous daffodil *Narcissus triandrus*. *Heredity*, **96**, 262–70.

Hodgins KA and Barrett SCH (2006b). Female reproductive success and the evolution of mating type frequencies in tristylous populations. *New Phytologist*, **171**, 569–80.

Lau P and Bosque C (2003). Pollen flow in the distylous *Palicourea fendleri* (Rubiaceae): an experimental test of the disassortative pollen flow hypothesis. *Oecologia*, **135**, 593–600.

Lloyd DG and Webb CJ (1992a). The evolution of heterostyly. In SCH Barrett, ed. *Evolution and function of heterostyly*, pp. 151–78. Springer-Verlag, Berlin, Germany.

Lloyd DG and Webb CJ (1992b). The selection of heterostyly. In SCH Barrett, ed. *Evolution and function of heterostyly*, pp. 179–208. Springer-Verlag, Berlin, Germany.

Lloyd DG, Webb CJ, and Dulberger R (1990). Heterostyly in species of *Narcissus* (Amaryllidaceae) and *Hugonia* (Linaceae) and other disputed cases. *Plant Systematics and Evolution*, **172**, 215–27.

Medrano M, Herrera CM, and Barrett SCH (2005). Herkogamy and mating patterns in the self-compatible daffodil *Narcissus longispathus*. *Annals of Botany*, **95**, 1105–11.

Orndruff R (1988). Distyly and monomorphism in *Villarsia* (Menyanthaceae): some evolutionary considerations. *Annals of the Missouri Botanical Garden*, **75**, 761–7.

Pailler T, Maurice S, and Thompson JD (2002). Pollen transfer patterns in a distylous plant with overlapping pollen-size distributions. *Oikos*, **99**, 308–16.

Pérez R, Vargas P, and Arroyo J (2003). Convergent evolution of flower polymorphism in *Narcissus* (Amaryllidaceae). *New Phytologist*, **161**, 235–52.

Ritland K (1990). Inferences about inbreeding depression based on changes of the inbreeding coefficient. *Evolution*, **44**, 1230–41.

Sage TL, Strumas F, Cole WW, and Barrett SCH (1999). Differential ovule development following self- and cross-pollination: the basis of self-sterility in *Narcissus triandrus* (Amaryllidaceae). *American Journal of Botany*, **86**, 855–70.

Schou O and Philipp M (1984). An unusual heteromorphic incompatibility system. III. On the genetic control of distyly and self-incompatibility in *Anchusa officinalis* L. (Boraginaceae). *Theoretical and Applied Genetics*, **68**, 139–44.

Stehlik I, Casperson JP, and Barrett SCH (2006). Spatial ecology of mating success in a sexually polymorphic plant. *Proceedings of the Royal Society of London, Series B*, **273**, 387–94.

Thompson JD (2005). *Plant evolution in the Mediterranean*. Oxford University Press, Oxford, UK.

Thompson JD, Pailler T, Strasberg D, and Manicacci D (1996). Tristyly in the endangered Mascarene Island endemic *Hugonia serrata* (Linaceae). *American Journal of Botany*, **83**, 1160–7.

Thompson JD, Barrett SCH, and Baker AM (2003). Frequency-dependent variation in reproductive success in *Narcissus*: implications for the maintenance of stigma-height dimorphism. *Proceedings of the Royal Society of London, Series B*, **270**, 949–53.

Weller SG (1992). Evolutionary modifications to tristylous breeding systems. In SCH Barrett, ed. *Evolution and function of heterostyly*, pp. 247–72. Springer Verlag, Berlin, Germany.

Wolley-Dod C (1886). Polymorphism of organs in *Narcissus triandrus*. *Gardeners' Chronicle*, April, p. 468.

Yeo PF (1975). Some aspects of heterostyly. *New Phytologist*, **75**, 147–53.

PART 4

Floral diversification

The most outstanding feature of angiosperms is their extraordinary reproductive diversification. Flowers vary enormously in size, colour, texture, and shape and probably exhibit greater structural variation than the equivalent reproductive organs of any other group of organisms. Floral diversity is the most prominent characteristic of the angiosperm radiation and therefore understanding the ecological and evolutionary processes responsible continues to be a major theme in floral ecology, and evolutionary biology in general. Although there seems little doubt that animal pollinators have played a causal role in angiosperm diversification, determining how often and under what ecological conditions pollinator-mediated selection shapes floral evolution represents a more difficult problem to solve. Linking this pollinator-mediated floral divergence to speciation represents a further challenge.

The final section of *Ecology and Evolution of Flowers* considers the problem of the evolution and adaptive basis of floral diversification and its consequences for speciation and species richness of lineages. The chapters in this section illustrate that tackling these issues requires a variety of approaches from quantitative genetics to comparative biology and analyses at both microevolutionary and macroevolutionary time scales. The authors illustrate that research on plant–pollinator interactions continues to provide some of the most compelling examples of the evolution of adaptation by natural selection. However, in outlining future research they also highlight some of the difficulties that will need to be overcome if we are to obtain a comprehensive understanding of the ecology and evolution of floral diversification.

The beginning to any research program on floral diversification requires an understanding of the genetic architecture of floral traits and measurements of the direction and strength of natural selection acting on those traits. In Chapter 14 Jeffrey Conner reviews evidence showing that significant genetic variance for floral traits is commonplace in plant populations thus providing the necessary raw material for natural selection. However, he makes a point of alerting us to deficiencies in studies of phenotypic selection because of the common failure to measure selection on male function. Genetic correlations among floral traits are near ubiquitous because of the integrated nature of the flower. Conner illustrates by artificial selection experiments in wild radish that these correlations may often not be strong enough to act as severe constraints on the independent evolution of individual traits and supports this inference by a comparative analysis of floral traits in Brassicaceae.

Linking microevolutionary processes operating at the intra-specific level to macroevolutionary patterns represents one of the most difficult challenges in modern evolutionary biology, including studies on floral diversification. In Chapter 15, Carlos Herrera, María Clara Castellanos, and Mónica Medrano point out that for this challenge to succeed a more rigorous approach to intra-specific studies is required. Few investigations of floral evolution are conducted in a broad geographical context and even fewer demonstrate local adaptive differentiation in response to spatial variation in pollinators through measurements of phenotypic selection. They outline a five-step protocol which can be used to identify geographical differentiation in floral traits resulting from local selection by pollinators and illustrate this approach through their field studies on *Lavandula*. Use of their procedure will be valuable for understanding how geographical variation becomes converted into inter-specific differentiation as a result of the evolution of reproductive isolation

Identifying the mechanisms driving speciation in plants is crucial for understanding the process of floral diversification. Speciation if often associated with evolutionary changes to reproductive traits and at least two hypotheses can explain these shifts. Divergence can arise through reinforcement to promote reproductive isolation and limit hybridization, or, alternatively differentiation may be the result of pollinator-mediated selection. In Chapter 16, Steven Johnson argues that too much emphasis has been placed on identifying mechanisms of reproductive isolation in sympatry, with not enough consideration given to the geographical factors promoting phenotypic divergence and local adaptation. He proposes that a better appreciation of the importance of pollinator-driven speciation will come from investigating pollinator-geographical mosaics and, echoing the sentiments of Herrera and colleagues, the spatially variable patterns of phenotypic selection that occur within them. This strongly selectionist view involves an inextricable link between floral adaptation and speciation in animal-pollinated plants.

Angiosperm clades differ considerably in species richness and this has stimulated efforts to identify plant traits that might be associated with rates of diversification and extinction. The availability of large-scale molecular phylogenies of angiosperms facilitates these types of comparative analyses. Although several non-reproductive traits have been linked to diversification, not surprisingly most attention has focused on reproductive characters. In Chapter 17, Scott Hodges and colleagues use a recent supertree of angiosperm phylogeny to investigate the association between rates of diversification and biotic pollination, dioecious sexual system, floral zygomorphy and nectar spurs. They find evidence in three of the four traits for a relation with species richness among angiosperm lineages, and propose functional explanations to account for these patterns. As more well-resolved phylogenies of angiosperm lineages become available, these approaches will undoubtedly be used to look at a wider range of reproductive traits and their correlated evolution.

A characteristic feature of many angiosperm taxa is the relatively weak reproductive isolation that occurs among closely related species. Natural hybridization is a relatively common phenomenon providing opportunities for the generation of novel floral diversity with potential consequences for adaptation and speciation. In animal-pollinated species, floral traits should influence both the formation of hybrids and their subsequent fitness as a result of pollinator visitation and behaviour. Surprisingly, given the long-standing interest in plant hybridization, very little is known about floral biology of hybrid zones. In the final chapter of this section Diane Campbell and George Aldridge remedy this deficiency by addressing the extent to which floral traits influence hybridization and reproductive isolation. Based on their empirical studies of *Ipomopsis* hybrid zones, and simulation models of pollinator visitation and pollen dispersal, they conclude that ethological isolation plays a more important role than mechanical isolation. They also show that differences in inter-specific pollen dispersal resulting from variation in pollinator behaviour likely account for the contrasting levels of hybridization among sites. These studies illustrate how investigations of the floral biology of hybrid zones offer unique opportunities to examine the functional significance of novel floral variation in an experimental context.

Selected key references

Barrett SCH, Harder LD, and Worley AC (1996). The comparative biology of pollination and mating in flowering plants. *Philosophical Transactions of the Royal Society of London, Series B*, **351**, 1271–80.

Dodd ME, Silvertown J, and Chase MW (1999). Phylogenetic analysis of trait evolution and species diversity among angiosperm families. *Evolution*, **53**, 732–44.

Kingsolver JG, Hoekstra HE, Hoekstra JM, *et al.* (2001). The strength of phenotypic selection in natural populations. *American Naturalist*, **157**, 245–61.

Rieseberg LH and Wendel J, eds (2004). Plant speciation. *New Phytologist*, **161** (Special Issue).

Schemske DW and Bradshaw HD Jr (1999). Pollinator preferences and the evolution of floral traits in monkeyflowers (*Mimulus*). *Proceedings of the National Academy of Sciences of the United States of America*, **96**, 11910–5.

Stebbins GL (1970). Adaptive radiation of reproductive characteristics in angiosperms. I Pollination mechanisms. *Annual Review of Ecology and Systematics*, **1**, 307–26.

CHAPTER 14

Ecological genetics of floral evolution

Jeffrey K. Conner

Kellogg Biological Station and
Department of Plant Biology, Michigan State University, East Lansing, MI, USA

Outline

The diversity of floral forms in nature can be explained largely as adaptations to the diversity of biotic and abiotic selective agents with which different plant species interact. Ecological genetics is the study of the process of adaptation, and therefore is an ideal approach to understanding floral adaptations. Here I review work on selection and genetic variance and covariance of floral traits, as these are the principal determinants of adaptive evolution. Early work focused on simple floral polymorphisms, because they were amenable to study, but more recent work has used their simplicity to understand the genetic mechanisms underlying adaptation in unprecedented detail. Because most floral adaptations are not simple polymorphisms, I also review studies that have measured selection and the **G** matrix (additive genetic variances and covariances among traits) for quantitative floral traits. I present new results from my research group on highly correlated traits in wild radish flowers, showing how these traits can evolve independently despite the constraint caused by the genetic correlation. A study of the same traits across the Brassicaceae suggests that macroevolution may be guided by the **G** matrix, but that independent evolution of highly correlated traits can also occur during these longer periods. I close by reviewing some topics for future study that have been opened up by recent technical advances, and which have the potential to expand our understanding of the mechanisms of floral adaptation greatly.

14.1 Introduction

Natural selection acts on phenotypic variation, and if some of this phenotypic variation is due to underlying genetic variation, then adaptation through genetic change can occur in the population. However, the rate of adaptive evolution can be constrained by either a lack of genetic variance for the adaptive trait or genetic correlations between the adaptive trait and other traits under selection. Ecological genetics focuses on the process of adaptive evolution, especially the direction and strength of natural selection and the nature of genetic variation and covariation underlying adaptive phenotypic traits (Conner and Hartl 2004).

Ecological genetics provides an ideal approach to understanding key factors underlying floral evolution. Flowers affect fitness directly through their role as the organs of mating and sexual reproduction in angiosperms. The primary selective agent on floral traits is clearly the pollen vector, usually animals or wind, although secondary selective agents, such as herbivores or abiotic factors, also influence floral evolution (Galen 1999; Chapter 7). Different floral organs (e.g., petals, stamens, pistil) may be under selection to work together as a functional unit (functional integration), which may alter correlations among floral parts (Armbruster 1991; Conner and Via 1993; see below), or conversely genetic correlations may constrain independent evolution of different floral parts (see below; Chapter 7).

In this chapter, I review current understanding of floral adaptations and suggest some lines of inquiry that will be particularly fruitful in the future. I first discuss floral colour polymorphisms, because simple polymorphisms were the first traits studied by ecological geneticists and their simplicity enabled recent detailed genetic analysis. However, most floral adaptations involve quantitative floral traits, so I next consider the genetic variance and covariance of these more complex, continuous traits. I then briefly review studies of natural selection on quantitative traits and identity continuing gaps in our knowledge. Next I present an overview of my current research, which illustrates the twin roles of covariance in integrating traits and constraining independent trait evolution. Finally, I discuss some directions for future analysis of the ecological genetics of floral traits that hold considerable promise for new insights into the process of floral adaptation.

14.2 Simple polymorphisms: floral colour

Early studies of ecological genetics focused on simple polymorphisms: traits that involve a few discrete types. These traits are more amenable to study than continuous traits, because the discrete variation arises from allelic differences at only one or at most a few gene loci, and the traits are not strongly affected by the environment. Classic examples of this research include industrial polymorphism in peppered moths and other insects, banding patterns in snails, and heterostyly in primrose and other plants (Ford 1975; Chapter 13). Here I will briefly discuss studies of another simple polymorphism, flower colour, both for historical context and because this work has progressed recently in important new directions. Because the genetics of polymorphic traits are simple, past work has focused on selection by pollinators and the maintenance of polymorphism, as selection favouring one form over another tends to eliminate variation. These twin themes of pollinator-mediated selection and genetic variation recur throughout this chapter. More recent work on floral polymorphisms has exploited the genetic simplicity of these traits to reveal the genetic mechanisms of adaptation in unprecedented detail; the genetic mechanisms underlying more complex adaptations are an important area for future research (Section 14.6).

Most, but not all, studies of selection on floral colour have found clear evidence for selection by pollinator preferences, but the form of these preferences varies. Both bumble bees and hummingbirds prefer the common blue flowers over rare white forms in *Delphinium nelsonii,* probably because the white flowers lack nectar guides, which increases the pollinators' handling time per flower (Waser and Price 1981). This pollinator preference probably explains the greater female fitness (seed production) of blue-flowered plants. Cabbage butterflies also prefer pigmented yellow flowers over white in wild radish (*Raphanus raphanistrum*), but this preference does not cause differences in female fitness (Fig. 14.1a and b; Stanton *et al*. 1986). This result suggests that fruit production is limited by resources other than pollen import (e.g., light, water, or soil nutrients). However, yellow-flowered plants sired more seeds than their white counterparts, showing that increased pollinator visitation enhanced male fitness (Fig. 14.1c). This was one of the first studies of floral evolution to measure seed siring success, which is crucial, but often neglected, in studies of floral evolution (see Section 14.4).

As in wild radish and *Delphinium*, pollinators of morning glories (*Ipomoea purpurea*) prefer pigmented flowers and discriminate against white flowers; however, in this species the selection is frequency dependent. Specifically, bumble bees prefer blue and pink flowers when they are common (greater than 75% combined), but exhibit no preference between white and pigmented flowers when they are roughly equally abundant (Epperson and Clegg 1987). In contrast to the previous examples, pollinators do not seem to discriminate between white and blue flowers in *Linanthus parryae* in the Mojave Desert (Schemske and Bierzychudek 2001). Nevertheless these morphs experience selection, and this selection fluctuated between years at one site: white-flowered plants produced more seeds during years with high rainfall, whereas blue-flowered plants produced more seeds during drier years. The cause of this selection remains unclear, although it is probably related to availability of essential cations in the soil

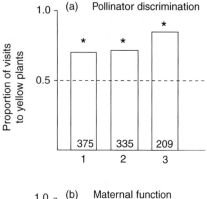
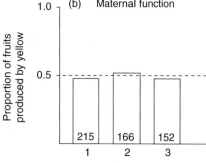
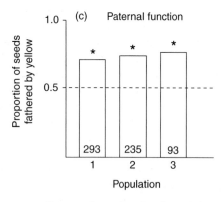

Figure 14.1 Pollinator preference for yellow-flowered plants over white-flowered plants in three populations of *Raphanus raphanistrum* (a) has no effect on female fitness (fruit production, b), but does enhance male fitness (siring success, c). An asterisk indicates a significant deviation from equal performance by the two colour morphs. Reprinted, with permission, from Stanton *et al.* 1986.

functions. Floral pigments serve various functions in plants, such as protection from UV radiation and from herbivores (Chapter 7).

The strong selection often reported in these studies raises the question of how genetic variation is maintained, because selection will tend to eliminate the polymorphisms. The authors of these studies postulate different mechanisms in each case, encompassing most of the major mechanisms that are thought to maintain genetic variation in general. Waser and Price explained the maintenance of the polymorphism in *Delphinium* as a balance between new mutations for white flowers and selection against the white morph. In wild radish, differences in colour preference among pollinators may maintain the polymorphism, as a series of studies of wild radish in England found that butterflies, syrphid flies, and honey bees preferred yellow flowers, whereas most bumble bees preferred white flowers (Kay 1978). Similarly, the fluctuating selection between wet and dry years in *Linanthus* could maintain the colour polymorphism in that species (Turelli *et al.* 2001). Finally, the negative frequency-dependent selection in morning glories would automatically maintain polymorphism.

Recent molecular studies of floral pigments in morning glory have revealed important new insights into the molecular mechanisms generating phenotypic variation and the biochemical and genetic basis of adaptation. One set of studies revealed that the mutations underlying the genetic variation in flower colour were caused mostly by transposable elements (Clegg and Durbin 2000), rather than by simple DNA base-pair substitutions. Other work examined different *Ipomoea* species in which red flowers evolved as part of an evolutionary switch from bee to hummingbird pollination. Zufall and Rausher (2004) determined that this switch involved inactivation of an enzyme in the pathway to blue pigments, which then shunts substrates along a different metabolic pathway, resulting in the synthesis of red pigments. Rarely is the genesis of an adaptive trait understood in this detail; knowledge of the molecular genetics of adaptation will go a long way towards improving our limited understanding of how adaptation occurs (Phillips 2005).

(D. W. Schemske personal communication). Thus, in this species selection on flower colour does not involve direct selection based on pollinator preference, but rather is indirect, caused by pleiotropic effects of the pigment gene on other plant

14.3 Ecological genetics of quantitative traits

Simple floral polymorphisms have provided and will continue to provide important insights into floral adaptation; however, most floral traits are not simple polymorphisms, but rather vary continuously. Examples include the sizes and shapes of petals and corolla tubes, the lengths of the stamens and pistils and the resulting placements of the anthers and stigmas, volume of nectar produced, and the sizes and numbers of ovules and pollen grains. Such continuous phenotypic distributions result from the action of more gene loci and/or stronger environmental effects compared with simple polymorphisms. Although both aspects of quantitative traits complicate their analysis, understanding floral adaptation has progressed significantly during the past 25 years, in part because of the availability of new theoretical, statistical, and molecular techniques, with the promise of even greater advances during the next decade (see also Chapter 2).

Two closely related equations encapsulate the key factors determining the adaptive evolution of quantitative traits:

$$R = h^2 S, \qquad (1)$$

$$\Delta \bar{z} = \mathbf{G}\beta. \qquad (2)$$

Both R and $\Delta \bar{z}$ represent the change in the mean phenotype from one generation to the next, that is, short-term adaptive evolution (Falconer and Mackay 1996; Conner and Hartl 2004). The magnitude of phenotypic evolution depends on the product of the amount of genetic variation for the trait, represented by heritability, h^2, or the genetic variance/covariance matrix, $\mathbf{G}$, and the strength of phenotypic selection, represented by the selection differential, S, or the selection gradient, β. Therefore, directional selection can cause extensive phenotypic evolution (a change in mean) between generations if selection is strong (large S or β), which occurs when the environment changes (including changes caused by human activity). The magnitude of this evolution also depends on the standing genetic variation for the trait under selection (h^2 or $\mathbf{G}$). Note that these equations focus on the change in the mean of a trait due to directional selection only, and do not model changes in trait variance caused by stabilizing and disruptive selection (cf. Chapter 13).

These equations differ in scope: the first considers evolution in one trait only, whereas the second is multivariate, considering multiple traits simultaneously. The multivariate approach offers two crucial advantages. First, the univariate selection differential estimates the combined selection on the trait, including both direct, adaptive selection and indirect, non-adaptive selection due to phenotypic correlations among traits. In contrast, the selection gradient estimates direct, adaptive selection, because it is based on multiple regression, so the effects of correlations among measured traits are removed. In fact, a well-conducted study estimating selection gradients is one of the best ways to determine what traits are adaptive in an undisturbed natural population (Conner and Hartl 2004).

The second advantage of the multivariate approach is that $\mathbf{G}$ includes both the genetic variance (as does the heritability) and genetic covariances among traits. The rate and short-term direction of adaptive evolution depend not only on selection and genetic variance, but also on genetic covariances. Genetic correlations (the standardized version of genetic covariances) result from pleiotropy (one gene locus affects variation in multiple traits) or gametic-phase disequilibrium (non-random association between alleles at distinct loci, each of which affect variation in one trait of interest). Such correlations have two important consequences for adaptive evolution. First, selection on one trait causes evolution in all other traits with which it is correlated genetically. This correlated response to selection can either accentuate or slow adaptive evolution. In the latter case, a genetic correlation acts as an evolutionary constraint, because correlated responses are not necessarily adaptive. For example, in scarlet gilia (*Ipomopsis aggregata*) direct selection favours increases in both corolla length and the proportion of time in the pistillate phase (when the stigma is receptive), but a negative genetic correlation between these traits slows the expected increases (Campbell 1996). Second, genetic correlations can themselves be adaptations, resulting from past selection for functional integration among traits that act together. For example, stabilizing selection on anther exsertion (the projection of the anthers beyond the opening of the corolla tube) through differences in male fitness in wild radish (Fig. 14.2; Morgan and Conner 2001) increases the correlation and integration between the filaments and corolla tube. Specifically, pollinators remove more pollen from flowers with anthers of intermediate height relative to the opening of the corolla tube than from flowers with lower or higher anthers. Thus, filament and corolla tube lengths are functionally integrated, because their relative lengths determine anther exsertion, which in turn affects successful pollination and subsequent male fitness.

Three approaches are used to estimate heritabilities, genetic correlations, and their unstandardized counterparts, genetic variances and covariances (the **G** matrix). The most common method involves the mating designs that plant and animal breeders have used for decades, such as offspring–parent regression, sibling analysis, and diallel mating crosses (Falconer and Mackay 1996; Conner and Hartl 2004). These techniques use controlled crosses to create sets of individuals of known genetic relationship, and then regression or analysis of variance to estimate genetic variances and covariances. For example, Campbell (1996) used both offspring–father regression and half-sibling analysis to demonstrate significant heritability for corolla length, width, and the positions of anthers and stigma of scarlet gilia, as well as genetic correlations among most traits. Many studies have used mating designs to estimate genetic variances and covariances for floral traits (e.g., Shore and Barrett 1990; Mazer and Schick 1991; O'Neil and Schmitt 1993; Conner et al. 2003a; Caruso 2004). These studies have found that significant genetic variance for, and covariance among, floral traits is very common in natural populations of a wide variety of plant species.

A newer approach uses molecular markers to estimate the relatedness of individuals in a population, instead of controlled crosses that create known relatedness (Ritland 2000; Thomas et al. 2000). This procedure allows estimation of quantitative genetic parameters for undisturbed natural populations and species for which controlled

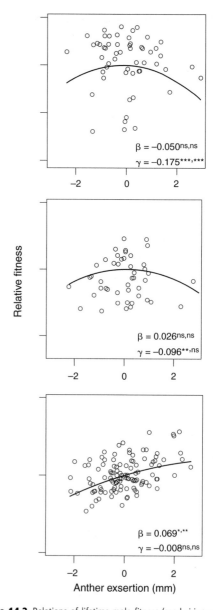

Figure 14.2 Relations of lifetime male fitness (seed siring success) to anther exsertion in wild radish during three field seasons. β and γ are the linear and quadratic selection gradients, respectively, and asterisks represent the outcomes of chi-square and simulation tests (*P < 0.05, **P < 0.01, ***P < 0.005). Adapted from Morgan and Conner 2001.

crosses are difficult, although it has not been developed and tested as thoroughly as standard crossing methods. Using a marker-based method, Ritland and Ritland (1996) reported significant heritability for both flower size and number in two natural populations of yellow monkeyflower, *Mimulus guttatus*. In contrast, Ritland and Ritland detected significant heritability only for flower size with more traditional offspring–parent regression methods, and all heritability estimates were lower. This contrast suggests that controlled crosses underestimate true heritability under natural conditions and/or that the marker-based method overestimates heritability. Clearly, the marker-based method requires more testing (but see Thomas *et al.* 2000).

The third general approach for estimating genetic variances and correlations involves artificial selection, which has been practiced by humans for millennia, resulting in all our domesticated plants and animals. To perform artificial selection, the investigator measures the phenotypic trait of interest on a controlled population, and selects individuals with the most extreme trait (in both directions, e.g., large and small) to breed the next generation. This process is typically continued for several generations. A significant difference between the lines selected in different directions provides good evidence for genetic variance for the trait in the original population. In addition, significant differences in any other traits that were not subject to selection demonstrates their genetic correlation with the selected trait. In essence, the investigator applies a known strength of selection (S) and measures the phenotypic change or response to this selection (R or $\Delta \bar{z}$) to infer the magnitude of heritability and parts of the **G** matrix (see eqs 1 and 2). Artificial selection is most advantageous as a direct test of whether a given trait can evolve in response to a known strength of selection. However, artificial selection can be time and labour intensive, it provides the genetic variances of only the selected trait(s) (usually only one) and covariances with that trait, rather than the entire **G** matrix, and the species used must be easily maintained and crossed (see Conner 2003 for further discussion).

To date, relatively few papers have reported the use of artificial selection to understand ecological genetics of floral traits. One of the first was by Huether (1968), who both increased and decreased petal number of *Linanthus androsaceus* during five generations of selection, even though five petals are diagnostic for the Polemoniaceae. Similarly, Karoly and Conner (2000) decreased the height difference between the short and long stamens of *Brassica rapa* during only three generations; this trait is diagnostic of another large family, the Brassicaceae. Finally, Stanton and Young (1994) changed the ratio of petal size and pollen production in *Raphanus sativus* with a single generation of selection, even though these two traits are subject to a strong positive genetic correlation, which should oppose such a response to selection. Taken together, these studies suggest weak evolutionary constraints on adaptive floral evolution, at least in the short term.

A few studies have used artificial selection to test for a trade-off between flower size and number that is expected if flower production draws from a limited resource pool. Worley and Barrett (2000) selected for larger and smaller flowers, as well as increased flower production per inflorescence, in *Eichhornia paniculata*. Interestingly, the predicted negative correlated responses occurred only in lines selected for smaller flowers, resulting in a correlated increase in flower number. Delph *et al.* (2004) also used artificial selection to study flower number–size trade-offs, but with the additional goal of understanding genetic correlations between the sexes in a dioecious plant, *Silene latifolia*. In this species, males produce more, smaller flowers than females. To reduce this sexual dimorphism, Delph *et al.* selected for larger flowers in males or smaller flowers in females and observed correlated responses between the sexes. Female flower size increased in the lines in which males were selected, and male flower size decreased in lines in which the females were selected. In addition, flower number exhibited negative correlated responses, decreasing under selection for larger flowers and vice versa.

Additional evidence for negative genetic correlations and possible constraints is evident from responses to selection on male versus female function within hermaphroditic flowers. Selection for production of more ovules and anthers in separate lines of *Spergularia marina* caused direct responses by each trait and negative correlated responses in the other trait, indicating a negative genetic correlation (Mazer *et al.* 1999). In contrast, Sarkissian and Harder (2001) reported significant direct responses to three generations of selection for large and small pollen grains in *Brassica rapa*, and a *positive* correlated response for ovule number. However, pollen number exhibited a negative response to selection on pollen size, indicating a negative genetic correlation and possible trade-off between pollen size and number in this species.

To summarize, studies of floral traits using different methods have commonly found significant genetic variance for floral traits, so that they can respond to selection. Genetic correlations among floral traits are also very common, which should constrain adaptive evolution, although these constraints might not be particularly strong (also see below). Consequently, additional measurements of genetic variation and covariance by themselves are no longer very useful, although such studies remain useful in the context of more comprehensive studies of adaptation and constraint.

14.4 Natural selection on floral traits

Selection has been measured for diverse floral traits in a wide variety of plant species (see Table 15.2 and associated electronic materials). Many authors have measured phenotypic selection on continuous floral traits (e.g., Schemske and Horvitz 1989; Johnston 1991; Caruso 2000; Maad and Alexandersson 2004; Chapter 15), and a few have combined such studies with estimates of genetic variances in the field for the same traits (Campbell 1996; Galen 1996). These studies often, but not always, detect strong directional selection on floral traits caused by pollinators, which demonstrates that quantitative floral traits are commonly adaptations for successful pollination. Stabilizing and disruptive selection seem to be less common, but this may be due to the greater statistical power needed to detect these forms of selection. These general results are mirrored in studies of natural selection in general (Kingsolver *et al.* 2001).

Despite many studies, important aspects of selection on floral traits remain poorly understood. Importantly, most studies focus exclusively on measures of female fitness, such as fruit and seed production, ignoring or estimating male fitness inadequately. The importance of male fitness for floral evolution has been recognized for more than 25 years (Willson and Price 1977), based on Bateman's principle (Bateman 1948), which predicts stronger selection on floral traits through differences in male fitness than through differences in female fitness. This prediction applies when the availability of resources other than pollen, such as light, water, or soil nutrients, limit female fertility, whereas male fitness depends on success in pollen export and ovule fertilization. Stanton *et al.*'s (1986) analysis of selection on flower colour in wild radish, discussed above, supported this prediction. How commonly Bateman's principle applies to plants has been subject to debate, as it depends largely on how commonly female fitness is pollen limited (Wilson *et al.* 1994; Larson and Barrett 2000; Ashman *et al.* 2004; Chapter 4). However, this debate does not diminish the relevance of male fitness in floral adaptation, because half of all genes are transmitted through pollen to the next generation.

Early attempts to measure male fitness focused on pollen removal from flowers, first in milkweeds and orchids, which package pollen in pollinia (e.g., Willson and Price 1977; Queller 1983; Nilsson 1988), and later in species with granular pollen, particularly after automated particle counters made such measurements more practical and accurate (see Galen and Stanton 1989; Harder and Barrett 1993). However, Harder and Thomson (1989), using a combination of theoretical and empirical approaches, showed that many floral traits may promote male siring success by *reducing* the number of pollen grains removed by each pollinator and instead placing pollen on more pollinators (Thomson and Thomson 1992; see also Stanton 1994). Therefore, pollen removal may often be a misleading proxy for male fitness.

For example, consider my research group's study of the effects of differences in anther height on both pollen removal and siring success in wild radish (Conner *et al.* 2003b). Like most members of the Brassicaceae, wild radish has four long and two short stamens, producing two anther heights within each flower. Using both experimental manipulation and natural variation, we found maximal single-visit pollen removal from flowers with the least difference in anther height. However, molecular genetic paternity analysis (see below) revealed stabilizing selection around the prevailing population mean difference in anther height during one year, perhaps reflecting a balance between the conflicting effects of increased pollen removal by individual pollinators and increased numbers of pollinators transporting pollen. In contrast, the same study found non-significant directional selection for *increased* differences during two other years. Thus, pollen removal was not a good guide to the pattern of selection based on siring success.

Surprisingly, few studies have measured selection on floral traits through seed siring success, even though the importance of male fitness in floral function and evolution has been long appreciated and the variety of genetic techniques, including AFLP, that enable measurement of siring success continues to increase. Stanton *et al.*'s (1986) work on floral colour polymorphism, discussed above, was among the first to do this. Their work was facilitated by the dual role of the floral colour trait as the trait of interest and as a genetic marker (inherited in a simple Mendelian fashion), so that the seed siring success of the yellow-flowered morphs could be measured directly by simply counting yellow-flowered offspring. In studies of quantitative traits, which do not exhibit simple Mendelian inheritance, molecular markers that are independent for the traits of interest are needed. Most studies of selection on floral traits through seed siring success have used allozymes; however, newer DNA-based markers, such as AFLP and microsatellites (see Conner and Hartl 2004 for an overview), reveal more variation than allozymes (increasing the power of inference) and can be used on virtually any species, with little development time in the case of AFLP.

New methods of analysis also increase the power to detect selection through differences in male fitness. For example, in a study of selection through male function on floral traits of *R. raphanistrum*, I used molecular genetic markers to genotype all possible fathers, mothers, and a sample of offspring to estimate seed siring success for each potential father. Initially, I used these data to estimate paternal fitness in a selection gradient analysis (Conner *et al.* 1996b); however, both steps in this two-step estimation process (first siring success and then the selection gradients) are subject to error, which can reduce the statistical power for detecting selection. Indeed, this indirect approach to measure selection found little evidence for selection on floral traits (Conner *et al.* 1996b). In contrast, reanalysis of the same data by direct estimation of the selection gradients from the molecular marker data and the measurements of the phenotypic traits of interest (Morgan and Conner 2001; Chapter 2) detected significant selection on all three measured traits during several years, although the form of selection varied among years for flower size and anther exsertion (Fig. 14.2). The stabilizing selection on anther exsertion during 1991 and 1992 suggests that an intermediate anther position was most effective in placing pollen on the pollinators, but it is unclear why selection favoured the most exserted anthers during 1993. The estimates of the selection gradients from both analyses were very similar, but the direct method produced smaller standard errors, facilitating rejection of a hypothesis of no selection.

All studies to date that have measured selection through both female and male fitness have found contrasting patterns of selection (Conner *et al.* 1996a; Kobayashi *et al.* 1999; Elle and Meagher 2000; Morgan and Conner 2001; van Kleunen and Ritland 2004; Wright and Meagher 2004). For example, in our work on wild radish, flower size experienced positive directional selection through female fitness, but stabilizing selection through male fitness during 1992 (Conner *et al.* 1996a; Morgan and Conner 2001). The differences in selection through male versus female fitness are not always as predicted; for example, van Kleunen and Ritland (2004) found selection for increased anther length through female, but not male, fitness. Therefore, measurements of selection through female fitness alone are likely to be misleading concerning both selection through male function and the total selection acting on floral traits.

In addition to the need for more estimates of selection based on male fitness, two aspects of natural selection on floral traits are obvious candidates for more thorough study. First, selection is rarely, if ever, measured during the entire life cycle. This problem is most acute in perennials, for which a heavy investment in reproduction in one season might lead to a high fitness contribution during that season at the expense of a reduced lifespan (also see Chapter 3). Thus selection favouring one trait during one year may be opposed by selection favouring a different trait during subsequent years, so that measurement of selection during a single season may mislead. Even in annuals, selection may be complicated by the "invisible fraction" (Grafen 1988), whereby differential germination success and/or mortality before flowering can cause non-adaptive evolution of floral traits if they are genetically correlated with traits that promote survival to flowering. Selection due to the invisible fraction is very difficult to measure, because it occurs before the floral traits of interest are expressed, so they cannot be measured. Therefore, the invisible fraction continues to be little studied (but see Bennington and McGraw 1995) and it remains a difficult problem for estimating natural selection in the field.

The second poorly understood aspect of selection on floral traits involves the spatial and temporal variation in selection on a given trait (also see Chapter 15). Unless selection is measured during multiple years at multiple sites, an assessment of the strength and nature of selection may be incomplete. In the context of floral evolution, variation in selection is particularly likely for species served by diverse pollinators, as the composition of the pollinator fauna may vary spatially and temporally. For example, Schemske and Horvitz (1989) measured selection on corolla length in *Calathea ovandensis* during three years, but found significant selection for decreased corolla length during only the year when one of the most effective pollinator species was present in appreciable numbers. Similarly, scarlet gilia (*Ipomopsis aggregata*) experienced significantly stronger selection for increased corolla length in the presence of another flowering species that competes with it for pollinators than when the competitor was absent (Caruso 2000). Such spatial and temporal variation in selection is probably ubiquitous, but the frequency with which selection acts in opposite directions on the same trait from year to year or site to site remains poorly studied. Regardless of how often selection changes direction, spatial and temporal variation will cause single-year studies conducted at single sites to misrepresent the strength of selection experienced by a trait.

14.5 Independent evolution of correlated traits in radish

As noted in Section 14.3 genetic correlation is often invoked as a likely evolutionary constraint, but this hypothesis is rarely tested directly. As a detailed illustration of the quantitative genetics of floral traits, I now describe a test of this hypothesis for wild radish flowers conducted by my research group. We have focused particularly on the lengths of the filaments and corolla tube, because together they determine the positions of the anthers relative to the opening of the corolla tube (anther exsertion), which affects the contact between anthers and pollinators. The phenotypic and genetic correlations between these traits are very strong (typically between 0.8 and 0.9), and significantly exceeds the average correlation between other pairs of floral traits in wild radish and some other closely related species in the Brassicaceae (Conner and Sterling 1995; J.K. Conner, K. Karoly, C. Stewart, V. Koelling, A.K. Monfils, L.A. Prather and H. Sahli unpublished manuscript). In wild radish this correlation is stable among populations and environments (Conner and Sterling 1995; J.K. Conner, K. Karoly, C. Stewart, V. Koelling, A.K. Monfils, L.A. Prather and H. Sahli unpublished manuscript), perhaps because it results from pleiotropy or extremely tight linkage (Conner 2002). A strong, stable genetic correlation caused by pleiotropy should constrain the independent evolution of filaments and corolla tubes. Below I describe both microevolutionary and macroevolutionary tests of this hypothesis.

14.5.1 Microevolution

To test whether the filament–corolla tube correlation constrains the independent evolution of these traits in wild radish over the short term, the research groups of Jeffrey Conner and Keith Karoly selected for increased and decreased anther exsertion (two lines per direction) for five or six generations, while maintaining two randomly mated control lines. Exsertion was defined as filament length minus corolla tube length (Plate 5). Our goal was to select in the direction that should be least responsive to change, that is, perpendicular to the major axis of the correlation between filament and corolla tube (Fig. 14.3). The major axis of the correlation is the direction of greatest genetic variance in two dimensions, so selection in this direction should produce the most rapid evolutionary response. This has been called the genetic line of least resistance (Schluter 1996). Perpendicular to the major axis is the direction of least genetic variance, so selection should be most constrained in this direction. Note that Stanton and Young's (1994) experiment discussed above (Section 14.3) selected perpendicular to the major axis of the correlation between petal size and pollen production.

Figure 14.4 illustrates the vectors of selection that we applied and the evolutionary responses to this selection. These vectors are the bivariate selection differentials, which depict the strength and direction of selection for both traits simultaneously in two dimensions. Although we sought to select perpendicular to the major axis of genetic variation, we were not entirely successful, because the direction of the selection vector depends on the phenotypic variance in each trait during each generation and their relative means (recall that selection was applied to the difference between filament and corolla tube, rather than to the individual traits themselves). These phenotypic means and variances depend strongly on the environment, as illustrated by the randomly mated controls (Fig. 14.4). For example, selection seems to have increased flower size overall during the experiment and flower size fluctuated from generation to generation. However, similar changes in the control lines indicated that these patterns resulted from environmental differences in a glasshouse among generations, rather than genetic changes in response to selection. Regardless, selection for increased anther exsertion achieved this outcome relative to selection for reduced exsertion in both replicates

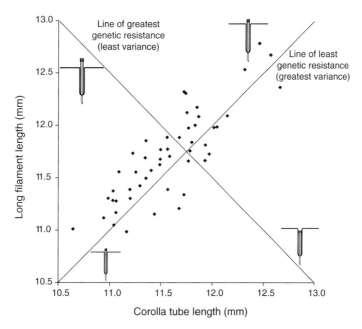

Figure 14.3 Genetic lines of least and greatest resistance. The points are means for filament and corolla tube length for half-sibling families of wild radish, *Raphanus raphanistrum* (data from Conner and Via 1993), so together they depict the additive genetic correlation between the two traits. The lines are the major and minor axes of this correlation, which correspond to the directions of greatest and least additive genetic variation in two dimensions. Selection along the major axis will produce the fastest evolutionary response, whereas selection along the minor axis will produce the slowest response; thus, they are known as the lines of least and greatest genetic resistance, respectively (Schluter 1996). The diagrams depict the relative sizes of the two traits at the four corners.

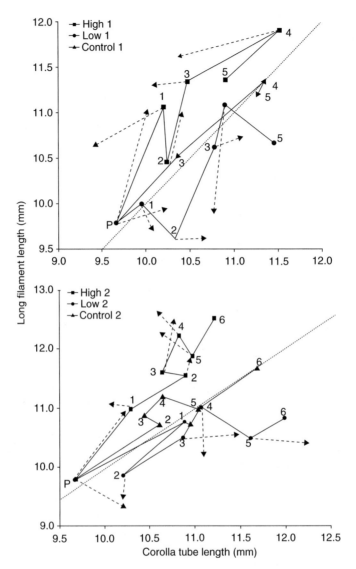

Figure 14.4 Bivariate selection differentials (dashed vectors) and overall changes in mean corolla tube and filament lengths (solid lines) of *Raphanus raphanistrum* during five and six generations of artificial selection in two replicate experiments. Each symbol represents the bivariate mean during the generation indicated by the adjacent number (P = parental), with squares indicating lines selected for increased exsertion, circles depicting lines selected for reduced exsertion, and triangles representing control lines. Control plants were not measured during the first two generations for replicate 1 and the first generation for replicate 2.

(Fig. 14.4), so that by the end of the experiment the lines had diverged perpendicularly to the major axis of genetic variation (see Plate 5 for examples).

Similarly, environmental differences among generations often affected the mean phenotype, despite the well-controlled glasshouse, which complicated assessment of the response to selection, especially for traits subject to genetic correlations. The genetic responses can be isolated partially from the influence of among-generation environmental differences by examining the phenotypic differences between the lines selected for increased and reduced anther exsertion (Fig. 14.5), which were raised together simultaneously in the same glasshouse.

ECOLOGICAL GENETICS OF FLORAL EVOLUTION

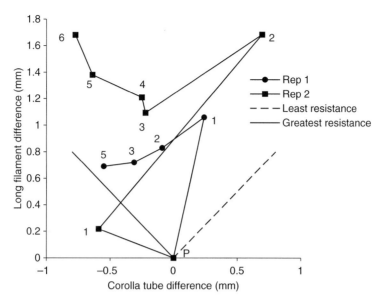

Figure 14.5 Mean differences in filament and corolla tube lengths between lines of *Raphanus raphanistrum* selected for high and low anther exsertion in two replicate experiments during five and six generations, respectively. Numbers beside each point indicate the respective generation (P = parental). Lines of greatest and least genetic resistance to selection, based on the genetic correlation between the traits, are plotted for reference. Results for generation 4 of replicate 1 are not plotted, because the selection lines were not grown simultaneously.

In replicate 1, anther exsertion changed considerably during the first generation, evolving along a trajectory roughly 30° from the line of least genetic resistance (represented by equal change for the two traits). This change in exsertion resulted from a large difference in filament length between lines and a much smaller difference in corolla tube length. During the next generation of selection the trajectory of evolution turned sharply, and further divergence between the lines selected for increased and reduced anther exsertion involved mainly evolution of tube length, with less change in filament length. Thus, replicate 1 never evolved along lines of maximum genetic resistance, that is, perpendicular to the major axis of variation, but rather at an angle to it.

Replicate 2 followed a somewhat different evolutionary trajectory. During the first generation of selection this replicate evolved roughly along the line of greatest genetic resistance (and perpendicular to the line of least resistance), as a result of both an increase in filament length and a decrease in tube length in the high lines relative to the low lines (Fig. 14.5). During the second generation, this replicate evolved parallel to the line of least resistance, resulting in a bivariate mean difference between lines similar to that in replicate 1 after one generation, with a large divergence in filament length and a smaller divergence in tube length between lines. As in replicate 1, the trajectory then turned sharply during the next generation and evolved in almost the opposite direction to that during generation two, again roughly parallel to the line of least resistance. Finally, the trajectory of divergence turned again and evolved along lines of greatest resistance during the final three generations.

The preceding results indicate that both traits evolved independently in both replicates, ultimately resulting in longer filaments and shorter corolla tubes in the lines selected for increased anther exsertion than in those selected for reduced exsertion. In the aggregate, anther exsertion evolved very nearly along the line of greatest resistance (especially in replicate 2), even though many individual segments of the trajectory deviated from this course. This outcome demonstrates that genetic correlations are not a strong constraint to independent evolution, at least over a few generations, and that net evolution can occur quite rapidly and predictably in the direction of greatest genetic resistance (least genetic variance) if selection is in this direction. However, the precise trajectories of evolution during each generation are less predictable.

14.5.2 Macroevolution

To test whether the type of independent evolution of filament and corolla tube observed in the preceding experiment is reflected in species divergence, Alan Prather, Anna Wiese, and I, along with members of my research group, measured floral traits in 23 species drawn broadly from the Brassicaceae, plus *Cleome spinosa*, representing an outgroup (Capparaceae). For these species, the mean lengths of the long filaments and corolla tubes have clearly evolved along lines of least genetic resistance (Fig. 14.6): most species cluster near the line of equality. Thus, most of the evolution in filament and corolla tube length in the Brassicaceae has resulted from changes in flower size, without much change in these traits relative to each other. Whether this isometry represents the influence of a genetic constraint is less clear. Schluter (1996) noted that isometry could occur if the line of least resistance also includes adaptive trait combinations. Accordingly, the observed isometry could reflect correlational selection for equal-length filament and corolla tubes within the Brassicaceae. We have evidence for this correlational selection in wild radish (Fig. 14.2; Morgan and Conner 2001), although whether this selection has occurred throughout such a large (>3000 species) and diverse family is unknown. The constraint is clearly not absolute, as very exserted anthers (filaments much longer than the corolla tube) have evolved in *Aethionema* and *Stanleya*, and highly inserted anthers (filaments shorter than corolla tube) have evolved in *Hesperis* and *Matthiola* (see Plate 5). These genera mirror the results of our artificial selection for increased and reduced

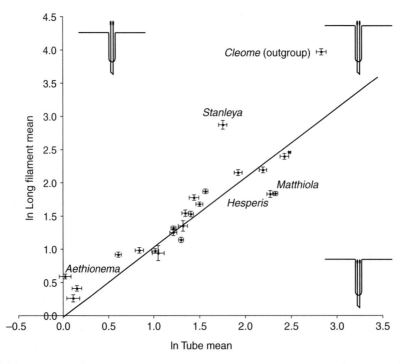

Figure 14.6 Macroevolution (mostly) along lines of least genetic resistance. Shown are ln-transformed mean (±95% confidence intervals) filament and corolla tube lengths of 23 species of Brassicaceae plus *Cleome* (Capparaceae). Some species were measured in the field, whereas others were measured in a glasshouse, so environmental differences are included. Schematic floral cross-sections depicting anther exsertion in the different regions of the graph are also shown.

anther exsertion, respectively. Overall, these comparative results suggest that genetic correlations among traits can guide the direction of interspecific phenotypic differentiation, but that these constraints can be overcome by selection.

14.6 Future directions

What directions might studies of the genetics of floral traits take in the near future? Here I highlight four areas:

Selection on floral traits through male fitness: More studies of selection on floral traits that incorporate lifetime male fitness, estimated through molecular genetic paternity analysis, are clearly needed. Because selection often acts differently through male than through female fitness (see Section 14.4 above), this information is critical to identify what floral trait values are adaptive and in what ways selection through male and female fitness conflicts. Although selection on floral traits through male fitness and functional gender have been major topics in plant evolutionary ecology for 25 years (Lloyd 1980; Lloyd 1984; Chapter 1), data that address these topics directly with reliable estimates of male fitness have been sorely lacking.

The genetic architecture of floral adaptations: The quantitative genetic techniques described in this chapter treat the genome as a black box, using the statistical abstractions of variance and covariance. This approach will continue to be extremely fruitful, especially to predict evolutionary change quantitatively using eqs 1 and 2 (Section 14.3), but it does not provide information on the details of genetic variances and covariances at the level of individual loci and alleles at those loci. The genetic architecture of a trait includes several interesting and fundamental issues. For example, is variation in a given trait governed by a few loci with large influence on phenotypic variance, many loci of small effects, or a mixture of both? The latter possibility seems most likely and is supported by many of the studies of quantitative trait loci (QTL) discussed below. If this result is confirmed, it implies that the initial stages of adaptation may be quite rapid, involving fixation of the large-effect loci, but the latter stages will be slower, relying instead on the greater number of loci with smaller phenotypic effect. Another key aspect of genetic architecture is whether genetic correlations between traits result from pleiotropy or gametic-phase disequilibrium (Conner 2002). Correlations caused by gametic-phase disequilibrium are unlikely to cause long-term evolutionary constraint; conversely, a finding of gametic-phase disequilibrium suggests that selection or some other evolutionary process is operating to maintain the disequilibrium. Finally, how important is epistasis i.e., interactions among gene loci, in adaptive evolution? Gene interactions are ubiquitous and they seem to be important in experimental evolution of microbes (Elena and Lenski 2001); however, their importance in adaptive evolution of more complex organisms and traits, such as floral traits, is largely unknown.

Mapping of QTL takes a first step towards characterizing genetic architecture. QTL mapping seeks to associate variance in complex phenotypic traits with molecular markers scattered throughout the genome. For floral traits this technique has been used mainly to map genes causing interspecific differences (e.g., Lin and Ritland 1997; Bradshaw *et al.* 1998). An exception was provided by Juenger *et al.* (2000), who mapped eight floral traits in the well-studied Landsberg *erecta* X Columbia recombinant inbred lines of *Arabidopsis thaliana*. They found 18 QTL that affected at least one trait, with 11 affecting more than one trait, indicating pleiotropy or fairly tight linkage. QTL mapping is only a first step in determining genetic architecture of complex traits, because the QTL identified are fairly large sections of chromosomes that can contain dozens or perhaps hundreds of genes. Because the *A. thaliana* genome has been sequenced, Juenger *et al.* could identify candidate genes (genes whose function has been identified in other studies) for some of these QTL. This is one approach that can facilitate the identification of the gene loci affecting a trait, and thus understanding of the genetic architecture of adaptive floral traits at the level of individual loci.

The relative roles of regulatory versus structural genes: A third fundamental question considers the degree to which variation in complex traits results from variation in structural genes (those coding for proteins) versus regulatory genes (those affecting the expression of other genes). Changes in regulatory genes can create dramatic and coordinated phenotypic change, as the effects of one regulatory change can cascade through a large metabolic pathway or regulatory network. Regulatory genes underlie major morphological adaptations in vertebrates (Abzhanov et al. 2004; Shapiro et al. 2004) and the evolution of regulatory elements has been proposed to be the main source of novel morphologies (Doebley and Lukens 1998). Although changes in regulatory genes may be particularly important in morphological evolution, changes in structural genes may be more important in simpler traits involving single molecules, such as resistance to insecticides and toxins (Geffeney et al. 2002; Ffrench-Constant et al. 2004) and pigmentation (Hoekstra and Nachman 2003). For example, recall that red flowers evolved in *Ipomoea* through changes in an enzyme-coding structural gene (Section 14.2; Zufall and Rausher 2004). Nevertheless, a substantial body of work demonstrates that regulatory genes control floral-organ identity (reviewed in Ng and Yanofsky 2001). Whether the same loci vary within natural populations, and are responsible for genetic variance in adaptive floral traits, is mostly unknown. The increasing availability of whole-genome microarrays and quantitative PCR are improving the ability to measure gene expression, so these questions are now more tractable.

Combined approaches: The largest advances in understanding the ecological genetics of flowers may result from studies that combine different approaches, such as QTL mapping and microarray analysis (Wayne and McIntyre 2002), artificial selection and QTL mapping (reviewed in Conner 2003), and quantitative genetics within species with phylogenetic studies (Schluter 1996; Baker and Wilkinson 2003; Fig. 14.6). These combinations will probably provide the most rapid progress towards some long-term goals in the ecological genetics of flowers, such as understanding complex floral adaptations at the level of allele frequencies at individual gene loci, and understanding the interplay between genetics and selection in creating floral diversification among species and even higher taxonomic groups. To date, these combined approaches have seldom been applied to floral traits, but advanced molecular genetic tools are currently under development for several plant species in addition to *A. thaliana*, including *Mimulus* (http://www.biology.duke.edu/mimulus/) and *Aquilegia* (S. A. Hodges personal communication), so that understanding of the genetic influences on floral adaptation is certain to continue evolving.

Acknowledgements

I thank my collaborators, Keith Karoly, Alan Prather, Anna Wiese, and members of our research groups for helping to generate many of the ideas and data presented here. I am also grateful to my current research group, Meghan Duffy, Frances Knapczyk, Cindy Mills, Neil Patel, Angela Roles, and Heather Sahli, for comments on an earlier version of this chapter. My research has been supported by the National Science Foundation under grants DEB 9796183, DEB 9903880, DEB 0108354, and DBI 9605168 and by the Cooperative State Research, Education, and Extension Service, US Department of Agriculture, under Agreement No. 2002*35320–11538.

References

Abzhanov A, Protas M, Grant BR, Grant PR, and Tabin CJ (2004). Bmp4 and morphological variation of beaks in Darwin's finches. *Science*, **305**, 1462–65.

Armbruster WS (1991). Multilevel analysis of morphometric data from natural plant populations: insights into ontogenetic, genetic, and selective correlations in *Dalechampia scandens*. *Evolution*, **45**, 1229–44.

Ashman TL, Knight TM, Steets JA, et al. (2004). Pollen limitation of plant reproduction: Ecological and evolutionary causes and consequences. *Ecology*, **85**, 2408–21.

Baker RH and Wilkinson GS (2003). Phylogenetic analysis of correlation structure in stalk-eyed flies (*Diasemopsis*, Diopsidae). *Evolution*, **57**, 87–103.

Bateman AJ (1948). Intra-sexual selection in *Drosophila*. *Heredity*, **2**, 349–68.

Bennington CC and McGraw JB (1995). Phenotypic selection in an artificial population of *Impatiens pallida*: The importance of the invisible fraction. *Evolution*, **49**, 317–24.

Bradshaw HD Jr, Otto KG, Frewen BE, McKay JK, and Schemske DW (1998). Quantitative trait loci affecting differences in floral morphology between two species of monkeyflower (*Mimulus*). *Genetics*, **149**, 367–82.

Campbell DR (1996). Evolution of floral traits in a hermaphroditic plant: field measurements of heritabilities and genetic correlations. *Evolution*, **50**, 1442–53.

Caruso CM (2000). Competition for pollination influences selection on floral traits of *Ipomopsis aggregata*. *Evolution*, **54**, 1546–57.

Caruso CM (2004). The quantitative genetics of floral trait variation in *Lobelia*: potential constraints on adaptive evolution. *Evolution*, **58**, 732–40.

Clegg MT and Durbin ML (2000). Flower color variation: A model for the experimental study of evolution. *Proceedings of the National Academy of Sciences of the United States of America*, **97**, 7016–23.

Conner JK (2002). Genetic mechanisms of floral trait correlations in a natural population. *Nature*, **420**, 407–10.

Conner JK (2003). Artificial selection: a powerful tool for ecologists. *Ecology*, **84**, 1650–60.

Conner JK and Hartl DL (2004). *A primer of ecological genetics*. Sinauer Associates, Sunderland, MA.

Conner JK and Sterling A (1995). Testing hypotheses of functional relationships: a comparative survey of correlation patterns among floral traits in five insect-pollinated plants. *American Journal of Botany*, **82**, 1399–406.

Conner JK and Via S (1993). Patterns of phenotypic and genetic correlations among morphological and life history traits in wild radish, *Raphanus raphanistrum*. *Evolution*, **47**, 704–11.

Conner JK, Rush S, and Jennetten P (1996a). Measurements of natural selection on floral traits in wild radish (*Raphanus raphanistrum*). I. Selection through lifetime female fitness. *Evolution*, **50**, 1127–36.

Conner JK, Rush S, Kercher S, and Jennetten P (1996b). Measurements of natural selection on floral traits in wild radish (*Raphanus raphanistrum*). II. Selection through lifetime male and total fitness. *Evolution*, **50**, 1137–46.

Conner JK, Franks R, and Stewart C (2003a). Expression of additive genetic variances and covariances for wild radish floral traits: comparison between field and greenhouse environments. *Evolution*, **57**, 487–95.

Conner JK, Rice AM, Stewart C, and Morgan MT (2003b). Patterns and mechanisms of selection on a family-diagnostic trait: Evidence from experimental manipulation and lifetime fitness selection gradients. *Evolution*, **57**, 480–6.

Delph LF, Gehring JL, Frey FM, Arntz AM, and Levri M (2004). Genetic constraints on floral evolution in a sexually dimorphic plant revealed by artificial selection. *Evolution*, **58**, 1936–46.

Doebley J and Lukens L (1998). Transcriptional regulators and the evolution of plant form. *The Plant Cell*, **10**, 1075–82.

Elena SF and Lenski RE (2001). Epistasis between new mutations and genetic background and a test of genetic canalization. *Evolution*, **55**, 1746–52.

Elle E and Meagher TR (2000). Sex allocation and reproductive success in the andromonoecious perennial *Solanum carolinense* (Solanaceae). II. Paternity and functional gender. *American Naturalist*, **156**, 622–36.

Epperson BK and Clegg MT (1987). Frequency-dependent variation for outcrossing rate among flower-color morphs of *Ipomoea purpurea*. *Evolution*, **41**, 1302–11.

Falconer DS and Mackay TFC (1996). *Introduction to quantitative genetics*, 4th edition. Longman, Harlow, UK.

Ffrench-Constant RH, Daborn PJ, and Le Goff G (2004). The genetics and genomics of insecticide resistance. *Trends in Genetics*, **20**, 163–70.

Ford EB (1975). *Ecological genetics*. Chapman and Hall, London.

Galen C (1996). Rates of floral evolution: adaptation to bumblebee pollination in an alpine wildflower, *Polemonium viscosum*. *Evolution*, **50**, 120–5.

Galen C (1999). Why do flowers vary. *Bioscience*, **49**, 631–40.

Galen C and Stanton ML (1989). Bumble bee pollination and floral morphology: Factors influencing pollen dispersal in the alpine sky pilot, *Polemonium viscosum* (Polemoniaceae). *American Journal of Botany*, **76**, 419–26.

Geffeney S, Brodie ED, and Ruben PC (2002). Mechanisms of adaptation in a predator-prey arms race: TTX-resistant sodium channels. *Science*, **297**, 1336–9.

Grafen A (1988). On the uses of data on lifetime reproductive success. In TH Clutton-Brock, ed. *Reproductive success*, pp. 454–71. University of Chicago Press, Chicago.

Harder LD and Barrett SCH (1993). Pollen removal from tristylous *Pontederia cordata*: effects of anther position and pollinator specialization. *Ecology*, **74**, 1059–72.

Harder LD and Thomson JD (1989). Evolutionary options for maximizing pollen dispersal of animal-pollinated plants. *American Naturalist*, **133**, 323–44.

Hoekstra HE and Nachman MW (2003). Different genes underlie adaptive melanism in different populations of rock pocket mice. *Molecular Ecology*, **12**, 1185–94.

Huether CA Jr (1968). Exposure of natural genetic variability underlying the pentamerous corolla constancy in *Linanthus androsaceus* ssp. *androsaceus*. *Genetics*, **60**, 123–46.

Johnston MO (1991). Natural selection on floral traits in two species of *Lobelia* with different pollinators. *Evolution*, **45**, 1468–79.

Juenger T, Purugganan M, and Mackay TFC (2000). Quantitative trait loci for floral morphology in *Arabidopsis thaliana*. *Genetics*, **156**, 1379–92.

Karoly K and Conner JK (2000). Heritable variation in a family-diagnostic trait. *Evolution*, **54**, 1433–8.

Kay QON (1978). The role of preferential and assortative pollination in the maintenance of flower colour polymorphisms. In AJ Richards, ed. *The pollination of flowers by insects*, pp. 175–90. Academic Press, New York.

Kingsolver JG, Hoekstra HE, Hoekstra JM, et al. (2001). The strength of phenotypic selection in natural populations. *American Naturalist*, **157**, 245–61.

Kobayashi S, Inoue K, and Kato M (1999). Mechanism of selection favoring a wide tubular corolla in *Campanula punctata*. *Evolution*, **53**, 752–7.

Larson BMH and Barrett SCH (2000). A comparative analysis of pollen limitation in flowering plants. *Biological Journal of the Linnean Society*, **69**, 503–20.

Lin J-Z and Ritland K (1997). Quantitative trait loci differentiating the outbreeding *Mimulus guttatus* from the inbreeding *M. platycalyx*. *Genetics*, **146**, 1115–21.

Lloyd DG (1980). Sexual strategies in plants III. A quantitative method for describing the gender of plants. *New Zealand Journal of Botany*, **18**, 103–8.

Lloyd DG (1984). Gender allocations in outcrossing cosexual plants. In R Dirzo and J Sarukhán, eds. *Perspectives on plant population ecology*, pp. 277–300. Sinauer Associates, Sunderland, MA.

Maad J and Alexandersson R (2004). Variable selection in *Platanthera bifolia* (Orchidaceae): phenotypic selection differed between sex functions in a drought year. *Journal of Evolutionary Biology*, **17**, 642–50.

Mazer SJ and Schick CT (1991). Constancy of population parameters for life-history and floral traits in *Raphanus sativus* L. II. Effects of planting density on phenotype and heritability estimates. *Evolution*, **45**, 1888–907.

Mazer SJ, Delesalle VA, and Neal PR (1999). Responses of floral traits to selection on primary sexual investment in *Spergularia marina*: The battle between the sexes. *Evolution*, **53**, 717–31.

Morgan MT and Conner JK (2001). Using genetic markers to directly estimate male selection gradients. *Evolution*, **55**, 272–81.

Ng M and Yanofsky MF (2001). Function and evolution of the plant MADS-box gene family. *Nature Reviews Genetics*, **2**, 186–95.

Nilsson LA (1988). The evolution of flowers with deep corolla tubes. *Nature*, **334**, 147–9.

O'Neil P and Schmitt J (1993). Genetic constraints on the independent evolution of male and female reproductive characters in the tristylous plant *Lythrum salicaria*. *Evolution*, **47**, 1457–71.

Phillips PC (2005). Testing hypotheses regarding the genetics of adaptation. *Genetica*, **123**, 15–24.

Queller DC (1983). Sexual selection in a hermaphroditic plant. *Nature*, **305**, 706–7.

Ritland K (2000). Marker-inferred relatedness as a tool for detecting heritability in nature. *Molecular Ecology*, **9**, 1195–204.

Ritland K and Ritland C (1996). Inferences about quantitative inheritance based on natural population structure in the yellow monkey flower, *Mimulus guttatus*. *Evolution*, **50**, 1074–82.

Sarkissian TS and Harder LD (2001). Direct and indirect responses to selection on pollen size in *Brassica rapa* L. *Journal of Evolutionary Biology*, **14**, 456–68.

Schemske DW and Bierzychudek P (2001). Perspective: evolution of flower color in the desert annual *Linanthus parryae*: Wright revisited. *Evolution*, **55**, 1269–82.

Schemske DW and Horvitz CC (1989). Temporal variation in selection on a floral character. *Evolution*, **43**, 461–5.

Schluter D (1996). Adaptive radiation along genetic lines of least resistance. *Evolution*, **50**, 1766–74.

Shapiro MD, Marks ME, Peichel CL, et al. (2004). Genetic and developmental basis of evolutionary pelvic reduction in threespine sticklebacks. *Nature*, **428**, 717–23.

Shore JS and Barrett SCH (1990). Quantitative genetics of floral characters in homostylous *Turnera ulmifolia* var. *angustifolia* Willd. (Turneraceae). *Heredity*, **64**, 1105–12.

Stanton ML (1994). Male–male competition during pollination in plant populations. *American Naturalist*, **144**, S40–68.

Stanton ML and Young HJ (1994). Selecting for floral character associations in wild radish, *Raphanus sativus* L. *Journal of Evolutionary Biology*, **7**, 271–85.

Stanton ML, Snow AA, and Handel SN (1986). Floral evolution: attractiveness to pollinators increases male fitness. *Science*, **232**, 1625–7.

Thomas SC, Pemberton JM, and Hill WG (2000). Estimating variance components in natural populations using inferred relationships. *Heredity*, **84**, 427–36.

Thomson JD and Thomson BA (1992). Pollen presentation and viability schedules in animal-pollinated plants: consequences for reproductive success. In R Wyatt, ed. *Ecology and evolution of plant reproduction*, pp. 1–24. Chapman and Hall, New York.

Turelli M, Schemske DW, and Bierzychudek P (2001). Stable two-allele polymorphisms maintained by fluctuating fitness and seed banks: Protecting the blues in *Linanthus parryae*. *Evolution*, **55**, 1283–98.

van Kleunen M and Ritland K (2004). Predicting evolution of floral traits associated with mating system in a natural plant population. *Journal of Evolutionary Biology*, **17**, 1389–99.

Waser NM and Price MV (1981). Pollinator choice and stabilizing selection for flower color in *Delphinium nelsonii*. *Evolution*, **35**, 376–90.

Wayne ML and McIntyre LM (2002). Combining mapping and arraying: an approach to candidate gene identification. *Proceedings of the National Academy of Sciences of the United States of America*, **99**, 14903–6.

Willson MF and Price PW (1977). The evolution of inflorescence size in *Asclepias* (Asclepiadaceae). *Evolution*, **31**, 495–511.

Wilson P, Thomson JD, Stanton ML, and Rigney LP (1994). Beyond floral Batemania: gender biases in selection for pollination success. *American Naturalist*, **143**, 283–96.

Worley AC and Barrett SCH (2000). Evolution of floral display in *Eichhornia paniculata* (Pontederiaceae): Direct and correlated responses to selection on flower size and number. *Evolution*, **54**, 1533–45.

Wright JW and Meagher TR (2004). Selection on floral characters in natural Spanish populations of *Silene latifolia*. *Journal of Evolutionary Biology*, **17**, 382–95.

Zufall RA and Rausher MD (2004). Genetic changes associated with floral adaptation restrict future evolutionary potential. *Nature*, **428**, 847–50.

CHAPTER 15

Geographical context of floral evolution: towards an improved research programme in floral diversification

Carlos M. Herrera, María Clara Castellanos, and Mónica Medrano

Estación Biológica de Doñana, CSIC, Sevilla, Spain

Outline

The diversification of animal-pollinated angiosperms is related to divergence in floral characteristics promoted by adaptations to different pollinators. According to prevailing evolutionary theory, this macroevolutionary pattern results from adaptive local or regional differentiation of pollination-related features in response to spatial divergence in pollinators. This crucial process links the micro- and macroevolution of floral adaptation, yet it has received much less attention than either floral diversification of species in a phylogenetic context, or pollinator-mediated phenotypic selection on pollination-related traits within populations. This chapter includes two components. We first use a literature survey to demonstrate that the study of plant–pollinator interaction in a geographical context is a relatively neglected element of research on floral diversification. In addition, the few studies that explicitly assess intraspecific variation in pollinators and pollination-related traits generally do not provide unequivocal evidence for a *causal* role of divergent selection from pollinators in intraspecific differentiation in floral traits. We then describe an analysis of regional variation in pollinators and corolla traits (upper lip and corolla tube length) of *Lavandula latifolia*, a Mediterranean evergreen shrub, which illustrates a five-step protocol for identifying geographical differentiation in floral traits driven by spatially variable selection from pollinators. Corolla traits, pollinator composition, and phenotypic selection on the upper corolla lip all vary geographically, and the morphological and pollination-related selection clines are closely congruent. Our results for this species implicate adaptive intraspecific floral differentiation in response to a cline in pollinator-mediated selection on pollination success, although confirmation of this conclusion awaits experiments to determine the genetic basis of floral variation.

15.1 Introduction

Since we view transpecific evolution as an extension of events at the species level, the foundation of most evolutionary theory rests upon inferences drawn from geographic variation or upon the verification of predictions made about it. Gould and Johnston (1972, p. 457)

The causal role played by animal pollinators in the extraordinary diversification of angiosperm flowers has figured prominently in plant biology since Darwin. The connection between floral diversity and divergence in pollination mechanisms of animal-pollinated lineages was recognized early in the history of evolutionary biology (Darwin 1862; Leppik 1957; Stebbins 1970). Several lines of evidence implicate animal pollinators in angiosperm diversification, including the fact that taxonomically distinctive traits primarily involve reproductive characters for animal-pollinated lineages, but not for abiotically pollinated taxa (Grant 1949); the temporal match in geological time between the radiations of angiosperms and major groups of animal pollinators (Grimaldi 1999); the frequent association between suites of floral traits and particular pollinator groups (Fenster et al. 2004); evidence of more rapid and/or extensive diversification in lineages of animal-pollinated plants (Eriksson and Bremer 1992; Ricklefs and Renner 1994; Dodd et al. 1999); and phylogenetic analyses showing that floral form has played a key role in the speciation of some animal-pollinated lineages (Graham and Barrett 2004; Sargent 2004; Chapter 17).

Recently, research on the adaptive origin of floral diversity in animal-pollinated angiosperms has generally adopted one of two approaches. On the one hand, and largely as a consequence of the increased availability of molecular phylogenies, a growing number of investigations have examined the ecological and pollination correlates of floral diversification in a phylogenetic context at the species level and above (Hapeman and Inoue 1997; Graham and Barrett 2004; Patterson and Givnish 2004; Chapter 17). On the other hand, many studies have assessed pollinator-mediated phenotypic selection on floral traits within populations by measuring the fitness consequences of floral variation that occurs naturally (Campbell et al. 1991; Herrera 1993; Maad 2000; Chapter 14) or has been induced artificially (Herrera 2001; Aigner 2004; Castellanos et al. 2004). The profusion of investigations adopting these approaches contrasts with the scarcity of studies of floral diversification that focus on intraspecific floral variation and its relation to geographic divergence in pollinators.

As summarized in Gould and Johnston's (1972) statement quoted at the beginning of this chapter, the hypothesis that macroevolutionary patterns represent the aggregate outcomes of microevolutionary processes at the intraspecific level is a central tenet of current evolutionary thought (Simpson 1953; Bock 1970). Local adaptation to contrasting pollination environments is an important component of adaptive floral diversification (e.g., Dilley et al. 2000; Patterson and Givnish 2004; Chapter 16). For this reason, studies of intraspecific geographical differentiation in floral traits and its potential relation to divergent selection from pollinators are crucial for understanding the linkage between the micro- and macroevolution of floral traits. Similar arguments have been raised by Barrett (1995; Barrett et al. 2001) in relation to the study of the evolution of plant mating systems. However, despite their interest and significance, relatively few studies have addressed the relation of intraspecific floral differentiation to geographically changing selection from pollinators, and most of these do not make convincing cases for pollinator-driven intraspecific differentiation, as discussed below.

In this chapter, we address the geographical context of floral evolution with a literature overview and a detailed, stepwise analysis of a case example. We begin by reviewing the relevant literature from two perspectives. First, we demonstrate that research on floral diversification has largely neglected the geographical context of plant–pollinator interactions. Then, we consider published evidence of intraspecific geographical differentiation in floral form and function and its relation to variation in pollinator faunas, highlighting some limitations that commonly hinder adaptive interpretations of observed patterns. Finally, we outline a relatively simple, stepwise protocol for identifying instances of geographical differentiation in floral traits driven by spatially variable selection from pollinators. We illustrate this approach with a study of geographical variation in the flowers and pollinators of *Lavandula latifolia*, a Mediterranean, evergreen shrub.

15.2 Representation of geographical variation in pollination studies

The neglect of geographical context by studies of floral diversification is evident from the remarkable scarcity of well-documented cases of pollinator-driven intraspecific geographical differentiation in floral form or function in recent books or reviews dealing with local differentiation in plants (Linhart and Grant 1996), ecological speciation (Levin 2000), or the geographical mosaic theory of plant–animal coevolution (Thompson 1994). To quantify this subjective impression, we conducted two literature surveys as described in the following two sections. First, we reviewed the literature looking for descriptions of geographic variation in floral characteristics and their pollinators. Next, we searched for studies that went beyond patterns and quantified processes, specifically, phenotypic selection on pollination-related plant traits. In both cases we were interested in evaluating the frequency of studies that considered geographical variation.

15.2.1 Patterns: how much attention has geographical variation in plant traits and pollinators received?

We screened the primary literature for papers describing both a plant species' pollinator fauna and one or more floral or plant traits putatively related to pollination. These studies were classified according to whether they provided data on geographical variation. Floral traits could be functional (e.g., dichogamy, floral longevity, nectar secretion rate) or structural (e.g., floral morphology, nectar composition, inflorescence height). We considered only studies conducted under natural field conditions, excluding studies performed in a glasshouse or in experimental plots or arrays, or that involved manipulated plant traits. The survey comprised articles published from 1995 until June 2005 that were accessible to us online; the starting year was later than 1995 for five journals with limited online availability. The journals screened and the first year reviewed (if different from 1995) were: *American Journal of Botany*, *Annals of Botany*, *Canadian Journal of Botany* (1998), *Ecography* (2000), *Ecological Monographs*, *Ecology*, *Evolution*, *International Journal of Plant Sciences*, *Journal of Evolutionary Biology*, *Oecologia* (1997), *Oikos* (2000), and *Plant Systematics and Evolution* (2001). These publications represent major outlets for pollination studies and thus likely provide a representative sample of published research in this field. We initially queried the ISI Web of Science database with the string *"pollinator or pollination biology or pollinated"* for each journal. The resulting articles ($N = 867$) were examined individually if the abstract indicated suitable content. Two reviewers performed the searches and classified the studies, one examining odd years and the other even years, to reduce possible biases.

Studies were classified according to whether they studied geographical variation in pollinator composition, abundance or visitation rates, and whether they studied geographical variation in plant traits (Table 15.1). By "geographical variation" we mean examination of at least two populations of the same plant species. We included plant species individually in the table, so that multi-species studies contributed more than one species. The upper-left cell in Table 15.1 includes single-site studies that reported only *quantitative* measures of plants and pollinators. This group excludes investigations that measured plant traits but mentioned only the main pollinators, and studies that quantified pollinator composition but provided simple descriptions of floral features. In contrast, for the upper-right and lower-left cells we relaxed the requirement that both plant traits and pollinator composition be measured quantitatively, because very few papers described variation in either plants or pollinators among sites, but quantified the other aspect in only one site. Also, because we were interested in studies that considered geographical aspects, we wanted to ensure that they all were included in the table. As a result of this procedure, the number of studies in the upper-left cell might be underestimated, but this conservative approach reinforces the conclusions drawn below. Finally, the lower-right cell in

Table 15.1 The incidence with which pollination-biology studies published during 1995–2005 in 12 ecological and botanical journals (see text for details) considered geographical variation in pollinator composition and pollination-related plant traits.

Sites studied for pollination-related plant traits	Sites studied for pollinator composition	
	1	>1
1	525 (79.1)	27 (4.1)
>1	62 (9.3)	50 (7.5)

Numbers in each cell represent the number of species considered, with the percentage of the overall total in parentheses. A list of the literature references used to construct this table is available upon request or in Electronic Appendix 15.1 (http://www.eeb.utoronto.ca/EEF/).

Table 15.1 includes studies that quantified both pollinator composition and pollination-related plant traits for more than one locality. Many of these papers did not compare localities (i.e., they were not testing for geographical variation explicitly), yet we adopted the conservative procedure of including them if findings for different populations were reported separately.

The final survey (Table 15.1) included 198 articles, which provided both pollinator and floral data for 664 plant species. The vast majority of species included in our sample provided information about pollinators and/or pollination-related traits for only one population. For only 7.5% of species were data on pollinator composition and pollination-related traits reported for multiple populations. Information on geographical variation was provided for an additional 13.4% of species, but it referred to either pollinators or plant traits alone, with information on plant traits being twice as common as that for pollinators. These results illustrate unequivocally that pollination biologists rarely consider the geographical context, even though our threshold for a study to qualify for "geographical variation" was quite liberal (number of populations > 1). Almost no studies would have been characterized as considering geographic variation if we had applied a slightly more restrictive threshold (e.g., number of populations > 3).

15.2.2 Processes: how much do we know about geographical variation in selection on pollination-related traits?

Our second literature survey considered studies of phenotypic selection (*sensu* Lande and Arnold 1983) on floral and other pollination-related traits. To make this search as comprehensive as possible, we did not limit the journals or years examined. We used a combination of sources to locate studies, including citations in review articles (e.g., Kingsolver *et al.* 2001) and searches of the ISI Web of Science. To be included, studies had to be conducted under natural pollination conditions and measure phenotypic selection on some character(s) hypothesized by the author(s) to be under pollinator-mediated selection. Glasshouse or flight cage studies were not considered. Selection had to be measured on traits with typical variation: artificially induced trait variation was acceptable only if it was kept within the range of phenotypic variation for the species. We included studies on both discrete (e.g., flower colour) and continuous (e.g., corolla size) traits. These criteria excluded studies using artificial conditions (e.g., controlled pollinator identity or extreme floral variation) to study phenotypic selection on plant traits, but we were more interested in studies of selection in the wild than in research designed to explore the mechanisms of selection. Likewise, we may have missed some studies of selection on modified floral or plant traits, because they often do not describe their results as "phenotypic selection." Because experimental studies are not generally replicated geographically, their exclusion should not bias our conclusions.

Results of our survey of phenotypic selection studies are summarized in Table 15.2, which includes data from 62 publications and 66 plant species. For only 39% of these species did the studies examine the possibility of geographical variation in selection by comparing phenotypic selection gradients among populations. However, despite this relative scarcity, the proportion of geographically informed studies was somewhat higher in this case than among the studies of general pollination biology surveyed in the preceding section (Table 15.1). This difference may

Table 15.2 Characteristics of published studies of phenotypic selection on pollination-related plant traits.

Type of pollination-related traits	Is phenotypic selection compared among populations?	
	No	Yes
Structural	25 (64.1)	14 (35.9)
Functional	5 (62.5)	3 (37.5)
Both trait types	10 (52.6)	9 (47.4)
Total	40 (60.6)	26 (39.4)

Numbers in each cell represent the number of studied species, with the corresponding percentages of the row total in parentheses. A list of the literature references used to construct this table is available upon request or in Electronic Appendix 15.1 (http://www.eeb.utoronto.ca/EEF/).

indicate that researchers who go beyond description of pollinators and plant traits to investigate the fitness consequences of floral variation under a particular pollination regime are more often aware of the importance of documenting variation in selective regimes among populations. However, this interpretation is contradicted by the fact that only 6 of the 26 geographically informed studies summarized in Table 15.2 quantified population differences in pollinators along with differences in phenotypic selection on plant traits. Therefore, phenotypic selection studies are not an exception to the predominant neglect of a geographical context in investigations of pollinator-mediated floral evolution.

15.3 Outcomes and limitations of geographically informed studies

15.3.1 Outcomes

This section summarizes the outcomes of the few studies in the preceding literature surveys that measured geographical variation in both plant traits and their pollinators (50 species from Table 15.1 plus 6 species from Table 15.2). We asked two questions for this subset of studies: (1) how often did pollinators *and* plant traits vary significantly among populations of the same species; and (2) when both plant traits and pollinators varied significantly, how often was the observed floral variation consistent with patterns expected from adaptive intraspecific diversification mediated by pollinators. To this end, we examined in detail studies in the lower-right cell of Table 15.1, and those in Table 15.2 that included information on pollinators, classifying them according to whether significant inter-population variation was found in floral traits, pollinator composition, or both. Populations were compared for only 33 species, and the outcomes of these studies are summarized in Table 15.3.

Plant–pollinator systems commonly vary geographically: 60.6% of the species included in Table 15.3 exhibit joint geographical variation in plant traits and pollinators. Many investigations published in journals or years not covered by our surveys also confirm the widespread occurrence of simultaneous geographical variation

Table 15.3 The incidence of significant geographical variation in pollinator faunas and pollination-related plant traits, based on the studies referred to in Tables 15.1 and 15.2.

Significant geographical variation in pollination-related plant traits?	Significant geographical variation in pollinators?	
	No	Yes
No	5 (15.1)	3 (9.1)
Yes	5 (15.1)	20 (60.6)

Numbers in each cell represent the number of species, with the percentage of the overall total in parentheses. A list of the literature references used to construct this table is available upon request or in Electronic Appendix 15.1 (http://www.eeb.utoronto.ca/EEF/).

in pollination-related traits and pollinator composition (e.g., Miller 1981; Armbruster 1985; Arroyo and Dafni 1995; Inoue *et al.* 1996; Boyd 2002; Malo and Baonza 2002). Studies of 13 of the 33 species included in Table 15.3 explicitly considered the association of floral variation or phenotypic selection on floral traits with variable pollinator faunas. In other words, less than half of these investigations were designed to assess whether geographical variation in floral traits was congruent with pollinator variation. Eight studies of seven species presented compelling evidence for congruent variation between plant traits and pollinator composition (Johnson and Steiner 1997; Gómez and Zamora 1999, 2000; Fausto *et al.* 2001; Totland 2001; Blionis and Vokou 2002; Elle and Carney 2003; Valiente-Banuet *et al.* 2004).

15.3.2 Limitations and a proposal

Except for two cases (see below), most studies included in Table 15.3 claiming that variation in pollinator faunas explained observed patterns of geographical variation in floral traits (or its lack thereof) relied entirely on correlative evidence. These studies described parallel spatial variation of floral traits and one or several aspects of the pollinator assemblage (e.g., taxonomic composition, abundance, mean body size) that may affect selection on the variable floral characters. In some cases, the correlative evidence for pollinator-driven intraspecific diversification is compelling. For instance, Valiente-Banuet et al. (2004) related variation in time of anthesis across the geographic range of a columnar cactus to the variable availability of bat pollinators. In areas where bats are migratory, flowers remain open and secrete nectar during the day, allowing diurnal and nocturnal visitors, whereas flowers are exclusively nocturnal where bats visit reliably. Correlative evidence has also been used to argue for uncoupled geographical variation between plant traits and pollinators, as in Herrera et al.'s (2002) study on variation of floral integration in the perennial herb, *Helleborus foetidus*, over the Iberian Peninsula. In our literature review, only studies by Gómez and Zamora (2000) and Totland (2001) assessed variation in plant traits and pollinators in conjunction with geographical variation in phenotypic selection.

If intraspecific variation reflects local adaptation, morphology or function should associate with those aspects of the environment that influence natural selection (e.g., Gould and Johnston 1972). However, the opposite need not be true, and character–environment correlations do not demonstrate a causal relation. Correlations linking geographical variation in flower traits with variation in pollinators of the sort often used, for example, to document "pollination ecotypes" (e.g., Robertson and Wyatt 1990; Arroyo and Dafni 1995; Johnson 1997) *suggest* only a plausible role of pollinators as agents of floral diversification. Floral traits could vary geographically for three reasons. First, floral traits could exhibit phenotypic plasticity in response to spatially variable environments. In this case the environmental factor(s) inducing floral variation (e.g., flower size) may also cause pollinator variation (e.g., species composition, mean body size). Second, floral variation among populations could reflect neutral phenotypic variation arising from genetic drift. In this scenario, floral variation would cause pollinator differences by "filtering out" available pollinators via, for example, morphological matching or differential exclusion, so that pollinator differences between populations are a proximate ecological consequence, rather than the ultimate evolutionary cause, of floral variation (i.e., an "ecological fitting" scenario sensu Janzen 1985). Finally, floral traits could vary geographically in response to divergent natural selection. Unequivocal demonstration of this process requires additional information on the crucial mechanism that differentiates it from the other two possible processes, namely evidence of spatially variable, pollinator-mediated selection on the floral traits involved. Therefore, in this respect studies of intraspecific floral adaptation conducted in a geographical context are no exception to the established principle that environment–trait correlations are the weakest and least conclusive evidence of natural selection (Lewontin 1974; Endler 1986).

Geographically informed studies of pollinator-driven intraspecific floral differentiation can be strengthened most simply by incorporating an explicit analysis of spatially heterogeneous selection. A study's ability to differentiate between phenotypic plasticity, neutral phenotypic variation and divergent natural selection, and thus reliably identify possible instances of pollinator-driven intraspecific diversification, will be enhanced considerably by the following five-step approach. Step 1 involves the usual practice of documenting geographical variation in pollinators. It must be stressed that, to allow for reliable geographical comparisons, pollinator composition studies should pay careful attention to sampling issues, as discussed in detail by Ollerton and Cranmer (2002) and Herrera (2005), for example. Step 2 tests whether geographically variable floral traits are subject to selection from pollinators. Step 3 examines whether the selection gradient on the floral traits is related to geographic variation in the pollinator fauna. Step 4 quantifies the spatial correlation between variable selection gradients and phenotypic values. Finally, step 5 determines whether population differences in floral traits have a genetic basis. Step 3 is the key component in this protocol. It represents an extended version of the "pollinator × floral-character interaction" approach suggested by Wilson and Thomson (1996) to account for pollinator-mediated floral divergence. It is also related to the ANCOVA-based phenotypic selection models proposed by Strauss et al. (2005) to test for differences in diffuse selection exerted on plants by different species groups of animals (see also Wade and Kalisz 1990). We will apply this five-step protocol in the following section to the study of clinal variation of *Lavandula latifolia* flowers and their pollinators.

15.4 A case study: clinal variation of *Lavandula latifolia* flowers and pollinators

Lavandula latifolia is a summer-flowering, insect-pollinated shrub of open woodlands in southern France and the eastern Iberian Peninsula (Fig. 15.1a). Flowers are hermaphroditic and self-compatible, but <4% of flowers set fruit in the absence of pollinators. More than 100 species of bees, flies and butterflies pollinate *L. latifolia* in southeastern Spain, so this species is an outstanding example of generalist pollination at the regional level (Herrera 1988, 2005). Below, we focus mainly on geographical variation in Hymenoptera and Lepidoptera, the two main groups of pollinators, whose proportions vary widely among *L. latifolia* populations. On average, Hymenoptera and Lepidoptera visitors differ in components of pollinating effectiveness, including flower visitation rate, frequency of pollen deposition on the stigmas, mean number of pollen grains left when deposition occurs, and the proportion of interfloral flights between flowers on different plants (Herrera 1987, 1989). Artificially induced variation in the relative abundance of major pollinator groups affects variable seedling recruitment prospects on a per-flower basis (Herrera 2000). Hymenoptera and Lepidoptera differ in morphology, foraging behaviour, thermal biology, and nutritional requirements, which presumably cause contrasting flower preferences and selection patterns. Therefore, the *L. latifolia*–Hymenoptera–Lepidoptera pollination system is characterized regionally by a combination of (1) non-equivalence of main pollinators in their potential fitness consequences for the plants; (2) possible differences among the main pollinators in flower selection; and (3) variation among populations in pollinator composition (Herrera 1988). This combination provides a suitable background for investigating the possibility of pollinator-driven geographical differentiation in pollination-related floral traits.

15.4.1 Methods

Floral form, pollinator composition, and the maternal component of pollination success, were studied concurrently during July–August 1996 on 300 *L. latifolia* plants from 15 widely spaced

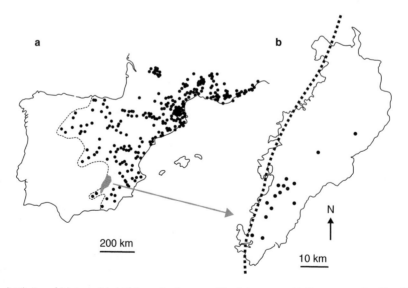

Figure 15.1 The distribution of (a) *Lavandula latifolia* on the European side of the western Mediterranean region (data from Upson and Andrews [2004] and Proyecto Anthos [http://www.programanthos.org]) and (b) the 15 populations of *L. latifolia* in Cazorla-Segura-Las Villas Natural Park considered in this chapter (dots). The dotted lines depict the western range limit of *L. latifolia*.

populations in the Sierras de Cazorla and Segura, around the southwestern limit of the species' range (Fig. 15.1b). Pollinator observations were repeated during 1997 in five populations. Twenty shrubs were marked at each site, and pollinators were observed on them between 0730 and 1230 h GMT. Four to six 3-min pollinator censuses were conducted on each plant (sample sizes shown in Electronic Appendix 15.2, http://www.eeb.utoronto.ca/EEF/). All flower visitors were identified to species and the number of flowers visited was recorded. Further details on pollinator observation methods are given by Herrera (2005). At each site, 20–25 open flowers were collected from each shrub during the afternoon of the corresponding pollinator census and stored in formaldehyde–acetic acid–ethyl alcohol solution. Flowers last for only 1.5–2.5 days and wither shortly after pollination (Herrera 2001, and unpublished), so pollen grains on the stigmas of afternoon-collected flowers could be related confidently to the activity of pollinators recorded during the preceding morning. For each flower, the lengths of the upper corolla lip and corolla tube (UL and CT hereafter, respectively; Fig. 15.2) were measured under a dissecting microscope using an ocular micrometer, and the numbers of pollen grains on the stigma and pollen tubes in the style were counted under an epifluorescence microscope (Herrera 2004).

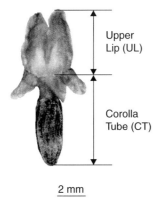

Figure 15.2 *Lavandula latifolia* flower in front view, showing the two measurements used to characterize floral morphology and symbols used in the text.

15.4.2. Step 1: Characterize geographical variation in pollinators

A total of 60 pollinator species (26 Lepidoptera, 23 Hymenoptera, and 11 Diptera) were recorded during the 1460 3-min observation periods at the 15 *L. latifolia* populations studied. The identity of the locally most important species of pollinators varied considerably among sites. Up to ten different taxa ranked among the two most important local pollinators at one site or another (Electronic Appendix 15.2): *Anthidiellum breviusculum* was one of the top two pollinators at nine sites, *Apis mellifera* at eight sites, *Macroglossum stellatarum* at four sites, *Bombus pascuorum* at three sites, *Bombus terrestris* at two sites, and *Ceratina* spp., *Anthophora quadrifasciata*, *Megachile pilidens* and *Lasioglossum* spp. at one site each. Only six of the 15 sites shared the same pair of top-two species (*Apis mellifera* plus *Anthidiellum breviusculum*).

Populations differed broadly in the relative contributions of Hymenoptera, Lepidoptera and Diptera to total floral visits (Fig. 15.3). Hymenoptera were the only or predominant ($>80\%$ of flower visits) visitors in six populations, Lepidoptera predominated in one population, and a variable mixture of Hymenoptera, Lepidoptera and Diptera occurred at the remaining seven sites. Diptera had minor importance in all sites and are not considered hereafter. Within populations, the relative occurrence of Lepidoptera tended to decline, and that of Hymenoptera to increase, from south to north ($r=0.514$, $N=15$, $P<0.05$ for Hymenoptera; $r=-0.477$, $N=15$, $P<0.10$ for Lepidoptera; correlations between latitude and population-level importance figures). This latitudinal trend is also evident for visits per plant (Fig. 15.4). Population differences in the proportion of flowers visited by the two major pollinator groups remained consistent between years in the five localities sampled during 1996 and 1997, as revealed by significant correlations between years for percent abundance of Hymenoptera ($r=0.903$, $N=5$, $P<0.05$) and Lepidoptera ($r=0.902$, $N=5$, $P=0.05$).

Populations differed also in pollinator species diversity, as measured by Shannon's diversity index for the proportional flower visitation data

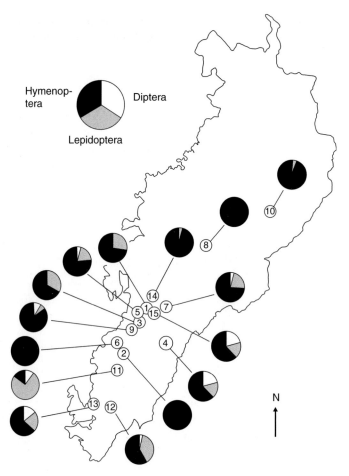

Figure 15.3 Geographical variation in the relative importance of the three main groups of pollinators of *Lavandula latifolia*, estimated by the proportion of total flower visits contributed. Localities are identified by numerals, as in Electronic Appendix 15.2 (http://www.eeb.utoronto.ca/EEF/). Locality names, geographical coordinates, and elevations are given in Herrera (2005: Table 2).

for each locality. Diversity correlated negatively and marginally significantly with latitude ($r = -0.498$, $N = 15$, $P < 0.10$). Pollinator abundance, measured as the mean number of flowers visited per 3-min period (all species combined), did not correlate significantly with latitude, for data from either populations ($r = 0.017$, $N = 15$, $P > 0.90$) or individual plants ($r = 0.010$, $N = 300$, $P > 0.80$).

15.4.3. Step 2: Demonstrate pollinator-mediated selection on floral traits

Phenotypic selection on floral morphology via its influence on the maternal component of pollination was assessed by fitting a generalized linear model to plant means ($N = 300$ plants), with pollen receipt per stigma (mean number of pollen grains; NPG) as the response variable, and the mean lengths of the UL and CT as independent variables. The response variable was ln-transformed and the analysis considered a negative binomial distribution of errors. Pollen receipt is a good surrogate of maternal fitness, as it correlates strongly with the number of pollen tubes in the style for the flowers sampled ($r = 0.660$, $N = 2987$, $P < 0.0001$; only flowers with NPG > 0 included), which in turn affects seed production per flower directly (CM Herrera unpublished data). Among-population variation in phenotypic selection on

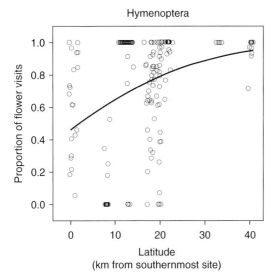

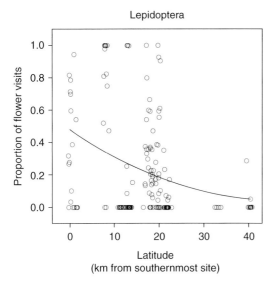

Figure 15.4 Latitudinal variation in the proportion of flower visits contributed by Hymenoptera and Lepidoptera to individual *Lavandula latifolia* plants. Each symbol corresponds to a different plant. Only plants with > 10 flower visits are used ($N = 161$). Logistic regressions are shown as solid lines (generalized $R^2 = 0.12$ and 0.10 for Hymenoptera and Lepidoptera, respectively; $P < 0.0001$ in both cases). A small random deviate was added to latitude data to reduce point overlap.

floral morphology was evaluated by testing the homogeneity of slopes of the relations of pollen receipt to the measures of floral morphology among populations with Population × UL and Population × CT interactions (e.g., Strauss *et al.*

2005; Rey *et al.* 2006). For simplicity, we focused only on directional selection gradients and did not assess quadratic terms in the phenotypic selection model, as this approach facilitates interpretation of population × trait interactions. Restriction of the analyses to directional selection is also justified in the present context, because directional selection seems to play the central role in phenotypic diversification at the species level and above (Rieseberg *et al.* 2002). Nevertheless, the model that we used to test geographical heterogeneity in selection could be extended easily to accommodate tests of heterogeneity in disruptive/stabilizing selection (Strauss *et al.* 2005; Rey *et al.* 2006).

This analysis revealed significant directional phenotypic selection on floral morphology through female function. Pollen receipt varied significantly among plants with the mean length of the UL ($F_{1,255} = 20.52$, $P < 0.0001$), but not with the mean length of CT ($F_{1,255} = 3.61$, $P > 0.05$). The relation between pollen receipt and length of UL differed significantly among populations (Population × UL interaction, $F_{14,255} = 2.44$, $P < 0.01$), demonstrating population differences in the nature of pollination-mediated phenotypic selection on this trait. Similar variation among populations was not evident for length of CT (Population × CT interaction, $F_{14,255} = 1.25$, $P > 0.1$). Consequently, we observed significant phenotypic selection only for the length of the UL and this selection varied among populations.

15.4.4 Step 3: Assess geographical divergence in selection

To examine whether the observed variation in selection gradients for the length of the UL has a geographic component, we assessed their correlation with latitude. Generalized linear models were fitted to plant means data separately for each population, and the standardized regression coefficients for UL length obtained from these models (b_{UL}'s hereafter) used as surrogates for phenotypic selection coefficients. b_{UL} increases significantly with latitude ($r_s = 0.671$, $N = 15$, $P < 0.01$: Fig. 15.5), demonstrating a geographical gradient in directional selection on that floral trait over the relatively restricted latitudinal range considered.

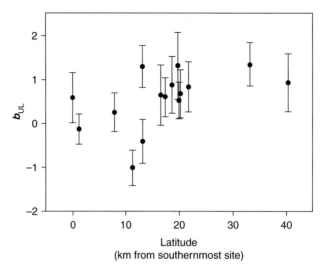

Figure 15.5 Latitudinal cline in pollination-mediated phenotypic selection on the upper corolla lip of *Lavandula latifolia* flowers. Dots represent the phenotypic selection coefficients ($b_{UL}\pm SE$) estimated separately for each locality.

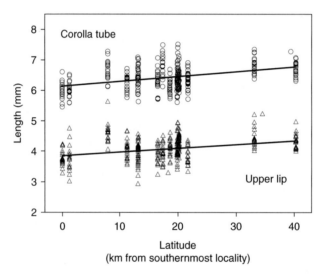

Figure 15.6 Clinal variation in lengths of upper lip (UL) and corolla tube (CT) of *Lavandula latifolia* flowers across the 40-km wide latitudinal range studied. Symbols represent means for individual plants (circles, CT; triangles, UL; $N = 300$ plants). Solid lines are least-squares linear regressions (CT: $F_{1,298} = 57.18$, $R^2 = 0.16$, $P < 0.0001$; UL: $F_{1,298} = 32.04$, $R^2 = 0.10$, $P < 0.0001$).

15.4.5 Step 4: Evaluate the match between divergent selection and phenotypic divergence

Corolla size varied gradually with latitude among the *L. latifolia* populations studied. Population means ranged between 3.7–4.6 mm and 5.9–6.8 mm for UL length and CT length, respectively (Electronic Appendix 15.2), with significant differences among populations ($F_{14,285} = 15.26$ and 15.03, for UL and CT, respectively; $P < 0.001$). Plant means for UL and CT increase significantly from southern to northern locations (Fig. 15.6), indicating a latitudinal cline in corolla size over the geographical range studied. The cline is rather steep, as denoted by average ($\pm SE$) gradients of 0.012 ± 0.0021 mm·km^{-1} and 0.016 ± 0.0021 mm·km^{-1} for UL and CT, respectively, as estimated from the slopes of linear regressions in Fig. 15.6. These gradients represent changes in population means of about 0.30% and 0.35% per km for UL and CT, respectively. The morphological cline and the pollination-related selection cline are therefore closely congruent, with the largest corollas in populations where directional selection favouring large corollas is strongest.

15.4.6 Step 5: Genetic basis of population differences in floral traits

A rigorous demonstration of adaptive, pollinator-driven, regional differentiation of *L. latifolia* flowers finally requires demonstrating that observed population differences in corolla traits have a genetic basis, rather than resulting from plastic responses to some variable environmental factor. This analysis could be accomplished with common-garden experiments or reciprocal transplants (reviewed by Schluter 2000); however, we have not performed such experiments, nor have most previous investigations relating geographic variation in floral traits to differences in pollinator composition and/or selection patterns. Clearly, such studies, which assess an essential component of natural selection, warrant consideration when planning future investigations.

Circumstantial evidence suggests a genetic component to variation among *L. latifolia* populations. Indirect evidence suggests that regional variation in corolla size is not a plastic response to changing abiotic environment. Soil nutrient availability and water stress can induce plastic variation in corolla size (Villarreal and Freeman 1990; Frazee and Marquis 1994; Galen *et al.* 1999); however, soil nutrient properties do not vary latitudinally among the 15 sites considered in our study of *L. latifolia* (C. M. Herrera, unpublished data). Total annual rainfall does vary latitudinally across the study region ($r = -0.559$, $P < 0.001$; mean rainfall data from $N = 40$ weather stations), but the relation is negative and thus contradicts the expected effects of water stress on latitudinal variation in corolla size. In the absence of relevant environmental variation, we expect that the variation in corolla size that we observed for *L. latifolia* has a genetic component, as has been observed for other plant species (Worley and Barrett 2000; Galen and Cuba 2001; Lendvai and Levin 2003).

15.4.7 Interpretation and caveats

Adaptive clines are maintained by the opposing interplay between the diversifying effect of variable selection along an environmental gradient and the "homogenizing" effect of gene flow (Slatkin 1985). Although we have demonstrated spatially variable selection, which is consistent with observed phenotypic variation, the observed clinal divergence in floral traits could partly reflect neutral phenotypic differentiation among populations under restricted gene flow and isolation by distance (Endler 1977). Data on amplified fragment length polymorphism (AFLP) markers for three of the *L. latifolia* populations studied here (Populations 1, 8, and 13 in Fig. 15.3) militate strongly against this possibility. These data reveal significant, but quantitatively modest genetic differentiation along the latitudinal range examined ($G_{ST} = 0.062$; 95% credible interval = 0.052–0.072), which is considerably smaller than the phenotypic differentiation for the floral traits that we examined (0.35–0.50; estimated using Spitze's [1993] formula for Q_{ST}: CM Herrera and P Bazaga unpublished data). If the AFLP markers used are effectively neutral and observed phenotypic differentiation reflects mainly genetic differences, these preliminary results support our interpretation that variable selection, rather than genetic drift, is the main factor maintaining the cline in corolla size (see Merilä and Crnokrak 2001). That CT length varies latitudinally even in the absence of demonstrable phenotypic selection may reflect a correlated response to selection on UL length resulting from the close integration between the two traits (Herrera 2001; see Chapter 14). The similarity in selection gradients for the two traits (Fig. 15.6) supports this interpretation. Therefore, our findings for *L. latifolia* are interpreted most reasonably as indicating a consistent latitudinal gradient in pollinator-mediated selection on corolla lip length through its effects on pollen import, resulting in adaptive intraspecific differentiation in the form of a latitudinal cline in corolla size. However, confirmation of this conclusion awaits common-garden or transplant experiments (step 5 above).

We have not investigated the proximate mechanisms whereby geographical variation in pollinators governs geographically variable selection on corolla size. As noted earlier, individual species and major groups of *L. latifolia* pollinators differ in incidence and the amount of pollen they deposit on stigmas, and they probably also respond differently to variations in corolla size. Therefore, population estimates of phenotypic selection on floral traits should reflect, in intricate ways, the differences among individual pollinating species in floral preferences, pollen dispersal, local abundance and flower visitation rates (see Eckhart 1991, 1992). With a taxonomically diverse pollinator assemblage, such as that of *L. latifolia*, dissecting the proximate mechanisms involved and the contribution of individual pollinator taxa to observed variation in selection may prove intractable. More positively, our results for *L. latifolia* are among the few to date providing empirical support for diffuse phenotypic selection on a plant trait exerted collectively by a multi-species animal assemblage (Strauss *et al.* 2005). This is an important result, as it suggests that adaptive floral divergence may not require specialization on particular pollinators as traditionally implied (e.g., Stebbins 1970). As shown here, taxonomically diverse pollinator assemblages, despite heterogeneity in pollinating characteristics, may collectively exert net selection on floral traits that, if spatially variable, may promote floral divergence.

15.5 Concluding remarks: towards an improved research programme in floral diversification

Inquiries into intraspecific diversification in floral traits mediated by divergent selection from pollinators represent a subclass of investigations on local adaptation, i.e., adaptive microevolutionary change. Nevertheless, in contrast with the voluminous literature on local adaptation in physiological, morphological or life history traits of plants (Linhart and Grant 1996; Jonas and Geber 1999, and references therein), our literature surveys found few substantiated studies of local adaptation in floral traits. One reason for this scarcity seems to be that pollination biologists have not always considered geographic variation to be important, as illustrated by scarcity of geographically informed investigations in our literature survey. Knowledge of the interaction of most plants with pollinators is based on single local snapshots of a process that varies among populations. Another reason for the rarity of geographic studies of selection on floral traits is that making a compelling case for pollinator-driven adaptive floral diversification is not easy, as illustrated by our *L. latifolia* study.

Little is known of patterns and processes related to intraspecific floral diversification (also see Chapter 16), so we largely focussed on how it should be studied, rather than on how it operates. Our literature review demonstrates that plant–pollinator interaction in a geographical context is a relatively neglected element of research on floral diversification. Furthermore, the few studies that address intraspecific variation explicitly generally provide ambiguous evidence for a *causal* role of variable selection by pollinators in generating intraspecific differentiation in floral traits, as most rely on correlative evidence alone, which provides the weakest support for adaptive interpretations. Research on floral diversification would benefit from both increased awareness of the central significance of incorporating the geographical context in studies of plant–pollinator interactions and, perhaps more importantly, reduced use of character–pollinator correlations to judge the occurrence of pollinator-mediated intraspecific diversification.

The five-step protocol, exemplified above for *L. latifolia*, may help circumvent some of the most obvious limitations of the few earlier studies on intraspecific floral variation. Particularly, we consider the demonstration of selection (step 2) and its geographical variation (step 3) essential to any investigation of the *current* adaptive value of intraspecific floral diversification. However, three aspects should be considered in relation to the phenotypic selection analyses involved in these steps. Firstly, although we considered only selection through the female function, steps 2 and 3 should also ideally assess possible selection through the male function, which may or may not mirror selection through female function (Conner *et al.* 1996; Maad 2000; Chapter 14). Secondly, the regression analysis promoted by Lande and Arnold (1983) and implemented in steps 2 and 3 above is not the only way to determine whether pollinators mediate selection on floral traits and document geographical variation. Other approaches to selection analysis, such as those based on path analysis and structural equations modelling, would be equally useful (e.g., Gómez 2000; Rey *et al.* 2006). Thirdly, contemporary measures of selection on floral traits provided by phenotypic selection analyses may provide limited insight into the adaptive origin of floral diversification in species where diversification occurred during past ecological scenarios, promoted by selective regimes different from those operating currently (see Herrera 1996 for some examples and a general discussion on "history-laden" versus "nonhistorical" approaches to the study of floral adaptations). In the Mediterranean Basin, where species originating prior to the appearance of Mediterranean climate conditions coexist with recent lineages evolved under current ecological conditions (e.g., Herrera 1992, Verdú *et al.* 2003), this limitation probably applies more importantly to phenotypic selection analyses of species that evolved before the appearance of Mediterranean climate conditions (e.g., *Viola cazorlensis*; Herrera 1990, 1993) than to those of species evolved under current ecological conditions (e.g., *Lavandula*).

The protocol that we propose focuses on the stepwise testing of an explicit *a priori* prediction: if variable pollinators are a major influence on floral diversification, *then* geographic variation in the abundance of pollinators with different floral preferences and pollinating quality (step 1) should impose geographic variation in selection on floral traits (steps 2 and 3), eventually causing phenotypic floral divergence (step 4) with a genetic basis (step 5). The sequence of steps of the proposed protocol reverses the inferential *a posteriori* approach typically applied to test links between intraspecific floral diversification and pollinator variation. The traditional approach can proceed beyond correlative evidence only with difficulty, leaving little room for incorporating explicit cause–effect hypotheses about selection, and it is susceptible to ad hoc hypothesis accommodation and hypothesis fudging (sensu Lipton 2005). In contrast, the approach illustrated here for *L. latifolia* tests the central elements of adaptive interpretations of floral diversification explicitly in a stepwise manner, running from putative causes to purported effects, and is thus less susceptible to accommodation and fudging. There are reasons for predictions counting more than accommodations (Lipton 2005) and also, therefore, for preferring a prediction-based, deductive logic when assessing the role played by pollinators in intraspecific floral diversification.

Acknowledgements

C Alonso, MC Anstett, A Benavente, D Ramírez, R Requerey, L Sánchez, and A Tíscar assisted with the *Lavandula* study in the field and/or the laboratory. P Bazaga shared unpublished genetical data, and two reviewers provided valuable suggestions. MCC was supported by a postdoctoral grant from the Ministerio de Educación y Ciencia, and MM by a I3P Programme postdoctoral contract from the Consejo Superior de Investigaciones Científicas.

References

Aigner PA (2004). Floral specialization without trade-offs: optimal corolla flare in contrasting pollination environments. *Ecology*, **85**, 2560–9.

Armbruster WS (1985). Patterns of character divergence and the evolution of reproductive ecotypes of *Dalechampia scandens* (Euphorbiaceae). *Evolution*, **39**, 733–52.

Arroyo J and Dafni A (1995). Variations in habitat, season, flower traits and pollinators in dimorphic *Narcissus tazetta* L. (Amaryllidaceae) in Israel. *New Phytologist*, **129**, 135–45.

Barrett SCH (1995). Mating-system evolution in flowering plants: micro- and macroevolutionary approaches. *Acta Botanica Neerlandica*, **44**, 385–402.

Barrett SCH, Dorken ME, and Case AL (2001). A geographical context for the evolution of plant reproductive systems. In J Silvertown and J Antonovics, eds. *Integrating ecology and evolution in a spatial context*, pp. 341–64. Blackwell Science, Oxford.

Blionis GJ and Vokou D (2002). Structural and functional divergence of *Campanula spatulata* subspecies on Mt Olympos (Greece). *Plant Systematics and Evolution*, **232**, 89–105.

Bock WJ (1970). Microevolutionary sequences as a fundamental concept in macroevolutionary models. *Evolution*, **24**, 704–22.

Boyd A (2002). Morphological analysis of Sky Island populations of *Macromeria viridiflora* (Boraginaceae). *Systematic Botany*, **27**, 116–26.

Campbell DR, Waser NM, Price MV, Lynch EA, and Mitchell RJ (1991). Components of phenotypic selection: pollen export and flower corolla width in *Ipomopsis aggregata*. *Evolution*, **45**, 1458–67.

Castellanos MC, Wilson P, and Thomson JD (2004). 'Antibee' and 'pro-bird' changes during the evolution of hummingbird pollination in *Penstemon* flowers. *Journal of Evolutionary Biology*, **17**, 876–85.

Conner JK, Rush S, Kercher S, and Jennetten P (1996). Measurements of natural selection on floral traits in wild radish (*Raphanus raphanistrum*). II. Selection through lifetime male and total fitness. *Evolution*, **50**, 1137–46.

Darwin CR (1862). *The various contrivances by which British and foreign orchids are fertilised by insects*. Murray, London.

Dilley JD, Wilson P, and Mesler MR (2000). The radiation of *Calochortus*: generalist flowers moving through a mosaic of potential pollinators. *Oikos*, **89**, 209–22.

Dodd ME, Silvertown J and Chase MW (1999). Phylogenetic analysis of trait evolution and species diversity variation among angiosperm families. *Evolution*, **53**, 732–44.

Eckhart VM (1991). The effects of floral display on pollinator visitation vary among populations of *Phacelia linearis* (Hydrophyllaceae). *Evolutionary Ecology*, **5**, 370–84.

Eckhart VM (1992). Spatio-temporal variation in abundance and variation in foraging behavior of the pollinators of gynodioecious *Phacelia linearis* (Hydrophyllaceae). *Oikos*, **64**, 573–86.

Elle E and Carney R (2003). Reproductive assurance varies with flower size in *Collinsia parviflora* (Scrophulariaceae). *American Journal of Botany*, **90**, 888–96.

Endler JA (1977). *Geographic variation, speciation, and clines*. Princeton University Press, Princeton, New Jersey.

Endler JA (1986). *Natural selection in the wild*. Princeton University Press, Princeton, New Jersey.

Eriksson O and Bremer B (1992). Pollination systems, dispersal modes, life forms, and diversification rates in angiosperm families. *Evolution*, **46**, 258–66.

Fausto JA, Eckhart VM, and Geber MA (2001). Reproductive assurance and the evolutionary ecology of self-pollination in *Clarkia xantiana* (Onagraceae). *American Journal of Botany*, **88**, 1794–800.

Fenster CB, Armbruster WS, Wilson P, Dudash MR, and Thomson JD (2004). Pollination syndromes and floral specialization. *Annual Review of Ecology, Evolution and Systematics*, **35**, 375–403.

Frazee JE and Marquis RJ (1994). Environmental contribution to floral trait variation in *Chamaecrista fasciculata* (Fabaceae: Caesalpinoideae). *American Journal of Botany*, **81**, 206–15.

Galen C and Cuba J (2001). Down the tube: pollinators, predators, and the evolution of flower shape in the alpine skypilot, *Polemonium viscosum*. *Evolution*, **55**, 1963–71.

Galen C, Sherry RA, and Carroll AB (1999). Are flowers physiological sinks or faucets? Costs and correlates of water use by flowers of *Polemonium viscosum*. *Oecologia*, **118**, 461–70.

Gómez JM (2000). Phenotypic selection and response to selection in *Lobularia maritima*: importance of direct and correlational components of natural selection. *Journal of Evolutionary Biology*, **13**, 689–99.

Gómez JM and Zamora R (1999). Generalization vs. specialization in the pollination system of *Hormathophylla spinosa* (Cruciferae). *Ecology*, **80**, 796–805.

Gómez JM and Zamora R (2000). Spatial variation in the selective scenarios of *Hormatophylla spinosa* (Cruciferae). *American Naturalist*, **155**, 657–68.

Gould SJ and Johnston RF (1972). Geographic variation. *Annual Review of Ecology and Systematics*, **3**, 457–98.

Graham SW and Barrett SCH (2004). Phylogenetic reconstruction of the evolution of stylar polymorphisms in *Narcissus* (Amaryllidaceae). *American Journal of Botany*, **91**, 1007–21.

Grant V (1949). Pollination systems as isolating mechanisms in angiosperms. *Evolution*, **3**, 82–97.

Grimaldi D (1999). The co-radiations of pollinating insects and angiosperms in the Cretaceous. *Annals of the Missouri Botanical Garden*, **86**, 373–406.

Hapeman JR and Inoue K (1997). Plant-pollinator interactions and floral radiation in *Platanthera* (Orchidaceae). In TJ Givnish and KJ Sytsma, eds. *Molecular evolution and adaptive radiation*, pp. 433–54. Cambridge University Press, Cambridge.

Herrera CM (1987). Components of pollinator quality: comparative analysis of a diverse insect assemblage. *Oikos*, **50**, 79–90.

Herrera CM (1988). Variation in mutualisms: the spatio-temporal mosaic of a pollinator assemblage. *Biological Journal of the Linnean Society*, **35**, 95–125.

Herrera CM (1989). Pollinator abundance, morphology, and flower visitation rate: analysis of the quantity component in a plant-pollinator system. *Oecologia*, **80**, 241–8.

Herrera CM (1990). The adaptedness of the floral phenotype in a relict endemic, hawkmoth-pollinated violet. 2. Patterns of variation among disjunct populations. *Biological Journal of the Linnean Society*, **40**, 275–91.

Herrera CM (1992). Historical effects and sorting processes as explanations for contemporary ecological patterns: character syndromes in Mediterranean woody plants. *American Naturalist*, **140**, 421–46.

Herrera CM (1993). Selection on floral morphology and environmental determinants of fecundity in a hawk moth-pollinated violet. *Ecological Monographs*, **63**, 251–75.

Herrera CM (1996). Floral traits and plant adaptation to insect pollinators: a devil's advocate approach. In DG Lloyd and SCH Barrett, eds. *Floral biology. Studies in floral evolution of animal pollinated plants*, pp. 65–87. Chapman & Hall, New York.

Herrera CM (2000). Flower-to-seedling consequences of different pollination regimes in an insect-pollinated shrub. *Ecology*, **81**, 15–29.

Herrera CM (2001). Deconstructing a floral phenotype: do pollinators select for corolla integration in *Lavandula latifolia*? *Journal of Evolutionary Biology*, **14**, 574–84.

Herrera CM (2004). Distribution ecology of pollen tubes: fine-grained, labile spatial mosaics in southern Spanish Lamiaceae. *New Phytologist*, **161**, 473–84.

Herrera CM (2005). Plant generalization on pollinators: species property or local phenomenon? *American Journal of Botany*, **92**, 13–20.

Herrera CM, Cerdá X, García MB, Guitián J, Medrano M, Rey PJ, and Sánchez-Lafuente AM (2002). Floral integration, phenotypic covariance structure and pollinator variation in bumblebee-pollinated *Helleborus foetidus*. *Journal of Evolutionary Biology*, **15**, 108–21.

Inoue K, Maki M, and Masuda M (1996). Evolution of *Campanula* flowers in relation to insect pollinators in islands. In DG Lloyd and SCH Barrett, eds. *Floral biology. Studies in floral evolution of animal pollinated plants*, pp. 377–400. Chapman & Hall, New York.

Janzen DH (1985). On ecological fitting. *Oikos*, **45**, 308–10.

Johnson SD (1997). Pollination ecotypes of *Satyrium hallackii* (Orchidaceae) in South Africa. *Botanical Journal of the Linnean Society*, **123**, 225–35.

Johnson SD and Steiner KE (1997). Long-tongued fly pollination and evolution of floral spur length in the *Disa draconis* complex (Orchidaceae). *Evolution*, **51**, 45–53.

Jonas CS and Geber MA (1999). Variation among populations of *Clarkia unguiculata* (Onagraceae) along altitudinal and latitudinal gradients. *American Journal of Botany*, **86**, 333–43.

Kingsolver JG, Hoekstra HE, Hoekstra JM, and Berrigan D (2001). The strength of phenotypic selection in natural populations. *American Naturalist*, **157**, 245–61.

Lande R and Arnold SJ (1983). The measurement of selection on correlated characters. *Evolution*, **37**, 1210–26.

Lendvai G and Levin DA (2003). Rapid response to artificial selection on flower size in *Phlox*. *Heredity*, **90**, 336–42.

Leppik EE (1957). Evolutionary relationship between entomophilous plants and anthophilous insects. *Evolution*, **11**, 466–81.

Levin DA (2000). *The origin, expansion, and demise of plant species*. Oxford University Press, Oxford.

Lewontin RC (1974). *The genetic basis of evolutionary change*. Columbia University Press, New York.

Linhart YB and Grant MC (1996). Evolutionary significance of local genetic differentiation in plants. *Annual Review of Ecology and Systematics*, **27**, 237–77.

Lipton P (2005). Testing hypotheses: prediction and prejudice. *Science*, **307**, 219–21.

Maad J (2000). Phenotypic selection in hawkmoth-pollinated *Platanthera bifolia*: targets and fitness surfaces. *Evolution*, **54**, 112–23.

Malo JE and Baonza J (2002). Are there predictable clines in plant-pollinator interactions along altitudinal gradients? The example of *Cytisus scoparius* (L.) Link in the Sierra de Guadarrama (Central Spain). *Diversity and Distributions*, **8**, 365–71.

Merilä J and Crnokrak P (2001). Comparison of genetic differentiation at marker loci and quantitative traits. *Journal of Evolutionary Biology*, **14**, 892–903.

Miller RB (1981). Hawkmoths and the geographic patterns of floral variation in *Aquilegia caerulea*. *Evolution*, **35**, 763–74.

Ollerton J and Cranmer L (2002). Latitudinal trends in plant-pollinator interactions: are tropical plants more specialised? *Oikos*, **98**, 340–50.

Patterson TB and Givnish TJ (2004). Geographic cohesion, chromosomal evolution, parallel adaptive radiations, and consequent floral adaptations in *Calochortus* (Calochortaceae): evidence from a cpDNA phylogeny. *New Phytologist*, **161**, 253–64.

Rey PJ, Herrera CM, Guitián J, Cerdá X, Sánchez-Lafuente AM, Medrano M, and Garrido JL (2006). The geographic mosaic in pre-dispersal interactions and selection on *Helleborus foetidus* (Ranunculaceae). *Journal of Evolutionary Biology*, **19**, 21–34.

Ricklefs RE and Renner SS (1994). Species richness within families of flowering plants. *Evolution*, **48**, 1619–36.

Rieseberg LH, Widmer A, Arntz AM, and Burke JM (2002). Directional selection is the primary cause of phenotypic diversification. *Proceedings of the National Academy of Sciences of the United States of America*, **99**, 12242–5.

Robertson JL and Wyatt R (1990). Evidence for pollination ecotypes in the yellow-fringed orchid, *Platanthera ciliaris*. *Evolution*, **44**, 121–33.

Sargent RD (2004). Floral symmetry affects speciation rates in angiosperms. *Proceedings of the Royal Society of London, Series B*, **271**, 603–8.

Schluter D (2000). *The ecology of adaptive radiation*. Oxford University Press, Oxford.

Simpson GG (1953). *The major features of evolution*. Columbia University Press, New York.

Slatkin M (1985). Gene flow in natural populations. *Annual Review of Ecology and Systematics*, **16**, 393–430.

Spitze K (1993). Population structure in *Daphnia obtusa*: quantitative genetic and allozymic variation. *Genetics*, **135**, 367–74.

Stebbins GL (1970). Adaptive radiation of reproductive characteristics in angiosperms, I: Pollination mechanisms. *Annual Review of Ecology and Systematics*, **1**, 307–26.

Strauss SY, Sahli H, and Conner JK (2005). Toward a more trait-centered approach to diffuse (co)evolution. *New Phytologist*, **165**, 81–9.

Thompson JN (1994). *The coevolutionary process*. University of Chicago Press, Chicago.

Totland O (2001). Environment-dependent pollen limitation and selection on floral traits in an alpine species. *Ecology*, **82**, 2233–44.

Upson T and Andrews S (2004). *The genus* Lavandula. Timber Press, Portland, Oregon.

Valiente-Banuet A, Molina-Freaner F, Torres A, Arizmendi MC, and Casas A (2004). Geographic differentiation in the pollination system of the columnar cactus *Pachycereus pecten-aboriginum*. *American Journal of Botany*, **91**, 850–5.

Verdú M, Dávila P, García-Fayos P, Flores-Hernández N, and Valiente-Banuet A (2003). 'Convergent' traits of mediterranean woody plants belong to pre-mediterranean lineages. *Biological Journal of the Linnean Society*, **78**, 415–27.

Villarreal AG and Freeman CE (1990). Effects of temperature and water stress on some floral nectar characteristics in *Ipomopsis longiflora* (Polemoniaceae) under controlled conditions. *Botanical Gazette*, **151**, 5–9.

Wade MJ and Kalisz S (1990). The causes of natural selection. *Evolution*, **44**, 1947–55.

Wilson P and Thomson JD (1996). How do flowers diverge? In DG Lloyd and SCH Barrett, eds. *Floral biology. Studies on floral evolution in animal-pollinated plants*, pp. 88–111. Chapman & Hall, New York.

Worley AC and Barrett SCH (2000). Evolution of floral display in *Eichhornia paniculata* (Pontederiaceae): direct and correlated responses to selection on flower size and number. *Evolution*, **54**, 1533–45.

CHAPTER 16

Pollinator-driven speciation in plants

Steven D. Johnson

School of Biological and Conservation Sciences, University of KwaZulu-Natal, Pietermaritzburg, South Africa

Outline

Speciation is often linked closely to evolutionary shifts in reproductive traits. In plants, these shifts have been viewed as a consequence of either direct adaptation to locally effective pollinators or selection for traits that impart reproductive isolation from congeners. Available evidence suggests that plant fitness is much more likely to be limited by pollinator availability or pollinator effectiveness than by hybridization. Thus, selection probably favours floral traits that alleviate pollen limitation or promote pollen dispersal more strongly than traits that prevent matings between congeners (reproductive isolating mechanisms). Although floral traits sometimes function as isolating mechanisms, this effect has probably not been a major influence in their evolution. Shifts between pollinators, a signal feature of many angiosperm radiations, occur largely because of spatial and temporal heterogeneity in pollinator availability, and result in the profound morphological changes that characterize speciation. This concept of pollinator-driven speciation links adaptation and speciation seamlessly and thus is essentially Darwinian in philosophy.

16.1 Introduction

Floral diversity is the outstanding characteristic of the angiosperm radiation. In a broad sense, abundant evidence now confirms that pollinators play a major role in this diversification (Dodd *et al.* 1999; Chapter 17). Yet, as we will see, opinion is divided about the importance of pollinator-driven speciation, the process itself, or how it should be studied. In this chapter I briefly review the development of ideas about pollinator-driven speciation, then focus on evidence for the process at several spatial scales and stages of divergence, and finally suggest a framework for the further development of this research programme.

Defining the nature of species has proved to be one of the most intractable of all debates in biology. This issue cannot be side-stepped, as speciation can be defined only in relation to its products—species. Darwin (1859) viewed speciation as little more than profound morphological change caused by natural selection. In fact, he doubted whether species were real in any sense of having properties different from intra-specific taxa. In Darwin's view, adaptation and speciation link seamlessly. Thus, in this tradition, the primary agenda of speciation research is to discover the selective factors and processes that lead to divergence among populations.

The Evolutionary Synthesis during the mid-twentieth century emphasized that species in the Darwinian sense would be imperilled, but for the presence of barriers to gene flow from congeners (Mayr 1942, 1963; Dobzhansky 1951). Phenotypic discontinuities between species were attributed primarily to the existence of these "isolating barriers" (Chapter 18). Thus, although morphology-based taxonomic systems continued to provide practical means of recognizing species, isolating barriers were considered the underlying *sine qua non* of a "good" species. This biological species concept (BSC) has been adopted almost universally by zoologists, but is still treated with scepticism by many botanists. It is neither possible nor desirable to deal here with all the objections, other than to note that various alternative species concepts have been formulated to address specific shortcomings of the BSC (e.g., difficulties in classifying asexual taxa and the arbitrary role of biological isolating barriers for allopatric taxa). The debate over the BSC is by no means over, despite recent claims by some of its more doctrinaire adherents (Coyne and Orr 2004). If the BSC is adopted, speciation can be studied as the process that produces isolating barriers. This research programe has led to remarkable insights into both the mechanisms and evolution of isolating barriers (Coyne and Orr 2004; Chapter 18).

The past few years have seen a resurgence of interest in ecological speciation (Schluter 2000). After decades of investigating models of speciation through genetic drift, biologists seem largely to agree that natural selection is the primary factor driving speciation (Rieseberg et al. 2002; Coyne and Orr 2004; Waser and Campbell 2004). Several lines of evidence support this viewpoint, including QTL sign tests, experimental translocations, and trait–environment correlations (Rieseberg et al. 2002, 2004). Intriguingly, evidence also indicates that isolating barriers arise easily as a pleiotropic consequence of adaptive divergence (Coyne and Orr 2004). Thus, Darwin was generally correct about selection being the main process leading to speciation, and Mayr was correct about one of its emergent properties—reproductive isolation.

Should studies of speciation then focus on the factors that promote divergence among populations, or on the evolution of isolating mechanisms? The answer, of course, is both, but the pendulum may have swung too far, so that factors promoting divergence have been neglected in favour of an emphasis on isolating mechanisms. Indeed, some works written in the context of the BSC give the unfortunate impression that natural selection is uninteresting for speciation unless it results in reproductive isolation (Schluter 2000; Coyne and Orr 2004). The elegant series of studies of *Mimulus cardinalis* and *Mimulus lewisii* by Schemske and his coworkers (Schemske and Bradshaw 1999; Bradshaw and Schemske 2003; Ramsey et al. 2003) are much-lauded as examples of how speciation occurs in plants, yet the geographical factors that drove divergence of these two species have yet to be identified. That a trait plays a role in reproductive isolation in a zone of contact is significant, but this observation need not explain its initial evolution. Factors promoting phenotypic divergence should remain the primary focus of speciation research: after all, it is the diversity of forms that characterizes organic diversity and, therefore, the primary pattern requiring explanation.

Grant and Grant (1965) first attempted to develop a conceptual model of pollinator-driven speciation in their landmark publication on pollination in the phlox family (Polemoniaceae). They outlined a simple scenario in which the relative abundance of two pollinators, A and B, varied spatially, such that species A visited one population of a plant species most frequently, whereas species B was the dominant pollinator in a second population. This contrasting pollination environment would promote divergence, whereby plants adapt to pollinators A and B, respectively. Grant and Grant used several cases of pollinator-linked "racial" differentiation in species of the Polemoniaceae to support this scenario. Speciation was considered an extension of this process, such that "when the specialization for different classes of pollinators approach or reach a stage of mutual exclusiveness, these differences contribute to the reproductive isolation between the species involved" (Grant and Grant 1965, p. 164). Stebbins (1970) subsequently expanded this perspective, emphasizing five key principles: (1) the most effective pollinator principle, (2) the significance of character syndromes, (3) selection along lines of least resistance, (4) transfer of function via an intermediate stage of double function, and (5) reversals of evolutionary trends.

Despite the intrinsic logic of the Grant–Stebbins conceptual model, empirical evidence for pollinator-driven speciation at all scales (except perhaps broad-scale radiation of lineages: Chapter 17) remains extremely fragmentary. The relative importance of the process in plants is also uncertain (see Section 16.7). For species that have undergone cryptic chromosomal changes, such as allopolyploidy, adaptation may have had little or nothing to do with their speciation (reviewed by Grant 1977), whereas for others speciation appears to be a direct consequence of adaptation to either biotic or abiotic environments, or both. In the following sections, I review both the theory and evidence for pollinator-driven speciation critically, and then use examples of divergence of intra-specific taxa and sister species in a geographical pollinator mosaic to support a Darwinian view of speciation which focuses on the process of phenotypic divergence.

16.2 Why flowers evolve

Biologists commonly argue that flowers evolve under selection for isolating barriers (Levin 1971; Goldblatt and Manning 1996; Jones 2001). This notion, that loss of fitness through hybridization is important in the evolution of reproductive traits, traces back to Wallace, and was formalized in Dobzhansky's concept of reinforcement (see Section 16.9). Although highly contentious as a theory (see Section 16.9), reinforcement continues to be invoked to explain the evolution of reproductive characters, especially in the systematics literature (cf. Goldblatt and Manning 1996). Another common explanation for floral differences among sympatric or parapatric congeners is competition for pollinators resulting in character displacement (cf. Macior 1982; Armbruster et al 1994). However, as argued below, selection for reproductive performance (efficient receipt and export of pollen) in environments in which pollinator availability is limited, rather than isolation from congeners or reduction of inter-specific competition, is likely to explain the vast majority of floral shifts (see West-Eberhard 1983 for a similar perspective on animal speciation).

Hundreds of ecological studies demonstrate that receipt of *compatible* pollen is the major proximate limitation on seed production (Burd 1994; Ashman et al. 2004). Therefore, selection should favour traits that alleviate this bottleneck to fitness (Johnston 1991; Chapters 2, 4, and 10). Indeed, the strength of selection through female function varies positively with the intensity of pollen limitation (Ashman and Morgan 2004). Interestingly, and in contradiction to theory (Chapter 10), shifts between pollinators, rather than selfing, may be the most common evolutionary outcome of pollen limitation. This outcome is exemplified by the orchids, a huge family in which pollinator shifts are common, and autonomous self-pollination rare (~5% of taxa), despite near-ubiquitous pollen limitation (Tremblay et al. 2005). There are two possible reasons for this pattern: pollinator shifts in certain environments may ameliorate pollen limitation without the cost of inbreeding depression, or for some plants, such as those in self-incompatible or dioecious lineages, a pollinator shift may be subject to fewer or weaker phylogenetic constraints than is the evolution of selfing.

The recent finding by Vamosi et al. (2006) of a relation between regional plant species richness and pollen limitation is particularly intriguing. Although the causal basis, if any, of this relation remains to be established, it hints at the possibility of a feedback loop, whereby pollen limitation, perhaps present initially because of an abiotic factor that limits insect biomass such as low soil nutrient levels, promotes pollinator-driven adaptive speciation, which in turn leads to higher biodiversity and smaller plant populations, and thus more pollen limitation.

Selection on floral traits through male function is much harder to measure, but is expected to be important, particularly when plants compete for access to ovules (Bell 1985; Chapters 2, 4, and 14). One unresolved issue is the extent to which variance in the expected positive relation between the removal of pollen and its successful export (which is apparently weak or not detectable in some species; Johnson et al. 2005) influences the strength of selection through male function (Broyles and Wyatt 1995; Queller 1997).

That plants can experience strong selection for traits that confer more efficient pollination is now abundantly clear (Campbell 1989; Galen 1989; Conner and Rush 1997; Alexandersson and Johnson 2001; Chapter 14). Unfortunately, most studies of selection on floral traits have considered single populations without any comparative or geographical component (see Chapter 15). Thus the role of selection on floral divergence has to be inferred indirectly. Furthermore, many of these studies provide inadequate data on the mechanism of selection, such as the fit between pollinators and flowers, thus making interpretation of the environmental basis of evolution difficult.

Flowers probably evolve particularly rapidly during shifts between pollination systems (Hodges and Arnold 1994a; Kay et al. 2005). Stebbins (1970) recognized that these shifts often involve relatively few floral traits, which are often correlated, giving rise to a distinctive pattern of floral syndromes (e.g., Armbruster 1993; Johnson et al. 2002a; Fenster et al. 2004; Goldblatt et al. 2004). In some cases, variation in a single trait may be enough to precipitate a pollinator shift. In a study of *Mimulus cardinalis* and *M. lewisii*, Bradshaw and Schemske (2003) showed how yellow carotenoid pigments, which are controlled by a single QTL, dramatically increase visitation by hummingbirds relative to bees. They speculated that increased relative abundance of hummingbirds in a region could favour yellow–orange mutants of a normally pink-flowered *Mimulus*, leading to rapid allele substitution in the population, thus initiating a speciation event consistent with insect-to-bird shifts evident in the phylogeny of *Mimulus* (Beardsley et al. 2003).

Even though the selective basis of floral evolution during shifts between different pollinators is almost self-evident, many plant lineages have undergone significant floral evolution without shifts in pollinators. Examples include *Pedicularis*, *Aconitum*, and *Delphinium* pollinated largely by bumble bees (cf. Macior 1982) and *Disperis* pollinated by oil-collecting bees (Steiner 1989). Floral evolution in these lineages often involves functional transitions that affect pollen placement on the same pollinators, such as on the dorsal versus ventral surfaces of bees in *Pedicularis* (Macior 1982). Most authors have invoked competitive interactions with congeners to explain these transitions (Macior 1982; Armbruster et al. 1994). Genetic drift too may play some role in these transitions, if peaks in the adaptive landscape are separated by relatively shallow valleys.

16.3 The geographical pollinator mosaic

Pollinators are distributed unevenly in time and space. Indeed, like all animals, pollinators have restricted ranges, which are often determined by both physical factors, such as altitude, temperature, and rainfall, and by biotic factors, such as vegetation structure and availability of flowering plants (Chapters 6 and 15). The landscape in which plants evolve presents a geographical mosaic of pollinators that is hard to visualize, but is no less important for plant evolution than the geographical mosaics of soils, climate, and herbivores. Grant and Grant (1965) referred to this mosaic as the "pollinator climate." The extent to which the geographical pollinator mosaic influences plant evolution depends on its stability in time and space, the sharpness of its boundaries, and the extent to which plants are generalized or specialized in their pollination systems. Like the Grants, Waser (2001, p. 327) emphasized that gradients in a pollinator mosaic will often be characterized by "quantitative differences among populations in the relative abundance of different pollinators, rather than from qualitative turnover in pollinators" (see also Aigner 2005).

The geographical pollinator mosaic is the basis for all allopatric and parapatric divergence in pollination systems. Yet, spatial and temporal variation in pollinator faunas remains poorly documented (see Chapter 15). Grant was keenly aware of its importance, and emphasized the significance of patterns in the distributions of hummingbirds, hawk moths, bumble bees, and flies within North America for plant evolution (Grant and Grant 1965; Grant 1983, 1994a). Unfortunately, much of this perspective seems to have been lost with the more recent emphasis on studies at single localities and reductionist approaches to studying pollination systems in general. Darwin himself was guilty on this score, as his famous orchid studies were conducted mostly at a single locality near his house. Despite being invaluable in showing the functional significance of floral traits, these studies did not provide the same insights into diversification that were afforded by Darwin's comparative studies of animals in their geographical context.

The existence of a pollinator mosaic can be illustrated graphically using data on the distribution of long-proboscid flies in South Africa. Figure 16.1 shows the main spatial and temporal components of this mosaic. This particular mosaic is significant for floral evolution, because plants tend to become involved in highly specialized relations with long-proboscid flies (Goldblatt and Manning 2000; Johnson and Steiner 2000). As individual fly species vary extensively in proboscis length, behaviour,

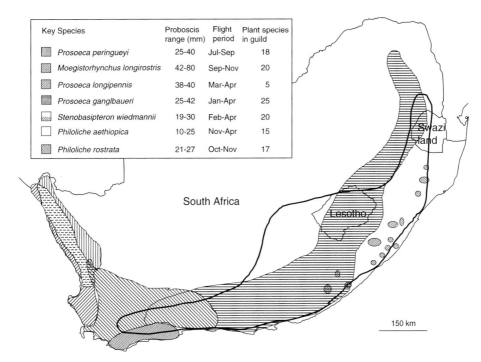

Figure 16.1 Distribution of long-proboscid fly species in southern Africa, illustrating a geographical pollinator mosaic. Note that the mosaic also involves a strong temporal dimension. The approximate number of plant species in each pollination guild associated with a fly species is given in the legend. *Philoliche* spp. belong to the Tabanidae, whereas the remaining species are nemestrinids. Based on information in Goldblatt and Manning (2000), Potgieter and Edwards (2005), and unpublished data of the author.

colour preferences, and flight period, local adaptation of a plant species for pollination by a particular fly species causes distinct changes in floral morphology and phenology (Johnson and Steiner 1997; Goldblatt and Manning 2000). Examples of speciation events linked to shifts across boundaries in this mosaic are discussed in Section 16.6.

Elevational differences in pollinator faunas have been documented in several studies (Cruden 1972; Arroyo *et al.* 1982; Warren *et al.* 1988). In the most widely cited example, Cruden (1972) showed that bees became less important pollinators and birds increased in importance with increasing elevation in Mexico. Similarly, studies in alpine regions of several continents have shown that flies become relatively common, but bees diminish in importance, as altitude increases (Muller 1880; Arroyo *et al.* 1982; Primack 1983; Warren *et al.* 1988), although Galen (1989) found the opposite trend with visitors to *Polemonium viscosum* in a North American alpine zone.

At an even finer scale, pollinator abundance can vary spatially between sites separated by only hundreds or tens of metres (see also Chapter 8). The South African butterfly *Aeropetes tulbaghia* seldom ventures far from steep rocky habitats, so that pollination success of one of its host plants, the orchid *Disa uniflora*, can vary three- or fourfold between sites a few hundred metres apart, a pattern which is consistent from year to year (Johnson and Bond 1992). Abundance of insect pollinators also varies according to the spatial distribution of primary nectar sources, which can have major implications for spatial variation in pollination success of less abundant or non-rewarding plants (Laverty 1992; Johnson *et al.* 2003).

Are biotic selection environments sufficiently stable in time and space to be important for floral character evolution and, ultimately, speciation? Mosaic stability probably varies in direct proportion to scale. Subcontinental distribution patterns of pollinators are probably stable for long enough to influence speciation. On the other hand, current mosaics in parts of the world that have been subjected to frequent climatic perturbation, such as Europe and northern North America, may not reflect those present during speciation. Johnson and Steiner (2000) suggested that the instability of pollination mosaics in postglacial environments may account for the relative lack of floral specialization evident in pollination systems in the northern temperate zone. This link between mosaic stability and specialization was supported by Valiente-Banuet's (2004) demonstration that populations of the cactus *Pachycereus pecten-aboriginum* are specialized for bat pollination in southern Mexico, where bats are resident, and more generalist in northern Mexico, where bats are migrant.

16.4 Pollination ecotypes

Evolutionary biologists have long recognized the value of geographical variation in traits *within* a species as evidence of the selective factors that promote divergence. Grant and Grant (1965, p. 162) argued that cases of intra-specific variation allow adaptive radiation to be studied as a "process rather than as an historical event." In many plant species, floral morphology varies geographically, allowing systematists to recognize subspecies or races. In many of these cases the intra-specific taxa probably reflect the evolution of pollination ecotypes—forms adapted to the local pollinator fauna. A careful observer may even recognize intra-specific geographical variation in floral traits that has not been recognized formally by taxonomists. The latter is particularly likely when the variation involves traits, such as flower colour or scent, that are not preserved in herbarium specimens (Pellmyr 1986). Species complexes in which species boundaries have been difficult to resolve taxonomically are likely to be among the most rewarding subjects for studies of incipient speciation.

In principle, sister-species comparisons can be similarly helpful for drawing inferences about the factors driving speciation. However, the time since divergence for sister species is likely to be greater than for intra-specific taxa, so whether current selective regimes are similar to those at the time of divergence is less certain. Sister species show changes in distribution that are proportional to the time since divergence (Barraclough *et al.* 1998), and may not even be closely related if many related species have become extinct. Use of intra-specific taxa allows researchers to side-step some of these problems, or at least reduce them to a manageable level.

Species in which intra-specific floral variation has been attributed to a geographical pollinator mosaic include *Dalechampia scandens* (Euphorbiaceae: resin-collecting bees: Armbruster 1985), *Lapeirousia pyramidalis* (Iridaceae: hawk moths and long-proboscid flies: Goldblatt and Manning 1996), *Gilia leptantha* (Polemoniaceae: bees and bee flies: Grant and Grant 1965), the *Disa draconis* complex (Orchidaceae: different long-proboscid flies: Johnson and Steiner 1997), *Satyrium hallackii* (Orchidaceae: bees and hawk moths: Johnson 1997), *Platanthera ciliaris* (Orchidaceae: butterflies of varying tongue length: Robertson and Wyatt 1990), *Macromeria viridiflora* (Boraginaceae: short- and long-billed hummingbirds: Boyd 2004), and *Cimifuga simplex* (Ranunculaceae: bees and butterflies: Pellmyr 1986). Several other studies have sought to link variation in floral form to a geographical mosaic of pollinators, but have not found clear patterns. Examples include *Aquilegia caerulea* (Ranunculaceae: Miller 1981), *Echinocereus coccineus* (Cactaceae: Scobell and Scott 2002), and *Helleborus foetidus* (Ranunculaceae: Herrera *et al.* 2002).

The first level of analysis of ecotypes involves an attempt to correlate traits with some aspect of the environment, in this case the pollinator fauna. As Herrera (Chapter 15) noted, trait–environment correlations comprise only partial evidence for adaptation. Nevertheless, these correlations are an essential starting point (Schluter 2000) and have scarcely been considered on any significant scale for floral traits.

Ecotype studies commonly fail to show that pollinators vary independently of the plant species being studied. In most studies the gradient is inferred from the spectrum of animals captured on flowers of the plant of interest. This correlative approach introduces uncertainty into the causal relations, as the visitor spectrum may reflect foraging preferences for particular floral traits, rather than floral adaptation to pollinator characteristics. Ideally, pollinator gradients should be established from data that are independent of the plant being studied, perhaps based on distribution records from systematic treatments of the animals, or from independent ecological surveys. Relatively few studies have verified pollinator mosaics using independent data (cf. Johnson 1997; Johnson and Steiner 1997; Valiente-Banuet *et al.* 2004).

Most of the aforementioned ecotype studies fail to test whether there is a genetic basis to the observed differences in floral traits among populations. Although floral traits are widely regarded as being less plastic than vegetative traits—hence the confidence placed by systematists in floral traits for classification—elimination of non-genetic effects through common garden or reciprocal transplant experiments should be a basic requirement in any thorough study of local adaptation (Kawecki and Ebert 2004; Chapter 15).

Perhaps the most serious problem with almost all existing studies of pollination ecotypes is that the variation is attributed to local adaptation, without direct experimental verification. Only a few studies have involved phenotypic selection experiments over a range of sites using either existing trait variation (Totland 2001; Herrera *et al.* 2002; Chapter 15), or manipulation of traits to recreate putative ancestral phenotypes (Johnson and Steiner 1997).

16.5 The scale of gene flow in plants

According to the BSC, a species consists of populations linked by gene flow and can thus evolve as a unit (*sensu* Morjan and Rieseberg 2004), whereas more or less impermeable isolating barriers block gene flow between species. Ehrlich and Raven (1969) mounted perhaps the most serious challenge to the BSC, arguing that gene flow between populations is too limited for species to evolve as cohesive evolutionary units. Whether species are units linked by gene flow or largely evolutionarily independent populations linked by common descent has implications for the development of models of pollinator-driven speciation. In particular, the extent of gene flow, as mediated by pollinators and seed-dispersal agents, determines the geographical scale of evolutionary diversification, the need for isolating barriers between parapatric taxa, and the extent to which species-level traits are fixed in small founder populations or diffuse across broad geographical ranges.

Much new information on gene flow has come to light since Ehrlich and Raven's paper. Several examples demonstrate extreme, long-distance gene flow in plants, such as wind-assisted vectoring of fig pollen by wasps up to 15 km (Nason *et al.* 1998). However, a recent survey of genetic studies of 289 plant species revealed that in almost 50% of these species less than one immigrant typically enters a population per generation (Morjan and Rieseberg 2004). Clearly gene flow in these species is insufficient to prevent genetic divergence through genetic drift and local selection (also see Chapter 12). However, Morjan and Rieseberg (2004) point out that "creative" flow of favourable alleles could result in cohesion among populations of species with low overall levels of gene flow.

Typical immigration rates, as estimated by genetic data from an arbitrary sample of populations of a species, may obscure the importance of major disjunctions in limiting gene flow. As pollen- or seed-mediated gene flow between populations is highly unlikely beyond 30 km (a rule of thumb suggested by Coyne and Orr 2004), the status of tens of thousands of plant species as cohesive evolutionary units must be called into question, as disjunctions on this scale are common. The extent of isolation is underscored by a simple analysis of the distribution records for the southern African orchids which found disjunctions >100 km in more than half of the 458 species (Fig. 16.2). These disjunctions mostly involve ancient geographical features (mountain ranges, dry valleys, different soil types) and are thus unlikely to be transient or anthropogenic in origin.

Rather than species evolving as cohesive units linked by persistent gene flow, species-level traits probably evolve during initial divergence, or shortly thereafter, and thus are represented in various populations primarily by common descent from the original founders. These traits are probably maintained largely by co-adaptation of gene complexes and stabilizing selection (Raven 1976).

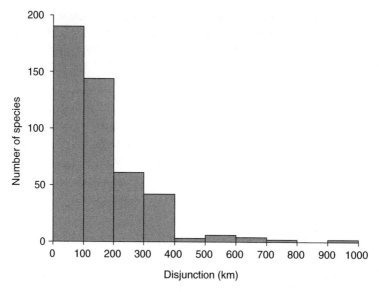

Figure 16.2 The frequency distribution of disjunctions in the ranges of the 458 orchid species that occur in southern Africa. The median disjunction distance is 128 km.

Conversely, populations probably diverge primarily in response to a change in selective regime, rather than interruption of gene flow. Furthermore, parent populations need not undergo evolutionary changes; thus paraphyly is both predicted and commonly observed at the species level (Rieseberg and Brouillet 1994).

16.6 Geographical modes of pollinator-driven speciation

The three main geographical modes of speciation are *allopatric*, whereby divergence involves populations that are not linked by gene flow; *parapatric*, in which divergence occurs between adjacent habitats in the face of some gene flow; and *sympatric*, in which populations diverge in situ through an instantaneous genetic barrier to gene flow or extremely assortative mating. All three modes may apply to pollinator-driven speciation, although only conventional allopatric speciation is strongly supported empirically.

Allopatric speciation is amply supported by the frequent allopatry of sister species and the tendency for range overlap to increase after divergence (Barraclough and Vogler 2002). Selection resulting in local adaptation is increasingly recognized as a significant influence on allopatric divergence, in contrast to previous models which emphasized neutral change resulting from genetic drift. In the context of pollinator-driven speciation without reinforcement, such divergence occurs in response to landscape-level changes in the pollinator fauna. Indeed, studies of sister taxa demonstrate that the mosaic of long-proboscid fly pollinators in southern Africa (Fig 16.1) has promoted allopatric plant speciation. Typically, one member of a species pair is pollinated by a long-proboscid fly species, whereas the other member, in a different geographical area, is pollinated by either a different fly species (Plate 6a and b) or other pollen vectors, such as hawk moths (Plate 6 c–f; Goldblatt and Manning 1996; Johnson and Steiner 1997; Johnson et al. 1998, 2002b; Goldblatt et al. 2001, 2004). In some cases, ancestral pollination systems can be inferred from phylogenies. For example, in *Zaluzianskya* (Scrophulariaceae), pollination by long-proboscid flies in *Zaluzianskya microsiphon* (Plate 6d) is a derived condition, representing a shift from hawk-moth pollination (Johnson et al. 2002a), whereas in *Lapeirousia* (Iridaceae), the reverse shift (fly to moth) occurred during the evolution of

Lapeirousia pyramidalis (Goldblatt and Manning 1996). Although reciprocal translocations and phenotypic selection experiments have yet to be performed to verify the adaptive basis of trait divergence in most of these instances, the observed patterns are consistent with the Grant–Stebbins model of allopatric pollinator-driven speciation.

Parapatric speciation, whereby divergence of contiguous populations occurs despite some gene flow, has been linked closely to ecological selection and reinforcement. Inferences of parapatric speciation based on the proximity of extant sister taxa are always dubious, as distributions can change substantially after speciation (Barraclough and Vogler 2002). Goldblatt and Manning (1996) argued for parapatric speciation in the radiation of the southern African iris genus *Lapeirousia*. Sister taxa of *Lapeirousia* typically occupy closely abutting habitats that differ chiefly in substrate. These taxa also tend to differ in their modes of pollination, which Goldblatt and Manning interpreted as the outcome of selection for isolating barriers at the contact zone. However, if the pollinator fauna itself differs between substrates, the linkage of pollinator and soil specialization in *Lapeirousia* may be simply be a case of parallel adaptations to the local environment. Whether these adaptations occurred in the face of some gene flow or in fully allopatric populations is thus almost impossible to gauge.

Sympatric speciation has always been controversial and can be excluded as a possibility in almost all cases of pollinator-driven speciation. Although a single individual can theoretically found a new lineage in a population of self-compatible plants, it is extremely unlikely that a mutant plant, no matter how novel its floral traits, would attract a completely different set of pollinators compared with plants in the rest of the population, and thus result in a new species.

One special case that may involve sympatric speciation deserves consideration, namely, sexual deception in orchids (Schiestl and Ayasse 2002; Schiestl *et al.* 2003; Mant *et al.* 2005). In these orchids, which attract male Hymenoptera chiefly by imitating the sex pheromones of female insects, a mutant with a novel fragrance could attract a different and non-overlapping set of pollinators (Schiestl and Ayasse 2002; Mant *et al.* 2005). However, despite common belief that sexually deceptive orchids attract male Hymenoptera of a specific species, molecular data show that gene flow between sexually deceptive orchid species is reasonably common (Soliva and Widmer 2003). Thus even in sexually deceptive orchids, speciation probably usually occurs allopatrically within a geographical mosaic of Hymenoptera species in response to selection favouring a shift to locally effective pollinators. Given that these orchids occasionally attract more than one pollinator species, this shift could occur according to Stebbin's principle of "transfer of function via an intermediate stage of double function."

16.7 Identifying pollinator-driven speciation

As we have seen, speciation can be pollinator-driven, but it can also result from adaptations to different abiotic environments (cf. Baldwin 1997; Verboom *et al.* 2004). How can the relative importance of these factors be gauged? Stebbins (1970) and Carson (1985) argued that the driving factors of speciation in plants could be identified by the phenotypic signal within lineages. In particular, speciation caused by local adaptation in a mosaic of physical (growth-influencing) environments should be reflected in vegetative diversification (cf. Baldwin 1997), whereas local adaptation in a mosaic of pollinator environments should involve floral diversification (cf. Johnson *et al.* 1998). Using the Cape flora as an example, Johnson (1996) identified several genera with very little vegetative diversification, yet considerable floral variation, and another set of genera that show the reverse pattern. Subsequent studies have confirmed that genera in the former category are indeed characterized by radiation of specialized pollination systems, including *Lapeirousia* (Goldblatt and Manning 1996), *Gladiolus* (Goldblatt *et al.* 2001), and *Disa* (Johnson *et al.* 1998).

Grant (1949) earlier showed that floral traits tend to comprise a much larger proportion of taxonomic characters in plant families characterized by specialized animal pollination than in families characterized by "promiscuous" or wind-pollination systems. Grant interpreted these results in the context of a discussion of the importance of ethological and mechanical isolating barriers in plants, but the same data also clearly link specialized pollination systems and floral diversification.

16.8 Pollinators and reproductive isolation

Regardless of one's view of species, isolating barriers in one form or another are essential to prevent most species from undergoing genetic dissolution (Mayr 1942; Coyne and Orr 2004). The mechanisms of pollinator-mediated isolation between congeners are well documented (Grant 1994b: Chapter 18) and do not need repeating here. However, I will consider two outstanding issues: the extent to which pollinator-mediated isolation is of general importance for the maintenance of species integrity, and recent claims that widespread generalization in pollination systems renders strict ethological isolation unlikely (cf. Waser 1998, 2001).

As several authors have noted (Coyne and Orr 2004; Rieseberg et al. 2004), occasional hybridization in a contact zone does not threaten a species' existence. Examination of almost all classic examples of pollinator-mediated isolation between closely related taxa reveals that habitat differences are also a significant barrier to hybridization (Hodges and Arnold 1994b; Goldblatt and Manning 1996; Campbell et al. 1997; Goulson and Jerrim 1997; Ramsey et al. 2003). For example, in *Lapeirousia*, pollinator shifts have occurred in four terminal sister pairs, three of which have also undergone shifts in substrate (Goldblatt and Manning 1996). Even orchids, which are notorious for their lack of postzygotic isolating barriers, offer few documented examples of fully sympatric sister taxa that owe their existence to differences in pollination systems (cf. Steiner et al. 1994). Such examples are probably rare because divergence of pollination systems typically occurs allopatrically and is thus likely to be associated with some form of habitat specialization that lessens the likelihood of secondary contact (cf. Patterson and Givnish 2003).

Although species integrity as a whole may seldom depend on pollinator-mediated isolation, this form of isolation undoubtedly contributes significantly to the coexistence of related taxa in zones of range overlap, and thus helps to sharpen species boundaries. There is now ample evidence for the efficacy of pollinator-mediated isolating barriers, particularly when both mechanical and ethological barriers operate simultaneously (Fulton and Hodges 1999; Ramsey et al. 2003; Ippolito et al. 2004). Many of these studies show that primary pollinators are largely responsible for assortative mating in the contact zones, whereas other flower-visiting animals, such as pollen-collecting bees or herbivorous beetles, cause much of the illegitimate pollen flow (Steiner et al. 1994; Fulton and Hodges 1999). However, these disassortative mating events seldom result in species dissolution, especially when other postzygotic barriers exist.

Waser (1998, 2001) has argued that pollination systems are seldom sufficiently specialized to result in complete ethological isolation. This has led him to question the traditional view of pollinator-driven speciation in which selection on floral traits leads to speciation because of the pleiotropic consequences for reproductive isolation (cf. Grant and Grant 1965; Bradshaw and Schemske 2003). However, ethological isolation is not central to a Darwinian view of speciation, which places primary importance on divergence, or one in which species maintenance depends less on isolating barriers and more on stabilizing selection. In a sharp departure from the BSC, Carson (1985, p. 380) reasoned that "the integrity of either a plant or an animal species is maintained not by ad hoc mechanisms, but primarily by selection that serves to maintain and sharpen the adaptive norm that characterizes species."

Hybrid zones continue to attract much research attention, largely because of their relevance for testing the efficacy of isolating barriers (Chapter 18). Rieseberg and others have also argued that hybrid zones allow the flow of important beneficial mutations between species through introgression (Rieseberg et al. 2004; Seehausen 2004). In this sense, hybrid zones might actually be coalfaces of speciation, but not in the sense of the reinforcement scenario proposed by Wallace and Dobzhansky in which hybridization results invariably in significant fitness losses.

16.9 Reinforcement of isolating barriers

Reinforcement of isolating barriers after secondary contact between diverging forms is one of the most controversial ideas in evolutionary biology. Even the *Drosophila* geneticist Hampton Carson believed that "selection is rarely directed towards reinforcing reproductive isolation. Rather, it will maximize fitness by making reproduction more efficient through mate recognition and sexual selection" (Carson 1985, p. 380). One problem with reinforcement is the lack of a mechanism by which traits selected under reinforcement can spread to populations away from the contact zone, as their effect on fitness is limited to this zone. Divergence of floral traits in allopatric populations also shows that reinforcement is not required for their evolution. Is there evidence that reinforcement plays a role in floral evolution? Such evidence would include exaggerated differences between traits in contact zones. In a classic study, Levin and Kerster (1967) showed that *Phlox pilosa*, which is usually pink-flowered, tends to have white flowers in contact zones with its pink-flowered congener *Phlox glaberrima*. This shift in flower colour was later shown to reduce hybridization effectively (Levin and Schaal 1970). Other than this example, there has been little subsequent support for reinforcement in plants.

Jones (2001) proposed a special scenario of reinforcement based on the foraging of individual pollinators. Noting that individual pollinators, especially Hymenoptera, often forage preferentially on particular forms of a species, she argued that this behaviour could cause assortative mating among slightly diverged forms on secondary contact. This form of assortative mating can function as an isolating barrier in sympatric populations only if it reinforces differences that existed already, because of divergence in allopatry. Although good explanations for divergence in floral form without shifts in pollinators may be lacking, this scenario would require rather extraordinary levels of assortative mating to be effective and is thus open to the same criticisms that Waser (1998, 2001) levelled against ethological isolation in general.

16.10 Adaptive radiation

A link between pollinator shifts and diversification is evident from the increasing number of studies that map pollination systems onto phylogenies to interpret their evolution (reviewed by Weller and Sakai 1999; Wilson *et al.* 2006; Chapter 17). Schluter (2000) proposed that the term adaptive radiation should be used only when diversification in a lineage is characterized by common ancestry, phenotype–environment correlation, trait utility, and rapid speciation. Most of the above examples of floral radiations satisfy the first three criteria. The fourth, rapid speciation, is still difficult to assess without good calibration of molecular clocks (see Kay *et al.* 2005), but it seems likely to be general on account of the limited sequence variation recorded in many studies.

Biologists have tended to attribute adaptive radiations to features of the organisms themselves. This "key innovation" approach has been used to explain why some angiosperm lineages have radiated more than others (Chapter 17). For example, Cozzolino and Widmer (2005) attributed the explosive speciation of orchids to their tendency to lack floral rewards, a trait that promotes cross-pollination (cf. Johnson *et al.* 2004). Using the relative number of species in sister lineages as a measure of the rate of diversification, Hodges and coworkers (Hodges and Arnold 1995; Hodges 1997; Chapter 17) showed that the evolution of floral spurs contributes to higher rates of speciation. Thus floral spurs are considered a key innovation. Although Hodges strongly emphasized the role that spurs play in reproductive isolation, spurs could also promote diversification because of their role in the development of specialized pollination systems that are more likely to undergo adaptive shifts in a geographical pollinator mosaic. The relative importance of these two factors (diversification and isolation) would depend largely on the extent of range overlap. Similar problems of interpretation apply to Sargent's (2004) recent demonstration that lineages with zygomorphic flowers (and hence likely to be associated with specialized pollination systems) tend to be more species rich than lineages with actinomorphic flowers.

In some cases, floral radiation may be facilitated by key environments, rather than key innovations. This appears to be true in the species-rich Cape region of South Africa, where parallel radiations have occurred in several unrelated lineages, such as orchids, irises, and ericas (Goldblatt and Manning 1996; Johnson et al. 1998). Although it has been speculated that ecological gradients, including those involving pollinators (Fig. 16.1), are particularly steep in this region, the environmental basis for these parallel radiations is still much debated (Linder 2003). Similar explosive speciation of plant lineages in the northern Andes has been attributed to a combination of climatic fluctuations, geological uplifts, and opportunities for shifts between insect and hummingbird pollination (Kay et al. 2005)

16.11 Conclusions

The conceptual model of pollinator-driven speciation developed by Grant and Grant (1965) and Stebbins (1970) is broadly supported by a wide range of pattern and process studies, but certain of its components remain vague or contentious. To conclude, I have identified some aspects of the "Grant–Stebbins model" (as it is referred to in this chapter) that require either theoretical development or further experimental work, or a combination of both.

One of the most outstanding findings of plant reproductive ecology has been that animal pollination is an uncertain and inefficient process that often limits seed production in plants. Although neither the Grants nor Stebbins used the term pollen limitation, it was clear that this state is what they had in mind when they described plants that would benefit from a mutation that increased their attractiveness to pollinators. Pollen limitation appears to dictate the strength of selection on floral traits that influence both female and male fertility (Section 16.2; Chapter 4), but the link between pollen limitation and shifts in pollination modes, other than shifts to selfing, is poorly understood.

The Grant–Stebbins model predicts that a geographical mosaic of pollinator availability should be conducive to between-population diversification in floral traits, because of selection for traits that improve the "fit" between locally effective pollinators and flowers. However, few studies of "pollination ecotypes" provide compelling support for this prediction. More experimental process studies are needed, focusing on spatial variation in selection imposed on floral traits and its link to the geographical pollinator mosaic.

In the Grant–Stebbins model, pollinator shifts lead to mechanical or ethological floral isolation, and thus to the formation of biological species. How often do forms that have undergone a pollinator shift depend on floral isolating barriers, as opposed to geographical, edaphic, or postzygotic ones to prevent genetic dissolution through hybridization? More studies of the kind pioneered by Ramsey et al. (2003) are needed to evaluate the importance of floral isolation in relation to other isolating barriers. If geography turns out to be the most important barrier (i.e., secondary contact between sister taxa that have undergone shifts in pollination mode is rare), then a central tenet of the Grant–Stebbins model—the linkage of floral evolution and reproductive isolation—may need to be revised.

The radiation of the angiosperms remains one of the great puzzles of evolution, and will undoubtedly occupy the minds of biologists for centuries to come. Solving it requires a diversity of approaches from natural history to molecular biology. In my view it is time to move beyond the reductionist view of speciation as the acquisition of isolating mechanisms to consider more fully the environmental factors behind the evolution of floral diversity.

Acknowledgements

I am grateful to Allan Ellis, Bruce Anderson, Mark van Kleunen, Nick Waser, Timo van der Niet, Lawrence Harder, Spencer Barrett, and two anonymous reviewers for their valuable inputs during the writing of this chapter.

References

Aigner PA (2005). Variation in pollination performance gradients in a *Dudleya* species complex: can generalization promote floral divergence? *Functional Ecology*, **19**, 681–9.

Alexandersson R and Johnson SD (2001). Pollinator mediated selection on flower-tube length in a hawkmoth-pollinated *Gladiolus* (Iridaceae). *Proceedings of the Royal Society of London, Series B*, **269**, 631–6.

Archibald JK, Wolfe AD, and Johnson SD (2004). Hybridization and gene flow between a day-and night-flowering species of *Zaluzianskya* (Scrophulariaceae SS, tribe Manuleeae). *American Journal of Botany*, **91**, 1333–44.

Armbruster WS (1985). Patterns of character divergence and the evolution of reproductive ecotypes of *Dalechampia scandens* (Euphorbiaceae). *Evolution*, **39**, 733–52.

Armbruster WS (1993). Evolution of plant pollination systems: hypotheses and tests with the neotropical vine *Dalechampia*. *Evolution*, **47**, 1480–505.

Arroyo MTK, Primack R, and Armesto J (1982). Community studies in pollination ecology in the high temperate Andes of central Chile. I. Pollination mechanisms and altitudinal variation. *American Journal of Botany*, **69**, 82–97.

Ashman T-L and Morgan MT (2004). Explaining phenotypic selection on plant attractive characters: male function, gender balance or ecological context? *Proceedings of the Royal Society of London, Series B*, **271**, 553–9.

Ashman T-L, Knight TM, Steets JA, et al. (2004). Pollen limitation of plant reproduction: Ecological and evolutionary causes and consequences. *Ecology*, **85**, 2408–21.

Baldwin BG (1997). Adaptive radiation of the Hawaiian silversword alliance: congruence and conflict of phylogenetic evidence from molecular and non-molecular investigations. In TJ Givnish and KJ Sytsma, eds. *Molecular evolution and adaptive radiation*, pp. 103–28. Cambridge University Press, New York.

Barraclough TG and Vogler AP (2002). Detecting the geographical pattern of speciation from species-level phylogenies. *American Naturalist*, **155**, 419–34.

Barraclough TG, Vogler AP, and Harvey PH (1998). Revealing the factors that promote speciation. *Philosophical Transactions of the Royal Society of London, Series B*, **353**, 241–9.

Beardsley PM, Yen A, and Olmstead RG (2003). AFLP phylogeny of *Mimulus* section Erythranthe and the evolution of hummingbird pollination. *Evolution*, **57**, 1397–410.

Bell G (1985). On the function of flowers. *Proceedings of the Royal Society of London, Series B*, **224**, 223–65.

Boyd EA (2004). Breeding system of *Macromeria viridiflora* (Boraginaceae) and geographical variation in pollinator assemblages. *American Journal of Botany*, **91**, 1809–13.

Bradshaw HD Jr and Schemske DW (2003). Allele substitution at a flower colour locus produces a pollinator shift in monkeyflowers. *Nature*, **426**, 176–8.

Broyles SB and Wyatt R (1995). A reexamination of the pollen-donation hypothesis in an experimental population of *Asclepias exaltata*. *Evolution*, **49**, 89–99.

Burd M (1994). Bateman's principle and plant reproduction: the role of pollen limitation in fruit and seed set. *Botanical Review*, **60**, 83–139.

Campbell DR (1989). Measurements of selection in a hermaphroditic plant: variation in male and female pollination success. *Evolution*, **43**, 318–35.

Campbell DR, Waser NM, and Meléndez-Ackerman EJ (1997). Analyzing pollinator-mediated selection in a plant hybrid zone: hummingbird visitation patterns on three spatial scales. *American Naturalist*, **149**, 295–315.

Carson HL (1985). Unification of speciation theory in plants and animals. *Systematic Botany*, **10**, 380–90.

Conner JK and Rush S (1997). Measurements of selection on floral traits in black mustard, *Brassica nigra*. *Journal of Evolutionary Biology*, **10**, 327–35.

Coyne JA and Orr HA (2004). *Speciation*. Sinauer, Sunderland, MA.

Cozzolino S and Widmer A (2005). Orchid diversity: an evolutionary consequence of deception? *Trends in Ecology and Evolution*, **20**, 487–94.

Cruden RW (1972). Pollinators in high-elevation ecosystems: relative effectiveness of birds and bees. *Science*, **176**, 1439–40.

Darwin CR (1859). *On the origin of species by means of natural selection or the preservation of favoured races in the struggle for life*. John Murray, London.

Dobzhansky T (1951). *Genetics and the origin of species*, 3rd edn. Columbia University Press, New York.

Dodd ME, Silvertown J, and Chase MW (1999). Phylogenetic analysis of trait evolution and species diversity variation among angiosperm families. *Evolution*, **53**, 732–44.

Ehrlich PR and Raven PH (1969). Differentiation of populations. *Science*, **165**, 1228–31.

Fenster CB, Armbruster WS, Wilson P, Dudash MR, and Thomson JD (2004). Pollination syndromes and floral specialization. *Annual Review of Ecology, Evolution and Systematics*, **35**, 375–403.

Fulton M and Hodges SA (1999). Floral isolation between *Aquilegia formosa* and *Aquilegia pubescens*. *Proceedings of the Royal Society of London, Series B*, **266**, 2247–52.

Galen C (1989). Measuring pollinator-mediated selection on morphometric floral traits: bumblebees and the alpine sky pilot, *Polemonium viscosum*. *Evolution*, **43**, 882–90.

Goldblatt P and Manning JC (1996). Phylogeny and speciation in *Lapeirousia* subgenus *Lapeirousia* (Iridaceae: Ixioideae). *Annals of the Missouri Botanical Garden*, **83**, 346–61.

Goldblatt P and Manning JC (2000). The long-proboscid fly pollination system in southern Africa. *Annals of the Missouri Botanical Garden*, **87**, 146–70.

Goldblatt P, Bernhardt P, and Manning JC (2000). Adaptive radiation of pollination mechanisms in *Ixia* (Iridaceae: Crocoideae). *Annals of the Missouri Botanical Garden*, **87**, 564–77.

Goldblatt P, Manning JC, and Bernhardt P (2001). Radiation of pollination systems in *Gladiolus* (Iridaceae: Crocoideae) in southern Africa. *Annals of the Missouri Botanical Garden*, **88**, 713–34.

Goldblatt P, Nanni I, Bernhardt P, and Manning JC (2004). Floral biology of *Hesperantha* (Iridaceae: Crocoideae): how minor shifts in floral presentation change the pollination system. *Annals of the Missouri Botanical Garden*, **91**, 186–206.

Goulson D and Jerrim K (1997). Maintenance of the species boundary between *Silene dioca* and *S. latifolia* (red and white campion). *Oikos*, **79**, 115–126.

Grant V (1949). Pollination systems as isolating mechanisms in flowering plants. *Evolution*, **3**, 82–97.

Grant V (1977). *Plant speciation*. Columbia University Press, New York..

Grant V (1983). The systematic and geographical-distribution of hawkmoth flowers in the temperate North-American flora. *Botanical Gazette*, **144**, 439–49.

Grant V (1994a). Historical development of ornithophily in the western North American flora. *Proceedings of the National Academy of Science of the United States of America*, **91**, 10407–11.

Grant V (1994b). Modes and origins of mechanical and ethological isolation in angiosperms. *Proceedings of the National Academy of Sciences of the United States of America*, **91**, 3–10.

Grant V and Grant KA (1965). *Flower pollination in the Phlox family*. Columbia University Press, New York.

Hapeman JR and Inoue K (1997). Plant-pollinator interactions and floral radiation in *Platanthera* (Orchidaceae). In TJ Givnish and KJ Sytsma, eds. *Molecular evolution and adaptive radiation*, pp. 433–54. Cambridge University Press, New York.

Herrera CM, Cerda X, Garcia MB, et al. (2002). Floral integration, phenotypic covariance structure and pollinator variation in bumblebee-pollinated *Helleborus foetidus*. *Journal of Evolutionary Biology*, **15**, 108–21.

Hodges SA (1997). Floral nectar spurs and diversification. *International Journal of Plant Sciences*, **158**, S81–8.

Hodges SA and Arnold ML (1994a). Columbines: a geographically widespread species flock. *Proceedings of the National Academy of Sciences of the United States of America*, **91**, 5129–32.

Hodges SA and Arnold ML (1994b). Floral and ecological isolation between *Aquilegia formosa* and *Aquilegia pubescens*. *Proceedings of the National Academy of Sciences of the United States of America*, **91**, 2493–6.

Hodges SA and Arnold ML (1995). Spurring plant diversification: are floral nectar spurs a key innovation? *Proceedings of the Royal Society of London, Series B*, **262**, 343–8.

Ippolito A, Fernandes GW, and Holtsford TP (2004). Pollinator preferences for *Nicotiana alata*, *N. forgetiana*, and their F-1 hybrids. *Evolution*, **58**, 2634–44.

Johnson SD (1996). Pollination, adaptation and speciation models in the Cape flora of South Africa. *Taxon*, **45**, 59–66.

Johnson SD (1997). Pollination ecotypes of *Satyrium hallackii* (Orchidaceae) in South Africa. *Botanical Journal of the Linnean Society*, **123**, 225–35.

Johnson SD and Bond WJ (1992). Habitat dependent pollination success in a cape orchid. *Oecologia*, **91**, 455–6.

Johnson SD and Steiner KE (1997). Long-tongued fly pollination and evolution of floral spur length in the *Disa draconis* complex (Orchidaceae). *Evolution*, **51**, 45–53.

Johnson SD and Steiner KE (2000). Generalization versus specialization in plant pollination systems. *Trends in Ecology and Evolution*, **15**, 190–3.

Johnson SD, Linder HP, and Steiner KE (1998). Phylogeny and radiation of pollination systems in *Disa* (Orchidaceae). *American Journal of Botany*, **85**, 402–11.

Johnson SD, Edwards TJ, Carbutt C, and Potgieter C (2002a). Specialization for hawkmoth and long-proboscid fly pollination in *Zaluzianskya* section *Nycterinia* (Scrophulariaceae). *Botanical Journal of the Linnaean Society*, **138**, 17–27.

Johnson SD, Peter CI, Nilsson LA, and Ågren J (2003). Pollination success in a deceptive orchid is enhanced by co-occurring rewarding magnet plants. *Ecology*, **84**, 2919–27.

Johnson SD, Peter CI, and Ågren J (2004). The effects of nectar addition on pollen removal and geitonogamy in the non-rewarding orchid *Anacamptis morio*. *Proceedings of the Royal Society of London, Series B*, **271**, 803–9.

Johnson SD, Neal PR, and Harder LD (2005). Pollen fates and the limits on male reproductive success in an orchid population. *Biological Journal of the Linnean Society*, **86**, 175–90.

Johnston MO (1991). Natural selection on floral traits in two species of *Lobelia* with different pollinators. *Evolution*, **45**, 1468–79.

Jones KN and Reithel JS (2001). Pollinator-mediated selection on a flower color polymorphism in experimental populations of *Antirrhinum* (Scrophulariaceae). *American Journal of Botany*, **88**, 447–54.

Jones NJ (2001). Pollinator-mediated assortative mating: causes and consequences. In L Chittka and JD Thomson, eds. *Cognitive ecology of pollination*, pp. 259–73. Cambridge University Press, Cambridge.

Kawecki TJ and Ebert D (2004). Conceptual issues in local adaptation. *Ecology Letters*, **7**, 1225–41.

Kay KM, Reeves PA, Olmstead RG, and Schemske DW (2005). Rapid speciation and the evolution of hummingbird pollination in neotropical *Costus* subgenus *Costus* (Costaceae): evidence from nrDNA ITS and ETS sequences. *American Journal of Botany*, **92**, 1899–910.

Laverty TM (1992). Plant interactions for pollinator visits: a test of the magnet species effect. *Oecologia*, **89**, 502–8.

Levin DA (1971). The origin of reproductive isolating mechanisms in flowering plants. *Taxon*, **20**, 91–113.

Levin DA and Kerster HW (1967). Natural selection for reproductive isolation in *Phlox*. *Evolution*, **21**, 242–50.

Levin DA and Schaal BA (1970). Corolla color as an inhibitor of interspecific hybridization in *Phlox*. *Bulletin of the Torrey Botanical Club*, **104**, 273–83.

Linder HP (2003). The radiation of the Cape flora, southern Africa. *Biological Reviews*, **78**, 597–638.

Macior LW (1982). Plant community and pollinator dynamics in the evolution of pollination mechanisms in *Pedicularis* (Scrophulariaceae). In JA Armstrong, JM Powell, AJ Richards, eds. *Pollination and evolution*, pp 29–45. Royal Botanic Gardens, Sydney.

Mant J, Bower CC, Weston PH, and Peakall R (2005). Phylogeography of pollinator-specific sexually deceptive *Chiloglottis* taxa (Orchidaceae): evidence for sympatric divergence? *Molecular Ecology*, **14**, 3067–76.

Mayr E (1942). *Systematics and the origin of species*. Columbia University Press, New York.

Mayr E (1963). *Animal species and evolution*. Belknap Press, Cambridge, MA.

Miller RB (1981). Hawkmoths and the geographical patterns of floral variation in *Aquilegia caerulea*. *Evolution*, **35**, 763–74.

Morjan CL and Rieseberg LH (2004). How species evolve collectively: implications of gene flow and selection for the spread of advantageous alleles. *Molecular Ecology*, **13**, 1341–56.

Muller H (1880). The fertilisers of alpine flowers. *Nature*, **21**, 275.

Nason JD, Herre EA, and Hamrick JL (1998). The breeding structure of a tropical keystone plant resource. *Nature*, **391**, 685–7.

Patterson TB and Givnish TJ 2003. Geographic cohesion, chromosomal evolution, parallel adaptive radiations, and consequent floral adaptations in *Calochortus* (Calochortaceae): evidence from a cpDNA phylogeny. *New Phytologist*, **161**, 253–64.

Pellmyr O (1986). Three pollination morphs in *Cimicifuga simplex*; incipient speciation due to inferiority in competition. *Oecologia*, **68**, 304–7.

Potgieter CJ and Edwards TJ (2005). The *Stenobasipteron wiedemanni* (Diptera, Nemestrinidae) pollination guild in eastern Southern Africa. *Annals of the Missouri Botanical Garden*, **92**, 254–67.

Primack RB (1983). Insect pollination in the New Zealand mountain flora. *New Zealand Journal of Botany*, **21**, 317–33.

Queller D (1997). Pollen removal, paternity, and the male function of flowers. *American Naturalist*, **149**, 585–94.

Ramsey J, Bradshaw HD Jr, and Schemske DW (2003). Components of reproductive isolation between the monkeyflowers *Mimulus lewisii* and *M. cardinalis* (Phrymaceae). *Evolution*, **57**, 1520–34.

Raven PH (1976). Systematics and plant population biology. *Systematic Botany*, **1**, 284–316.

Rieseberg LH and Brouillet L (1994). Are many plant species paraphyletic? *Taxon*, **43**, 21–32.

Rieseberg LH, Widmer A, Arntz AM, and Burke JM (2002). Directional selection is the primary cause of phenotypic diversification. *Proceedings of the National Academy of Sciences of the United States of America*, **99**, 12242–5.

Robertson JL and Wyatt R (1990). Evidence for pollination ecotypes in the yellow-fringed orchid, *Platanthera ciliaris*. *Evolution*, **44**, 121–33.

Sargent RD (2004). Floral symmetry affects speciation rates in angiosperms. *Proceedings of the Royal Society of London, Series B*, **271**, 603–8.

Schemske DW and Bradshaw HD Jr (1999). Pollinator preference and the evolution of floral traits in monkeyflowers (*Mimulus*). *Proceedings of the National Academy of Science of the United States of America*, **96**, 11910–5.

Schiestl FP and Ayasse M (2002). Do changes in floral odor cause speciation in sexually deceptive orchids? *Plant Systematics and Evolution*, **234**, 111–9.

Schiestl FP, Peakall R, Mant JG, et al. (2003). The chemistry of sexual deception in an orchid-wasp pollination system. *Science*, **302**, 437–8.

Schluter D (2000). *The ecology of adaptive radiation*. Oxford University Press, Oxford.

Scobell SA and Scott PE (2002). Visitors and floral traits of a hummingbird-adapted cactus (*Echinocereus coccineus*) show only minor variation along an elevational gradient. *American Midland Naturalist*, **147**, 1–15.

Seehausen O (2004). Hybridization and adaptive radiation. *Trends in Ecology and Evolution*, **19**, 198–207.

Soliva M and Widmer A (2003). Gene flow across species boundaries in sympatric, sexually deceptive *Ophrys* (Orchidaceae) species. *Evolution*, **57**, 2252–61.

Stebbins GL (1970). Adaptive radiation of reproductive characteristics in angiosperms. I. Pollination mechanisms. *Annual Review of Ecology and Systematics*, **1**, 307–26.

Steiner KE (1989). The pollination of *Disperis* (Orchidaceae) by oil-collecting bees in southern Africa. *Lindleyana*, **4**, 164–83.

Steiner KE, Whitehead VB, and Johnson SD (1994). Floral and pollinator divergence in two sexually deceptive South African orchids. *American Journal of Botany*, **81**, 185–94.

Totland O (2001). Environment-dependent pollen limitation and selection on floral traits in an alpine species. *Ecology*, **82**, 2233–44.

Tremblay RL, Ackerman JD, Zimmerman JK, and Calvo RN (2005). Variation in sexual reproduction in orchids and its evolutionary consequences: a spasmodic journey to diversification. *Biological Journal of the Linnean Society*, **84**, 1–54.

Valiente-Banuet A, Molina-Freaner F, Torres A, Del Coro Arizmendi A, and Casas A (2004). Geographic differentiation in the pollination system of the columnar cactus *Pachycereus pecten-aboriginum*. *American Journal of Botany*, **91**, 850–5.

Vamosi JC, Knight TM, Steets JA, Mazer SJ, Burd M, and Ashman T-L (2006). Pollination decays in biodiversity hotspots. *Proceedings of the National Academy of Sciences of the United States of America*, **103**, 956–61.

Verboom GA, Linder HP, and Stock WD (2004). Testing the adaptive nature of radiation: growth form and life history divergence in the African grass genus *Ehrharta* (Poaceae: Ehrhartoideae). *American Journal of Botany*, **91**, 1364–70.

Warren SD, Harper KT, and Booth GM (1988). Elevational distribution of insect pollinators. *American Midland Naturalist*, **120**, 325–30.

Waser NM (1998). Pollination, angiosperm speciation, and the nature of species boundaries. *Oikos*, **82**, 198–201.

Waser NM (2001). Pollinator behaviour and plant speciation: looking beyond the "ethological isolation" paradigm. In L Chittka and JD Thomson, eds. *Cognitive ecology of pollination: animal behavior and floral evolution*, pp. 318–35. Cambridge University Press, Cambridge.

Waser NM and Campbell DR (2004). Ecological speciation in flowering plants. In U Dieckmann, M Doebeli, JAJ Metz, and D Tautz, eds. *Adaptive speciation*, pp. 264–77. Cambridge University Press, Cambridge.

Weller SG and Sakai AK (1999). Using phylogenetic approaches for analysis of plant breeding system evolution. *Annual Review of Ecology and Systematics*, **30**, 167–99.

West-Eberhard MJ (1983). Sexual selection, social competition, and speciation. *Quarterly Review of Biology*, **58**, 155–83.

Wilson P, Castellanos MC, Wolfe AD, and Thomson JD (2006). Shifts between bee- and bird-pollination in penstemons. In NM Waser and J Ollerton, eds. *Plant-pollinator interactions: from specialization to generalization*, pp. 47–68. Chicago University Press, Chicago.

CHAPTER 17

Floral characters and species diversification

Kathleen M. Kay, Claudia Voelckel, Ji Y. Yang, Kristina M. Hufford, Debora D. Kaska, and Scott A. Hodges

Department of Ecology, Evolution and Marine Biology, University of California, Santa Barbara, CA, USA

Outline

The burgeoning of phylogenetic information during the past 15 years has focused much interest on whether specific features of clades enhance or hinder the evolution of species diversity. In the angiosperms many of the traits thought to affect clade diversity are floral in nature, because of their association with reproduction and thus species isolation. Therefore, we briefly review mechanisms by which floral traits can affect diversification. We then consider the possible influences of four specific traits by comparing the species diversity of a clade possessing a trait with that of its sister clade that lacks the trait. Clearly, this approach requires correct identification of sister groups, so that changes in phylogenetic reconstruction can have profound effects on these analyses. Here we use a recent supertree analysis of the angiosperms, which includes nearly all described families, along with other phylogenies to reexamine a number of floral traits thought to affect diversification rates. In addition, because many of the previous analyses employed a statistical test that has since been shown to be misleading, we use a suite of signed-rank tests to assess associations with diversification. We find statistical support for the positive effect of animal pollination and floral nectar spurs and a negative effect of dioecious sexual system on diversification, as proposed previously. However, our results for the effect of bilaterally symmetric flowers on species diversity are equivocal. We discuss several factors that will aid in future analyses and the need for both more detailed phylogenetic analyses and more studies on floral biology.

17.1 Introduction

The angiosperms are the most abundant and diverse group of plants on Earth today. Since their first appearance in the fossil record during the early Cretaceous (ca. 130 Ma, Crane *et al.* 2004), they have colonized almost every habitat on the planet, and now number approximately 260,000 extant species (Soltis and Soltis 2004). These myriad species vary impressively in morphology, life history, chemistry, and reproductive biology. Especially striking is the floral diversification, which fossils show began among early angiosperms (Friis *et al.* 2000), and therefore must have occurred concurrently with their radiation and rise to ecological dominance. Flowers exhibit an amazing variety of sizes, shapes, colours, arrangements, scents, rewards, and sexual systems, from the tiny self-fertilizing flower of *Arabidopsis thaliana* to the intricate flowers of *Ophrys* orchids, which mimic a female mate for an unsuspecting male wasp, to the enormous putrid inflorescence of the corpse flower, *Amorphophallus titanium.*

Which factors promote angiosperm diversification, especially the role of floral traits and sexual systems, is an enduring question and its resolution is a central goal of plant evolutionary biology. Darwin puzzled over the apparent sudden appearance of the angiosperms in the fossil record, clearly finding it a challenge to his view of "extremely gradual evolution" and forever labelled the phenomenon with his famous quotation as "an abominable mystery" (Darwin 1903). Knowledge of the timing and pace of angiosperm diversification has progressed considerably with both the discovery of fossil flowers (e.g., Dilcher and Crane 1984; Crane et al. 1995; Gandolfo et al. 1998) and molecular-based phylogenies (e.g., Qiu et al. 2000; Zanis et al. 2002). These findings have revealed an increasingly detailed view of patterns of floral diversification.

During the past 10 years or so, phylogenies have been used extensively to assess whether species diversity occurs non-randomly among clades (Sims and McConway 2003) and whether particular traits may be responsible for these patterns. Although studies have reported that changes in angiosperm diversity correlate with several traits, including the rates of molecular evolution (Barraclough et al. 1996; Barraclough and Savolainen 2001), latex and resin canals (Farrell et al. 1991), herbaceous growth habit (Dodd et al. 1999), and climbing habit (Gianoli 2004), floral traits have been implicated most commonly. This apparent evolutionary importance of floral traits is perhaps not surprising, because a correlation between the rise and diversification of angiosperms and the diversification of pollinating insects has long been recognized (Crepet 2000). As much of this volume attests, aspects of both sexual system and floral morphology can affect how a plant reproduces and with which other plants it mates. Thus, these traits are natural subjects for investigating their effects on species diversity.

Comparative studies have identified several floral characters that affect rates of angiosperm diversification, including animal pollination (Dodd et al. 1999), floral nectar spurs (Hodges and Arnold 1995; Hodges 1997a, b), bilateral symmetry (Sargent 2004), and a dioecious sexual system (Heilbuth 2000) (Plate 7). However, most previous analyses of the effects of floral traits on species diversity either used now-discredited statistical methods (Dodd et al. 1999; Hodges and Arnold 1995; Hodges 1997a, b) or relied largely on angiosperm phylogenies that lacked representatives of many families and were constructed mostly with plastid-gene sequences (Dodd et al. 1999; Heilbuth 2000; Sargent 2004). Clearly, incorrect statistical techniques can cause misinterpretations about diversification hypotheses. The limited taxon sampling in phylogenies can both lead to errors in inferring sister-group relationships and the timing of the origin of a trait of interest and reduce the sample of replicate origins of a key trait. Finally, because plastids do not undergo recombination, plastid genes are inherited essentially as a single locus, so that sequences of different genes provide limited independent phylogenetic information. Consequently, phylogenetic information from plastid genes should be combined with data from other loci for a robust phylogeny based on multiple independent lines of evidence. These problems all call for a reanalysis of the role of floral traits in angiosperm diversification.

Here we reanalyse four purported floral correlates of angiosperm diversity—animal pollination, floral nectar spurs, bilateral symmetry, and the dioecious sexual system—using both more complete phylogenetic analyses and appropriate statistical tests. We use a recently constructed and nearly comprehensive supertree of angiosperm families derived from 46 source trees (Davies et al. 2004), along with other phylogenies at lower taxonomic levels, and family circumscriptions consistent with APGII (2003). For each character we identify phylogenetically independent contrasts and compare the species richness of the sister clades composing each contrast. We discuss our findings in light of hypotheses for how these traits affect diversification, and suggest avenues for future work to clarify the mechanisms responsible for any correlations.

17.2 How might floral traits affect diversification?

For a trait to affect diversification rates, it must influence the probability of speciation, extinction, or both. Speciation involves the evolution of reproductive isolation and is generally initiated by geographical isolation (Coyne and Orr 2004).

Therefore, a trait that promotes the colonization of new habitats, limits dispersal between populations, or increases the propensity or ability to mate with phenotypically similar individuals is a good candidate for a trait that might affect diversification by increasing speciation. Certain traits may affect speciation, but in almost all cases these effects are thought to be coincidental to adaptation to local conditions or genetic divergence in isolation. Thus, although natural selection may cause evolution in these traits, they are not selected for reproductive isolation or higher diversification *per se* (Chapter 16). Conversely, some traits may be selected in certain environments and during short periods, but predispose their possessors to a higher chance of extinction. A trait that leads a population to experience greater demographic stochasticity or lower adaptive genetic diversity (see Chapter 2) would be a good candidate for a trait affecting diversification through extinction. Unfortunately, without a detailed phylogeny and fossil record, speciation and extinction are exceedingly difficult to tease apart. Thus analysis of diversification requires a hypothesis for how a trait affects either speciation, extinction, or both, which can be tested more directly than by simply assessing its association with the overall diversification rate.

Animal pollination, bilateral floral symmetry, and nectar spurs have all been suggested to enhance speciation rates for seemingly similar reasons, namely, their likely effects on the specificity of mating among plants. Animal pollination is one of the most striking features of angiosperms, with plants using an incredible array of insects, birds, and mammals to disperse pollen. Because successful pollen transfer is so important to fitness, pollinators exert selection on floral traits (reviewed in Fenster *et al.* 2004 and most chapters in this volume). Spatial and temporal variation in pollinator assemblages can promote evolutionary divergence in floral traits among populations (Chapters 8, 15, and 16), and plant populations adapted to different suites of pollinators may be less likely to mate with each other (Thompson 1994). This avenue for reproductive isolation is not available to abiotically pollinated lineages, which depend on wind or water to transfer pollen. Dodd *et al.* (1999) compared the diversity of sister clades using the methods of Slowinski and Guyer (1993) and found a strong overall pattern that animal pollination was associated with more rapid diversification than abiotic pollination. This finding is bolstered by an extensive body of empirical research on the role of plant–pollinator interactions in speciation (reviewed in Stebbins 1974; Grant 1981; Coyne and Orr 2004; Chapter 16).

Similarly, floral nectar spurs may further promote specialization on different pollinators; affecting reproductive isolation, and thus diversification (Hodges and Arnold 1995; Hodges 1997a, b). The presentation of nectar at the base of a relatively long, thin tube requires a match between the pollinator and the floral morphology, limiting the number of pollinating species that can manipulate the flower successfully. Hodges and Arnold (1995) and Hodges (1997a, b) found an association between the evolution of floral nectar spurs and higher diversification in both a comparative study among angiosperms and a detailed study of columbines (*Aquilegia*).

In contrast, bilateral floral symmetry, or zygomorphy, may affect diversification somewhat differently. Compared with radially symmetric, or actinomorphic, flowers, zygomorphy constrains the orientation of pollinators while they visit flowers, thereby enhancing the precision of pollen exchange between pollinators' bodies and the sexual organs of flowers (Neal *et al.* 1998; Sargent 2004). This increased precision could affect reproductive isolation if it promotes specialization by different pollinators on different types of zygomorphic flowers or if flowers diverge in the location of pollen placement on a pollinator (Chapter 16). Zygomorphy may also limit the number or type of pollinating species that manipulate a species' flowers effectively, which may increase the variance in pollinator assemblages and hence selection on floral traits across the landscape. Although examples of pollen placement affecting reproductive isolation between species visited by the same pollinator are known (Brantjes 1982; Grant 1994; Kay 2006; Chapter 16), the importance of such shifts in speciation remains to be clarified. Nevertheless, in a sister-group study among angiosperms, Sargent (2004) found accelerated diversification in lineages with bilaterally symmetric flowers.

Conversely, a trait may increase the chances of extinction, even if it is favoured by selection in the short term (Chapter 2). The dioecious sexual system, with separate male and female individuals, may be an example of this. Dioecy has evolved multiple times and occurs in approximately 6% of angiosperm species (Renner and Ricklefs 1995). In a comparative study across angiosperms, Heilbuth (2000) found a striking association between dioecy and lower species richness of sister clades. Dioecy may retard diversification for several reasons. Because they lack the reproductive assurance of being able to self-pollinate, dioecious plants may have a higher risk of dying without reproducing and may have a lower colonization ability (Baker 1954; Bawa 1980; Chapter 12). However, self-incompatible species should be subject to the same constraint and have lower species diversity than their self-compatible sister taxa, but no such association has been found (Heilbuth 2000). Vamosi and Otto (2002) proposed that differential selection on male and female flowers can lead one sex to become more showy than the other (typically male), resulting in poor pollination during years of low pollinator abundance. Dioecious plants can also suffer increased variance in both pollination and seed dispersal, because, unlike hermaphrodites, not every individual is a potential mate and only females disperse seeds (Heilbuth et al. 2001; Wilson and Harder 2003). Because reproductive success varies nonlinearly with pollination and seed dispersal, this increased variance can reduce the average reproductive performance of dioecious species relative to that of otherwise similar hermaphroditic species (Wilson and Harder 2003).

17.3 Common tests for key innovations

Discovery of the mechanisms by which particular traits influence speciation and extinction is fundamental, but phylogenetically based comparative studies are necessary to identify the importance of a trait to angiosperm diversification in general. Many traits can influence diversification in some circumstances, but certain traits have been suggested to act as key innovations, allowing their possessors to diversify rapidly and create new niches. Such effects should be relatively consistent across lineages in which they evolve. Ideally, identification of key innovations requires knowledge of the evolutionary relationships among taxa and the timing of all critical events, such as speciation, extinction, and the origins of the trait of interest. Unfortunately, barring an exceptionally detailed fossil record, these factors are usually incompletely known, dictating the use of less powerful statistical tests for an association between a trait and diversification rate.

The simplest technique for testing whether a trait alters diversification compares the numbers of species in two sister taxa differing in the trait of interest. By definition, sister groups are the same age, so any difference in species numbers must be a result of differences in rates of speciation and/or extinction. These differences are compared to a null model of equal diversification to determine whether they are sufficiently large to indicate a change in the diversification rate with the origin of the putative key innovation or whether they occurred stochastically during speciation and extinction (Sanderson and Donoghue 1994, 1996). Such an analysis can be implemented with only a rudimentary phylogeny showing sister-group relationships, and may therefore be feasible in many diverse and poorly characterized lineages; however, it has low statistical power and can detect only extremely large differences in diversification rates (Sanderson and Donoghue 1996).

More powerful inferences can be drawn for traits that evolve repeatedly, which can provide replicated evidence for changes in diversification. Most simply, numbers of species between pairs of sister clades can be compared with a sign test. However, the sign test ignores the magnitude of differences in species numbers, and thus provides limited statistical power. Consequently, the sign test should be used only when the relative sizes and not the species numbers of sister groups are known. Perhaps the most commonly used method has been that of Slowinski and Guyer (1993), which compares the difference in species richness between individual sister groups with a null model based on random speciation and extinction, and then combines probabilities from multiple comparisons. However, several researchers have identified severe shortcomings with this method (e.g., de Queiroz 1998; Goudet 1999; McConway and Sims 2004). Vamosi and Vamosi (2005) recently reviewed and compared statistical tests for sister-group comparisons and showed clearly that the Slowinski–Guyer method is prone to type I errors, because a few large differences in species numbers can result in a significant test statistic, regardless of the direction or magnitude of the remaining contrasts. This problem is especially severe for datasets in which some sister-group comparisons have large differences in species counts that favour the hypothesis, but other sister group comparisons have large differences in the opposite direction. Such a dataset results in a U-shaped frequency distribution of the proportion of species in each sister-group pair possessing the trait of interest. In these cases, the Slowinski–Guyer method can give the nonsensical result that the trait both promotes and retards diversification significantly. For these reasons, Vamosi and Vamosi (2005) recommended against the use of the Slowinski–Guyer method, and suggested more suitable, less biased techniques. They also recommended that plots of the data accompany any statistical tests, making it possible to check visually for data with a U-shaped frequency distribution.

Instead of the Slowinski–Guyer test, contrasts between sister clades can be analysed using a Wilcoxon signed-rank test. Vamosi and Vamosi (2005) reviewed various methods for calculating the contrasts to be tested (also see Isaac *et al.* 2003). "Simple" contrasts based on the absolute difference in species numbers between sister clades may seem straightforward, but do not account for the overall species richness of the pair and can be misleading. For example, Vamosi and Vamosi showed that simple contrasts of 1020 versus 1010 species and 20 versus 10 species result in the same test statistic, but represent two intuitively contrasting cases. Two alternative methods of calculating the contrasts avoid this problem. "Proportional" contrasts are calculated as the proportion of all species in the sister group represented by the clade possessing the trait of interest minus 0.5, so that it ranges from −0.5 to +0.5. For example, the proportional contrast between clade A with 10 species and clade B with 5 species shown in the hypothetical phylogeny of Fig. 17.1 equals 0.167. This test is prone to errors if the proportion of species with the trait of interest for each contrast has a U-shaped distribution. In this case, tests for effects of either character state can result in significant test statistics. In contrast, this approach applied to a data set with an L-shaped frequency distribution, in which most of the large contrasts in species counts fall in the same direction, will yield a significant result for only one of the character states. Finally, "log" contrasts compare sister-group diversity based on the ratio of the log number of species in the larger group to that for the smaller group. In this case, the contrast between A and B in Fig. 17.1 equals 1.43. Log contrasts may favour small or young sister groups, and therefore should be used cautiously if replicate sister groups differ systematically in phylogenetic age according to the direction of their contrast. Tests based on log contrasts yield the same result if a specific character state promotes or retards diversification; however, the direction of the effect can be identified from a plot of the contrast distribution (Vamosi and Vamosi 2005). Isaac *et al.* (2003)

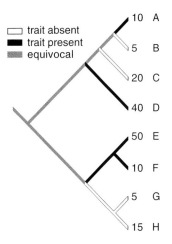

Figure 17.1 Hypothetical phylogenetic tree illustrating the mapping of both the character state for a trait (present, absent or equivocal) and the species numbers for each clade (A–H) at the tips of each branch. The construction of specific contrasts is explained in the text.

used a simulation study to examine the performance of these three contrast methods and recommended using either log contrasts, when sister groups are of similar age, or proportional contrasts.

Note that all techniques based on the Wilcoxon signed-rank test necessarily rely on phylogenies, which in themselves constitute hypotheses. Thus, a sister-group analysis should ideally incorporate the confidence in each phylogenetic hypothesis. Of particular concern are large-scale phylogenies with limited sampling and ability for analysis of statistical support. Incomplete sampling can impact the nature of the comparisons profoundly. For instance, in Fig. 17.1, A and B are sister clades and A, possessing the trait of interest, is also more species rich. However if B was not sampled for the phylogenetic analysis, one would erroneously conclude that A and C are sister clades and that the clade lacking the trait is more diverse. Studies based on comparisons of multiple sister groups implicitly assume that such errors are not biased in one direction or another.

17.4 Methods

Our general approach in reanalysing the effects of the four key traits on diversification was to review the plant groups identified in previous studies, identify new groups that could be added, update the taxonomic circumscriptions to reflect APGII (2003), and update the phylogenetic information according to Davies et al. (2004) and other available phylogenies for lower taxonomic ranks. Independent contrasts between groups possessing and lacking the trait of interest were identified starting at the tips of the phylogeny and proceeding towards the root. Nested contrasts were removed from higher-level contrasts, so that no group was used in more than one contrast. For example, in Fig. 17.1 we would calculate a contrast between A and B, and then remove that contrast from the tree and calculate another contrast between C and D. We used the consensus tree presented by Davies et al. (2004) for family-level and higher contrasts. This tree includes several unresolved nodes, which causes uncertainty about the appropriate sister group. In such cases, we used the species counts from the clades that would be most conservative with regard to the hypothesis of contrasting diversification rates (i.e., the results are biased against finding an effect). Because families are arbitrary constructs and do not represent a well-defined evolutionary unit, we used total species numbers in our higher-level contrasts, instead of averaging across the species richness of the constituent families, as is typically done in nested contrasts of continuous variables. For example, in the contrast of E and F versus G and H in Fig. 17.1, we would contrast species counts of 60 and 20, rather than the average counts of 30 and 10.

17.4.1 Trait datasets

We first compared species richness between sister clades with biotic and abiotic pollination at the family level and higher. We used the data of Dodd et al. (1999), with species counts taken from Davies et al. (2004). Additional pollination information was obtained from Watson and Dallwitz (2005), or from literature searches on the ISI Web of Science using the word "pollination" and the family name in the topic field. Pollination mode for each family was coded as either primarily biotic, primarily abiotic, both modes present, or unknown, and we excluded families in the latter two categories from analyses.

To assess the role of zygomorphy in diversification, we expanded the dataset constructed by Sargent (2004). Character state determinations were taken from Sargent (2004), Watson and Dallwitz (2005), Takhtajan (1997), and Mabberley (1997). Families were considered zygomorphic if they were described as primarily having zygomorphic, bilaterally symmetrical, irregular or bilabiate corollas, whereas actinomorphic families were described as having radially symmetrical, polysymmetric, or regular corollas. Only animal-pollinated families are considered in this analysis, because the hypothesis for how floral symmetry affects diversification depends on plant–pollinator interactions. To be conservative in finding an effect, Sargent (2004) subtracted actinomorphic genera from zygomorphic families, but did not subtract zygomorphic genera from actinomorphic families. As this method could bias the results, we did our analysis both with and without these subtractions.

For nectar spurs, many contrasts between spurred and non-spurred groups occur within families. Therefore we searched the literature to find as many spurred taxa as possible, regardless of rank, and to determine their putative sister clade. Spurred lineages surveyed previously by Hodges (1997a, b) and Hodges and Arnold (1995) were reviewed for more recent evidence regarding sister-group relationships. In addition, we searched the literature for phylogenetic data identifying sister groups for additional spurred lineages that were unavailable in previous analyses. We excluded groups possessing nectarless spurs, which may not function in the hypothesized manner, and groups with flowers described as only saccate. We included the Marcgraviaceae, which does not have spurs within flowers, but rather highly modified floral bracts that form elaborate extrafloral nectaries (Ward and Price 2002), which pollinators probe to access to nectar in a similar manner to probing nectar spurs. We also considered whether each group is zygomorphic or actinomorphic (e.g., Plate 7d versus 7e) to test whether spurs correlate with diversity for the subset of instances in which nectar spurs evolved independently of floral symmetry.

For our analysis of dioecy, we reviewed the dioecious taxa identified by Heilbuth (2000). Lineages were considered dioecious if most or all of the species exhibit separate sexes on different individual plants, whereas lineages were considered non-dioecious if most or all of the species exhibit both sexes on the same individual plants. To avoid inflating the number of species in the non-dioecious sister group, we either subtracted any dioecious genera from the non-dioecious families, or subtracted the estimated number of dioecious species, if this information was available. Information on dioecious genera was taken from Mabberley (1997), Takhtajan (1997), and the database of Renner and Ricklefs (1995).

17.4.2 Analyses

For each trait of interest, we first constructed a frequency distribution of the proportion of species from each sister group possessing the trait of interest to examine qualitatively whether the data exhibited a U-shaped distribution. We then performed one-tailed Wilcoxon signed-rank tests on simple, proportional and log contrasts (see Section 17.3). For datasets including monotypic groups, we added one to all species numbers before log-transformation. For each contrast, we assigned a positive to contrasts matching our hypothesis and a negative to those opposing it. We excluded cases for which the focal and sister clades have equal species numbers, because they are uninformative.

17.5 Results

17.5.1 Pollination mode

Of 379 families included in this analysis, we identified 39 with abiotic pollination, 202 with animal pollination, 17 with both modes present, and 121 for which there is insufficient information. For these data we found 16 independent contrasts between pollination modes. Because the animal-pollinated clade contained more species than the abiotically pollinated clade for 11 of the 16 contrasts (Electronic Appendix 17.1, http://www.eeb.utoronto.ca/EEF/), the frequency distribution of the proportion of biotically pollinated species in sister groups was strongly L-shaped (Fig. 17.2a). Indeed, regardless of the contrast measure used, animal pollination seems to promote significantly higher diversification (Table 17.1). A notable exception to this pattern is the contrast between the animal-pollinated Bromeliaceae and an abiotically pollinated clade including the Poaceae, Juncaceae, and Typhaceae.

17.5.2 Floral symmetry

We found 22 independent contrasts in floral symmetry among animal-pollinated angiosperms (Electronic Appendix 17.2). Only 16 of these contrasts involve the predicted higher species richness in clades with asymmetric flowers and the frequency distribution of the proportion of species in sister groups with asymmetric flowers is distinctly U-shaped, with most contrasts being either strongly positive or strongly negative (Fig. 17.2b).

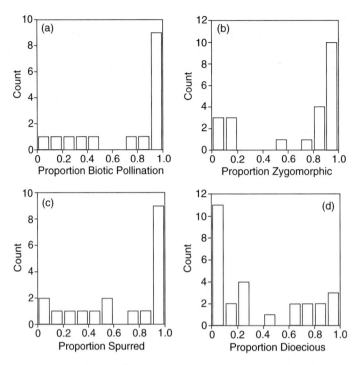

Figure 17.2 Frequency distributions of the proportion of species in a sister group represented by the clade exhibiting (a) biotic pollination, (b) zygomorphic floral symmetry, (c) presence of floral nectar spurs, and (d) dioecious sexual system.

Table 17.1 Results of Wilcoxon signed-ranks tests for the effects of four reproductive characters on diversification.

Character	Method of Ranking		
	Difference in species numbers	Proportion	Log (high)/ log (low)
Biotic pollination	$P=0.033$	$P=0.005$	$P=0.005$
Zygomorphy	$P=0.035$	$P=0.009$	$P=0.010$
Spurs	$P=0.137$, ns $(P=0.010)$	$P=0.019$ $(P=0.007)$	$P=0.007$ $(P=0.007)$
Dioecy	$P=0.019$	$P=0.025$	$P=0.020$

For each character, we considered three methods for calculating the difference in diversity between clades, as described in the text. For floral nectar spurs, the results in parentheses represent tests that considered only comparisons for which both sister groups have the same floral symmetry. Probabilities represent the results of one-tailed tests for each comparison, because test comparisons addressed a specific directional hypothesis.

The statistical tests for proportional and log contrasts are both significant (Table 17.1); however, the U-shaped distribution makes the statistical results suspect. Furthermore, the sister group for one contrast, with 10 zygomorphic families including Marantaceae (total of 1549 spp., Electronic Appendix 17.2), was unresolved and could involve either the relatively species-rich Arecaceae (2500 spp.), or the relatively species-poor Rapataceae (80 spp.), or both combined. In our analysis (Electronic Appendix 17.2) we used the Rapataceae as the sister group, as this relationship supported the hypothesis of greater diversity in the zygomorphic clade; however, use of either other possible sister group reduces statistical support for zygomorphic flowers promoting diversification and amplifies the U-shaped distribution.

17.5.3 Floral nectar spurs

We found 16 independent origins of floral nectar spurs for which the sister group can be identified (Electronic Appendix 17.3). For 12 cases, the spurred group includes more species than its sister clade. A one-tailed Wilcoxon signed-rank test of simple contrasts rejected an association higher diversification with nectar spurs, whereas tests of proportional and log contrasts are highly significant (Table 17.1). Furthermore, the proportion of species with nectar spurs in sister groups has a distinctly L-shaped frequency distribution (Fig. 17.2c) that is consistent with spurs promoting diversification. Restriction of the dataset to consider only comparisons for which the sister groups have the same floral symmetry resulted in 10 independent comparisons, of which 9 have more species in the spurred group compared with its sister clade. Each of the signed-rank tests for this restricted analysis detected significantly greater diversity in spurred clades (Table 17.1).

17.5.4 Dioecy

We identified 29 independent contrasts in sexual system among angiosperms (Electronic Appendix 17.4). For 18 contrasts, the dioecious clade has lower species richness, whereas the opposite is true for 9 contrasts. Two comparisons, Barbeyaceae versus Dirachmaceae, and Myricaceae versus Juglandaceae, involve equivalent species numbers between sexual systems, and we excluded them from the analysis. The frequency distribution for the proportion of species in sister groups with a dioecious sexual system is L-shaped (Fig. 17.2d), with dioecious clades being less diverse than hermaphroditic clades. Regardless of the contrast measure used, dioecy is associated with significantly lower diversification rates (Table 17.1).

17.6 Discussion

With one exception, our reanalysis supports previous findings that the evolution of floral traits can alter subsequent species richness within clades. Like previous analyses, we found that the evolution of animal pollination (Dodd *et al.* 1999) and floral nectar spurs (Hodges and Arnold 1995; Hodges 1997a, b) enhanced species diversification, whereas the evolution of dioecy retarded diversification (Heilbuth 2000). In contrast, our results cast some doubt on Sargent's (2004) conclusion that the evolution of bilaterally symmetric flowers affects diversification. The overall similarity of our results to previous analyses occurred despite our use of a different phylogenetic tree, which represents angiosperm families much more completely, and new statistical methods. The general robustness of these results to new analyses provides strong support to the conclusion that a variety of floral traits thought to affect the likelihood of speciation or extinction contribute to species diversification. However, note that every trait that we considered had contrasts that span the full range of outcomes (Fig. 17.2). Thus, the effect of any of these traits on diversification is likely to be context dependent, with other factors influencing specific cases. Also, the general pattern found for any trait need not explain the true causal factor for diversity in any specific contrast, even those strongly supporting the general trend. As we emphasize below, even when a multiple sister-group analysis supports a hypothesis for diversification, these correlations should represent starting points of more thorough phylogenetic and population analyses.

In contrast to the expectation that changes in phylogenetic reconstruction should not favour one hypothesis or another, the difference between our results and those of Sargent (2004) suggest that even a strong association between the evolution of trait and subsequent diversity should be treated cautiously. Such caution is especially necessary in the absence of additional data supporting the functional hypothesis (see below). In this specific case, we found that although most contrasts support the hypothesis that the evolution of zygomorphic flowers enhances diversification, a substantial number of the contrasts support the exact opposite conclusion.

The results for flower symmetry should be treated prudently, as zygomorphy may influence diversification in some instances, but not others. As noted in Section 17.2, zygomorphy may influence reproductive isolation by constraining the orientation of pollinators during flower visitation, encouraging precise placement of pollen on their bodies. However, some species with actinomorphic flowers may have other traits that restrict the position of pollinators while they visit flowers, such as inflorescence architecture and flower orientation. For example, bees visiting flowers arranged in a raceme inflorescence (e.g., *Chamerion angustifolium*) do so primarily from the bottom towards the top and so approach each flower from a similar angle (Routley and Husband 2003), especially if the flower face is roughly vertical. Similarly, hummingbirds visiting *Aquilegia formosa* do so by probing the nectar spurs in a precise manner causing their chin to brush against the anthers and stigmas. Thus, not all zygomorphic taxa may have more precise pollen placement or pollinator specificity than their actinomorphic sister groups. We encourage studies concerning whether the evolution of zygomorphy enhances diversity in specific cases and especially studies of how zygomorphy may enhance pollinator and pollen placement specificity and reproductive isolation, which remain uncertain, as it has never been tested explicitly.

Speculation about how a trait affects rates of species diversification points to a general problem with tests such as those performed here. Associations of diversification with specific traits are, of course, just associations. Any additional factor that co-varies with the trait of interest could be the true causal factor for increased diversification, even if it is not recognized. Thus, although identification of characters that associate significantly with diversification is an important first step, additional analyses of how a particular trait affects speciation or extinction are essential for testing whether a true causal relationship exists.

One such analysis requires more detailed phylogenetic sampling for specific examples of the origin of a trait. For instance, von Hagen and Kadereit (2003) examined a detailed phylogeny for *Halenia*, which possesses floral nectar spurs. Simple sister-group analysis shows that *Halenia* has many more species than its non-spurred sister group; however, von Hagen and Kadereit (2003) showed that diversification did not follow the evolution of nectar spurs immediately. Rather, diversification seems to have increased after the invasion of South America by a subclade of the genus. Although contradicting an immediate diversification effect of nectar spurs, this pattern is consistent with a general hypothesis of a key innovation (Simpson 1953), which considers two factors, the origin of the trait and the ecological context in which it evolves. If the evolution of nectar spurs promotes diversity by facilitating pollinator transitions, this role can be played only in the presence of a diverse pollinator fauna. Perhaps *Halenia* encountered a sufficient pollinator diversity for nectar spurs to affect diversification only after invading South America (von Hagen and Kadereit 2003). Unfortunately, little is known about the pollination biology of this group, so that this latter hypothesis remains untested.

A second type of phylogenetic analysis explores how a trait may affect diversification. For instance, the hypothesized effects of both animal pollination and floral nectar spurs on diversification involve an increased likelihood of transitions to novel pollinators, thereby promoting reproductive isolation and thus speciation (Hodges and Arnold 1995; Dodd *et al.* 1999). Therefore, a species-level phylogeny should reveal frequent transitions to novel pollinators, especially for recent radiations for which extinction is less likely to influence clade diversity. Unfortunately, few species-level phylogenies are currently available for entire groups, particularly for those in which a trait correlated with species diversification has evolved recently (although see Beardsley *et al.* 2003; Kay 2005; Whittall 2005). In addition, although transitions between major pollinator types (e.g., bee and hummingbird) are most convincing, transitions between different species within a major pollinator type may also provide reproductive isolation. Thus, detailed knowledge of pollination in multiple species will be needed for a full analysis.

Lack of phylogenetic information also restricted our analyses because the sister-group relationships could not be determined for many groups. This difficulty is especially problematic for our review of floral nectar spurs, because spurs are commonly generic, rather than family, traits and genus or species-level phylogenetic information is comparatively rare. Consequently, we had to exclude many groups from our analyses, reducing the power of our tests. In addition to reducing the number of comparisons available for simple tests, such as those described here, this lack of phylogenetic information precludes more detailed analyses needed to tease apart the effects of multiple characters.

Many characters probably affect the rate of either speciation or extinction. For instance, Vamosi and Vamosi (2004) performed nested analyses of the effect of dioecy, while also controlling for woody versus herbaceous growth habit, tropical versus temperate distribution, and fleshy versus dry fruits. Based on the detailed analysis of dioecious genera and their non-dioecious sister lineages, Vamosi and Vamosi found significantly slower diversification in dioecious groups, which is ameliorated somewhat for clades with a tropical distribution and/or fleshy fruits. They attributed these results to opposing effects on extinction, with dioecy increasing the risk of extinction, whereas fleshy fruits, a tropical distribution, and possibly woody growth reduce the risk of extinction. This untangling of multiple factors was possible only with a large number of sister-group comparisons derived from a fairly comprehensive character database and phylogenetic information below the family level.

We attempted a similar analysis for floral nectar spurs, because 6 of 16 sister-group comparisons coincide with a change from actinomorphic to zygomorphic flowers. Comparisons within symmetry classes found even stronger support for the association of spurs with higher diversification rates (Table 17.1). Although these specific tests involve fewer comparisons, they are more robust, because a potentially confounding factor has been excluded.

More detailed phylogenetic information would enhance the study of diversification in several other ways. With more complete taxon sampling and information on the timing of lineage divergence, more powerful methods can be used to detect changes in diversification rates. Sanderson and Donoghue (1994) developed a maximum-likelihood approach that employs the diversities of three branches of a clade and determines which models of changes in diversification best fit these diversities. Wollenberg *et al*. (1996) proposed a method for comparing branching patterns in empirical trees to those generated by a stochastic model of speciation and extinction. Ree (2005) also proposed using stochastic models of speciation, but allowed for uncertainty in the tree topology and for multiple gains and losses of the putative key innovation. As yet, such analyses tend to be performed on specific groups for which the necessary phylogenetic information is available (e.g., spur evolution in *Halenia* described above). More detailed tests, such as these, performed on multiple groups that have evolved the same trait would be especially useful for exposing how particular traits affect the timing and tempo of diversification.

Last, tests of a key-innovation hypothesis can focus on whether a particular trait actually affects speciation and/or extinction. For example, Fulton and Hodges (1999) showed that aspects of nectar spurs in *Aquilegia* (spur length and orientation) affect pollinator visitation and pollen removal (and therefore, presumably, pollen dispersal) and Hodges *et al*. (2004) showed that nectar spur colour affects pollinator visitation. Such studies link variation in nectar spurs directly to pollinator behaviour and reproductive isolation. Several other studies suggest that nectar spurs affect pollinator visitation or pollen dispersal, including studies of orchids (Nilsson 1988; Johnson and Steiner 1997) and *Epimedium* (Suzuki 1984). More such studies are needed, particularly between sister species, to test fully how proposed key innovations affect speciation or extinction. Studies such as these are particularly amenable in hybrid zones (Chapter 18) between species that differ in the trait of interest.

Our review also revealed the remarkable lack of knowledge about the floral biology of most plant species and, therefore, the need for more studies of floral biology. This lack of information hindered our tests of hypothesized associations of floral traits with diversification rates. For example, we found no information on the dominant mode of pollination for over 30% of plant families (121 out of 379). This paucity of information probably resulted in fewer independent contrasts in our data set, inaccurate estimates of species numbers in some sister clades, and misidentification of some sister-group relationships. Although these problems need not have biased our tests in favour of increased diversification, we note again how new information can alter the interpretation of associations with diversity, as we found for zygomorphy. The Eriocaulaceae illustrate this point. Although this family is listed as having either biotic or abiotic pollination (Watson and Dallwitz 2005), the pollination system of the family was first reported only recently, for two species of *Syngonanthus* (Ramos et al. 2005). This study found clear evidence of animal pollination in these species, despite contrary predictions of other authors, and suggests that other species that have been considered abiotically pollinated in the family may actually be animal pollinated as well. Future studies targeting families with poorly known pollination systems and sexual systems will be especially fruitful avenues for research.

We conclude that there is strong evidence that a number of floral characters have affected the species diversity of many angiosperm lineages. However, determining how these characters may have stimulated these changes remains elusive, and these characters can explain only some of the numerous shifts in diversification rates that have been detected during angiosperm evolution (Davies et al. 2004). Thus, other floral characters must also be considered with comparative, detailed phylogenetic, and population studies. For example, exploration of the effects of other sexual systems on diversification will probably be a fruitful avenue of research. Many authors have suggested that the evolution of self-pollination is an evolutionary dead-end leading to extinction (Barrett et al. 1996; Schoen et al. 1997) and self-incompatibility would be expected to have similar effects to dioecy, though no effect has been detected in sister-group analysis (Heilbuth 2000). Other floral structures and features that may enhance specific pollinator visitation and therefore specialization include the evolution of tubular flowers, and specific floral attractants and rewards, such as fragrances and oils. Thus, future comparative research aimed at understanding the evolutionary dominance and diversity of the angiosperms will provide many fruitful avenues for investigating the ecology and evolution of flowers.

Acknowledgements

We wish to thank Lawrence Harder and Spencer Barrett for putting together this volume and giving us the opportunity to contribute. We also thank them and Tim Holtsford for careful reading and suggestions that improved our manuscript, along with an anonymous reviewer who provided thorough comments. We gratefully acknowledge grant support from the NSF (EF-0412727) and a National Parks Ecological Research Fellowship to KMK.

References

Andersson L and Andersson S (2000). A molecular phylogeny of Tropaeolaceae and its systematic implications. *Taxon*, **49**, 721–36.

APGII (2003). An update of the Angiosperm Phylogeny Group classification for the orders and families of flowering plants: APG II. *Botanical Journal of the Linnean Society*, **141**, 399–436.

Baker HG (1954). Race formation and reproductive method in flowering plants. *Evolution*, **7**, 114–43.

Barraclough TG and Savolainen V (2001). Evolutionary rates and species diversity in flowering plants. *Evolution*, **55**, 677–83.

Barraclough TG, Harvey PH, and Nee S (1996). Rate of *rbcL* gene sequence evolution and species diversification in flowering plants (angiosperms). *Proceedings of the Royal Society of London, Series B*, **263**, 589–91.

Barrett SCH, Harder LD, and Worley AC (1996). The comparative biology of pollination and mating in flowering plants. *Philosophical Transactions of the Royal Society, Series B*, **351**, 725–33.

Bawa KS (1980). Evolution of dioecy in flowering plants. *Annual Review of Ecology and Systematics*, **11**, 15–39.

Beardsley PM, Yen A, and Olmstead RG (2003). AFLP phylogeny of *Mimulus* section *Erythranthe* and the evolution of hummingbird pollination. *Evolution*, **57**, 1397–410.

Brantjes NBM (1982). Pollen placement and reproductive isolation between two Brazilian *Polygala* species (Polygalaceae). *Plant Systematics and Evolution*, **141**, 41–52.

Caddick LR, Rudall PJ, Wilkin P, Hedderson TAJ, and Chase MW (2002). Phylogenetics of Dioscoreales based on combined analyses of morphological and molecular data. *Botanical Journal of the Linnean Society*, **138**, 123–44.

Coyne J and Orr HA (2004). *Speciation*. Sinauer, Sunderland, MA.

Crane PR, Friis EM, and Pedersen KR (1995). The origin and early diversification of angiosperms. *Nature*, **374**, 27–33.

Crane PR, Herendeen PS, and Friis EM (2004). Fossils and plant phylogeny. *American Journal of Botany*, **91**, 1683–99.

Crepet WL (2000). Progress in understanding angiosperm history, success, and relationships: Darwin's abominably "perplexing phenomenon." *Proceedings of the National Academy of Sciences of the United States of America*, **97**, 12939–41.

Darwin CR (1903). *More letters from Charles Darwin*, Appleton, New York.

Davies TJ, Barraclough TG, Chase MW, Soltis PS, Soltis DE, and Savolainen V (2004). Darwin's abominable mystery: insights from a supertree of the angiosperms. *Proceedings of the National Academy of Sciences of the United States of America*, **101**, 1904–9.

de Queiroz A (1998). Interpreting sister-group tests of key innovation hypotheses. *Systematic Biology*, **47**, 710–18.

Dilcher DL and Crane PR (1984). *Archaeanthus*—an early angiosperm from the Cenomanian of the western interior of North America. *Annals of the Missouri Botanical Garden*, **71**, 351–83.

Dodd ME, Silvertown J, and Chase MW (1999). Phylogenetic analysis of trait evolution and species diversity variation among angiosperm families. *Evolution*, **53**, 732–44.

Farrell BD, Dussourd DE, and Mitter C (1991). Escalation of plant defense—do latex and resin canals spur plant diversification? *American Naturalist*, **138**, 881–900.

Fenster CB, Armbruster WS, Wilson P, Dudash MR, and Thomson JD (2004). Pollination syndromes and floral specialization. *Annual Review of Ecology, Evolution and Systematics*, **35**, 375–403.

Friis EM, Pedersen KR, and Crane PR (2000). Reproductive structure and organization of basal angiosperms from the early Cretaceous (Barremian or Aptian) of western Portugal. *International Journal of Plant Sciences*, **161**, S169–82.

Fulton M and Hodges SA (1999). Floral isolation between *Aquilegia formosa* and *Aquilegia pubescens*. *Proceedings of the Royal Society of London, Series B*, **266**, 2247–52.

Gandolfo MA, Nixon KC, Crepet WL, Stevenson DW, and Friis EM (1998). Oldest known fossils of monocotyledons. *Nature*, **394**, 532–3.

Ghebrehiwet M (2000). Taxonomy, phylogeny and biogeography of *Kickxia* and *Nanorrhinum* (Scrophulariaceae). *Nordic Journal of Botany*, **20**, 655–89.

Ghebrehiwet M, Bremer B, and Thulin M (2000). Phylogeny of the tribe Antirrhineae (Scrophulariaceae) based on morphological and *ndhF* sequence data. *Plant Systematics and Evolution*, **220**, 223–39.

Gianoli E (2004). Evolution of a climbing habit promotes diversification in flowering plants. *Proceedings of the Royal Society of London, Series B*, **271**, 2011–15.

Goudet J (1999). An improved procedure for testing the effects of key innovations on rate of speciation. *American Naturalist*, **153**, 549–55.

Grant V (1981). *Plant speciation*, 2nd edition. Columbia University Press, New York.

Grant V (1994). Mechanical and ethological isolation between *Pedicularis groenlandica* and *P. attollens* (Scrophulariaceae). *Biologisches Zentralblatt*, **113**, 43–51.

Heilbuth JC (2000). Lower species richness in dioecious clades. *American Naturalist*, **156**, 221–41.

Heilbuth JC, Ilves KL, and Otto SP (2001). The consequences of dioecy on seed dispersal: modeling the seed shadow handicap. *Evolution*, **55**, 880–8.

Hodges SA (1997a). Floral nectar spurs and diversification. *International Journal of Plant Sciences*, **158**, S81–8.

Hodges SA (1997b). Rapid radiation due to a key innovation in *Aquilegia*. TJ Givnish and KJ Sytsma, eds. *Molecular evolution and adaptive radiation*, pp. 391–405. Cambridge University Press, New York.

Hodges SA and Arnold ML (1995). Spurring plant diversification: are floral nectar spurs a key innovation? *Proceedings of the Royal Society of London, Series B*, **262**, 343–8.

Hoot SB, Kadereit JW, Blattner FR, Jork KB, Schwarzbach AE, and Crane PR (1997). Data congruence and phylogeny of the Papaveraceae sl based on four data sets: *atpB* and *rbcL* sequences, *trnK* restriction sites, and morphological characters. *Systematic Botany*, **22**, 575–90.

Isaac NJB, Agapow PM, Harvey PH, and Purvis A (2003). Phylogenetically nested comparisons for testing correlates of species richness: a simulation study of continuous variables. *Evolution*, **57**, 18–26.

Jobson RW and Albert VA (2002). Molecular rates parallel diversification contrasts between carnivorous plant sister lineages. *Cladistics*, **18**, 127–36.

Jobson RW, Playford J, Cameron KM, and Albert VA (2003). Molecular phylogenetics of Lentibulariaceae inferred from plastid *rps16* intron and *trnL-F* DNA sequences: implications for character evolution and biogeography. *Systematic Botany*, **28**, 157–71.

Johnson SD and Steiner KE (1997). Long-tongued fly pollination and evolution of floral spur length in the *Disa draconis* complex (Orchidaceae). *Evolution*, **51**, 45–53.

Karehed J (2001). Multiple origin of the tropical forest tree family Icacinaceae. *American Journal of Botany*, **88**, 2259–74.

Kay KM (2005). Rapid speciation and the evolution of hummingbird pollination in neotropical *Costus* subgenus *Costus* (Costaceae): evidence from nrDNA ITS and ETS sequences. *American Journal of Botany*, **92**, 1899–910.

Kay KM (2006). Reproductive isolation between two closely related hummingbird-pollinated Neotropical gingers. *Evolution*, **60**, 538–52.

Kim YD, Kim SH, Kim CH, and Jansen RK (2004). Phylogeny of Berberidaceae based on sequences of the chloroplast gene *ndhF*. *Biochemical Systematics and Ecology*, **32**, 291–301.

Liden M, Fukuhara T, Rylander J, and Oxelman B (1997). Phylogeny and classification of Fumariaceae, with emphasis on *Dicentra* s l, based on the plastid gene *rps16* intron. *Plant Systematics and Evolution*, **206**, 411–20.

Mabberley DJ (1997). *The plant book: a portable dictionary of the higher plants*. Cambridge University Press, Cambridge.

McConway KJ and Sims HJ (2004). A likelihood-based method for testing for nonstochastic variation of diversification rates in phylogenies. *Evolution*, **58**, 12–23.

Neal PR, Dafni A, and Giurfa M (1998). Floral symmetry and its role in plant–pollinator systems: terminology, distribution, and hypotheses. *Annual Review of Ecology and Systematics*, **29**, 345–73.

Nilsson LA (1988). The evolution of flowers with deep corolla tubes. *Nature*, **334**, 147–9.

Olmstead RG, dePamphilis CW, Wolfe AD, Young ND, Elisons WJ, and Reeves PA (2001). Disintegration of the Scrophulariaceae. *American Journal of Botany*, **88**, 348–61.

Patterson TB and Givnish TJ (2004). Geographic cohesion, chromosomal evolution, parallel adaptive radiations, and consequent floral adaptations in *Calochortus* (Calochortaceae): evidence from a cpDNA phylogeny. *New Phytologist*, **161**, 253–64.

Qiu YL, Lee J, Bernasconi-Quadroni F, *et al.* (2000). Phylogeny of basal angiosperms: analyses of five genes from three genomes. *International Journal of Plant Sciences*, **161**, S3–27.

Ramos COC, Borba EL, and Funch LS (2005). Pollination in Brazilian *Syngonanthus* (Eriocaulaceae) species: evidence for entomophily instead of anemophily. *Annals of Botany*, **96**, 387–97.

Renner SS and Ricklefs RE (1995). Dioecy and its correlates in the flowering plants. *American Journal of Botany*, **82**, 596–606.

Routley MB and Husband BC (2003). The effect of protandry on siring success in *Chamerion angustifolium* (Onagraceae) with different inflorescence sizes. *Evolution*, **57**, 240–8.

Sargent RD (2004). Floral symmetry affects speciation rates in angiosperms. *Proceedings of the Royal Society of London, Series B*, **271**, 603–8.

Schoen DJ, Johnston MO, L'Heureux A-M, and Morsolais J (1997). Evolutionary history of the mating system in *Amsinckia* (Boraginaceae). *Evolution*, **51**, 1090–9.

Simpson GG (1953). *The major features of evolution*. Columbia University Press, New York.

Sims HJ and McConway KJ (2003). Nonstochastic variation of species-level diversification rates within angiosperms. *Evolution*, **57**, 460–79.

Slowinski JB and Guyer C (1993). Testing whether certain traits have caused amplified diversification—an improved method based on a model of random speciation and extinction. *American Naturalist*, **142**, 1019–24.

Soltis PS and Soltis DE (2004). The origin and diversification of angiosperms. *American Journal of Botany*, **91**, 1614–26.

Stebbins GL (1974). *Flowering plants: evolution above the species level*. Belknap Press, Cambridge, MA.

Sun Y, Fung KP, Leung PC, and Shaw PC (2005). A phylogenetic analysis of *Epimedium* (Berberidaceae) based on nuclear ribosomal DNA sequences. *Molecular Phylogenetics and Evolution*, **35**, 287–91.

Suzuki K (1984). Pollination system and its significance on isolation and hybridization in Japanese *Epimedium* (Berberidaceae). *Botanical Magazine-Tokyo*, **97**, 381–96.

Sytsma KJ, Morawetz J, Pires JC, *et al.* (2002). Urticalean rosids: circumscription, rosid ancestry, and phylogenetics based on *rbcL*, *trnL-F*, and *ndhF* sequences. *American Journal of Botany*, **89**, 1531–46.

Sytsma KJ, Litt A, Zjhra ML, *et al.* (2004). Clades, clocks, and continents: historical and biogeographical analysis of Myrtaceae, Vochysiaceae, and relatives in the Southern Hemisphere. *International Journal of Plant Sciences*, **165**, S85–105.

Takhtajan A (1997). *Diversity and classification of flowering plants*. Columbia University Press, New York.

Thompson JN (1994). *The coevolutionary process*. University of Chicago Press, Chicago.

Vamosi JC and Otto SP (2002). When looks can kill: the evolution of sexually dimorphic floral display and the extinction of dioecious plants. *Proceedings of the Royal Society of London, Series B*, **269**, 1187–94.

Vamosi SM and Vamosi JC (2005). Endless tests: guidelines for analysing non-nested sister group comparisons. *Evolutionary Ecology Research*, **7**, 567–79.

von Hagen KB and Kadereit JW (2003). The diversification of *Halenia* (Gentianaceae): ecological opportunity versus key innovation. *Evolution*, **57**, 2507–18.

Ward NM and Price RA (2002). Phylogenetic relationships of Marcgraviaceae: insights from three chloroplast genes. *Systematic Botany*, **27**, 149–60.

Watson L and Dallwitz MJ (2005). The families of flowering plants: descriptions, illustrations, identification, and information retrieval. http://delta-intkey.com.

Whittall JB (2005). Ecological speciation and convergent evolution in the North American columbine radiation (*Aquilegia*, Ranunculaceae). Ph.D. Thesis University of California, Santa Barbara.

Wilson WG and Harder LD (2003). Reproductive uncertainty and the relative competitiveness of simultaneous hermaphroditism versus dioecy. *American Naturalist*, **162**, 220–41.

Wollenberg K, Arnold J, and Avise JC (1996). Recognizing the forest for the trees: testing temporal patterns of cladogenesis using a null model of stochastic diversification. *Molecular Biology and Evolution*, **13**, 833–49.

Yuan YM, Song Y, Geuten K, *et al.* (2004). Phylogeny and biogeography of Balsaminaceae inferred from *ITS* sequences. *Taxon*, **53**, 391–403.

Zanis MJ, Soltis DE, Soltis PS, Mathews S, and Donoghue MJ (2002). The root of the angiosperms revisited. *Proceedings of the National Academy of Sciences of the United States of America*, **99**, 6848–53.

CHAPTER 18

Floral biology of hybrid zones

Diane R. Campbell[1] and George Aldridge[2]

[1] Department of Ecology and Evolutionary Biology, University of California, Irvine, CA, USA
[2] Rocky Mountain Biological Laboratory, Crested Butte, CO, USA

Outline

Hybridization between closely related species is relatively common in angiosperms and can create a natural hybrid zone. We review recent experimental studies of floral biology in pairs of hybridizing species, emphasizing comparisons of the floral morphology and nectar rewards of hybrid plants with that of their progenitors, and quantifying the influence of these floral traits on pollinator behaviour and pre-zygotic and post-zygotic reproductive isolation. Floral traits of hybrids can be intermediate or transgressive. Floral differences between species, which in one case are attributable to particular chromosomal regions, can have differing impacts on interspecific flights by pollinators. A simulation model of mating in a hybrid zone between *Ipomopsis aggregata* and *Ipomopsis tenuituba* shows that behavioural responses by hummingbird and hawk-moth pollinators affect pre-zygotic ethological isolation more strongly than mechanical isolation. We apply this model to compare contact sites between these species which differ greatly in the frequency of natural hybrids. Striking differences in hawk-moth behaviour between the sites generated large differences in the rate of interspecific pollen movement, potentially explaining the dissimilar frequencies of hybrids. Although floral traits influence both the formation and fitness of hybrids primarily through effects on pollinators, impacts on plant enemies also need consideration. Recent research has revealed much about how floral traits influence pollinator visitation, but mainly for systems with hummingbird versus insect pollinators. Such studies should be extended to other pollinators. Further research is also needed on how floral traits influence pollen dispersal and other post-visitation events that impact reproductive isolation, and their genetic basis in natural populations.

18.1 Introduction

Although no record of the word "hybrid" exists before the early seventeenth century (Oxford English Dictionary 1971), ancient Greek mythology was populated by mixed creatures, such as the chimaera, a composite of a lion, goat, and dragon, which were considered unnatural and monstrous. During the early twenty-first century, "hybrid" is more likely to connote an innovative and synergistic invention than a monstrosity. The evolutionary biologist's view of hybrids has changed similarly. Based largely on animal studies, natural hybrids were traditionally considered as either evolutionary dead-ends, occurring rarely and usually being sterile, or indicating a breakdown in reproductive isolation caused by disturbance, often attributed to human activity (Mayr 1963). However, evolutionary biologists now ascribe to hybrids a much wider range of evolutionary roles, including introgression of genes from another species, as stressed originally by Anderson (1949); persistence of stable hybrid zones (Barton and Hewitt 1985); introduction of novel genetic variation that can increase adaptation to a particular environment (Anderson and Stebbins 1954; Kim and Rieseberg 1999; Martinsen *et al.* 2001);

promotion of speciation through reinforcement (Servedio and Noor 2003); and production of a new hybrid species (Rieseberg et al. 1996; Rieseberg 1997).

Flowering plants hybridize commonly. Based on analysis of five floras, Ellstrand et al. (1996) estimated that hybrids comprise 9–22% of the total species in a particular region. More recently, Mallet (2005) reported that 25% of vascular plant species in the UK hybridize. Many of the observations or specimens undoubtedly come from hybrid zones, or areas of mixing between two related species that include some viable hybrids. Such a hybrid zone may be a result of secondary contact or ongoing speciation in sympatry or parapatry. Hybrid zones can form between native species, as for example between *Iris fulva*, *Iris brevicaulis*, and *Iris hexagona* in Louisiana (Arnold 1994), or between a recently introduced and a native species, as between two species of cordgrass (*Spartina*) in San Francisco Bay (Ayres et al. 1999).

Hybrid zones often present a striking profusion of flower morphologies. These floral traits can profoundly influence the degree of reproductive isolation between the species, by altering any of several steps in pre-zygotic and post-zygotic isolation (Fig. 18.1). Studies of *Phlox* provided an early example: plants of *Phlox pilosa* with white corollas set a lower percentage of hybrid seed than did plants with pink corollas that were similar to the congener *Phlox glaberrima* (Levin and Schaal 1970). Perhaps the best-known mechanism for such pre-zygotic isolation depends on behaviour of animal pollinators; the floral morphology and rewards offered by each species may attract its own type of pollinator, which tends to move more often between flowers of the same species (ethological isolation, Fig. 18.1a). As outlined by Verne Grant (1949), seasonal and mechanical isolation (differences in floral structure restrict interspecific pollen movement: Fig. 18.1b) can also prevent interspecific mating, the deposition of heterospecific pollen on stigmas. Furthermore, recent studies have shown associations between certain floral traits and post-mating isolation that influence production of hybrid seed (Fig. 18.1c).

Floral traits may also influence post-zygotic reproductive isolation by altering several components of hybrid fitness (Fig. 18.1d–f). Divergent selection on alternative floral morphologies would coincide with low hybrid fitness, and could be important in driving floral diversification (Chapter 15). This impact on post-zygotic events has received less study than that of pre-zygotic events;

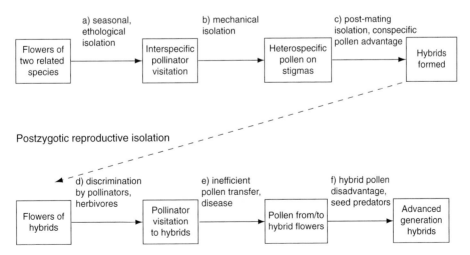

Figure 18.1 Steps in pre-zygotic and post-zygotic isolation that can be influenced by floral traits. The step from "hybrids formed" to "flowers of hybrids" represents survival and is shown by a dashed arrow as it is influenced indirectly via correlations with other traits.

however, in principle, floral traits could influence pollination success of hybrids directly, through both male and female functions (Chapter 2), and affect several components of fitness through interactions with floral herbivores, nectar thieves, and seed predators (Chapter 7).

The resulting fitness of hybrids relative to the parental species is critically important to evolution in hybrid zones and the origin and maintenance of species (Hatfield and Schluter 1999). Natural selection against intrinsically unfit hybrids balanced by some gene flow between the species (and hence incomplete pre-zygotic reproductive isolation) can maintain a spatially stable hybrid zone (Barton and Hewitt 1985). In this case, distinct floral morphologies and species could be maintained at opposite ends of a cline. Certain types of environmentally dependent selection on hybrids can also lead to a stable zone (reviewed in Arnold 1997). Alternatively, a hybrid zone might expand due to neutral mixing of genes following secondary contact between two species. If genes for floral traits are neutral, distinct floral morphologies would not persist. Finally a hybrid zone can shrink or disappear if reinforcement of reproductive isolation by selection against interspecific mating decreases hybridization. Hybridization between invasive exotic and native species is of special concern as a cause of extinction and genetic assimilation of endangered plant species (Levin et al. 1996).

In this chapter, we explore the influences of floral traits on reproductive isolation in hybrid zones. We begin by examining the morphology of hybrid flowers, asking whether flowers of hybrids are intermediate or transgressive in phenotype, and how this trait distribution depends on the genetic architecture of floral morphology. We then examine effects of floral traits on pre-mating reproductive isolation (steps a and b in Fig. 18.1), especially that resulting from plant–pollinator interactions, and on post-mating reproductive isolation (step c). We find that despite early recognition of mechanisms of pre-zygotic isolation, surprisingly few studies have quantified the relative importance of various steps in the process. In addition to identifying specific forms of reproductive isolation that require more study, we derive a model of pollinator behaviour to explain mating patterns and quantify impacts of ethological, mechanical, and post-mating isolation in hybrid zones between *Ipomopsis aggregata* and *Ipomopsis tenuituba*. We then examine more briefly recent studies exploring how floral traits influence hybrid fitness and thus post-zygotic isolation (steps d–f in Fig. 18.1). We conclude with recommendations for future study of reproductive isolation and hybridization.

18.2 Genetic architecture of species differences; what do hybrid flowers look like?

The distribution of floral morphologies in hybrid zones depends on two major factors. First, because of the underlying genetic architecture, hybrids need not be intermediate in floral morphology or other traits between the parental species, but instead can be transgressive, or more extreme than either parent (Rieseberg and Ellstrand 1993). Second, the distribution may reflect past selection for hybrids with particular phenotypes.

Hybrids necessarily have intermediate phenotypes only if they are F_1 individuals, as in a *Rhododendron* hybrid zone (Milne et al. 2003), and the traits measured vary solely due to additive effects of genes. In contrast to this very restrictive situation, many hybrid zones are dominated by hybrids from later generations, as in the Louisiana irises for which no F_1 individuals have been found in the wild (Hodges et al. 1996). Even among F_1 individuals, variation due to dominance causes the phenotypic mean to deviate from that of the parents. Such effects are well known in agriculture, as many crop-development programmes take advantage of heterosis shown by F_1 individuals, whose vigour exceeds that of the parents. Furthermore, second generation (F_2) hybrids formed by crossing two F_1 individuals often exhibit transgressive traits (reviewed by Rieseberg et al. 1999a).

Transgressive traits can arise from epistasis, in which trait values deviate from the summed effects at two loci. Such epistasis can produce hybrid breakdown due to Dobzhansky–Muller incompatibilities (Rhode and Cruzan 2005), as found in monkey flowers (*Mimulus guttatus* and *Mimulus nasutus*; Fishman and Willis 2001). Although rarely documented, epistasis could also generate transgressive floral traits.

Second, transgression can arise from complementary effects of additive alleles at multiple loci. If one species has an overall higher phenotypic value than the species with which it hybridizes, because the positive effects of alleles at some loci outweigh the negative effects of alleles at other loci, a recombinant F_2 individual might by chance inherit all positive alleles and be transgressive with a yet higher phenotypic value. Recent studies using QTL mapping have suggested that complementary genes are the major cause of transgression in plants (Rieseberg *et al.* 1999a).

Note that transgressive phenotypes are not an exception, but occur frequently in crosses between plant species. For example, 44% of traits reviewed by Rieseberg *et al.* (1999a: 58% if intraspecific crosses are included) exhibited transgression, although floral traits were not identified specifically. An example of transgressive floral traits comes from QTL mapping of species differences between the monkey flowers *Mimulus cardinalis* and *Mimulus lewisii* (Bradshaw *et al.* 1998). Several floral traits in the F_2 generation exhibited transgression, notably the size-related characters of corolla width and petal width, and nectar volume. Characters related to flower colour showed little transgressive segregation, which might reflect the influence of only a few major QTLs. In contrast, F_2 progeny of the columbines *Aquilegia formosa* and *Aquilegia pubescens* were intermediate for all five floral traits measured (Hodges *et al.* 2002). The possibility of transgressive segregation calls into question the use of morphological indices by themselves to assess the frequency of plant hybrids, but such indices can be valuable in combination with controlled crosses that characterize the trait values of F_1 and F_2 individuals to determine which traits are transgressive (Section 18.4.1).

18.3 Floral traits and the frequency of mating between species and hybrids

Production of hybrids is the converse of reproductive isolation. Therefore, studying the causes of hybrid production leads to an examination of forms of potential reproductive isolation and their roles in speciation. We begin with pre-mating isolation (Fig. 18.1a and b), which can occur through seasonal, ethological, or mechanical mechanisms (Grant 1949, 1994). Seasonal isolation refers to phenological differences in flowering that prevent mating (Husband and Schemske 2000; Chapter 8). Ethological isolation occurs when behavioural preferences of pollinators for particular floral traits restrict their movement between flowers of different species. Complete ethological isolation results if pollinators of different types (e.g., bumble bees and hummingbirds) have narrow and non-overlapping preferences. Such strong preferences are envisioned in the textbook concept of "pollinator syndromes" in which suites of floral traits lead to visitation by a particular type of pollinator (Faegri and van der Pijl 1966; Baker and Hurd 1968); for example, "hummingbird flowers" are red with broad corolla tubes. Alternatively, mechanical isolation results when differences in floral structures restrict interspecific pollen movement. For example, the positions of stigma and anthers may cause pollen grains from two species to be carried on different parts of a pollinator's body. Morphological adaptation to a specific pollinator could evolve in response to selection exerted by pollinators that visit a species most frequently and effectively ("the most effective pollinator": Stebbins 1970).

18.3.1 Ethological isolation

Although some degree of ethological isolation can be inferred from observations in natural hybrid zones, quantifying the process requires observations at experimental arrays of plants that provide simultaneous choice to pollinators. Observations of plants growing *in situ* are usually not sufficient, as biased visitation to one species can result instead from spatial separation of the species. Randomization of the locations of plants in an array also allows assessment of the relative importance of pollinator preference for floral traits of one plant species versus constancy, whereby individual pollinators visit both species but make long bouts of visits to one species before switching to the other (Waser 1986). Table 18.1 summarizes recent experimental array studies of pollinator visitation between potentially hybridizing species, using either randomized or alternating positions of species (the latter can underestimate ethological isolation). Some arrays used just the two parental species, illustrating what might happen during initial contact between two species, whereas others also included hybrids, illustrating the potential for interspecific pollinator movement in established hybrid zones. For each case, we calculated reproductive isolation due to ethological isolation from the relative frequency of pollinator movement between plants of different species (Ramsey et al. 2003): $RI = 1 - $ (heterospecific transitions/conspecific transitions). Not all of the studies reported the necessary information, instead focusing on the demonstration of pollinator preference. However, estimates of ethological RI varied across the entire range from 0 (random visitation) to 1 (complete pollinator specialization; Table 18.1).

Many of these studies focused on situations with one hummingbird and one insect-pollinated species. In all such cases, the two major types of pollinators showed contrasting preferences, although they varied so much in strength that RI still ranged from 0 to 1, suggesting that the variation is not explained simply by pollinator type. In arrays of red-flowered *I. fulva* and blue-flowered *I. brevicaulis*, along with F_1 and backcross hybrids, hummingbirds preferred *I. fulva* flowers and bumble bees preferred F_1s (Wesselingh and Arnold 2000). However, both pollinator types visited nearest neighbouring plants nearly 80% of the time, generating zero or weak ethological isolation. In a similar experiment involving blue-flowered *I. hexagona* instead of *I. brevicaulis*, bumble bees preferred *I. hexagona* and hummingbirds preferred *I. fulva*, but both moved frequently between a parental species and F_1 individuals (Emms and Arnold 2000). These backcross movements were more common than heterospecific movements,

Table 18.1 Experimental array studies of ethological isolation between closely related species due to pollinator visit transitions.

Species	Hybrids included?	Pollinators	Floral traits selected	Pre-zygotic ethological RI	Source
Iris brevicaulis, I. fulva	Yes	Bumble bees, hummingbirds		0–weak	Wesselingh and Arnold (2000)
Iris hexagona, I. fulva	Yes	Bumble bees, hummingbirds		0–0.69[a]	Emms and Arnold (2000)
Mimulus lewisii, M. cardinalis	Yes	Bees, hummingbirds	Colour (bees), nectar (birds)	Strong, 0.976[b]	Schemske and Bradshaw (1999)
Aquilegia pubescens, A. formosa	No	Hawk moths, hummingbirds	Orientation (moths)	0.85–1.00[c]	Fulton and Hodges (1999)
Nicotiana alata, N. forgetiana	Some arrays	Hawk moths, hummingbirds		0.59	Ippolito et al. (2004)
Ipomopsis tenuituba, I. aggregata	Some arrays	Hawk moths, hummingbirds	Colour (birds), width (both)	0.50, 0.30–0.96[d]	Campbell et al. (2002), Aldridge and Campbell, unpublished and this study
Baptisia leucophaea, B. sphaerocarpa	Some arrays	Bumble bees		0.48–0.84	Leebens-Mack and Milligan (1998)
Asclepias spp.	No	Generalist insects		0.39–1.00[e]	Kephart and Theiss (2003)

Ethological RI was calculated as 1 − (heterospecific plant transitions/conspecific transitions), unless otherwise noted.
[a] Value depends on site.
[b] The estimate of 0.976 is based on frequencies of foraging bouts that included one versus both species in an area of natural sympatry (Ramsey et al. 2003).
[c] Estimates based on frequencies of foraging bouts.
[d] Value depends on site.
[e] Value depends on pollinator type.

suggesting some ethological isolation between the species, but little pollinator discrimination against hybrids (step d in Fig. 18.1)

Only a few studies have used phenotypic or genetic manipulation to demonstrate effects of particular floral traits on pollinator visitation within and between species (Table 18.1). One case involved two *Mimulus* species (Schemske and Bradshaw 1999) that show strong ethological isolation in nature (Table 18.1). The pink-flowered *M. lewisii* has a wide corolla, produces little nectar, and is predominantly bee pollinated, whereas the red-flowered *M. cardinalis* has a narrower corolla, secretes more nectar, and is hummingbird pollinated. Including artificially produced hybrids in the array allowed investigation of how major QTLs for floral traits generate ethological isolation. The QTL marker genotype for petal carotenoid concentration dramatically affected visitation by bees, but not by hummingbirds. Conversely, the QTL for nectar production affected only hummingbird visitation.

In columbines (*Aquilegia*), flower orientation contributes significantly to ethological isolation (Fulton and Hodges 1999). *Aquilegia formosa* has pendent, red and yellow flowers with short nectar spurs, whereas *A. pubescens* has upright, pale flowers with long spurs. The two species produce hybrid zones in the southern Sierra Nevada Mountains of California. In an array of *A. pubescens* in which half of the flowers were tied to make the flowers pendent and thus given the trait of the other parental species, hawk moths visited upright flowers more than 10 times as often as pendent flowers. Manipulation of this one trait reproduced the strong preference for *A. pubescens* exhibited by hawk moths at arrays of the two parental species (Fulton and Hodges 1999).

Many pollinators may respond to suites of floral characters. An example of interacting effects of floral traits comes from hybrid zones between *Ipomopsis aggregata* and *I. tenuituba*. Hummingbirds prefer to visit the red-flowered *I. aggregata* when given simultaneous choice in arrays, visiting them 3–4 times as often as *I. tenuituba* (Campbell *et al.* 1997). However, other floral traits are also involved, as hummingbirds preferentially visit flowers with wide corolla tubes (Campbell *et al.* 2002) and can rapidly learn to associate white, instead of red, with high nectar reward. Manipulation of flower colour in arrays showed that the combination of traits found naturally in *I. aggregata* induces higher hummingbird preference than occurs when flowers differ only in colour, suggesting that multiple characters contribute to the rate of hybridization (Melendez-Ackerman and Campbell 1998).

The presence of hybrids could lead to high interspecific pollen transfer, even though pollinators seldom move between species when hybrids are absent. Experiments that contrast responses to arrays of only the parental species and those including hybrids can evaluate the importance of this "hybrid bridge" (Leebens-Mack and Milligan 1998). Studies of *Nicotiana alata*, *Nicotiana forgetiana*, and their F_1 hybrids in Brazil provide a good example (Ippolito *et al.* 2004). In experimental plots with only plants of the two species, hummingbirds visited *N. forgetiana* exclusively, and hawk moths strongly preferred *N. alata*, in agreement with their pollinator syndromes. However, when plots contained the two species and F_1 individuals, hummingbirds visited both species, so that their movements could have produced interspecific gene flow, and backcrossing was also possible. This finding suggests that even if F_1 hybrids arise rarely due to pre-zygotic ethological isolation, their presence can accelerate further gene flow between species.

18.3.2 Mechanical isolation

Mechanical isolation is often suggested if pollen of two species is carried on different parts of pollinators' bodies, if a pollinator fails to contact reproductive parts of one species due to poor fit to the flower, or if hybrid production differs from interspecific pollinator movement (Macior 1965; Kephart and Theiss 2003; Ippolito *et al.* 2004). However, quantifying mechanical isolation requires tracking the movement of pollen grains within and between flower species. These patterns of pollen movement depend on the number of flowers displayed on a plant and on pollen carryover, the pattern of flower-to-flower pollen transfer (Chapter 5).

Few studies of pollen carryover have included related plant species (Levin and Berube 1972; Stucky 1985). One exception based on pollen dispersal within a species showed an intriguing asymmetry in pollen transfer by hummingbirds and bumble bees visiting *Penstemon* (Castellanos *et al.* 2003). Because they carried pollen to a longer sequence of recipient flowers, birds could pollinate the bee-syndrome *Penstemon strictus* flowers nearly as well as bees, whereas bees pollinated the bird-syndrome *Penstemon barbatus* poorly, suggesting that mechanical isolation would be asymmetric. However, as far as we are aware, the measurements of pollen dispersal between anthers and stigmas of two related flower species by individual pollinators that are necessary to quantify mechanical isolation are available only from studies of *Ipomopsis*. On a per-visit basis, hummingbirds transfer little pollen from the highly exserted anthers of *I. aggregata* subsp. *formosissima* to the inserted stigma of *Ipomopsis arizonica*, but much pollen in the opposite direction, demonstrating asymmetric mechanical isolation between these species visited by the same hummingbird pollinators (Wolf *et al.* 2001). Similarly, hummingbirds transfer about 35% as much pollen from a donor flower of *I. tenuituba* to a recipient flower of *I. aggregata* as they do to another conspecific flower [mechanical RI = 1 − (transfer to heterospecific/transfer to conspecific) = 0.65], with no demonstrable mechanical isolation in the opposite direction (Campbell *et al.* 1998).

18.4 The relative importance of ethological and mechanical isolation

Studies of *I. aggregata*, *I. tenuituba*, and their hybrids allow the opportunity to quantify mechanical isolation and compare it with the strength of ethological isolation. This system offers the further advantage that hybridization varies across the geographical range (Grant and Wilken 1988). Comparison of ethological isolation at contact sites with and without large numbers of hybrids, while controlling for other features of the species, provides a powerful assessment of its role in determining the frequency of hybrid formation.

18.4.1 Floral traits in *Ipomopsis aggregata* and *I. tenuituba* and pollinator behaviour

To set the stage for these comparisons, we first describe in more detail differences in flowers between the two species and hybrids. *I. aggregata* subsp. *aggregata* has red flowers with relatively short, wide corolla tubes, high nectar production, and slightly inserted to exserted reproductive organs. *Ipomopsis tenuituba* has white to pale pink flowers with a long, narrow corolla tube, low nectar production, and strongly inserted reproductive organs (Plate 8). Flowers of both species are protandrous, and the plants are self-incompatible. The most common pollinators are hummingbirds and hawk moths. Hybrid zones form frequently, but not always, where the species come into contact in the western mountains of the USA (Grant and Wilken 1988).

To determine whether floral traits of hybrids are intermediate or transgressive to those of the parental species, we analysed seven traits of screenhouse-grown F_2 hybrids between *I. aggregata* subsp. *aggregata* and *I. tenuituba* subsp. *tenuituba* collected near a hybrid zone at Poverty Gulch, CO (PG). The mean phenotype was intermediate or similar to one parent for corolla size, nectar volume, and flower colour (assessed by optical density), and significantly exceeded that of both parents (positively transgressive) only for style length (Fig. 18.2). Next we constructed a composite index of floral morphology by finding the canonical discriminant function (CDF) that maximized the difference between the two species ($P < 0.001$). The CDF correlated strongly and positively with corolla width, nectar volume, and colour ($r = 0.87$, 0.97, and 0.74, respectively), and negatively with corolla length ($r = -0.69$). Based on this CDF, the mean phenotype for F_2 individuals (6.8) was intermediate between those for the parental species (11.5 and 0.9), suggesting that a composite index can be useful as a hybrid index, even when an individual trait may not be intermediate.

As described in Section 18.3.1, some of these floral traits influence pollinator behaviour in a way that impacts ethological isolation. That ethological isolation is evident in a reduction in hybrid formation. Painting all flowers red to eliminate the difference in flower colour between the plant species reduced the percentage of seeds formed that were conspecific, as shown by multilocus paternity analysis (Campbell 2004). This trait manipulation alone caused RI, based on the relative formation of

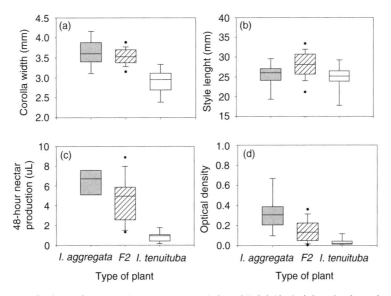

Figure 18.2 Representative floral traits for *Ipomopsis aggregata*, *I. tenuituba* and F$_2$ hybrids. Each box plot shows the 10th, 25th, 50th, 75th, and 90th percentiles (and 5th and 95th where the number of plants permitted). Traits were measured using methods described in Campbell *et al.* (2002), with two to five flowers measured per plant and averaged prior to analysis. For corolla width (a) the F$_2$ differs from *I. tenuituba* (*t*-test, $P < 0.0001$). For traits shown in (b), (c), and (d), both parental species differed from the F$_2$ (all $P < 0.05$), but only style length (b) is transgressive. Corolla length, and maximum and minimum stamen lengths (not shown) were also included in the canonical discriminant analysis.

hybrid and conspecific seed, to drop from 0.81 to 0.60 in arrays visited by hummingbirds. Whereas hummingbirds preferentially visit flowers with traits characteristic of *I. aggregata* (Campbell *et al.* 2002), hawk moths (usually rare at Poverty Gulch) slightly prefer plants with narrow corolla tubes, typical of *I. tenuituba* (Campbell *et al.* 1997).

To quantify the effects of the ethological isolation on pollen transfer, Campbell *et al.* (2002) set up arrays with equal numbers of *I. aggregata*, *I. tenuituba*, and hybrids (F$_1$ or F$_2$). The pollen in dehiscing anthers was marked with different colours of dye, so that it was possible to estimate interspecific pollen movement to stigmas simultaneously with interspecific flights by pollinators. Dye placed on anthers of one of the parental species reached heterospecific stigmas 7% as often as conspecific ones (Fig. 18.3a). This difference resulted partly from ethological isolation, as hummingbirds starting at one of the parental species flew to a heterospecific plant only 50% as often as they flew to a conspecific (Fig. 18.3b). However, ethological isolation was far from complete, and the rate of backcrossing exceeded interspecific pollen movement, as estimated by dye dispersal (Fig. 18.3a).

18.4.2 Simulation model of ethological and mechanical isolation

To determine the relative importance of ethological versus mechanical isolation in reducing interspecific pollen movement, Campbell *et al.* (2002) constructed a simulation model in which these parameters could be varied realistically. We review that study of a single hybrid zone and then apply a modification of the model to a new investigation of how geographical variation in pollinator behaviour influences pre-zygotic reproductive isolation.

The model simulated the movement of pollen grains carried by pollinators in an array of plants consisting of equal numbers of the two parental species and hybrids. Ethological isolation was determined by the sequence of plants visited by a pollinator, the usual method of measurement (Table 18.1). Visit sequences sampled with replacement from foraging bouts observed at

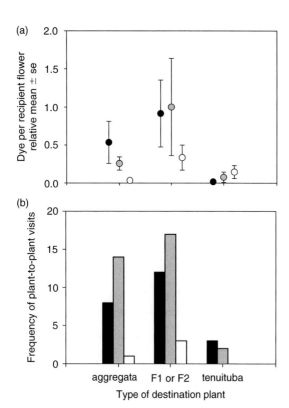

Figure 18.3 Relative amounts of (a) pollen transfer as estimated by dye particles and (b) frequency of visits by hummingbirds in experimental arrays with equal numbers of *Ipomopsis aggregata*, *I. tenuituba* and F_1 or F_2 hybrids. Responses are shown for the nine combinations of transfer or visit transition, with the type of the source plant indicated by shading (*I. aggregata*, closed symbol or bar; *I. tenuituba*, open symbol or bar; F_1 or F_2 hybrids, grey symbol or bar) and the type of the destination plant identified on the abscissa. Adapted from Campbell et al. (2002). © 2002 by The University of Chicago.

flower, and the probability of pollen deposition on a flower.

Mechanical isolation was incorporated by varying the probability of pollen deposition on a stigma as a function of the type of pollen donor and flower recipient. The simulation incorporated estimates of the natural incidence of mechanical isolation from measurements of pollen transfer between hand-held flowers during visitation by captive hummingbirds (Campbell et al. 1998). Birds removed pollen from one of three types of donor flowers (*I. aggregata*, *I. tenuituba*, or hybrid) and then visited a long series of emasculated recipient flowers including all three types. The amounts of pollen deposited on the series of recipients fitted well to a model of pollen carryover in which each pollen grain has the same probability of being deposited on a flower's stigma ($P_{deposit}$), resulting in exponential decay in the amount deposited on subsequently visited stigmas (see Chapter 5). When a hummingbird removed pollen from an *I. tenuituba* donor, $P_{deposit}$ was higher for subsequent visits to conspecific flowers than for visits to *I. aggregata* (0.182 versus 0.064; Campbell et al. 1998). For all nine combinations of pollen donor and recipient, this probability of pollen deposition varied from 0.064 to 0.226 (for full parameter set see Campbell et al. 2002). A constant $P_{deposit}$ for all combinations of donor and recipient simulated no mechanical isolation.

To assess the impacts of ethological and mechanical isolation, Campbell et al. (2002) ran four types of simulations: presence or absence of each type of isolation in a crossed design. Pollinators moved either randomly between plants (no ethological isolation) or by sampling from observed visitation sequences. The probability of pollen deposition ($P_{deposit}$) was either constant (no mechanical isolation) or depended on the combination of donor and recipient flower. With both ethological and mechanical isolation, the simulation predicted relative amounts of pollen dispersal between various combinations of flower types well (compare Fig. 18.3a with Fig. 18.4d), and the fit improved further when the simulation included the few visitation sequences by bees (Campbell et al. 2002). Note that the distribution of predicted pollen transfer (Fig. 18.4d) differed from that of

arrays in the field were used to simulate natural levels of ethological isolation, whereas random movements between plants simulated no ethological isolation. Patterns of pollen movement cannot be obtained simply from combining such measurements of ethological isolation with mechanical isolation, as they also depend on the floral display. Thus pollen export and import by a plant depended also on field measurements of the number of flowers displayed, the proportion of flowers in female or male phase, the probability that a pollinator visits a given flower once at a plant, the amount of pollen it picks up at a male-phase

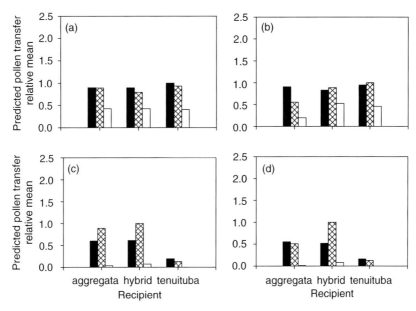

Figure 18.4 Effects of ethological and mechanical isolation on predicted pollen export per flower from *Ipomopsis aggregata* (filled bars), *I. tenuituba* (open bars) and F_1 or F_2 hybrids (hatched bars) in a simulation model of a community with the three classes of plants at equal frequencies. The four cases illustrate the effects of: (a) random visitation sequences (no ethological isolation) and a constant probability ($P_{deposit}$) of 0.135 that a pollen grain on a pollinator is deposited on a flower during an individual flower visit (no mechanical isolation); (b) random visitation sequences (no ethological isolation) and $P_{deposit}$ determined by the particular combination of pollen donor and recipient (mechanical isolation); (c) observed bird visitation sequences (ethological isolation) and constant $P_{deposit}$ (no mechanical isolation); and (d) observed bird visitation sequences (ethological isolation) and variable $P_{deposit}$ (mechanical isolation). Each panel is based on 200 replicates. Adapted from Campbell et al. (2002). © 2002 by The University of Chicago.

pollinator visits (Fig. 18.3b). For example, 52% of all pollinator movements, but only 42% of predicted pollen transfer, were of a backcross type. In the null model with no ethological or no mechanical isolation (Fig. 18.4a), interspecific pollen transfer equalled conspecific pollination. Introducing ethological isolation reduced the relative interspecific pollen transfer (Fig. 18.4c) to 38% that of conspecific transfer (RI = 0.62). In contrast, inclusion of variable pollen carryover, and hence the influence of mechanical isolation (Fig. 18.4b), reduced interspecific pollination to only 84% of conspecific pollination (RI = 0.16). Thus ethological isolation appears to be more effective than mechanical isolation in this system.

18.4.3 Geographical variation in pre-mating isolation

To identify the reproductive isolating mechanisms that best explain geographical variation in hybrid formation, we compared the subalpine site described above (PG) with a sage-oak site where natural hybrids between the same *Ipomopsis* species are mostly absent (Grizzly Ridge, CO [GR]; Aldridge 2005). Both sites have the same pollinators, but hummingbirds at GR visit only *I. aggregata*, and hawk moths visit only *I. tenuituba* in natural populations, suggesting complete ethological isolation, rather than the partial isolation seen at PG (G. Aldridge and D.R. Campbell unpublished manuscript). This difference could reflect dissimilar floral morphology, as the species are more divergent for corolla width at GR (mean = 2.8 versus 4.3 mm for *I. tenuituba* and *I. aggregata*) than at PG (2.9 versus 3.6 mm), or it could result from contrasting pollinator behaviour between sites. To test these hypotheses, we set up experimental arrays of potted *I. aggregata* and *I. tenuituba* plants in the four combinations of site of origin and site of observation (PG or GR; G. Aldridge and D.R. Campbell unpublished manuscript). Hummingbirds preferred *I. aggregata* in all situations, although their preference was lower for plants from PG. In contrast, hawk moths exhibited both strong preference for *I. tenuituba* (one-sample t-test, $P < 0.001$) and constancy (G-test, $P < 0.001$) only for plants from GR and observed at GR, with 95% of plant-to-plant movements being conspecific. In contrast, the proportion of conspecific movements by hawk moths foraging on GR plants translocated to PG fell below the random expectation of 50%. Based on the combined visits from birds and moths, the intensity of ethological isolation depended on both plant traits (compare Fig. 18.5a with Fig. 18.5b) and differences in pollinator behaviour between sites (compare closed and open bars). The difference between sites may relate partly to warmer temperatures at GR, which allow some nocturnal foraging, in contrast to the largely diurnal foraging at the high-elevation PG site (Campbell *et al.* 1997). Under low light, the difference in flower colour may be more visible, causing hawk moths to restrict their visits to the white-flowered species, in keeping with the hawk-moth pollination syndrome. These site-to-site differences underscore the importance of considering the "pollinator geographical mosaic"

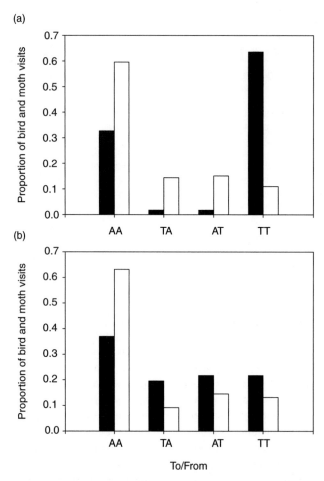

Figure 18.5 Effects of sites of observation and origin of *Ipomosis aggregata* and *Ipomosis tenuituba* plants on the proportions of pollinator movements in experimental arrays at Grizzly Ridge (GR, filled bars) and Poverty Gulch (PG, open bars). AA, movements to *I. aggregata* from *I. aggregata*, AT to *I. aggregata* from *I. tenuituba*, TA to *I. tenuituba* from *I. aggregata*, and TT to *I. tenuituba* from *I. tenuituba*. Comparisons of panels (a) plants originating from Grizzly Ridge and (b) plants originating from Poverty Gulch illustrate the effect of site of origin, whereas comparisons within a panel illustrate the effect of site of observation. Based on 686 plant-to-plant movements by hummingbirds and hawk moths.

(Chapter 15) and show that geographical differences in pollinator behaviour can appear even in the absence of spatial variation in the pollinator community.

We now explore the importance of this variation in ethological isolation to the production of hybrids by altering the simulation model to reflect the differences in plant-to-plant movements among the four array types. To incorporate these new data, we modified Campbell et al.'s (2002) model to allow changes in positions of plants within the arrays between foraging bouts. The number of flowers per plant and the probability of visiting a particular flower were altered to reflect our new field observations. The latter parameter varied significantly among the four array types; both plant origin and site of study influenced the proportion of flowers visited on an inflorescence (main effects in two-way ANOVA, both $P < 0.001$), so we used separate values for each situation. Visitation sequences were obtained by sampling with replacement from the observed sequences by hawk moths and hummingbirds. No data are available on dispersal of pollen by hawk-moths, let alone how it changes between conspecific and heterospecific visits, but Hodges' (1995) study of *Mirabilis* suggested extensive carryover, with $P_{deposit} = 0.05$. We ran the model with four scenarios for $P_{deposit}$: (1) fixed at 0.135 (the overall estimate for hummingbirds in Campbell et al. 1998); (2) fixed at 0.05; (3) 0.135 for hummingbird sequences, but 0.05 for hawk-moth sequences; and (4) variable pollen deposition for hummingbird sequences, reflecting natural mechanical isolation, but 0.05 for hawk-moth sequences.

With $P_{deposit}$ fixed at 0.135, the model predicted that just 4% of pollen transfer would be heterospecific at GR with GR plants presented in a mixture of the two species, compared with 29% at PG with PG plants (Fig. 18.6). Thus both plant differences and site differences acting in combination could account for a seven-fold difference in interspecific pollen movement between these two natural situations. These figures are similar to the percentage of interspecific plant-to-plant flights (4% and 24%, respectively, in Fig. 18.5), indicating that ethological isolation alone primarily explains the lack of hybrids at GR; although note that predicted pollen transfer at PG is higher than the observed interspecific visitation.

Predicted heterospecific transfer was greatest for PG plants at GR (40%), suggesting that overall differences in ethological isolation could produce a 10-fold difference in the rate of hybridization.

Variation in pollen carryover within limits indicated by the available data had relatively little impact on heterospecific pollen transfer. A reduction of $P_{deposit}$ from 0.135 to 0.05, thus increasing carryover, increased the relative frequency of heterospecific transfer slightly (Fig. 18.7). This change in $P_{deposit}$ also reduced total pollen transfer, as more pollen remained on a pollinator's body when it left the simulated array. A 10-fold increase in $P_{deposit}$ to 0.5 lowered the frequency of heterospecific transfer from 32% to 27% for PG plants at PG (results not shown), but in all cases heterospecific transfer remained higher than the percentage of heterospecific visit transitions (24%). Addition of the measured mechanical isolation for hummingbirds had little impact on the percentage of heterospecific pollen transfer (compare grey and hatched bars in Fig. 18.7). In sum, the model illustrates some differences between heterospecific pollinator movements and heterospecific pollen transfer. However, ethological isolation due to the visitation behaviour of hawk moths at GR would remain a powerful reproductive isolating mechanism over a wide range of pollen carryover and probably explains most of the reduction in hybrids at that site.

18.5 Post-mating isolation

Even if pollen is transferred between species, formation of hybrids may be reduced if heterospecific pollen germinates poorly on stigmas or few pollen tubes grow down styles (step c in Fig. 18.1). In at least one case, this post-mating isolation acts in combination with mechanical isolation to prevent hybridization in nature; *I. arizonica* pollen performs poorly on *I. aggregata* stigmas (Wolf et al. 2001) and hummingbirds move little pollen in the reverse direction (Section 18.3.2). Heterospecific pollen that fertilizes when present by itself on a stigma may perform poorly when competing with conspecific pollen (Darwin 1859; Arnold 1997). Examples of such a conspecific pollen advantage include the Louisiana irises (Carney et al. 1996),

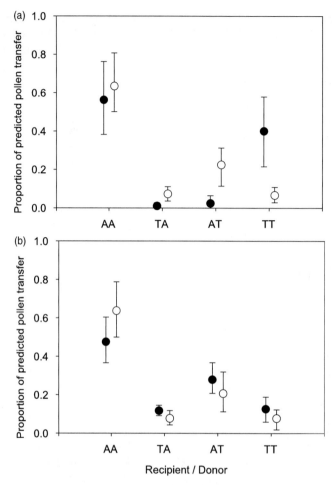

Figure 18.6 Mean (±95% confidence interval) proportion of pollen transfer per flower predicted for visitation sequences observed for *Ipomopsis aggregata* and *I. tenuituba* plants originating from (a) Grizzly Ridge (GR, filled symbols) and (b) Poverty Gulch (PG, open symbols: see Fig. 18.5) in 100 replicates of a simulation model. Comparisons within a panel illustrate the effect of site of observation. $P_{deposit} = 0.135$. Number of flowers displayed = 13 for *I. aggregata* and 12 for *I. tenuituba*. Probability of visiting a flower once at a plant = 0.31 and 0.35 for plants at GR and PG in (a), and 0.34 and 0.50 in (b). Proportion of flowers in female phase = 0.44 for *I. aggregata* and 0.23 for *I. tenuituba* (based on Campbell et al. 2002). Pollen removal per visit to a male-phase flower = 1713 grains for *I. aggregata* and 482 grains for *I. tenuituba*.

two *Brassica* species (Hauser et al. 1997), *Helianthus* (Rieseberg et al. 1995), *Piriqueta* (Wang and Cruzan 1998), and *Senecio chrysanthemifolius* (Chapman et al. 2005).

Whereas post-mating isolation is well documented, less is known about particular floral traits involved in natural contact sites. Pollen-tube growth rate has been associated with both the pollen source and characteristics of the recipient pistil (e.g., Kerwin and Smith-Huerta 2000). Asymmetries in the advantage of one species may reflect a tendency for species with longer styles to have faster-growing pollen tubes (Arnold 1997). In addition, crosses between species with different mating systems often show unilateral interspecific incompatibility in which pollen from the self-incompatible species fertilizes the self-compatible species, but the reciprocal cross fails (e.g., Harder et al. 1993). Such asymmetries could produce an asymmetrical pattern of hybridization and help to strengthen reproductive isolation (Brandvain and Haig 2005).

Incorporation of pollen competition into the model described in Section 18.4.3 allowed

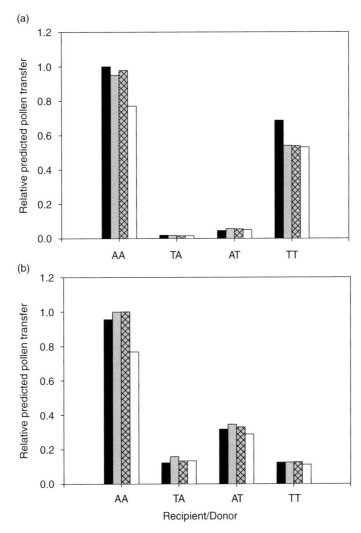

Figure 18.7 Effects of pollen carryover by hummingbirds and hawk moths on predicted pollen transfer per flower in mixed arrays of *Ipomopsis aggregata* (A) and *I. tenuituba* (T) for (a) visitation sequences for Grizzly Ridge (GR) plants observed at GR, and (b) Poverty Gulch (PG) plants observed at PG. Pollen transfer is expressed relative to the highest mean. Filled bars: $P_{deposit} = 0.135$. Shaded bars: $P_{deposit} = 0.135$ for hummingbird sequences and $P_{deposit} = 0.05$ for hawk-moth sequences. Hatched bars: $P_{deposit} = 0.133, 0.115, 0.064$, and 0.182 for A to A (*I. aggregata* to *I. aggregata*), A to T (*I. aggregata* to *I. tenuituba*), T to A, and T to T transitions in hummingbird sequences, and $P_{deposit} = 0.05$ for hawk-moth sequences. Open bars: $P_{deposit} = 0.05$.

quantitative evaluation of the impact of post-mating isolation. At the GR site, *I. aggregata* has a longer style than *I. tenuituba*. When equal amounts of pollen of the two species were placed together on stigmas, *I. aggregata* sired 70–80% of the seeds produced by recipient plants of either species (Aldridge and Campbell 2006). Thus, rather than a conspecific pollen advantage, there was a unilateral advantage to *I. aggregata*. Incorporating this threefold seed siring advantage in the simulation model has its greatest impact on the relative expected production of the reciprocal hybrids, altering the mix from 29% of F_1 individuals with a *tenuituba* mother and *aggregata* father (TA; GR plants at GR in Fig. 18.6a) to 56% in this direction of hybridization.

Whereas most studies of pre-zygotic isolation in angiosperms have focused on particular mechanisms, such as ethological isolation, an alternative approach uses genetic markers to determine the mating patterns in natural hybrid zones. Rieseberg et al. (1998) used a multilocus allozyme analysis of the progeny of open-pollinated maternal families and showed that hybrid plants were more likely to be fertilized by pollen from Helianthus petiolaris than from H. annuus. This approach may be particularly useful for wind-pollinated species. In principle, genetic markers could also be used to ascribe the likelihood of paternity to individual plants in natural populations (Smouse et al. 1999), and thereby examine how patterns of mating depend on floral traits. If floral traits are critical to reducing hybridization, such an analysis would show that mating occurs more often between plants of similar phenotype.

18.6 Floral traits and the fitness of hybrids

In addition to affecting hybrid formation, floral traits can influence hybrid fitness and thus the level of post-zygotic reproductive isolation (Fig. 18.1d–f). Floral traits play this role most directly by affecting pollen export and import by hybrids. Low pollination success of hybrids could cause either low female fitness (seed production) or low male fitness (seeds sired). David Lloyd (1980) first articulated in detail that the functional gender of individual hermaphroditic plants varies within populations, depending on their relative success as female and male parents. Male fertility is usually difficult to measure in natural plant populations, because of the large number of potential fathers for each seed (Meagher 1986), and this has rarely been attempted for natural hybrids (Melendez-Ackerman and Campbell 1998). However, continuing advances in molecular markers, such as microsatellites, and statistical analysis of male fertility (Smouse et al. 1999; Morgan and Conner 2001) make this increasingly feasible (Chapter 14).

Only a few studies present data on the pollination success of hybrid flowers and even fewer on how floral traits are involved. In the Louisiana irises, F_1 individuals received pollinator visits by bees and hummingbirds at a rate intermediate to that of the two parental species (Emms and Arnold 2000). Visitors to Nicotiana arrays exhibited less discrimination for or against hybrids than for the parental species (Ippolito et al. 2004). In experimental arrays of Ipomopsis, per-flower hummingbird visitation to hybrids ranged between years from intermediate between the two parental species I. aggregata and I. tenuituba to exceeding that of both parents (Campbell et al. 2002). In the latter case, pollen receipt as estimated by dyes followed the general pattern in pollinator visitation (Fig. 18.3). In a separate experiment, hybrids produced and sired intermediate numbers of seeds to those of the parental species (Melendez-Ackerman and Campbell 1998). These results do not follow the common expectation that hybrids have low pollination success, but they involve cases in which natural hybridization is relatively frequent.

In principle, if differences in floral traits impose strong ethological isolation, hybrids intermediate for those traits will receive few pollinator visits. This poor performance of hybrids constitutes disruptive or divergent selection on floral morphology, in which the low hybrid fitness represents a form of post-zygotic reproductive isolation. Thus, the same set of traits (and presumably genes) under divergent selection would also produce pre-zygotic reproductive isolation. In this "single-variation" model (Rice and Hostert 1993), sympatric or parapatric speciation occurs relatively easily (Rice 1987). A potential example involves corolla width in Ipomopsis, as plants with intermediate corolla widths receive less combined visitation by hummingbirds and hawk moths (Campbell et al. 1997). Whether the species difference in corolla width also enforces pre-zygotic isolation through assortative mating remains unknown.

The steps between pollinator visitation of hybrids and production of advanced generation hybrids (Fig. 18.1e and f) have rarely been studied in natural hybrid zones. In studies of efficiency of pollen transfer involving I. aggregata and I. tenuituba, hybrids between the two species actually received more pollen per visit than did the parental species (Campbell et al. 1998). This pattern reflected stabilizing selection on the position of the stigma relative to the opening of the corolla tube, which is intermediate in hybrids, probably due to frequent misses of exserted stigmas by hummingbirds and inability of hummingbirds to insert their bill fully into narrow *I. tenuituba* flowers. Once pollen is deposited or exported from hybrids, its final success can depend on pollen interactions. Hybrid pollen between these two species of *Ipomopsis* is at a disadvantage in competing with pollen of the two species to fertilize ovules (Campbell *et al.* 2003).

Although floral traits are typically assumed to evolve in response to pollinators, they could also have other fitness consequences for hybrids. Selection by nectar thieves and by floral herbivores that damage petals or reproductive parts is well documented (Chapter 7) and provides a potential mechanism for reducing pollinator visitation to hybrids (step d in Fig. 18.1). Similarly, floral traits can influence fruit or seed predation, potentially reducing the production of viable seed by hybrids (step f). A more complex interaction occurs when pollinators are vectors for sexual disease, as the same floral trait can increase pollen transfer and spore transfer, causing sterility (Elmqvist *et al.* 1993). In some cases, the responses of pollinators and non-pollinators to the same floral trait lead to antagonistic selection on floral colour, shape, or size (Chapter 7). Despite being well documented in single-species populations, the influence of these effects on hybrid fitness is largely unknown and warrants more study. Hybrid zones between *I. aggregata* and *I. tenuituba* provide a potential example of antagonistic selection between hummingbird pollinators and seed predators. The anthomyiid fly *Hylemya* sp. (= *Delia*) lays eggs on the inside of the sepals, and the larvae consume seeds before pupating (Brody 1992). Oviposition rates correlated positively with corolla width in experimental arrays that included both plant species and hybrids (Campbell *et al.* 2002). This trait also increases hummingbird visitation, so damage by the seed predator and pollination can correspond closely, leading to opposing selection.

Floral traits can also be associated with aspects of survival, such as the ability to handle drought stress or resistance to vegetative herbivory. Usually a pleiotropic effect on a vegetative trait is involved, for example the multiple effects of the anthocyanin pathway on flower colour and vegetative traits (Chapter 7). In this case, indirect selection of the genetically correlated trait must be considered to predict the evolutionary fate of the floral trait (Lande and Arnold 1983, Chapter 14). In one example, the white allele that affects flower colour and pollinator visitation in the morning glory *Ipomoea purpurea* (Epperson and Clegg 1987) also controls stem colour (Schoen *et al.* 1984) and resistance to herbivory by tortoise beetles (Simms and Bucher 1996). Again, little is known about such indirect selection or genetic correlations between floral traits and vegetative traits in hybrid zones.

18.7 Conclusions and future directions

The studies reviewed above represent considerable progress during the past decade in understanding of the influences of floral traits on hybridization and reproductive isolation. However, further work is needed along several lines. First, even in the best-studied natural systems the influences of particular floral traits on reproductive isolation have seldom been quantified, largely because the data are restricted to a single step in the process. Most investigators have focused on pollinator visitation or post-mating isolation, and only rarely have all or most steps involved in reproductive isolation been examined (Campbell 2004), making it hard to assess their relative importance. Pollinator visitation is a topic with which to begin, because complete ethological isolation precludes opportunities for reproductive isolating mechanisms that act later. However, even when focusing on this initial step, pollinator transitions between species, rather than pollinator preference alone, must be measured.

Second, in systems with any breakdown of ethological isolation, later-acting processes that may influence pollen transfer, interspecific gene flow, or hybrid fitness should also be measured. For example, even if a pollinator visits a second species only occasionally during a long foraging bout, many plants could receive some heterospecific pollen if pollen carryover is extensive. Our simulation models illustrate that the details of pollen carryover affect the correspondence between interspecific pollinator visitation and heterospecific pollen transfer. This link between pollinator visitation patterns and pollen movement warrants more study, especially for whole pollinator assemblages.

Third, the extent of interspecific gene flow can depend strongly on environmental characteristics of the contact site, and on the presence versus absence of hybrids, with pollinators showing stronger preference when the parental species only are present. In a study of *Ipomopsis* we found as much difference in ethological isolation between two contact sites visited by the same pollinators (0.30–0.96) as could be explained by pollinator type in the broader survey of studies (Table 18.1).

Fourth, molecular genetic approaches offer increasing promise for further understanding of how particular traits influence components of reproductive isolation. Initial studies of the genetic basis of reproductive isolation have proceeded by QTL analyses of floral traits in a glasshouse setting (Bradshaw *et al.* 1998), combined with separate field experiments to assess pollinator responses. In experimental arrays of *Mimulus*, in which the normal genetic associations between flower colour and nectar were disrupted, bee visitation depended on the presence/absence of an allele controlling carotenoid pigments in the petals (Schemske and Bradshaw 1999). Genetic mapping could also be used in natural hybrid zones to test more directly for QTLs that influence reproductive isolation. In one example, Rieseberg *et al.* (1999b) employed a large number of mapped RAPD markers to examine associations with reduced pollen fertility of natural sunflower hybrids. Use of such methods in natural hybrid zones will become easier as more species-diagnostic markers are developed. However, a simpler approach could be applied that involves planting a QTL mapping population of offspring of known parentage in the field and examining correlations of genetic markers specific to the original parents with floral traits and visitation by specific pollinators, or even patterns of pollen transfer (e.g., using dyes to mark pollen). Such field data could allow mapping of chromosomal blocks associated with particular floral traits and their influences in restricting pollination by a particular type of animal.

Finally, most of the more detailed studies of pre-mating isolation have focused on systems involving divergence of insect and hummingbird-pollinated species. The specific influences of floral traits on hybridization in such situations probably do not apply generally in other pollination systems. Therefore studies of other pollination systems, including those involving multiple types of insect pollinators, or animal versus wind-pollination, are essential to identify general principles of the floral biology of hybridization.

Acknowledgements

We thank Matthew Crawford, Greg Pederson, and Holly Prendeville for assistance in the field, and Lawrence Harder, Spencer Barrett, and two anonymous reviewers for comments on the manuscript. The National Science Foundation (USA) provided financial assistance through grants NSF DEB-9806547 to D. Campbell and NSF DEB-0206279 to D. Campbell and G. Aldridge.

References

Aldridge G (2005). Variation in frequency of hybrids and spatial structure among *Ipomopsis* contact sites. *New Phytologist*, **167**, 279–88.

Aldridge G and Campbell DR (2006). Asymmetrical pollen success in *Ipomopsis* (Polemoniaceae) contact sites. *American Journal of Botany*, **93**, 903–9.

Anderson E (1949). *Introgressive hybridization*. Wiley, New York.

Anderson E and Stebbins GL (1954). Hybridization as an evolutionary stimulus. *Evolution*, **8**, 378–88.

Arnold ML (1994). Natural hybridization and Louisiana irises. Defining a major factor in plant evolution. *BioScience*, **44**, 141–7.

Arnold ML (1997). *Natural hybridization and evolution*. Oxford University Press, Oxford.

Ayres DR, Garcia-Rossi D, Davis HG, and Strong DR (1999). Extent and degree of hybridization between exotic (*Spartina alterniflora*) and native (*S. foliosa*) cordgrass (Poaceae) in California, USA determined by random amplified polymorphic DNA. *Molecular Ecology*, **8**, 1179–86.

Baker HG and Hurd PD (1968). Intrafloral ecology. *Annual Review of Entomology*, **13**, 385–415.

Barton NM and Hewitt GM (1985). Analysis of hybrid zones. *Annual Review of Ecology and Systematics*, **16**, 113–48.

Bradshaw HD Jr, Otto KG, Frewen BE, McKay JK, and Schemske DW (1998). Quantitative trait loci affecting differences in floral morphology between two species of monkeyflower (*Mimulus*). *Genetics*, **149**, 367–82.

Brandvain Y and Haig D (2005). Divergent mating systems and parental conflict as a barrier to hybridization in flowering plants. *American Naturalist*, **166**, 330–8.

Brody AK (1992). Oviposition choices by a pre-dispersal seed predator (*Hylemya* sp.). II. A positive association between female choice and fruit set. *Oecologia*, **91**, 63–7.

Campbell DR (2004). Natural selection in *Ipomopsis* hybrid zones: implications for ecological speciation. *New Phytologist*, **161**, 83–90.

Campbell DR, Waser NM, and Melendez-Ackerman EJ (1997). Analyzing pollinator-mediated selection in a plant hybrid zone: hummingbird visitation patterns on three spatial scales. *American Naturalist*, **149**, 295–315.

Campbell DR, Waser NM, and Wolf PG (1998). Pollen transfer by natural hybrids and parental species in an *Ipomopsis* hybrid zone. *Evolution*, **52**, 1602–11.

Campbell DR, Waser NM, and Pederson GT (2002) Predicting patterns of mating and potential hybridization from pollinator behavior. *American Naturalist*, **159**, 438–50.

Campbell DR, Alarcon R, and Wu CA (2003). Reproductive isolation and hybrid pollen disadvantage in *Ipomopsis*. *Journal of Evolutionary Biology*, **16**, 536–40.

Carney SE, Hodges SA, and Arnold ML (1996). Effects of differential pollen tube growth on hybridization in the Louisiana Irises. *Evolution*, **50**, 1871–8.

Castellanos MC, Wilson P, and Thomson JD (2003). Pollen transfer by hummingbirds and bumblebees, and the divergence of pollinator modes in *Penstemon*. *Evolution*, **57**, 2742–52.

Chapman MA, Forbes DG, and Abbott RJ (2005). Pollen competition among two species of *Senecio* (Asteraceae) that form a hybrid zone on Mt. Etna, Sicily. *American Journal of Botany*, **92**, 730–5.

Darwin CR (1859). *On the origin of species*. John Murray, London.

Ellstrand NC, Whitkus R, and Rieseberg LH (1996). Distribution of spontaneous plant hybrids. *Proceedings of the National Academy of Sciences of the United States of America*, **93**, 5090–3.

Elmqvist T, Liu D, Carlsson U, and Giles BE (1993). Anther-smut infection in *Silene dioicea*—variation in floral morphology and patterns of spore deposition. *Oikos*, **68**, 207–16.

Emms SK and Arnold ML (2000). Site-to-site differences in pollinator visitation patterns in a Louisiana iris hybrid zone. *Oikos*, **91**, 568–78.

Epperson BK and Clegg MT (1987). First-pollination primacy and pollen selection in the morning glory, *Ipomoea purpurea*. *Heredity*, **58**, 5–14.

Faegri K and van der Pijl L (1966). *The principles of pollination ecology*. Pergamon, Oxford.

Fishman L and Willis JH (2001). Evidence for Dobzhansky–Muller incompatibilities contributing to the sterility of hybrids between *Mimulus guttatus* and *M. nasutus*. *Evolution*, **55**, 1932–42.

Fulton M and Hodges SA (1999). Floral isolation between *Aquilegia formosa* and *Aquilegia pubescens*. *Proceedings of the Royal Society of London, Series B*, **266**, 2247–52.

Grant V (1949). Pollination systems as isolating mechanisms in angiosperms. *Evolution*, **3**, 82–97.

Grant V (1994) Modes and origins of mechanical and ethological isolation in angiosperms. *Proceedings of the National Academy of Sciences of the United States of America*, **91**, 3–10.

Grant V and Wilken DH (1988). Natural hybridization between *Ipomopsis aggregata* and *I. tenuituba* (Polemoniaceae). *Botanical Gazette*, **149**, 213–21.

Harder LD, Cruzan MB, and Thomson JD (1993). Unilateral incompatibility and the effects of interspecific pollination for *Erythronium americanum* and *Erythronium albidum* (Liliaceae). *Canadian Journal of Botany*, **71**, 353–8.

Hatfield T and Schluter D (1999). Ecological speciation in sticklebacks: environment-dependent hybrid fitness. *Evolution*, **53**, 866–73.

Hauser TP, Jorgensen, RB, and Ostergard H (1997). Preferential exclusion of hybrids in mixed pollinations between oilseed rape (*Brassica napus*) and weedy *B. campestris* (Brassicaceae). *American Journal of Botany*, **84**, 756–62.

Hodges SA (1995). The influence of nectar production on hawkmoth behavior, self pollination and seed production in *Mirabilis multiflora* (Nyctaginaceae). *American Journal of Botany*, **82**, 197–204.

Hodges SA, Burke JM, and Arnold ML (1996). Natural formation of *Iris* hybrids: experimental evidence in the establishment of hybrid zones. *Evolution*, **50**, 2504–9.

Hodges SA, Whittall JB, Fulton M, and Yang JY (2002). Genetics of floral traits influencing reproductive isolation between *Aquilegia formosa* and *Aquilegia pubescens*. *American Naturalist*, **159**, S51–60.

Husband BC and Schemske DW (2000). Ecological mechanisms of reproductive isolation between diploid and tetraploid *Chamerion angustifolium*. *Journal of Ecology*, **88**, 689–701.

Ippolito A, Fernandes GW, and Holtsford TP (2004). Pollinator preferences for *Nicotiana alata*, *N. forgetiana*, and their F-1 hybrids. *Evolution*, **58**, 2634–44.

Kephart S and Theiss K (2003). Pollinator-mediated isolation in sympatric milkweeds (*Asclepias*): do floral morphology and insect behavior influences species boundaries? *New Phytologist*, **161**, 265–77.

Kerwin MA and Smith-Huerta NL (2000). Pollen and pistil effects on pollen germination and tube growth in selfing and outcrossing populations of *Clarkia tembloriensis* (Onagraceae) and their hybrids. *International Journal of Plant Sciences*, **161**, 895–902.

Kim S and Rieseberg LH (1999). Genetic architecture of species differences in annual sunflowers: implications for adaptive trait introgression. *Genetics*, **153**, 965–77.

Lande R and Arnold SJ (1983). The measurement of selection on correlated characters. *Evolution*, **37**, 1210–26.

Leebens-Mack J and Milligan BG (1998). Pollination biology in hybridizing *Baptisia* (Fabaceae) populations. *American Journal of Botany*, **85**, 500–7.

Levin DA and Berube DE (1972). *Phlox* and *Colias*—efficiency of a pollination system. *Evolution*, **26**, 242–7.

Levin DA and Schaal BA (1970). Corolla color as an inhibitor of interspecific hybridization in *Phlox*. *American Naturalist*, **104**, 273–83.

Levin DA, Francisco-Ortega J, and Jansen RK (1996). Hybridization and the extinction of rare plant species. *Conservation Biology*, **10**, 10–6.

Lloyd DG (1980). The distributions of gender in four angiosperm species illustrating two evolutionary pathways to dioecy. *Evolution*, **34**, 123–34.

Macior (1965). Insect adaptation and behavior in *Asclepias* pollination. *Bulletin of the Torrey Botanical Club*, **92**, 114–26.

Mallet J (2005). Hybridization as an invasion of the genome. *Trends in Ecology and Evolution*, **20**, 229–37.

Martinsen GD, Whitham TG, Turek RJ, and Keim P (2001). Hybrid populations selectively filter gene introgression between species. *Evolution*, **55**, 1325–35.

Mayr E (1963). *Animal species and evolution*. Harvard University Press, Cambridge, MA.

Meagher TR (1986). Analysis of paternity within a natural population of *Chamaelirium luteum*. I. Identification of male parentage. *American Naturalist*, **128**, 199–215.

Melendez-Ackerman EJ and Campbell DR (1998). Adaptive significance of flower color and inter-trait correlations in an *Ipomopsis* hybrid zone. *Evolution*, **52**, 1293–303.

Milne RI, Terzioglu S, and Abbott RJ (2003). A hybrid zone dominated by fertile F1s: maintenance of species barriers in *Rhododendron*. *Molecular Ecology*, **12**, 2179–29.

Morgan MT and Conner JK (2001). Using genetic markers to directly estimate male selection gradients. *Evolution*, **55**, 272–81.

Oxford English Dictionary (1971). Oxford University Press, Oxford.

Ramsey J, Bradshaw HD Jr, and Schemske DW (2003). Components of reproductive isolation between the monkeyflowers *Mimulus lewisii* and *M. cardinalis* (Phrymaceae). *Evolution*, **57**, 1520–34.

Rhode JM and Cruzan MB (2005). Contributions of heterosis and epistasis to hybrid fitness. *American Naturalist*, **166**, E124–39.

Rice WR (1987). Speciation via habitat specialization: the evolution of reproductive isolation as a correlated character. *Evolutionary Ecology*, **1**, 301–14.

Rice WR and Hostert EE (1993). Laboratory experiments on speciation—what have we learned in 40 years? *Evolution*, **47**, 1637–53.

Rieseberg LH (1997). Hybrid origins of plant species. *Annual Review of Ecology and Systematics*, **28**, 359–89.

Rieseberg LH and Ellstrand NC (1993). What can molecular and morphological markers tell us about plant hybridization? *Critical Reviews in Plant Sciences*, **12**, 213–41.

Rieseberg LH, Desrochers AM, and Youn SJ (1995). Interspecific pollen competition as a reproductive barrier between sympatric species of *Helianthus* (Asteraceae). *American Journal of Botany*, **82**, 515–19.

Rieseberg LH, Sinervo B, Linder CR, Ungerer MC, and Arias DM (1996). Role of gene interactions in hybrid speciation: evidence from ancient and experimental hybrids. *Science*, **272**, 741–5.

Rieseberg LH, Baird SJE, and Desrochers AM (1998). Patterns of mating in wild sunflower hybrid zones. *Evolution*, **52**, 713–26.

Rieseberg LH, Archer MA, and Wayne RK (1999a). Transgressive segregation, adaptation and speciation. *Heredity*, **83**, 363–72.

Rieseberg LH, Whitton J, and Gardner K (1999b). Hybrid zones and the genetic architecture of a barrier to gene flow between two sunflower species. *Genetics*, **152**, 713–27.

Schemske DW and Bradshaw HD Jr (1999). Pollinator preference and the evolution of floral traits in monkeyflowers (*Mimulus*). *Proceedings of the National Academy of Sciences of the United States of America*, **96**, 11910–5.

Schoen DJ, Giannasi DE, Ennos RA, and Clegg MT (1984). Stem color and pleiotropy of genes determining flower color in the common morning glory. *Journal of Heredity*, **75**, 113–6.

Servedio MR and Noor MAF (2003). The role of reinforcement in speciation: theory and data. *Annual Review of Ecology and Systematics*, **34**, 339–64.

Simms EL and Bucher MA (1996). Pleiotropic effects of flower-color intensity on herbivore performance on *Ipomoea purpurea*. *Evolution*, **50**, 957–63.

Smouse PE, Meagher TR, and Kobak CJ (1999). Parentage analysis in *Chamaelirium luteum* (L.) Gray (Liliaceae): why do some males have higher reproductive contributions? *Journal of Evolutionary Biology*, **12**, 1069–77.

Stebbins GL (1970). Adaptive radiation of reproductive characteristics in angiosperms. I. Pollination mechanisms. *Annual Review of Ecology and Systematics*, **1**, 307–26.

Stucky JM (1985). Pollination systems of sympatric *Ipomoea hederacea* and *I. purpurea* and the significance of interspecific pollen flow. *American Journal of Botany*, **72**, 32–43.

Wang J and Cruzan MB (1998) Interspecific mating in the *Piriqueta caroliniana* (Turneraceae) complex: effects of pollen load size and composition. *American Journal of Botany*, **85**, 1172–9.

Waser NM (1986). Flower constancy: definition, cause and measurement. *American Naturalist*, **127**, 593–603.

Wesselingh RA and Arnold ML (2000). Pollinator behaviour and the evolution of Louisiana iris hybrid zones. *Journal of Evolutionary Biology*, **13**, 171–80.

Wolf PG, Campbell DR, Waser NM, Sipes SD, Toler TR, and Archibald JK (2001). Tests of pre-and post-pollination barriers to hybridization between sympatric species of *Ipomopsis* (Polemoniaceae). *American Journal of Botany*, **88**, 213–9.

Glossary

Additive genetic variance (V_A): the magnitude of the phenotypic (and genotypic) variance that is due to additive effects of genes and that determines the degree to which the average phenotype of the parents is reflected in the average phenotype of their progeny.

Additive genetic variance–covariance matrix: see *G matrix*).

AFLP (amplified fragment length polymorphism): genetic markers detected by cleaving DNA with one or more restriction enzymes and then amplifying some of these fragments by PCR using primers with random nucleotide sequences.

Allee effect: a positive relation between density and recruitment, often resulting in a threshold density (extinction threshold) below which populations are driven deterministically to extinction.

Androdioecy: a sexual polymorphism in which populations are composed of hermaphroditic and male plants. Very rare in flowering plants.

Anisoplethy: unequal frequency of morphs in a population.

Anthocyanin: blue, violet, or red flavonoid pigment.

Apomixis: production of an unreduced embryo sac without fertilization of the egg cell. The resulting seed is genetically identical to the maternal parent. May or may not involve fusion between a sperm nucleus and polar nuclei to form endosperm.

Approach herkogamy: a flower's stigma(s) is exserted beyond its anthers, so that pollinators typically contact the stigma before the anthers as they enter a flower.

Assortative mating: non-random mating based on the characteristics of the partners. Includes *positive assortative mating*, whereby the traits of mating partners are more similar than expected from random pairing (e.g., flowering time), and *negative assortative mating* (or *disassortative mating*), whereby the traits of mating partners are less similar than expected from random pairing (e.g., heterostyly).

Autogamy: self-fertilization following the transfer of viable pollen between anthers and receptive stigmas within the same flower.

Autonomous apomixis: asexual seed production that occurs without pollination. Involves spontaneous embryo and endosperm development. Provides reproductive assurance.

Autonomous autogamy: self-fertilization following intra-floral self-pollination that occurs spontaneously without the activity of pollen vectors. Often arises from contact between dehiscing anthers and receptive stigmas at some time during floral development. Can occur before (prior autogamy), during (simultaneous autogamy), or after (delayed autogamy) the period during which outcross pollen is deposited on stigmas. All forms provide reproductive assurance.

β-diversity: the variety of species within a collection of communities, such as a landscape.

Chasmogamous: an open flower that is accessible to pollen carried by a vector. Usually contrasted with closed (*cleistogamous*) flowers.

Cleistogamous: a flower that never opens and self-pollinates autonomously.

Cline: a geographical gradient in phenotype among populations resulting from either phenotypic plasticity or a mixture of local adaptation and gene flow in response to an underlying environmental gradient.

Clonal reproduction: often refers to any form of asexual reproduction, but here refers only to asexual reproduction involving vegetative propagation of some part of the plant other than the ovule or embryo sac. Like other forms of asexual reproduction (see *apomixis*) progeny are genetically identical to the parent.

Constancy: an individual pollinator's fidelity to one plant species during a single foraging bout.

Correlated response to selection: an evolutionary change in an unselected trait caused by an additive *genetic correlation* between the unselected trait and a trait under selection.

Cosexuality: a condition in which a population comprises a single sexual class of hermaphrodites (cosexes) in which individuals reproduce equally as both maternal and paternal parents, on average. Because of the latter condition, hermaphroditic individuals in a population with females (*gynodioecy*) are not cosexual, because they contribute more genes to the next generation as males than as females. Cosexuality is the commonest sexual system in flowering plants.

Cost of meiosis: the reduced genetic contribution of a female to her offspring caused by "gene sharing" with an unrelated mate during sexual reproduction. An asexually reproducing female contributes all of each offspring's genes compared to only half of an offspring's genes if she reproduces sexually via mating with an unrelated individual. Biparental mating bears an equivalent cost relative to self-mating, although both of these processes are sexual and so involve meiosis.

Deme: a partially isolated subpopulation, connected to other such subpopulations by some dispersal.

Dichogamy: differences in the timing of anther dehiscence and stigma receptivity of flowers. Also occurs at the inflorescence or plant level and is common in flowering plants. Two types: *protandry*, with male function before female, and *protogyny*, with the reverse pattern. Generally reduces intrafloral self-pollination and can reduce between-flower self-pollination, if flowers are arranged so that pollinators tend to visit female-phase flowers before male-phase flowers.

Dicliny: refers to a species with separate pistillate and staminate flowers, either on the same plant (*monoecy*), gynomonoecy or on different plants (*dioecy*), gynodioecy.

Dioecy: a sexual polymorphism in which populations are composed of female and male plants, often differing also in secondary sex characters. Often associated with large plant size, fleshy fruits, small inconspicuous flowers, and abiotic pollination.

Distyly: the commonest form of *heterostyly*, involving two genetically determined floral morphs, one with long styles and short stamens, and the other with short styles and long stamens.

Emasculation: experimental removal of anthers from a hermaphroditic flower, rendering it functionally female and incapable of autonomous self-pollination.

Enantiostyly: a condition in which flowers have styles that bend either to the right or to the left.

Epistasis: interaction between the effects of alleles at multiple loci that produces a phenotype that differs from that expected solely from additive genetic effects.

Ethological isolation: behavioural preferences of pollinators for particular floral traits restrict their movement between flowers of different species.

Evolutionary stable strategy (ESS): a strategy that results in higher fitness than any other strategy when adopted by all members of a population. An ESS can be pure if all individuals adopt the same strategy, or mixed (polymorphic) if individuals adopt alternative strategies but none could realize higher fitness by changing strategies. A mixed ESS is maintained by negative frequency-dependent selection, resulting in an evolutionary stable state.

Floral design: characteristics of individual flowers including their size, structure, sex condition, colour, scent, nectar production, and degree of herkogamy and dichogamy.

Floral display: the number of open flowers on a plant and their arrangement within and among inflorescences. The important functional unit for pollination is usually daily inflorescence size.

Floral isolation: differences in floral traits prevent mating of related species by affecting either pollinator behaviour (*ethological isolation*) or the contact between pollinators and sexual organs (*mechanical isolation*).

Florivore: animals, usually insects, that feed on flowers.

Fluctuating asymmetry: random variation in bilateral symmetry among individuals caused by unequal growth of the two sides of a structure.

Functional gender: The relative (or absolute) contribution to fitness by an individual through female (ovules) and/or male (pollen) function. In hermaphrodites the "maleness" or "femaleness" of an individual.

Functional response: the effect of resource density on the rate at which individual organisms consume resources.

G matrix: a square matrix with additive genetic variances for the traits on the diagonal and additive genetic covariances between traits as off-diagonal elements (synonym: additive genetic variance–covariance matrix).

Gametic phase disequilibrium: a non-random association between alleles at different loci in a population as a result of linkage, non-random mating, gene flow, or recent evolution (selection or genetic drift) in the ancestral population. Gametic phase disequilibrium creates a *genetic correlation*.

Geitonogamy: self-pollination resulting from transfer of pollen between flowers on an individual. Common in mass-flowering species, genetically equivalent to autogamous selfing, and is probably a common cause of complete *pollen* and *seed discounting*.

Gender diphasy: Sequential hermaphroditism in which plants function entirely as females or entirely as males, depending on their size.

Gender strategies: concern the femaleness and maleness of individuals and reflect the relative contributions to fitness from maternal and paternal investment. Hermaphroditic, female, and male morphs are distinguished by their functional gender and co-occur in different sexual systems.

Genet: a genetic individual, which may be composed of multiple connected or independent physical individuals that share the same genotype. A synonym for "clone" and usually contrasted with *ramet*.

Genetic correlation (r_A): *genetic covariance* expressed relative to the additive genetic variation in both traits.

Genetic covariance: an absolute measure of the degree to which two traits are affected by the same genes (*pleiotropy*) or pairs of genes (*gametic phase disequilibrium*). Selection on one trait produces an evolutionary change in all traits that exhibit an additive genetic covariance with the selected trait.

Genetic load: the frequency of deleterious alleles in a population. Genetic load increases by the accumulation of mutations and is generally reduced by the selection associated with *inbreeding depression*.

Genetic structure: usually referring to the amounts and kinds of genetic variation within and among populations. Measures of population genetic structure commonly include the number of polymorphic loci (P), the number of alleles per locus (A) and average heterozygosity (H).

Genetic variance: the component of the variance in a phenotypic trait resulting from all genetic differences among individuals.

Geophyte: a herbaceous plant with an underground storage structure, such as a bulb, corm, or tuber, usually occupying a seasonal climate with a marked dry season.

Glucosinolate: (or mustard oil glycosides): a class of organic compounds that contain sulfur, nitrogen and a group derived from glucose and are characteristic of many plants in the Brassicaceae, and the Capparales. They are hydrolysed by myrosinase enzyme to produce glucose, sulfate and either a nitrile, isothiocyanate, thiocyanate, epithionitrile, oxazolidine-2-thione, or other products. Glucosinolates play a role in plant defense, and may also have allelopathic functions.

Gynodioecy: a sexual polymorphism in which populations are composed of hermaphroditic and female individuals. This sexual system is scattered among angiosperm families, but is considerably more common than *androdioecy*. Grades into *subdioecy*.

Heritability: the proportion of the total phenotypic variance that is due to genetic causes. Heritability measures the relative importance of genetic variance in determining phenotypic variance. Narrow-sense heritability (h^2) is the additive genetic variance divided by the phenotypic variance (V_A/V_P), whereas broad-sense heritability (H^2) is the total *genotypic variance* divided by the phenotypic variance (V_G/V_P).

Herkogamy: the spatial separation of dehiscing anthers and receptive stigmas within flowers. Common in flowering plants and, like dichogamy, generally reduces intra-flower self-pollination.

Heteromorphic incompatibility: a form of *self-incompatibility* that occurs in heterostylous populations in which there are two (dimorphic) or three (trimorphic) physiological mating types corresponding to the floral morphs (see *heterostyly*). Self and intra-morph pollinations are incompatible, whereas pollinations between anthers and stigmas of equivalent height are compatible.

Heterostyly: a sexual polymorphism in which populations are composed of two (*distyly*) or three (*tristyly*) floral morphs with reciprocal arrangements of anthers and stigmas (reciprocal herkogamy). Usually associated with *heteromorphic incompatibility*. The syndrome functions to promote more proficient pollen dispersal and to reduce selfing.

Homomorphic incompatibility: the most widespread form of *self-incompatibility* in which populations are composed of multiple mating groups that cannot be distinguished morphologically.

Hybrid zone: area of contact between two related species in which viable hybrids are present.

Ideal free distribution: a stable distribution of consumers that are both aware of the resource distribution (ideal) and free to move among discrete resource patches (or species). This distribution is realized when no consumer could improve its state by moving to an alternative resource patch.

Inbreeding depression: the reduction in viability and/or fertility of inbred offspring in comparison with those from outcrossed matings, or $1 - (w_i/w_x)$, where w_i and w_x are the fitnesses of inbred and outcrossed individuals, respectively. Results primarily from the expression of deleterious recessive alleles in homozygous genotypes. Inbreeding depression can occur throughout the life cycle and is expressed most strongly in outcrossing species.

Introgression: genetic assimilation of genes from another species through inter-specific hybridization.

Isoplethy: equal frequency of morphs in a population.

Late-acting self-incompatibility: a type of *self-incompatibility* expressed in the ovary, usually prior to fertilization. Post-zygotic self-incompatibility has been reported, but it can be difficult to distinguish from early-acting *inbreeding depression*.

Local mate competition: competition among siblings (usually brothers) for mating opportunities.

Local resource competition: competition among independent siblings for environmental resources.

Marginal fitness: the change in fitness associated with a small change in a trait or strategy. Formally, the first derivative of fitness with respect to a trait or strategy.

Mating: processes that result in the fertilization of ovules by male gametophytes. A key component of sexual reproduction.

Mating system: the mode of transmission of genes from one generation to the next through sexual reproduction. Important determinants of plant mating systems are the maternal (ovule) *selfing rate* and male siring success through pollen (male fertility).

Mechanical isolation: differences in floral structures, such as the positions of stigma and anthers, that restrict inter-specific hybridization.

Metapopulation: a population of populations more or less connected by gene flow and characterized by colonization–extinction dynamics.

Microarray: a technique for measuring the expression of many genes simultaneously.

Microsatellites: genetic markers consisting of repeated units 2–9 nucleotides long. Also called simple sequence repeats (SSR), simple sequence repeated polymorphisms (SSRP), or short tandem repeats (STR).

Monoecy: the condition in which individuals produce separate female and male flowers. Prevents intra-flower self-pollination and enables adjustment of female and male allocation to environmental conditions. Most commonly associated with *dichogamy*.

Morph: a genetically determined class of individuals with a discrete morphology, colour, strategy, etc. Usually used to distinguish one class of individuals from other classes.

Multivoltine: a species that undergoes more than one generation per year; usually used in reference to animals.

Numerical response: the relation of per capita population growth rate (via birth, rather than immigration) to resource density.

Oligolectic: refers to a specialized bee species that collects pollen from one or a few related plant species.

Outcrossing rate: the proportion of seeds produced by an individual or population that are cross-fertilized: the complement of the *selfing rate*. Typically used to refer to the female outcrossing rate, or the proportion of cross-fertilized seeds. An individual's female and male outcrossing rates differ when the numbers of outcrossed seeds that it produces and sires on other plants are not equal.

Ovule discounting: reduced female fertility caused by self-pollen tubes disabling some ovules.

Reported from species with ovarian self-incompatibility (see *late-acting self-incompatibility*).

Ovule limitation: a constraint on seed production that occurs when all ovules are fertilized but too few zygotes survive genetic death and predation to compete for maternal resources. Ovule limitation can be alleviated by increased ovule production.

Parapatric: populations with contiguous but non-overlapping distributions.

Paraphyly: a taxon or portion of a phylogenetic tree that is derived from a single ancestor but that does not include all the descendants of that ancestor.

Phenotypic selection: the association of differential survival and/or reproduction with phenotypic variation within a population. If the phenotypic variation depends partially on additive genetic differences among individuals, phenotypic selection results in natural selection.

Pleiotropy: the phenotypic effect of a gene on more than one trait.

Pollen allelopathy: chemicals exuded from heterospecific pollen on a stigma inhibit germination or tube growth of otherwise compatible conspecific pollen.

Pollen carryover: the residue of pollen from a specific donor flower on a pollinator as it visits a sequence of recipient flowers. Pollen carryover results in the dispersal of pollen from one flower to several (many) recipient flowers by an individual pollinator.

Pollen discounting: a loss in outcrossed siring success caused by self-pollination. Reduces the transmission advantage of selfing and, along with *inbreeding depression*, represents a major cost of selfing.

Pollen limitation: a reduction in potential seed production caused when some ovules remain unfertilized and too few embryos survive genetic death and predation to compete for maternal resources.

Pollination ecotype: a population (or populations) that is adapted to the local pollinator fauna and so differs in floral traits from other populations of the same species.

Pollination syndrome: a correlated suite of floral traits adapted to the morphology and behaviour of a specific class of pollen vector (e.g., bees, flies, butterflies, hawk moths, birds, bats, wind).

Protandrous: male before female. In the context of hermaphroditic species, protandrous indicates that an individual flower or plant functions as a male before functioning as a female.

Protogynous: female before male. In the context of hermaphroditic species, protogynous indicates that an individual flower or plant functions as a female before functioning as a male.

Pseudogamy: a form of *apomixis* (asexual seed production) in which endosperm development requires fusion between a sperm nucleus from a pollen grain and embryo-sac polar nuclei, even though the egg nucleus is unreduced and develops into an embryo without fertilization. Means "false marriage," in that pollination is required for seed production, but the pollen donor does not sire the plant arising from the seed.

Quantitative trait locus (QTL): a neutral genetic marker used to identify a chromosomal region affecting a quantitative trait. Used for QTL mapping studies.

Ramet: the unit of clonal growth in plants, often corresponding to a shoot. Ramets may be connected, but semi-autonomous, or if severed from the parental plant become physiologically independent.

Reciprocal transplant experiment: a method generally used to detect local adaptation. Genotypes from a set of k source populations are planted in all k populations. Local adaptation is apparent when genotypes from the local population outperform genotypes from all other populations (home-site advantage).

Reinforcement: selection against inter-specific mating, decreasing hybridization.

Reproductive assurance: an increase in seed production caused by self-fertilization when conditions for outcrossing are unfavourable because of an absence of mates or pollinators. Requires plants to be self-compatible and usually capable of some form of *autonomous autogamy*.

Reproductive compensation: expenditure of extra effort on reproduction to counteract reduced fecundity caused by the loss of gametes or dependent offspring.

Reproductive investment: allocation of limited resources to reproduction, rather than growth, maintenance, and storage for future expenses.

Resource limitation: a constraint on seed production that occurs when an ovary contains more embryos than can mature into seeds given the available maternal resources.

Ruderal: a species adapted to open habitats maintained by human disturbance. Often distinguished from agrestals, which are weeds of agriculture.

Seed discounting: reduced production of outcrossed seeds caused by self-fertilization, either because self-fertilization pre-empts ovules, or because selfed seeds consume maternal resources that would otherwise have been used to produce outcrossed seeds. Because seed discounting involves resources and ovules, it can occur within flowers or between flowers produced at different times on the same plant.

Selection differential (S): a combined measure of total directional phenotypic selection, both direct and indirect. The selection differential can be estimated as the difference in the mean of the selected group and the mean of the entire population before selection. The selection differential can also be estimated as the covariance of fitness and a trait, or as the slope of the relation between relative fitness and standardized trait values.

Selection gradient (β): a measure of direct selection on each trait after removing indirect selection from all other traits that are in the analysis.

Self-incompatibility: the inability of a fertile hermaphroditic plant to set abundant seed following self-pollination. Involves diverse physiological mechanisms that typically operate pre-zygotically. The most common anti-selfing mechanism in flowering plants.

Selfing rate: the proportion of an individual hermaphrodite's offspring produced by self-fertilization: the complement of the outcrossing rate. Often used to refer to the female selfing rate, or the proportion of self-fertilized seeds. An individual's female and male selfing rates differ when the numbers of outcrossed seeds that it produces and sires on other plants are not equal.

Sex allocation: investment of limited reproductive resources in female versus male function by hermaphroditic individuals.

Sex inconstancy: the production of the opposite gamete type by female or male individuals in species with gender dimorphism. Male sex inconstancy is more prevalent than female sex inconstancy and is especially common in subdioecious and some dioecious populations.

Sexual system: the particular deployment of sexual structures (stamens and pistil) within and among plants in a population and the influence of this variation on mating patterns. Can also include whether plants are self-compatible, or if self-incompatible (SI), the type of SI system. Examples of sexual systems include *cosexuality, gynodioecy, monoecy, dioecy, heterostyly,* and *self-incompatibility*.

Subdioecy: a sexual polymorphism in which populations are composed of hermaphroditic, female and male individuals.

Sympatric speciation: speciation involving populations occurring in the same location, so that the opportunity for interbreeding is present.

Transgene: a gene from one species that has been inserted into the genome of another species, usually by molecular genetic techniques.

Transgression: a hybrid individual possessing more extreme characteristics than either of its parents.

Tristyly: the rarest form of *heterostyly*, in which there are three floral morphs.

Univoltine: a species that undergoes one generation per year; usually used in reference to animals.

INDEX

Note: page numbers in **bold** refer to Glossary entries, and those in *italics* refer to Figures and Tables.

abiotic influences
  flower size 251, 289
  inbreeding depression 213
  nectar production 128
  pollen limitation 297
  pollination 244
  selection on floral traits 121, 131–2, 296, 303, 341
    flower size 129
    petal colour 125–6
    phenology 130, 140, *143–4*, 148, 155
    sexual systems 209–15, 217
abiotic pollination 121, 279
  floral diversification 313, 316–17, 322
  see also wind pollination
Acer species, anthocyanins 125
Aconitum species, floral evolution 298
actinomorphic flowers 305, 313, 316–17, 319–21, *Plate 7*
adaptive radiation 305–6
additive genetic variance **346**
additive genetic variance-covariance matrix
  (**G** matrix) 260, 263, 264, **346**
Aeropetes tulbaghia, geographical mosaics 299
AFLP (amplified fragment length polymorphism) 267, **346**
  Lavandula latifolia 289
agrestal species 350
Allee effect 33, 35, **346**
  effect of congeners 113
allocation strategies 23–4, 27, 41–57, 190
  D. G. Lloyd's work 9, 11, *12–13*, 14
  genetic determination 42
  pollinator attraction 27, 31–2
  size-dependency 53–5, 57
  see also sex allocation
allopatric speciation 298, 303–5
allopolyploidy 296
Alsinidendron species, ecological influences *210*, 214
Alstroemeria aurea, effect of introduced ungulates 170–1
altitude, effects on pollinator faunas 299
androdioecy 204, **346**
  in metapopulations *228, 229*, 230
angiosperm diversification 311–12
  effects of floral traits 312–14

  phylogenetic studies 314–22
anisoplethy 240, **346**
antagonistic selection on floral traits 121, *122*, 123, *124*, 126–7, 131–4, 148, 268, 341
  flower shape 126–7
  relative strengths of agents 131–2
Antennaria species, apomixis 198
anther exsertion, wild radish *264*, 267, 268–73
anther orientation 6
anthesis, effect of bat pollinators 283
anthocyanins 121, 122, 123, 125–6
  Raphanus sativus 12–14, *Plate 2*
Anthonomous weevils 129
anthropogenic disturbance *see* human habitat alteration
anti-selfing mechanisms 15
ants, effect on *Polemium viscosum* floral shape 126–7
Apis mellifera 165, 167–8, 175
apomixis 183, 184, 196–8, **346**
approach herkogamy *see* herkogamy
Aquilegia species
  A. caerulea 300
  A. canadensis
    emasculation studies *188, 189*–90
    florivory 130
  A. formosa
    ethological isolation *330*, 331
    hummingbird pollination 320
    hybridization 329
Arabidopsis species
  A. Formosa x *A. pubescens Plate 7*
  A. thaliana
    genetic studies of flowering phenology 155
    QTL studies 273
  flowering phenology, correlation with sex allocation 147
Arenaria uniflora, self-fertilization 108, 194
artificial selection 126, 153, 265, *270*, 274, *Plate 5*
Arum maculatum, flowering phenology 142
Asclepias species
  A. syriaca, sex allocation 48
  ethological isolation *330*
asexual reproduction 183, 184, 196–8

INDEX

assortative mating 130, 142, *240–1*, 246, *248–9*, 302, 304–5, 340, 241, 305, **346**
Astilbe biternata, female frequency associations 207
Astragalus scaphoides, synchronous flowering 141
asymmetrical mating *240*
  in *Narcissus* species 241–2, 244, 253
    N. assoanus 246
    N. triandrus 248–9
asymmetry
  in hybridization 338
  plant-pollinator interactions *173*
  pollen and ovule fates 76–7
asynchronous flowering 141, 142, 155
  see also sequential flowering
autogamy 151, 166, *185–7*, *188–9*, 190, 215, 216, **346**
  autonomous 171, *186–8*, 192–3, 205, **346**, *347*, 350
    Aquilegia canadensis 189–90
    Camissonia cheiranthifolia 192
    Collinsia verna 187–8
  of defense 195
  see also competing self-fertilization; delayed self-fertilization; facilitated self-fertilization; prior self-fertilization; self-pollination
autonomous apomixis **346**
autonomous simultaneous self-pollination *see* self-pollination

Baker, H. G. 2
Baker's Law 225
Banksia serrata, competition for pollination 145
Baptisia species, ethological isolation 330
barriers to gene flow 295
Bateman, A. J. 85–8, 97, 98, 251
Bateman's principle 14, 15, 43, 77, 266
Batesian mimicry 145
bat pollination
  effect on time of anthesis 283
  Pachycereus pecten-aroriginum 300
bees 65, 85–6, 106–7, *112*, 113, *114*, 165, 167, 192, 194, 242, 244–5, 298–300, 304, 319–30, *330–2*, 342, 349
  species richness
    correlation with floral species richness 107
    effect on coffee fruit production 167
  see also Apis mellifera; bumble bees; Hesperapis; Lasioglossum; Osmia; pollen-collecting bees
β-diversity 165, **346**
Bidens species
  environmental influences 215
  female frequency associations 207
bilateral floral symmetry *see* zygomorphy
biogeography 113–15, 124–5, *191–6*, 247–51, 278–91, 298–305, Plate 1, Plate 4
biological species concept (BSC) 295–6, 301

bird
  pollination 145, 229
  seed dispersal 147, 225
  see also hummingbird
bivariate selection differentials 269, *270*
Blandfordia grandiflora, competition for pollination 145
Brassica species
  B. napus, pollen-mediated gene dispersal 90
  B. rapa
    artificial selection 265
    negative genetic correlation 266
    post-mating isolation 338
bumble bees 29, 90, *122*, 126–7, 131, 145–6, 149, 151–3, 261–2, 298, *330*, 332

Calathea ovandensis, selection on corolla length 268
Calyptrogtyne ghiesbreghtiana, antagonistic selection on floral traits *122*, 127
calyx length, antagonistic selection 131
Camissonia cheiranthifolia
  evidence for reproductive assurance hypothesis 192–5
  floral morphology variation and self-incompatibility 191–2
  reciprocal transplant experiments 199
Cape flora, pollinator-driven speciation 303, 306
Castilleja linariaefolia, antagonistic selection on floral traits *122*, 131
cattle introduction, effects on *Alstroemeria aurea* 170–1, 174
chalcone synthase deficiency, *Ipomoea purpurea* 125–6
Chamaelirium luteum, variability of male reproductive success 86
Chamerion angustifolium, nectar volume 128
Charlesworth, D. 218
Charlesworth, D. and Charlesworth, B. 16, 37, 208
Charnov, E. L. 14, 23, 37, 57, 76
chasmogamous flowers **346**
  modes of self-pollination 6–7
chemical agents
  effects on plant attributes *162*, 163
  effects on pollinator attributes *164*
Chionohebe species, female frequency associations 207
Cimifuga simplex 300
Cirsium palustris, antagonistic selection on floral traits *123*
Clarkia species 111–15, *122–3*, 125–6, 129, *188*, 194, Plate 1
Claytonia virginica, antagonistic selection on floral traits *123*
cleistogamous 192, **346**
climate change, effect on flowering phenology 130
cline **346**
  clinal variation, *Lavandula latifolia* flowers and pollinators 284–90
clonal reproduction 196–8, 215, 225–6 240, **346**
coevolution
  plants and parasites 196
  plants and pollinators 32

coffee fruit production, relationship to bee species richness 167
co-flowering species 81–2, 107, 145
  patterns of selection 110–11
  pollinator sharing 102–3, 115
  coincident selection 121, 123, *124*
Collinsia verna, emasculation studies 187–*8*
colonization 223–4, 236
  metapopulations *185*, 224, 236, **349**
    evolution of dominant versus recessive traits 229–30
    evolution of geitonogamy 231–6
    sexual system evolution 226–9
  single-event, effect on sexual systems 224–6
community context of pollination 32, 102–16, 120–35, 139–40, 142–4, 148–*51*, 155, 160, *162–7*, 171, 173–5
comparative studies 109, 110, 112, 121, *169*, 184, 190, *241*, 272–3, 279–82, 311–22
competing self-fertilization 33–6, 68, 73, 205, *216*, 219
competition
  between embryos 67–8, 70
  between pollen tubes 69–70, 168–9
  caused by herbivory 146
  and community flowering phenology 143, 144–5
  for pollinators 297
  interspecific 107–8, 115, 116
  variability, effect on phenological evolution 148
competitive ability, association with petal colour 126
complementary genes 329
congeners
  competitive interactions 298
  effects on *Clarkia xantiana* 112–13
constancy 105, 107–8, 115, 116, 329, 336, **346**
constraint *124*, 127, 130, 131, 141, 142, 247
  allometric 55–6
  ecological 208, 218
  functional 72, 143
  genetic 126, 128, 134, 155, 198, 208, 218, 260–73
  mate limitation 231
  physiological 140
  phylogenetic 140, 143, 148, 297
  see also ovule limitation; pollen limitation; resource limitation
consumption rate 105
convergent evolution 110–11, 121, 142, 145, 242
  see also pollination syndrome
corolla
  tube length 263–4, 268–73, 285, 287–90, 332, Plate 5
  width *122*, 126, 131, *191*–2, 329, 331–5, 340–1
correlated response to selection 263, 265–6, 289, **347**
cosexuality 10, 47, **347**
  temporal displacement of male and female function 51–3
cost of meiosis 6, 184, 190, 196, **347**

costs of reproductive strategies 9–10, 27, 32, 71, 127–8, 141, 196–8, 210–12, 214, 218
  life-history trade-offs 27, 54, 57, 186, 190, 210
  self-pollination 7, 187–90, 192, 196–7, 199, 231, 234, 297
  see also cost of meiosis; inbreeding depression; ovule discounting; pollen discounting; seed discounting
Cotula species
  D. G. Lloyd's work 8–9, 225–6
  sexual systems 225–*6*
crop plants, apomixis 196
cross-fertilization 7, 66, *67*, 145
  see also competition, between pollen tubes; outcrossing, self-incompatibility
cross-pollination 6, 44, 62–5, 92, 166, 188, 219, 245
  disturbance effects 167–8
  evolution 71–3, 96–7
  role of floral traits 16–17, 219, 240, 245–6, 305
  see also pollen discounting; pollen dispersal; pollen export; pollen limitation
Cucurbita foetidissima, female frequency *208*
Cynoglossum officinale, competition for pollination 145

Dactylorhiza maculata, sex allocation 48
Dalechampia species 121, 125, 300
Daphne laureola 207,210
Darwin, C. R.
  angiosperm origin 312
  floral traits 14–15, 126, 181, 278
  heterostyly 15–17, 239
  influence on D. G. Lloyd 5–6
  orchids 75
  outcrossing versus selfing 6–7, 14–15, 181, 183–4
  pollen dispersal 83
  sexual systems 205, 209–*10*
  speciation 295–6, 298, 304
Datura stramonium, antagonistic selection on nectar volume *122*, 128
delayed self-fertilization 7, 34, *63–4*, 66, *67*, 186, 231
  Camissonia cheiranthifolia 192
  disadvantage of 76
  ecological consequences 35
  evolution of gynodioecy 205
  influence of variable pollination 36
  selection conditions *72*, 76
  see also self-pollination
Delphinium species
  D. nelsonii, flower colour 262
    pollinator preferences 261
  floral evolution 298
demes 227–30, 232, 235, **347**
density-dependence *see* population density
Dianthus sylvestris, seed predation 217
dichogamy 15, 72, 104, 147, 280, **347**

dichogamy (*cont.*)
   see also heterodichogamy; protandry; protogyny
dicliny 8, 9, 15, 129, **347**
   see also dioecy; gynodioecy; monoecy; subdioecy
diffusion-advection model, Morris 88–9, 97
Digitalis purpurea, antagonistic selection on floral traits *123*
dioecy 9–10, 182, 204, 225, 265, **347**
   association with herbivory 129, 131
   association with polyploidy 9
   effect on diversification 47, 312, 314, 317, *318*, 319, 321
   evolution 45–6, 50, 204–19, 225–6, 297
     importance of harsh environment 209–15
     influence of plant enemies 131, 215, *216*, 217
     role of pollinators 219
   flowering phenology 147
   in metapopulations 226–9
   see also dicliny; sex ratio
Diptera (flies)
   geographical pollinator mosaic 298, *299*, 300, 302
   pollination of *Lavandula latifolia* 285, *286*
directional selection *114*, 141, 146, 263, 266–7, 287, 289, 351
Disa species
   D. cooperi pollen fates 65
   D. draconis 300, Plate 6
   D. Scullyi Plate 6
   D. uniflora, geographical pollinator mosaics 299
disassortative mating 240–1, 304, 346
disjunctions, Southern African orchids 301, *302*
Dispersis species, floral evolution 298
disruptive selection 141, 147, 204, 246, 251, 263, 266, 287, 340
distyly *16*, 204, 230, 239, *240*, *243*, 245, **347**
   evolution *16*, 230, *243*
     from stigma-height dimorphism *16*, 242–3, 246
   mating patterns 17, *240*
   Narcissus species 241–3, 246
Dobzhansky-Muller incompatibilities 328–9
dominant alleles
   effect on rate of evolution 227, 229–30, 232, 235–6
   and flower colour, *Raphanus sativus* 132, Plate 2
   in stylar polymorphisms 245
Dromiciops australis, fruit dispersal 131
drought
   influence on fecundity *123*
   influence on flower size or shape *122*, 127, 131
   influence on nectar production 128
   influence on pollen quality 163
   influence on sex ratio *210*
   tolerance and flower colour 120, *121*, *123*, 125–6

Ecballium elaterium, ecological influences *210*
Echinocereus coccineus 300
Echium plantagineum, antagonistic selection on floral traits *123*

Echium vulgaris
   competition for pollination 145
   plant size, relationship to seed production 212, *213*
ecological genetics 260–74
   correlated traits, evolution in wild radish 268–73
   floral colour polymorphisms 261–2
   future directions 273–4
   metapopulation effects 227–36
   natural selection on floral traits 266–8
   quantitative traits 263–6
   see also population genetics; quantitative genetics
ecological speciation 296
Eichhornia paniculata
   flower number-size trade-offs 265
   tristyly 230, 236
emasculation **347**
   emasculation studies 187, *188*, *189*, 193, 198
embryo
   abortion 36, 67–72
   apomixis 197–8, 346, 350
   competition 67–73, 75–6
   fates 63, 66–70, 351
enantiostyly 231, **347**
environmental influences
   isolation of genetic responses 270
   key environments 306
   on evolution of dioecy 47, 209–*16*
   on flowering phenology 130, 140–4, 149, 152–4
   on fitness 30, 36, 196, 263
   on floral traits 127–8, 269–70, 283, 289
   on inbreeding depression 37, 196, 213
   on mating 8, 35, 47, 61, 129, 139, 193, 198, 214–15
   on mortality 33, 56, 126, 154
   on plant size 53–4
   on pollinators 32. 214–16
   on resource allocation 41–2, 56
   on sex allocation 46–7, 52–3, 213
   see also altitude; cline; competition; drought; florivores; habitat fragmentation; heat; herbivory; human habitat alteration; local adaptation; light intensity; magnet-species effect; mate limitation; mating environment; nectar robbers; plant communities; plant-pollinator interactions; pollination ecotype; pollination environment; resource availability; snowmelt gradients
ephemeral habitats
   association with apomixis 198
   association with self-fertilization 193
epistasis 223, 273, 328–9, **347**
Eriocaulaceae, pollination modes 321–2
Eritrichum aretioides, ecological influences *210*
Erysimum mediohispanicum, antagonistic selection on floral traits *122*, 127, 131

Erythronium grandiflorum, plant-pollinator interactions 29
Eschscholzia californica Plate 7
ethological isolation *see* reproductive isolation
euglossine bees
　aggregative responses 106
　population size fluctuations 106
evolution *see* floral diversification; genetic drift; mutation; natural selection
evolutionarily stable strategy (ESS) 12, 23, 25, 26–7, 28, 37, 41–2, **347**
　mixed mating 74
　self-fertilization 198
　sex allocation 44–6, 56
Evolutionary Synthesis 295
exclusion methods, genetic analysis 29
exotic plant introductions
　effects on plant attributes *162*, 163
　effects on pollinator attributes *164*
　see also invasive species
experimental array studies, pollinator preferences 329–30
exploitative competition 108, 116, 145
extinction
　population
　　Allee effect 33, 36, 113
　　facilitation and demographic rescue 108
　　metapopulation context 224, *227*, 230, 232, *233*, *234*, 235
　　pollinators 173, *174*
　　plants *174*
　　rates *227*, 232, *233*, *234*, *235*
　　relation to abundance 175
　　self-pollination 234
　　thresholds *33*, 35–6
　species
　　dioecy 321
　　effects of floral traits 312–14, 319–21
　　hybridization 328
　　rates 313–14, 321
　　self-pollination 322
　　woody growth 321

facilitated self-pollination *see* self-pollination
facilitation, interspecific 82, 108, 110–16, *144*–5, 148, 155, 167
fellfields, flowering phenology 149, *150*
female fertility 26, 28–9, 34, 43, 68, 198, 204–8, 213, 217, 229, 245–51, 261–2
　limitations 68–71, 77, 266, 340
　measurement 28–9
　see also fruit production; seed production
female fitness 27, 42–8, 131, 261, 266
　correlation with pollen receipt 17
　gain curve 43–57, *211*
　marginal fitness 26–7, 42, 45–7, 49, 52–5, 211

　role in selection 266–8, 287, 297
　selection gradient 27–8
female frequency
　dioecy *see* sex ratio
　gynodioecy 205–17, 228–30
　subdioecy 218
fig species, pollination by wasps 77
filament-corolla tube correlation, wild radish 268–73
fire
　effects on plant attributes *162*
　effects on pollinator attributes *164*
first flowering, time of 140
Fisher, R. A. 42
fitness *12–13*, 26–7, *30*, 42–3, *45–7*, *49*, 52, 55, 68, 72, 76, *123*, 131–2, 134, 146, 153, 249, 297, 305
　arithmetic mean 36
　costs of apomixis 196–7
　frequency dependence 23, 26
　gain curve 43, 56
　hybrids 29, 304, 327–8, 340–1
　influence of floral traits 12–13, 26, *30*, 43, 49, 50, 71, 107, 114, *124*–8, 194, 258, 260, *264*, 282
　marginal fitness *12–14*, 26, *44*, 46, 52, 54, **349**
　measurement 28–9, 56, 266
　selection gradient 27, 28, *288*
　self-fertilizing plants 6, 34–5, *122*, 132, 184–7, 195
　surface or landscape 28, 92
　trade-offs 31, 88, 135, 184, 268
　see also female fitness; hermaphrodite fitness; male fitness
floral design 121, *144*, 241–2, 248–9, 253, **347**
　see also anther; dichogamy; enantiostyly; floral morphology; flower colour; flower shape; flower size; herkogamy; heterostyly; nectar; stigma
floral display 88, 91, 140, **347**
　geitonogamy 8, 65–6, 74, 88, 92, 96, 141, 206, 215
　herbivory 131, 163
　influences on 66, 140, *162*, *164*, *214*–15
　pollen dispersal 85, 91–2, 96, 331, 334
　pollinator attraction 32, 91, 98, 140, 164, 166
　see also flower number; inflorescence traits; pollen discounting
floral diversification 257–8, 260, 279, 285–9, 311–12, 327
　allopatric divergence 279, 285–91, 301–3
　influence of pollinators 282–3, 289–91, 295, 303–6, 313, 317
　see also adaptive radiation; cline; local adaptation
floral isolation **347**
　see also ethological isolation; mechanical isolation; reproductive isolation
floral longevity 66, 140–1, 219
floral morphology 11, 127–8, 162, 166, 175, 191–2, 195, 214, 300, 327–8

floral morphology (*cont.*)
    hybrids 332–3
    interaction with pollinators 32, 104, 283, 313, 327
    Narcissus species 241–3, 246, 253
    selection 286–9
    see also anther; dichogamy; enantiostyly; floral design; flower shape; flower size; herkogamy; heterostyly; stigma
floral number 51, 55, 217
    geitonogamy 65, 166, 216
    pollen dispersal 43–4, 96, 166, 216, 331, 334
    pollinator behaviour 104
    selection *122*
    trade-off with flower size 265
    see also floral display; inflorescence size; pollen discounting
floral physiology
    disturbance effects *162*, *166*, 175
    nectar 32
    phenotypic cues 144
    see also self-incompatibility
floral traits
    adaptations to self-pollination 6
    antagonistic selection 122–3, *124*
    effects on diversification 312–14
        phylogenetic studies 314–22
    effects on pollination success 166
    effect of pollinator diversity 111
    genetic variance 257
    influence on hybrid fitness 340–1
    minimization of heterospecific pollen transfer 108
    natural selection 266–8, 297
    role in reproductive isolation 327–8
    selection
        role of flowering phenology 140
        role of male fitness 266–8, 273
floral visitation rate, influencing factors 107
florivores **347**
    effect on *Camissonia cheirathifolia* flower size *195*–6
    influence on floral evolution 127, 341
    influence on flowering phenology 130, 146
flower colour 125–6, 305
    antagonistic selection *122*, 132–4, 262, 341
    association with heat of drought tolerance 125–6
    herbivory 126–7, 134
    hybrids 329, 332
    mimicry 111, 145
    pollinator attraction 125, 132, 261, 321, *330*–1, 336, 342
    polymorphism 123, 125–6, 132–4, 261–2, 266–7
    Raphanus sativus 132–4
    see also anthocyanins
flowering phenology 82, 130–1, 139–56
    alpine plants 149–53
    Camissonia cheiranthifolia 193
    community phenology 142–3
    correlation with sex allocation 147
    effects on pollination success and seed quality 149, 166
    future research 153–5
    individual phenology 140–1
    influencing factors
        abiotic influences 130, 143–4, 162
        habitat disturbance *162*
        herbivory 130–1, 146, 163
        pollination conditions 144–6
        seed dispersal and germination 146–7
        sexual systems 147
    initiation cues 144
    limits on evolution 148
    non-pollinator selection 130–1
    phenological isolation 153, *154*
    population phenology 141–2
flowering rate 140–1
flower orientation, *Aquilegia* species *330*, 331
flower patch size
    and pollen-mediated gene dispersal 85, 91, 97
    pollinator preference 106
flower production rate 52
flower shape 108, 121–2, 126–7, 131, 163
flower size
    abiotic influences 127–8, 162, *214*–15
    association with pollinator fauna 251, 283
    correlation with flower number 265
    correlation with pollen production 269
    disturbance effects 162–3, 175
    effect on self-pollination 129, 183, 215
    environmental influences 269
    geographical variation *191*–4, 251, *288*–90
    heritability 264–5, 289
    herbivory 127, 163, *195*–6
    herkogamy 114–*15*, 195
    hybridization 329, 332
    macroevolution 242, 272
    non-pollinator selection 127–8
    phenology 147, 193
    pollinator preference 127
    polymorphism 263
    selection 128, 267, *269*, 288, 290, 341
    self-pollination 129, 192
    variation in *Camissonia cheiranthifolia* 191, 192–3, *195*–6
    variation in *Narcissus* 241–2, 251
    wild radish (*Raphanus raphanistrum*) 267
fluctuating asymmetry 163, **347**
Fragaria virginiana
    antagonistic selection on floral traits *122*, 127
    ecological influences *210*
    effects of herbivory 217, 218
    female frequency 207, *208*–9, 213
    plasticity in fruit set *212*
frequency dependence 23, 230

fitness 26
mating 23, 25–6, 141, 245
selection
    flower-colour polymorphism 261–2
    gynodioecy 205–7, *210*, *216*, 227
    heterostyly 182, 240, 248–50
fruit characteristics and dioecy 225, 321, 347
fruit dispersal 43, 131, 147
fruit herbivory (predation) 121–3, 126–7, 131, 146, 217, 341
fruit production (set) 141, *195*, 211–3, 261
    relation to pollination 89–91, 149, 167–8, *170*, 188, *262*, 284
    reproductive assurance *191*–3
    timing 51–3, 130–1, 142–7, *151*
    see also female fertility; seed production
functional gender 11, *49*, 226, 340, **347**
functional responses 105, 107, *109*–11, 113, 115, **347**
fungal effects on reproduction 132, 196, 217

gain curves 43–4, *45*, 46, 56
    budget effects of plant size 48–50
    direct effects of plant size 50–1
    influencing factors 216
    relationship to sexual system evolution 46–7
    see also female fitness; fitness; male fitness
gametic phase disequilibrium 263, 273, **348**
geitonogamy 7, 8, 15, 64, 65–6, 186, **348**
    effects of floral display 8, 66, 88, 166, 215
    and evolution of gynodioecy 215–16
    influence of individual flowering phenology 66, 141
    and mating-system evolution 74–6, *185*–6, 231–6
    pollen discounting 7, 74, 88, 215, 231
    see also self-pollination
gender 10–11, 340, **348**
    D. G. Lloyd's contributions 8–11, 225–6
    dimorphism *see* dioecy
    diphasy 49–51, **348**
    functional 11, *49*, **347**
    phenotypic *9*, 11, 47, *49*
    selection of 47–51, *122*, 214
    see also cosexuality; gynodioecy; hermaphrodite; sex allocation
gene flow
    intraspecific 85, 90, 95, 141–2, 152–3, 155, 227, 230, 289, 295, 301–3
        see also cline; metapopulation; reproductive isolation
    interspecific 328, 331, 341–2
        see also hybridization
generalist plant-pollinator interactions 284, 200, *330*
    resilience to disturbance 171–4
    selection 30–*1*, 37
generalist pollinators 106, 112, 143, 148, 155, 164, 167, *169*, 173, 225
genet **348**

genetically modified (GM) crops
    Brassica napus, pollinator-mediated gene dispersal 90
    confinement 85, 88, 89, 97
genetic bottleneck 223, 236
genetic constraint on selection 28, 120, 218, 260, 263, 265–6, 268–73
genetic correlation or covariance 28, 126, 194, 211–12, 218, 260, 263–73, 341, 347, **348**
genetic drift 224, 230–1, 283, 289, 298, 348
genetic load 36–7, 67, 70–1, 184, 232, **348**
    see also inbreeding depression
genetic markers 29, 56, 85–6, 90, 93–4, 188–9, 267, 340, 342
    see also AFLP; microsatellites; QTL
genetic transmission advantage of selfing 34–5, 56, 68, 71, 184
genetic variation 127–8, 167–8, 189, 196, 213, 218, 223, 228, 262–6, 269, 272, 274, 313, 326, 346, 348
    see also additive genetic variance; **G** matrix; heritability
Gentiana nipponica
    flowering phenology 150
    phenological isolation 153, *154*
geographical variation
    floral morphology 190–2, 195, 247, 251, 279–83, *288*, 300–1
    genetic 153–4, *228*, 248, Plate 4
    herbivory 195
    mating system 113, 184, 190, 192, 198, 227
    pollinators 113, 192, 251, *280*–3, *286*, 290, 333
    pollinator mosaic 297–300, 303, 305–6
    reproductive isolation 335–7
    selection 278–91
    see also allopatric speciation; biogeography; cline; local adaptation; pollination ecotypes
geophytes 50, 145, 244, **348**
Geranium species
    female frequency associations *207*
    G. sylvaticum
        antagonistic selection on gender *122*, 129, 131
        ecological influences *210*
Gilia leptantha 300
Gingidia species, female frequency associations *207*
Glechoma hederecea, female frequency associations *207*
glucosinolate 126, 133–4, **348**
G matrix 28, 260, 263–5, **348**
Grant, V. 2, 303–4, 327
grapefruit plantations, pollinator species richness *165*
gynodioecy 28, 182, 204, 218, **348**
    association with herbivory 127, 129, 131
    D. G. Lloyd's work 9, 10
    flowering phenology 147
    in metapopulations *228*, 230
    sex-differential plasticity 213
gynodioecy pathway 9, 204–9, *216*
    see also dicliny

habitat fragmentation
    effects on plants 162–3, 166, 168–9, 172–3
    effects on pollinators 106, 164, 173
Haldane's sieve 230–1
Halenia, nectar spurs 320
Hamilton, W. D., influence on D. G. Lloyd 17
hawk moth pollination *Plate 6*, *Plate 7*
    Ipomopsis species 332, 335–6
heat, influence on flower colour 121–3, 125–6, 134–5
    see also drought
Hebe species
    ecological influences *210*, 214
    female frequency associations *207*
Helianthus species
    H. annuus, florivory 130, 146
    post-mating isolation 338, 340
Helleborus foetidus, geographical variation 283, 300
herbivory
    influence on population dynamics 33
    influence on flowering phenology 130, 141, *143*, 146
    influence on nectar rewards 128–9
    influence on plant attributes 133, *162*–3, 170, 194–5, 327, 341
    influence on pollinator attributes 48, 162–4,
        Raphanus sativus 133–4
    influence on sex allocation 48, 208–9
    influence on sexual systems 55, 129, 194–5, 208–10, 215–17
    role in floral evolution 121–3, 126, 127, 131, 133–5, 341
    variation 52, 129, 148
heritability 37, 128, 263, **348**
    methods of estimation 264–5
herkogamy 15, 104, 113–15, 187, 189–92, 195, 244–6, **348**, *Plate 3*
    approach herkogamy 16, 242, *243*, 245, **346**
    Aquilegia canadensis 189–90
    Camissonia cheiranthifolia 192, 195
    Clarkia xantiana 113, *114*, *115*
    Narcissus 242, *243*–6
    reciprocal herkogamy 16, 240, 245–6, 349
hermaphrodite 25, 42, 181, 204
    D. G. Lloyd's terminology 10
    evolution 31, 45–8, 50, 56
    fertility 26, 42–3
    fitness 46–7
        in gynodioecious populations 10, 129, 131, 205–6, 208–13, 215–17
        in subdioecious populations 218
    flowering phenology and sex allocation 147
    herbivory 122, 129, 131, 208, 217
    resource limitation of seed production 209–13
    selection in metapopulations 227, 228, 229
    temporal displacement of male and female functions 51–3, 147

    see also androdioecy; cosexuality; gender; gynodioecy; self-fertilization; self-pollination; sex allocation; subdioecy
Hesperapis regularis, pollination of *Clarkia* species 113
heterodichogamy 147, 204
heteromorphic incompatibility *see* self-incompatibility
heterospecific pollination 107–8, 113, 116, 145, *152*, 166–7, *170*–1, 194, 327, 331–3, 337, 341–2
heterostyly 182, 253, **348**
    D. G. Lloyd's work 15–17
    evolution 15–17, *243*
    in *Narcissus* species 241, 246–52
    non-random mating 239–53
    see also distyly; heteromorphic incompatibility; tristyly
Heuchera grossularifolia, phenotypic selection gradient analysis 28
Holcus lanatus, antagonistic selection on floral traits *123*
homomorphic incompatibility *see* self-incompatibility
honey bees *see Apis mellifera*
human habitat alteration 81, 82, 159–76
    effects on plant attributes 162–4
    effects on pollination interaction networks 172–4
    effects on pollinator attributes 164–5
    general effects and reproductive consequences 161–2
    introduced cattle, effects on *Alstroemeria aurea* 170–1
    modulators of plant reproductive response 171–2
    relation of plant reproduction to modified pollination 168–9
    relation of pollination to modified plant attributes 165–7
    relation of pollination to modified pollinator attributes 167–8
hummingbird 298
    behaviour 106, 131, 146, 216, 298, 320, *330*–1, 333, 335, 340, 341
    pollination 332, *334*, 337, *339*, 341
hummingbird-pollinated plants 131, 142, 146, 148, 300, 306
    Aquilegia formosa 320
    Ipomopsis species *330*–41
hummingbird pollination syndrome 121, 129, 262, 298, 329, 331
Hybanthus prunifolius, synchronous flowering 141
hybridization 258, 304, 326, *Plate 8*
    hybrid zone 304, 326–42, **349**
    role in angiosperm evolution 297, 304, 327
    see also heterospecific pollination; reproductive isolation
Hylemya species, seed predation 341
Hyles lineata Plate 7
Hypericum perforatum 155

Iberis species *Plate 5*
ideal free distribution **349**

Impatiens pallida, flowering phenology 142
inbreeding 171
　biparental 96, 232
　see also self-fertilization
inbreeding depression 36–7, 96, 145, 166, 169, 184, 196, 349
　Aquilegia canadensis 189, 190
　Camissonia cheiranthifolia 195
　Collinsia verna 188, 190
　environmental influences 37, 196, 209, 213–14, *216–17*
　evolution of 184–5, 297
　　purging 186, 190, 197
　　relation to mutation rate 36, 184
　evolutionary influence 14, 88
　　apomixis 196
　　dioecy 47, 215, 227
　　geitonogamy 232, 233, 234, 236
　　gynodioecy 205–6, 207–8
　　mating systems 26, 34–6, 73, 75–6, 78, 88, 184, 187–8, 195–6, 205, 231, *233–4*
　　self-incompatibility 16
　Narcissus 244
　post-dispersal (late acting) 68, 73, 76, 78
　pre-dispersal (early acting) 67, 75
　see also genetic load; reproductive compensation
inflorescence traits 147–8, 265, 311, 347
　effect on herbivory 127
　effect on plant mating 75, 190, 231
　effect on pollinator behaviour 32, 104, 122, 127, 145, 319–20
　see also floral display; geitonogamy
intra-sexual selection 11, 14
introgression 304, 326, **349**
invasive species 155, 161, 223
　effects on native species 155, 175, 328
Ipomoea species
　I. purpurea
　　anthocyanins 125–6
　　pollinator preferences 261
　　selection on flower colour *122, 123*, 135, 341
　　molecular studies of floral pigments 262
Ipomopsis species
　hybrid zones 326, 340–1, 342, Plate 8
　　ethological isolation *330*, 331
　　pollen carryover 332
　　pollen competition 338–9
　　post-mating isolation 337
　　relative importance of ethological and mechanical isolation 332–7
　I. aggregata
　　antagonistic selection 121, *123*, 131
　　genetic correlation 263
　　seed predation 146
　　selection on corolla length 268

Iris species, hybrid zones 327, 328, 340
　ethological isolation *330*
　post-mating isolation 337
island colonization 224
　effects on sexual systems 224–5
Isomeris arborea, floral herbivory 127
isoplethy 240, **349**
　Narcissus assoanus 245

Jepsonia heteranda, pollen load studies 17

key innovations 304–6, 320–1
　identification 314–16

Lapeirousia species
　L. pyramidalis 300
　pollinator shifts 302, 304
　speciation 303
Lasioglossum pulliabre, pollination of *Clarkia* species 113
late-acting self-incompatibility *see* self-incompatibility
Lavandula latifolia 278
　clinal variation of flowers and pollinators 284–90
　distribution *284*
　floral morphology *285*
Leavenworthia species, D. G. Lloyd's Ph.D. thesis 4, 5–6, 8
Lepidoptera
　herbivores 141
　pollination 246, 284
　pollinators 242, 284–7
Lianthus parryae
　antagonistic selection on floral traits *123*
　petal colour 126
life-history 139, 224–6
　trade-off 27, 29, 56–7
　theory 24, 56
light availability 46, 261, 266, 336
　effects of habitat disturbance 162, 163, 164
　flowering phenology 140, 143–4
Lignocarpa species
　female frequency associations *207*
　L. carnosula, ecological influences *210*
Limnanthes douglasii, female frequency *208*
Linanthus androsaceus artificial selection 265
Linanthus parryae, selection on flower colour 261–2
Linaria triomithophora Plate 7
Lloyd, D. G. 1–2, *3*, 17–18, 37
　allocation strategies 11–14, 23, 46, 208, 340
　biographical sketch 2, 4–5
　Cotula species studies 225–6
　ecological perspective on mating systems 8
　education 2, 4
　floral mechanisms 14–17
　gender strategies 8–11
　mating systems 5–8, 77–8, 183, 190, 196, 233

Lloyd, D. G. (*cont.*)
  pollen and ovule fates 62, 65
  seed discounting 68
  self-fertilization 34–5, 36, 76, *185*, 186, 205
  local adaptation 81, 83, 85, 141–2, 145, 152, 155, 199, 218, 226, 230, 248, 279, 283, 290, 296, 299, 301–3, 326, 349, 350
    see also cline; geographical variation; pollination ecotype
  local mate competition 76, **349**
Lupinus species *Plate 7*
Lythrum salicaria, flowering phenology 142

MacArthur, R. H. 4
macroevolutionary patterns 257, 278, 279
  Brassicaceae flowers 272–3
Macromeria viridiflora 300
magnet-species effect *see* facilitation, interspecific
male fertility 11, 14, 26, 34, 42–3, 45, 68, 72, 85–7, 89, 91, 93–8, 166, 168, 261, 339, 340
  limitations 69, 77, 166, 180
  measurement 29, 56, 266–7, 340
  see also Bateman's principle; male fitness
male (paternal) fitness 15, 27, 29, 43, 47, 49, 184, 197, 212, 217, 261–2, *264*, 266
  gain curve 44, 46–8, 50–*1*, 56, 210, 211, *216*, 218
  marginal fitness 45, 48, 52–3, *55*, 211
  relation to pollen dispersal 44, 48, 50, 184, 266, 313, 340
  role in floral evolution 11, 266, 267, 273
  selection gradient 28, *264*, 267
male sterility 204–5, 210, 217, 219, 228, 230
  see also dioecy; gynodioecy
Mangelsdorf, P. 4
Marcgraviaceae, nectar spurs 317
marginal fitness **349**
  see also fitness, marginal
mate diversity 149, *151*
mate limitation 231, 239, 314
  selection of flowering synchrony 142
mating 67, 141, 239–41, **349**
  see also assortative mating; disassortative mating
mating asymmetry
  intermorph 240–53
  intersexual 76–7
    see also Bateman's principle
mating environment 139, 142, 147, 152, 163, 236, 241
  see also pollination environment
mating system **349**
  diversity 4, 5, 113–15, 187–96, 206, 227, 241, 244, 253
  ecological influences 6–8, 35, 69, 70, 113, 129, 144, 151, 185, 188–96, 231–6
  evolution 8, 34, 62, 68–78, 113, 151, 184–96, 199, 231–6
  genetic influences 66–75, 77, 184, 186
    see also inbreeding depression

    see also apomixis; assortative mating; disassortative mating; mixed mating; outcrossing; selfing
Matthiola species *Plate 5*
Maynard Smith, J. 23
Mayr, E. 296
mechanical isolation *see* reproductive isolation
Mercurialis annua, metapopulations 227, *228*
  metapopulations 152–3, 186, 224, *229*, 236, **349**
  evolution of dominant versus recessive traits 229–30
  evolution of geitonogamy 231–6
  sexual system evolution 226–9
microarrays 274, **349**
microsatellites 267, 340, **349**
"mid-domain" effect, flowering phenology 142–3
migration 182, 224, 226–7, *229*, *235*, 301
  see also gene flow
migratory birds, seed dispersal 147
mimicry 111, 129, 145–6
Mimulus species
  ethological isolation *330*, *331*, 342
  M. guttatus, heritability of floral size and number 264–5
  nectar rewards 129
  pollinator shift 121, 298
  studies of speciation 296
  transgression 328–9
Minuartia obtusiloba, ecological influences 210
mistletoe, flowering phenology 131, 146–7
mixed mating
  evolution 66, 74, 185
  in metapopulations 232–6
  role of pollen discounting 75–7
  role of reproductive assurance 187–90
  prevalence 77, 186, 214, 244
Moegistorynchus longirostris Plate 6
monoecy 9, 48, 142, 204, 212, 225–8, 231, 347, **349**
  see also dicliny
morph **349**
  flower colour 123–6, 132–4, 261–2, 267, *Plate 2*
  flower morphology 16, 132, 142, 182, 230, 239–53
  ratios 125, 132, 133, 230, 240–53, *Plate 4*
  see also dioecy; frequency dependence; gynodioecy; heterostyly; polymorphism
Müller, H. 6
Müllerian mimicry 145–6
multi-agent selection 121, 123–4, 135, 217
  on flowering phenology 140, 146
  Raphanus sativus 132–4
  relative strengths of agents 131–4
multivariate selection 27–9, 263, 268–73
multivoltine species **349**
mutation 34, 36–7, 184, 262
  see also genetic load; Haldane's sieve

Narcissus species
  asymmetrical mating 241–2, 253
  daffodil floral design 242
  floral diversity 239, 252
  floral morphology and sexual system diversity 242–3
  mating in dimorphic populations 244–6
  mating in monomorphic populations 243–4
  mating in trimorphic populations 246–52
  N. albimarginatus 246
  N. assoanus 241, 244–6, *247*, Plate 3
  N. longispathus 241, 243–4, 253, Plate 3
  N. triandrus 247, 253, Plate 3
    evolution of morph ratios 247–51, Plate 4
    paperwhite floral design 242
    variation and evolution of sexual organs 251–2
  reproductive attributes *241*
  stylar polymorphisms 182, *241*, 242, 247–8, 251–2
  "triandrus" floral design 242
Narthecium asiaticum, sex allocation 48
natural selection 23, 25, 27–8, 85, 223–4, 230, 236, 240, 251–2, 260, 263, 274, 289
  floral traits 42, 81, *103*, 108, 110–11, 114, 116, 121–31, 135, 141–2, 146, 148, 261, 304, 306, 313–14
  geographical variation 47, 148, 153, 251, 302–3
  in hybrid zones 327–9, 340–1
  mating system 113, 184–6, 194–7, 231–6, 244
  sexual systems 129–30, 204, 211–12, 218, 225, 227–9,
  see also adaptive radiation; directional selection; disruptive selection; evolutionarily stable strategy; fitness; frequency dependence, selection; genetic constraint on selection; local adaptation; phenotypic selection; selection differential; selection gradient; stabilizing selection
nectar
  effect on pollination 94, 166
  pollinator attraction 43, 104, 106–7, 122, 127–30, 167, 242, 299, 342
  pollinator energetics 32, 107
  production 32, 127–9, 151–2, 162–3, 195, 263, 280, 283, 329–33, 347
nectar guides 104, 108, 261
nectar robbers 121–3, 126, 128, 131, 341
nectar spurs
  as key innovation 305
  role in floral diversification 313, 317, *318*–19, 320, 321
Nemophila menziesii
  ecological influences *210*, 213
  floral herbivory 127
Nicotiana species
  effects of herbivory 128–9
  ethological isolation *330*, 331
  hybrid zones 340

non-adaptive evolution 268
non-pollinator selection 120–4, 134–5
  on flowering phenology 130–1
  on flower shape 126–7
  on floral size and display size 127–8
  on nectar 128–9
  on petal colour 125–6
  in *Raphanus sativus* 132–4
  relative strength 131–2
  on sexual systems 129–30
non-random mating 37, 239–40, 348
  see also assortative mating
numerical responses 105–7, 110, 112–13, **349**

Ochradenus baccatus
  ecological influences *210*
  female frequency *207*, 208
  flowering phenology 147
oligolectic bees 105, **349**
opium poppy, petal colour 126
orchids
  adaptive radiation 305
  Batesian mimicry 145
  pollen fates 65
  pollinator shifts 297
  reproductive isolation 304
  rewardless pollination 32
  sexual deception 303
Osmia lignaria, preferences 106–7
outcrossing 62–72, 190, 253
  advantages 14, 35, 66, 76, 88, 186
  characteristics 114–15, 152, 169, 181, 184
  disadvantages 76, 108, 129, 184, 187, 190, 192–4, 196, 205, 233, 235
  disturbance effects 167
  evolution 34, 36, 72, 74–6, 186, 196, 225, 231
  female outcrossing rate 92, 188, 214–15, 244, **349**
  floral traits 6, 9, 14–16, 43, 76–7, 92, 141, 147, 187, 244, 253
  Lloyd, D. G., contributions 5–9, 14
  requirements 23, 25, 81, 186
  see also cross-fertilization; cross-pollination; inbreeding, biparental; mating environment; mating system; mixed mating; ovule discounting; pollen discounting; seed discounting; self-incompatibility
ovule 62, 197, 346, **350**
  fates 61–3, 66–8, 71–2, 76–8, 248–9, 350
  production 11–13, 33, 42–3, 48, 54, *70*–1, 74, 76–7, 132, 205, 218, 263, 265–6
  see also cross-fertilization; seed production; self-fertilization
ovule discounting 68, 245–6, 249, **349–50**

ovule limitation
    of mating opportunities 44, 66, 168, *234–5*, 297
    of seed production 68–77, **350**

Pachycereus pecten-arboriginum, bat pollination 300
Pachycereus pringlei
    ecological influences *210*
    female frequency associations *207*
Panax trifolium, flowering phenology 147
parapatric 112, 297–8, **350**
    speciation 301–3, 340
paraphyly 302, **350**
paternity shadow 83
    estimation of 93–5
    evolutionary biology 95–7
Pedicularis species, floral evolution 298
Penny, D. 2
Penstemon species, pollen carryover 332
perenniality, consequences 37, 76, *185*, 197, 268
perennials
    association with dioecy 225
    measurement of selection 268
    resource allocation 42
    sex allocation 53–6
    size-dependence of reproductive effort 55–6
petal colour *see* flower colour
Peucedanum multivittatum
    flowering phenology *150*
    geographical isolation 153, *154*
Phacelia linearis, sex-differential plasticity 213
phenotypic gender *9*, 11, 47, *49*
phenotypic plasticity
    costs 211–12, 214
    flowering phenology 153
    and geographical variation 127, 218, 283, 289, 301, 346
    sex allocation 55, 210–14, *216*, 218
    see also cline; plant size
phenotypic selection 23, 26–8, 34, 37, **350**
    floral traits 131–4, 145–7, 261, 264, 266–8, 273, 279–91
    generalist versus specialist pollination 30–1, 92
    mating system 35–7, 71–6
    measurement 28–9, 263, 266–8, 286–9
    pollen dispersal 95–7
    sex allocation 44–56
    sexual system 46–7, 204
    variation 268, 279–84, 287–8, 301
    see also directional selection; disruptive selection; evolutionarily stable strategy; fitness; frequency dependence, selection; selection differential; selection gradient; stabilizing selection
Philipson, W. R. 2
Philoliche rostrata Plate 6
Phlox species
    hybrid zones 327

P. drummondii
    antagonistic selection on floral traits *122, 123*
    flower colour, association with competitive ability 126
    flowering phenology 141
    pollinator-driven speciation 296
    reinforcement 305
Phyllodoce species
    competition for pollination 151–2
    flowering phenology *150*
phylogenetic constraints 140, 143, 148, 297
phylogenetic studies *314–15*
    angiosperm diversity 279, 312, 315–16, 319–20
    dioecy 209
    Narcissus 242–3
    see also comparative studies
Piper arieianum, asynchronous flowering 141
Piriqueta species, post-mating isolation 338
Plantago coronopus, female frequency associations *207*
Plantago lanceolata
    seed quality 153–4
    sex-ratio variation 228
plant communities 102, 104, 115–16
    attributes, effects on pollination variables *166*
    Clarkia species studies 111–15
    effects of habitat disturbance *162*, 163–4
    flowering phenologies 142–3
    pollinator-community structure 107
    pollinator-mediated interspecific interactions 107–8
    pollinator responses *103*, 105–7
        visitation rate 108–11, *109*
    properties 104–5
    trait evolution 135
    see also community context of pollination
plant–pollinator interactions 26, 29–30, 37–8, 257
    in community context 102–3
    ecological dynamics 32–4
    effects of human habitat alteration 161, 172–4, 175
    effects of plant enemies 217
    effects of plant size 50
    geographical variation 282–4, 290–1
    Lavandula latifolia 284–90
    pollination as trade 32
    specialists and generalists 30–1
    see also bees; flower colour; flower number; flower shape; floral design; floral display; hummingbird; inflorescence traits; nectar; pollinator (all entries)
plant population dynamics 32–5, 68, 161, 173, 185, 199
plant size 42, 56–7, 155, *210*, **347**
    effect on reproductive error 55–6
    effect on seed production *211, 212*, 213
    effect on sex allocation 46, 47, 53–5, 210–13, *216*, 251
        budget effects 48–50, *51*

direct effects 50–1, 56
plant species richness, correlation with bee species richness 107
Platanthera ciliaris 300
pleiotropy 126, 205, 210–11, 219, 262–3, 268, 273, 296, 304, 341, 348, **350**
Polemoniaceae, pollen-driven speciation 296
Polemonium folioissimum, flowering phenology 146
Polemonium viscosum
    flower shape, antagonistic selection *122*, 126–7, 131
    pollinators 299
pollen
    distribution on pollinator's body 71, 95, 104, 145, 242, 298, 313, 319–20, 329, 331
    fates 61–6, 76–8
        export 29, *63*–5, 74–6, 196, 205, 249, 253, 297, 334–5, 340
        relation to removal 29, 44
        role of floral traits 11–12, 15, 44, 71–2, 107, 132, 141, 184, 341
        see also pollen dispersal
    losses 44, *63*–5, 75, 77, 91, 96, 107–8, 145, 195–6
    removal 14, 29, 44, 65, 166–7, 264, 266–7, 297, 321
    see also cross-pollination; male fertility; male fitness; self-pollination
    production 11–12, 29, 42–3, 45, 48, 54, 61, 76–7, 132, 147, 163, 208, *211*–12, *216*, 218, 227–8, 253, 265, 269
        size-number trade-off 13, 266
    quality 68–9, 76, 163, 166, 168
pollen allelopathy 145, 167, **350**
pollen carryover 86, 88, 95–8, 331–2, 334–5, 337, 339, 341–2, **350**
pollen-collecting bees 65, 104, 106–7, 130, 149, 167, 304, 349
pollen competition 7, 44, 48, 50, 66–7, 69–2, 75, 132, 166, 168, 249, 341
    Ipomopsis species 338–9
pollen discounting 7, *63*–5, 68, 71, 97, 166, 184, 214, 231, **350**
    Camissonia cheiranthifolia 195–6
    effect on mating-system evolution *72*–7, *185*–7, 190, 193, 195, 206, *233*–4, 236
    and evolution of dioecy 206, 207, 216
    in apomixis 197
    influence of floral design and display 15, 44, 166, 195–6, 215, 231, 246, 348
    influence of flowering phenology 66, 141
pollen dispersal 15, 33, 43, 175, 219, 313, *327*
    extent 44, 50, 69, 75, 83–98, 142, 152–3, 239
    pollen shadow 10, 93–4
    influencing factors 48, 50, 96, 104–5, 142, 214, 241, 246, 253, 321, 332
    Ipomopsis species 333–5, 337, 340

heterostylous species *16*–7, 132, *240*, 244–5, *248*–9, 251–3
Narcissus triandrus 249–50
see also heterospecific pollination
pollen export *see* pollen, fates
pollen import *63*–4, 68, 71, 112, 145, 167, *170*–1, 284, 290, 340
    role of floral traits 15, 107, 127–8, 141, 166, 219, 245, 286, 287, 289, 334
pollen limitation 68–9, 77, 127, 132, 145, 151–2, 168, 184, 187, 248, 266, 297, 306, **350**
    Antennaria species 198
    benefits of delayed self-pollination 76
    Clarkia xantiana 112–13
    effect of human habitat alteration 161, 168, *170*–1
    effect of pollen export improvements 69, 71, 72, 166
    gynodioecy 205–7, *216*–17, 219
    Narcissus species 245, 249
    and self-fertilization 16, 72–6, 184, 186–7, 190, 192
    see also reproductive assurance
pollen:ovule ratios *6*, 76–7, 132
pollen receipt *see* pollen import
pollen removal *see* pollen, fates
pollen-stigma interference, avoidance 15
pollen-tube growth 66, 337–8
pollen vectors 81, 144, 184, 260, 313, **350**
    role in self-pollination 7, 63–4, 76, 186
    see also bees; Diptera; hawk moth; hummingbirds; Lepidoptera; pollination syndromes
pollination 43, 62–6, 83–98, 115–16, 129, 132, 209–10
    disturbance effects 160–76
    evolution of 29, 121, 262, 298–9, 302–6, 312, 317
    effect on floral longevity 66, 141
    generalist versus specialist 30–1, 37, 169, 171–5, 298, 300, 304
    see also generalist plant-pollinator interactions
    hand-pollination 112, 132, 149, 152, 187, 192–3, 199, 244
    Narcissus species 242, 244–5
    quality 69, 166–8
    quantity 68–70, 131, 143, 145, 147, 166–7, 171, 184–200, 207, 214, 219, 231, 340
    role in mating 6–8, 71–6, 83, 239–53
    role of floral traits 11, 32, 126–8, 135, 163, 166, 168–9, 175, *240*–1, 246, 264, 266, 278–91, 297, 328, 342, 347
    see also flower colour; flower size; dichogamy; herkogamy; heterodichogamy; heterostyly; nectar; pollination syndromes; protandry; protogyny
    variable 35–6, 184–8, 190, 195, 198, 279–91, 299, 306, 314
    see also community context of pollination; cross-pollination; facilitation, interspecific; generalist plant-pollinator interactions;

pollination (cont.)
    geographical variation, pollinator mosaics;
    heterospecific pollination; pollen fates
pollination ecotypes 283, 300–1, 350
    see also geographical variation, pollinator mosaics
pollination environment 7, 36–7, 66, 81, 144, 164,
    186–7, 194–5, 214–15, *248*, 279, 296, 303
pollination syndromes 129, 298, 326, 329, 331, 336, **350**
pollinator abundance
    aggregative response 105–6, 107, 110, 112–13
    Camissonia cheiranthifolia 192
    disturbance effects 164, 168
    evolution of selfing 113, 183–200
        see also reproductive assurance
    floral visitation rate 107, 110, 112, 146, 164
    numerical response 105–7, 110
    pollen dispersal 91, 112, 167, 314
    pollination quality 167, 284
    pollinator specialization 30, 31, 296
    relation to resource abundance 6, 106–7, 113, 164,
        214, 299
    reproductive assurance 6, 35, 113–*14*, 183–200, 231
    selection on floral traits 127, 132, 280, 282, 290–1, 296,
        298
    variation 107, 110, 127, 214, 280, 282, 284–6, 291, 298,
        299
pollinator assemblage (community) 104, 107, 111, 148,
    282, 290, 313, 337, 342
    disturbance effects 160, *164*, 167, *173*, 175
    diversity (species richness) 103, 107, 111, 112, 120, *165*,
        167, 286
    see also geographical variation, pollinator mosaic
pollinator attraction *31*, 111, 130, 141, 145–6, 219, 306, 327
    effect of floral traits 98, 110, 121, 125–8, 131, 141, 147,
        152, 166, 195, 303
        see also floral design; floral display
    effect of herbivory 126–7, 130–1, 187, 194, 216
    investment in 27, 31, 43–4
    role in pollination 25, 71, 91, 98, 166
    see also ethological isolation; facilitation, interspecific;
        synchronous flowering
pollinator behaviour
    aggregation 106–7, 110, 113
    constancy (fidelity) 103, 105, 107–8, 115–16, 143–5,
        329, 336, **346**
    effect on paternity shadow 96–7
    preference 105, 107, 111, 121, *124*, 284, 305, 329, 332
        floral morphology 122, 126–7, 131, 311, 333
        floral display *122*, 152
        flower colour 122, 132–4, 261–2, 299, 331
        plant height 50, *122*
        plant species 145, 330–1, 335–6, 342
        rewards 110, *122*, *128*, 151–2, 167
    see also ethological isolation

visitation rate 30 105, 107, 108–13, 115, 122, 127, 132,
    141, 185, 192, 214–*16*, 244, 261, 284, 290, 298,
    321, 327, 330–1, 337, 340–2
    disturbance effects 164, 167, *170*, 171, *174*
    see also facilitation, interspecific
    see also functional response
pollinator-driven speciation 295–306
pollinator extinction 33, 174
pollinators
    effects of environmental conditions 214–15
    role in floral evolution 25, 120–1, 246, 251, 257, 260, 266,
        278–9, 282–3
    floral colour preferences 261–2
    Lavandula latifolia 286–7
    sexual system evolution 219
    strength as agents of selection 131–2, 134–5
    see also pollinator-driven speciation
    role in reproductive isolation, see ethological isolation
    see also plant-pollinator interactions
pollinator sharing 81–2, 102–3, 115–16
    and flowering phenology 143, 144–6, 148
    see also facilitation, interspecific
pollinator shifts 72, 121, 246, 253, 295, 297–8, 302–6
    see also pollination, evolution of
pollinia 65, 75
Polygonum persicaria, antagonistic selection on floral
    traits *123*
polymorphism 123–4, 142, 236, 239, 240, 263
    floral morphology *16*, 240–53
    flower colour *123*, 125–6, 261, 267, *Plate 2*
    genetic 29, 289, 346
    mating system 227, 231–4
    sexual system 147, 230
    see also androdioecy; dioecy; gynodioecy; heterostyly;
        morph; style-length polymorphism;
        subdioecy
polyploidy 9, 198, 296
Pontederia cordata, pollen load studies 17
population genetics 289, 313, 348
    genetic structure of populations 139, 141–2, 149
    metapopulations 153–5, *228*–36, 301
    see also gene flow
population size 186, 224
    disturbance effects 162–3, 166–7, *170*–1
    effect on fitness or population growth 33–6, 43, *185*,
        235
    effect on mating 10, 44, 85, 108–13, 141, 145, 148, 164,
        166–7, *170*–1, *185*, 246
    effect on resource consumption 105–6, 109–10
    see also Allee effect; extinction; functional response;
        numerical response
post-mating isolation see reproductive isolation
post-zygotic reproductive isolation see reproductive
    isolation

pre-zygotic reproductive isolation *327–8*, 330, 331, 333, 340
  in *Ipomopsis* species 332–7
  see also ethological isolation; mechanical isolation; post-mating isolation
Primula sieboldii, relative strengths of agents of selection 132
prior self-fertilization 7, *33–5*, *63–7*, 185–6, 192, 197, 205, 216, 245, 346
  see also self-pollination
Prosoeca gangbaueri Plate 6
protandry 113, 147, 245, **350**
  see also dichogamy
protogyny **350**
  see also dichogamy
pseudogamy 197, **350**

quantitative genetics 25–9, 263–6
  see also additive genetic variance; genetic correlation; genetic variation; **G** matrix; heritability
quantitative trait locus (QTL) 273–4, 298, 329, 331, 342, **350**

ramet 7, 48, 212–13, 215–16, 348, **350**
Ranunculus adoneus, geographical isolation 153
Raphanus raphanistrum (wild radish) Plate 5
  evolution of correlated traits 268, *269*
    macroevolution 272–3
    microevolution 268–72
  flower colour, pollinator preferences 261, *262*, 266
  male fitness 266–7
  sex allocation 48
Raphanus sativus
  antagonistic selection on floral traits *122*, 123
  artificial selection 265
  petal colour variation 126, 132–4, *133*, Plate 2
recessive alleles *see* dominant alleles
reciprocal herkogamy *see* herkogamy
reciprocal transplant experiments 199, 289, 301, **350**
regulatory genes 273–4
reinforcement 297, 303–5, *327–8*, **350**
reproductive assurance 6, 35, 66, 69, 113, 184–200, 314, 346, **350**
  asexual reproduction 196–8, 346
  intraspecific variation 190–6
  future research 199
  mixed mating 8, 184–5, 187–90
  relation to mode of self-pollination 7, 35, 76, *185*, 231
  role in metapopulation dynamics 225–6, 228, 234, 236
  see also Baker's law; self-fertilization; self-pollination
reproductive compensation 36, 70–1, 77, 205, **350**
reproductive effort (investment) 41–57, **350**
  and sex allocation 53–5
  size dependence 55–6

reproductive isolation 121, 258, 295–7, 304–6, 312–13, 319–21, 326–8, 329, 341
  ethological isolation 258, 304–6, *327–37*, 340–2, **347**
  in *Ipomopsis* species 332–7
  mechanical isolation 304, 306, *327*, 329, 331–2, **349**
  post-mating isolation *327–8*, 337–40
  post-zygotic isolation 304, 306, *327–8*
  see also gene flow; hybridization; reinforcement
resource allocation *see* allocation strategies
resource availability
  for plants 41, 129, 163–4, 214, 218, 265
    effect on pollinators 214–15
    effect on reproductive effort 53–6
    effect on sex allocation 42–57
    flowering phenology 140, 143–4, 148
    see also resource limitation
  for pollinators 104–10, 112–13
    see also functional response; pollinator behaviour, aggregation
resource limitation (of seed production) 13, *69–71*, 77, 168, 261, 266, **351**
  consequences for gynodioecy 209–13, *216*
  consequences for mating-system evolution 71–5
rewardless pollination 32, 167
  see also facilitation, interspecific; mimicry
Rhododendron species
  hybrid zones 328
  R. aureum, seed quality and quantity, seasonal differences 149, *151*
Rollins, R. 4
Rosmarinus officinalis, flower size variations 127
ruderal species **351**
  colonization 224

Sagittaria latifolia
  ecological influences *210*, 213
  plant size, relationship to female frequency 212
  sex allocation 48
Salvia pratensis, female frequency associations *207*
Satyrium hallackii 300
Scandia rosaefolia, female frequency associations *207*
Scheidea salicaria, female frequencies *208*
Schiedea species, ecological influences *210*, 214
seed
  abortion 48, 52, 67–8, 70, 245
  development 62–3, 66–8, 75
    in apomixis 197, 346, 350
  dispersal 153, 155, 301, 314
    influence of flowering phenology 130–1, 141, *143*, 146–7
    seed shadow 10, 93
  production 52, 54, *70*, *123*, 129, 131–2, 187, 194, *198*, 245, *247*, 261, 286, 340

seed (*cont.*)
  constraints *see* ovule limitation; pollen limitation; resource limitation
  disturbance effects 166, 168–71,
  effects of flowering phenology 146–7, 149, *151–2*
  gynodioecious populations 205–8, *211*, 213, *216*, 228–9
  herbivory effects 48, 134
    see also seed predation
  seed:ovule ratio 70
  size dependence 48, 54, 56, 212–13
  see also female fertility; reproductive assurance
  quality 73–4, 149, 153–5, 166, 168–9, 245
  size 7, 71, 168, *185*
seed discounting 7, 63–5, 68, 71, 97, 166, 184, 21, 231, **351**
  in apomixis 197
  Aquilegia canadensis 189–90
  Camissonia cheiranthifolia 195–6
seed limitation of populations 29, 108, 161
seedling 168
  establishment 33, 43, *63*, 68, 73–4, 76–8, 125, 134, 141, 147, 154, 284
  inbreeding depression 6, 73, 75–6
seed predation 62, 67, *122*, 126–7, 129, 131–2, 195, 208, 217, 341
  influence of flowering phenology 130, 141, *143*, 146
selection differential 263, 269, *270*, **351**
  see also female fitness; fitness; male fitness
selection gradient 27–9, *114*, 263–4, *267*, **351**
  see also female fitness; fitness; gain curves; male fitness
selective harvesting
  effects on plant attributes *162*, 163
  effects on pollinator attributes *164*
self-fertilization (selfing) 37, 56, 67, 69, 113, 129, 183, 184, 187, 189–90, 193, 225, 244
  associated floral traits 16, 76, 129, 132, 166, 192–3, 195, 246, 311
  and colonization 225, 226–7
  effect on population dynamics 33, 35
  and evolution of sexual systems 205–8, 216, 219, 226–7, 229
  incidence 56, 65, 114–*15*, *152*, 171, 184, 186–7, 190, 197, 214, *216*, 231, 246
    see also mixed mating; selfing rate
  influence of harsh environments 129, 193, 209, 214–15
  D. G. Lloyd's investigations 4–8, 10
  selection of 34–7, 72–7, 145, 151, 184–6, 194, 196, 199, 231–6, 297
  see also autonomous autogamy; reproductive assurance; self-pollination
self-incompatibility 15, 66, 142, *169*, 171–2, 225, **351**
  heteromorphic 66, 240–1, 252–3, **348**
  homomorphic 66, 152, 191–2, 230, **349**
  late-acting (ovarian) 67–8, 241, 243, 246, **349**

selfing rate 35, 56, *188–9*, 208, 215–*16*, 231, *234*, 246, 249, 253, **351**
self-pollination 8, 69, 71–2, 108, 122, 129, 132, 149, 160, 168, 184, 193, 215, 231, 243, 249, 322
  associated floral traits 6, 111, 113, 187, 192, 198, 244, 246, 314
  modes 6, *7*, *63*, *185*
    autonomous 62, *64*, 152, 186–90, 192–3, 214–15, 245, 297
    competing (D. G. Lloyd's definition) 7, 34–6, 68, 205, 219
      see also simultaneous autonomous self-pollination
    delayed 34–6, *64*–5, 72, 76, 186, 188–9, 205, 231
    prior 7, 34–6, *64*–5, 205, 245
  facilitated (D. G. Lloyd's definition) 7, 34, 63
  evolution *185*–7, 194
    see also simultaneous facilitated self-pollination
  simultaneous 7, 65, 205, 245
    autonomous 64, 68, 72–4
    facilitated
      intrafloral 63–*5*, 75–6
      evolution 72, *73*, 74
      incidence, *Disa cooperi* 65
      geitonogamy 7, 8, 15, 64, 65–6, 186, **348**
        effects of floral display 8, 66, 88, 166, 215
        and evolution of gynodioecy 215–16
        influence of individual flowering phenology 66, 141
        and mating-system evolution 74–6, *185*–6, 206, 231–6
        pollen discounting 7, 74, 88, 215, 231
    see also autogamy; pollen discounting; reproductive assurance
Senecio chrysanthemifolius, post-mating isolation 338
sequential flowering 107, 110, 142
  Clarkia species 113
sex allocation 41–7, 206–7, *216*, 218, 226, **351**
  correlation with flowering phenology 147
  effects of plant size 47–8, 55–6, 210–13
    budget effects 48–50, *51*
    direct effects 50–1
  individual variation 42
  in metapopulations 227, 228
  in perennial plants 53–6
  temporal displacement of male and female functions 51–3
  see also allocation strategies; gender; hermaphrodite
sex change *see* gender diphasy
sex inconstancy 9–10, 50, 226 229, **351**
sex ratios 42, 46, 205–6, 208–9, 212–13, 227–8
  D. G. Lloyd's work 9–10
sexual deception, orchids 303
sexual interference 15, 147, 245

sexual selection 11, 14, 24, 62, 77, 147, 305
  see also Bateman's principle
sexual systems 46–7, *144*, 147, 171–2, 236, *241*–3, 253, 311–12, **351**
  D. G. Lloyd contributions 9–11
  diversity in *Narcissus* species 241–3
  effects of single-event colonization 224–6
  evolution in metapopulations 226–9
  non-pollinator selection 129–30, 131
  see also androdioecy; cosexuality; dioecy; enantiostyly; gender; gynodioecy; heterostyly; self-incompatibility; stigma-height dimorphism
Sidalcea species
  female frequency *207*, 208
  S. hendersonii, ecological influences *210*
  S. malviflora, gynodioecy 228
Silene species
  female frequency associations *207*
  S. acaulis, ecological influences *210*
  S. dioica, antagonistic selection on floral size *122*, 127–8
  S. latifolia, flower number-size trade-offs 265
  S. vulgaris, gynodioecy 228
Simmondsia chinensis Plate 7
simultaneous self-fertilization *64*, 65, 67, 68, 72–4, 186
  Camissonia cheiranthifolia 192–3
  see also self-fertilization; self-pollination
single-event colonization 224, 236
  effects on sexual systems 224–6
sister taxa comparisons 300, 302, 314–16
  methods 316–17
size-dependent reproductive effort 55–6
size-dependent resource allocation 53–5, 57
size-number strategies 12, *13*–14, 29
Slowinski-Guyer test 313, 314–15
snowbeds, flowering phenologies 144, 149, *150*
snowmelt gradients
  as cause of phenological isolation 152–3, *154*
  comparative studies 140, 155
  interspecific competition for pollination 149, 151–2
Spartina species, hybrid zones 327
specialist plant-pollinator interactions 30–2, 106, 112–13, 194
  effects of human habitat disturbance 171–3, 175
speciation 31, 121, 125, 279, 312–14, 319–21, 327, 329, 340
  see also allopatric speciation; ecological speciation; parapatry; pollinator-driven speciation; sympatric speciation
species concept 295–6
species richness 107, 111
  of plants
    ecological 107, 167, 297
    phylogenetic 47, 314–17, 319
  of pollinators *see* pollinator assemblage
Spergularia marina, negative genetic correlation 265–6

Sprengel, C. K. 8, 120
stabilizing selection 30, 74, *264*, 266–7, 301, 304, 340
  flowering phenology 141, 142
stamen position, *Narcissus triandrus* 248, 252, 253
Stanleya pinnata Plate 5
Stebbins, G. L. 2, 6
  "most effective pollinator principle" 30, 92, 296, 329
  pollinator-driven speciation 296, 303, 306
stigma-height dimorphism *16*
  Narcissus species 241–3, 244–6, 253
stress *see* entries for harsh environments under: dioecy; inbreeding depression; self-fertilization
structural genes 273, 274
style length 122, 128, 332–3, 337–8, Plate 8
  variation, *Narcissus triandrus* 251–2
  see also heterostyly; stigma-height dimorphism
subdioecy 48, 49–50, 218, 348, **351**
symmetrical mating *240*
symmetric plant-pollinator interactions *173*
sympatric speciation 302, 303, 340, **351**
synchronous flowering 52, 141–2, *144*, 146, 153, 155
  in tropical forests 107, 143
  see also co-flowering species
Syngonanthus species, pollination mode 322

temperature, role in initiation of flowering 144
temporal displacement, male and female function 51–3
Thymus species
  female frequency associations 206, *207*
  T. vulgaris
    ecological influences *210*
    flowering phenology 147
    sex-ratio variation 228
time of first flowering 140
tobacco plants, effects of herbivory 128–9
"toxic" nectar 128–9
trade-offs 31
  life-history 27, 29, 53, 56–7
  sex function 27, 31, 218
  size-number 13–14, 29, 185–6, 265–6
  see also allocation strategies; ovule discounting; pollen discounting; seed discounting; sex allocation
transgenes 88, **351**
transgression **351**
  see hybridization
transplant experiments 183, 199, **350**
  Antennaria species 198
transport loss, pollen 44, *63*–5, 71–2, 75–7, 93, 96–7
Tristerix corymbosus, flowering phenology 131, 146
tristyly 239, 246–7, 349, **351**
  mating patterns *240*
  Narcissus species 241, *242*–3, 247–52
tropical forests, synchronous flowering 107, 143, 145

Umbelliferae, sex ratios 9–10
uniparental reproduction, evolution 183
  see also asexual reproduction; self-fertilization
univoltine **351**

Vaccinium hirtum, flowering phenology 146
vegetative diversification 303
vegetative growth 54, 148, 210
vegetative reproduction 196–7, 225, 346
vegetative tissues, anthocyanin content 125–6, 341
Veronica stelleri
  flowering phenology *150*
  phenological isolation 153, *154*
Vicia sepium, antagonistic selection on floral traits 123
visitation rates see pollinator behaviour

Wahlenbergia albomarginata, avoidance of sexual interference 15
water-pollination 44, 313
  see also abiotic pollination
Webb, C. 15, 16, 17
weevil herbivory
  effect on sex ratio 208
  Fragaria virginiana 217, 218
wild radish see *Raphanus raphanistrum*
wild strawberry see *Fragaria virginiana*
wind-pollination 303–4, 313
  dispersal models 84
  fitness effects of plant size 50
  flowering phenology 144
  grasses, anthocyanin polymorphisms 125
  male gain curve 44
  see also abiotic pollination
Wurmbea species
  W. biglandulosa, ecological influences *210*
  W. dioica
    ecological influences *210*, 213, 214, 215
    sex allocation 50

Yates, J. M. A. 16

Zaluzianskya species
  Z. microsiphon 302, *Plate 6*
  Z. natalensis Plate 6
zygomorphy 11, 258, 305, *Plate 7*
  role in floral diversification 313–14, 316, 317–*18*, 319–20
zygote fates 66–7